REPRODUCTOMICS

REPRODUCTOMICS

The -Omics Revolution and its Impact on Human Reproductive Medicine

Edited by

José A. Horcajadas

Department of Molecular Biology and Biochemical Engineering, Pablo de Olavide University; SINAE, Sevilla, Spain

Jaime Gosálvez

Department of Biology, University Autónoma of Madrid, Madrid, Spain

Academic Press is an imprint of Elsevier
125 London Wall, London EC2Y 5AS, United Kingdom
525 B Street, Suite 1650, San Diego, CA 92101, United States
50 Hampshire Street, 5th Floor, Cambridge, MA 02139, United States
The Boulevard, Langford Lane, Kidlington, Oxford OX5 1GB, United Kingdom

Notices
Knowledge and best practice in this field are constantly changing. As new research and experience broaden our understanding, changes in research methods, professional practices, or medical treatment may become necessary.

Practitioners and researchers must always rely on their own experience and knowledge in evaluating and using any information, methods, compounds, or experiments described herein. In using such information or methods they should be mindful of their own safety and the safety of others, including parties for whom they have a professional responsibility.

To the fullest extent of the law, neither the Publisher nor the authors, contributors, or editors, assume any liability for any injury and/or damage to persons or property as a matter of products liability, negligence or otherwise, or from any use or operation of any methods, products, instructions, or ideas contained in the material herein.

Library of Congress Cataloging-in-Publication Data
A catalog record for this book is available from the Library of Congress

British Library Cataloguing-in-Publication Data
A catalogue record for this book is available from the British Library

ISBN 978-0-12-812571-7

Publisher: John Fedor
Acquisition Editor: Peter B. Linsley
Editorial Project Manager: Megan Ashdown
Production Project Manager: Poulouse Joseph
Cover Designer: Matthew Limbert

Typeset by SPi Global, India

Contents

Contributors

Ashok Agarwal American Center for Reproductive Medicine, Cleveland Clinic, Cleveland, OH, United States

Signe Altmäe Department of Biochemistry and Molecular Biology, Faculty of Sciences, University of Granada, Granada, Spain; Competence Centre on Health Technologies, University of Tartu, Tartu, Estonia

Ester Anton Genetics of Male Fertility Group, Cell Biology Unit, Autonomous University of Barcelona, Cerdanyola del Vallès, Spain

Rosa Roy Barcelona Department of Biology, Autonomous University of Madrid, Madrid, Spain

Lluís Bassas Laboratory of Seminology and Embryology, Andrology Service, Fundació Puigvert, Barcelona, Spain

Davina Bonte Ghent-Fertility and Stem cell Team (G-FaST), Department for Reproductive Medicine, Ghent University Hospital, Ghent, Belgium

José P. Carrascosa Pablo de Olavide University, Sevilla, Spain

Petra De Sutter Ghent-Fertility and Stem cell Team (G-FaST), Department for Reproductive Medicine, Ghent University Hospital, Ghent, Belgium

Aysenur Ersoy American Center for Reproductive Medicine, Cleveland Clinic, Cleveland, OH, United States; Capa Medical School, Istanbul University, Istanbul, Turkey

Francisco J. Esteban Department of Experimental Biology, Faculty of Experimental Sciences, University of Jaén, Jaén, Spain

José Luís Fernández Unidad de Genética—Complejo Hospitalario Universitario A Coruña-INIBIC; Laboratorio de Genética Molecular y Radiobiología, Centro Oncológico de Galicia, A Coruña, Spain

A.F. Fernández Instituto de Investigación Sanitaria del Principado de Asturias (ISPA), Hospital Universitario Central de Asturias (HUCA), University Institute of Oncology of Asturias (IUOPA), Oviedo, Spain

M.F. Fraga Nanomaterials and Nanotechnology Research Center (CINN-CSIC), University of Oviedo, Principado de Asturias, Asturias, Spain

Iria González-Vasconcellos Unidad de Genética—Complejo Hospitalario Universitario A Coruña-INIBIC, A Coruña, Spain; Institute of Radiation Biology, Helmholtz Zentrum München, Deutsches Forschungszentrum für Gesundheit und Umwelt, Neuherberg, Germany

Jaime Gosálvez Department of Biology, University Autónoma of Madrid, Madrid, Spain

Ramesh Reddy Guggilla Ghent-Fertility and Stem cell Team (G-FaST), Department for Reproductive Medicine, Ghent University Hospital, Ghent, Belgium

Sezgin Gunes American Center for Reproductive Medicine, Cleveland Clinic, Cleveland, OH, United States; Department of Medical Biology, Ondokuz Mayis University, Samsun, Turkey

Samir Hamamah ART/PGD Department, Arnaud de Villeneuve Hospital, INSERM Unit 1203, Montpellier, France

Delphine Haouzi ART/PGD Department, Arnaud de Villeneuve Hospital, INSERM Unit 1203, Montpellier, France

Björn Heindryckx Ghent-Fertility and Stem cell Team (G-FAST), Department for Reproductive Medicine, Ghent University Hospital, Ghent, Belgium

Ralf Henkel American Center for Reproductive Medicine, Cleveland Clinic, Cleveland, OH, United States; Department of Medical Bioscience, University of the Western Cape, Bellville, South Africa

William V. Holt Academic Unit of Reproductive and Developmental Medicine, University of Sheffield, Sheffield, United Kingdom

José A. Horcajadas Department of Molecular Biology and Biochemical Engineering, Pablo de Olavide University; SINAE, Sevilla, Spain

Meritxell Jodar Molecular Biology of Reproduction and Development Research Group, August Pi i Sunyer Biomedical Research Institute (IDIBAPS), Clínic Foundation for Biomedical Research, University of Barcelona, Barcelona, Spain

Ahmet Kablan Department of Medical Biology, Ondokuz Mayis University; Nuclear Medicine, Ondokuz Mayis University, Samsun, Turkey

Neha Kumar Genetic Counselor, New York, NY, United States

Sara Larriba Molecular Genetics of Male Fertility, Human Molecular Genetics Group, Bellvitge Biomedical Research Institute (IDIBELL), Barcelona, Spain

Giulia Mariani Department of Biomedical, Experimental and Clinical Sciences, Division of Obstetrics and Gynecology, University of Florence, Florence, Italy

Swati Mishra Ghent-Fertility and Stem Cell Team (G-FAST), Department for Reproductive Medicine, Ghent University Hospital, Ghent, Belgium

Juan M. Moreno-Moya SINAE, Sevilla, Spain

Santiago Munne Yale University, New Haven, CT, United States

Sergio Oehninger The Jones Institute for Reproductive Medicine, Eastern Virginia Medical School, Norfolk, VA, United States

José Bellver Pradas Instituto Valenciano de Infertilidad, University of Valencia; Department of Pediatrics, Obstetrics and Gynecology, Faculty of Medicine, University of Valencia; IVI Foundation, Catedrático Agustín Escardino 9, Parc Científic Universitat Valencia, Valencia, Spain

Anaís García Rodríguez Department of Biology, Autonomous University of Madrid, Madrid, Spain

Sally Ann Rodríguez Certified Genetic Counselor, New York, NY, United States

Sarthak Sawarkar University of Kent, Canterbury, United Kingdom

Panagiotis Stamatiadis Ghent-Fertility and Stem cell Team (G-FaST), Department for Reproductive Medicine, Ghent University Hospital, Ghent, Belgium

Jasin Taelman Ghent-Fertility and Stem Cell Team (G-FAST), Department for Reproductive Medicine, Ghent University Hospital, Ghent, Belgium

R.G. Urdinguio Nanomaterials and Nanotechnology Research Center (CINN-CSIC), University of Oviedo, Principado de Asturias, Asturias; Instituto de Investigación Sanitaria del Principado de Asturias (ISPA), Hospital Universitario Central de Asturias (HUCA), University Institute of Oncology of Asturias (IUOPA), Oviedo, Spain

Margot Van der Jeught Ghent-Fertility and Stem Cell Team (G-FAST), Department for Reproductive Medicine, Ghent University Hospital, Ghent, Belgium

Eva Vargas Department of Experimental Biology, Faculty of Experimental Sciences, University of Jaén, Jaén, Spain

Xavier Vendrell Sistemas Genómicos Ltd, Valencia, Spain

Preface

The biological process by which any living organism is able to maintain its own specific status is known as reproduction. Reproduction represents the ability of the species to transmit the required information to the next generation in order to produce a new quasi-identical organism when compared with its predecessor. Reproduction and genetics therefore are linked concepts in science. While reproduction precedes the notion of genetics, both have "walked together" since William Bateson in 1905 coined the term genetics. In fact, the origins of the concept of phenotypic characteristic inheritance in association with reproduction might be considered ancient when compared with the era of molecular genetics we currently find ourselves in.

It is possible to find in the literature some references to the idea of linking reproduction and inheritance. While court physician to the Spanish caliphate, ca. 1000, Abū al-Qāsim gave the earliest known description of hemophilia linked to a hereditary nature, August Weismann, in 1883, anticipated the germ plasm theory of inheritance, where hereditary information was carried only in sperm and oocytes. With this theory, he challenged both Pangenesis and Lamarckism, as these concepts about how life and associated characters were transmitted to the offspring were predominant at that time. Gregor Mendel (1822–84) was the first author that, using an experimental design, described the existence of "heritable factors" that can be transmitted to the offspring. Most likely, this was the first recorded scientific contribution that connected reproduction and genetics.

To understand how reproduction and genetics have evolved in close relationship with the development of modern -omics technologies, we must acknowledge the "in situ hybridization of nucleic acid" as the primary precursor of the new genetics era. This technique has allowed significant development of most of the areas that are currently trying to ascertain the risks of undesirable inheritance associated with the use of assisted reproductive technology (ART). The first contributions to the concept of in situ hybridization of nucleic acids emerge from Ferruccio Ritossa and Sol Spiegelman when they published a paper in 1965, showing the possibility of immobilizing and denaturing DNA on a nitrocellulose filter and later hybridized radioactive-labeled RNA to the single-stranded DNA in what became known as DNA-RNA hybridization. The concept of in situ hybridization was later exploited by Joseph Gall and Mary Lou Pardue during the late 1960s when they hybridized DNA to DNA. The paper *Molecular hybridization of radioactive DNA to the DNA of cytological preparations* was published in 1969; 1969 was also the year that both authors showed, for the first time, the chromosomal localization of satellite DNA associated with the centromeres in mouse chromosomes using their DNA-DNA in situ hybridization protocol. Hence, this protocol can be regarded as the prototechnology for DNA mapping and identification of DNA allelic variations within a given genome.

In 1975, Sir Edwin Mellor Southern was able to identify specific sequences among DNA fragments that were separated by gel electrophoresis: the so-called Southern blot. The rationale of the Southern blot, as a molecular technique to localize and characterize DNA-specific DNA molecules, was later exported to protein (known as the Western blot) and RNA (Northern blot) characterization. Later, Southern founded a company (Oxford Gene Technology) in Oxford (the United Kingdom) to develop and establish a large part of the DNA microarray technology that today is a fundamental tool to nearly every element of genetics and the clear precursor of most -omics technologies.

It is interesting to note that while in all species reproduction has been linked to a slow process of evolutionary change producing major differentiations such as sexual or asexual reproduction, human beings have carried out a quick revolution in how the genetic information is transmitted to the progeny. The in vitro fertilization (IVF) that conceived birth of Louise Brown in 1978 was originally regarded by many with a certain level of fear and prejudice, and yet, IVF and other alternative strategies to produce human-influenced gamete syngamy have now become common and reliable procedures. In fact, the existing technology for ART is now challenging the "natural concept of infertility," because a naturally conceived infertile individual, with the help of ART, can now be considered as a "potentially assisted fertile individual."

The individual and social implications of fertility go beyond the natural need of reproduction emerging from any unfertile couple. The full range of technologies to overtake infertility is progressively being used by the industry. The last-generation technologies provide the proper environment to give pregnancy at the edge of not natural biological assumable conditions, such as postmenopausal situations. Gene editing is now available and thought to be of relatively low risk; but should gene editing be performed on human embryos? This is a hot spot that is clearly controversial, but science and the general public cannot ignore the possibility that genetically engineered babies with the aim of creating a perfect child currently do exist and will be available in the not-so-distant future.

The intention of this book is to summarize and contextualize the arrival of -omics revolution and how it will impact two connected areas of knowledge, "reproduction and genetics." The book not only covers a series of topics within the realm of -omics but also directs our attention to other areas of research that use equivalent technologies to understand the possible future of reproduction. In this sense, understanding the future in every field of science is always difficult without being conscious of an existing and motivating past.

Chapter 1 recognizes the important role that the "basic semen analysis" has played and is still playing to understand the primary causes of infertility associated with male factor. Before the use of an -omics approach, ejaculate characteristics such as volume, sperm concentration, sperm morphology, sperm viability, motility, and DNA quality need to be assessed and quantified. As useful as the spermiogram may be as a diagnostic, Prof. William Holt reminds us that "none of these parameters can really be regarded as simple." While measurement of seminal parameters is relatively easy to implement in the laboratory on a routine basis, these observations typically provide little insight with respect to the molecular basis of fertility and the differences and relative importance of these parameters between patients.

Understanding infertility makes more sense when we look for a genetic explanation behind it. In Chapters 2 and 3, the authors have reviewed some of the well-known genetic variations that can be associated with an infertile profile. However, establishing etiologic causes that might be directly associated with male infertility can be particularly challenging given that a high proportion of infertility associated with the male factor is defined as "idiopathic." Given that thousands of genes have now been reported as contributing in male reproduction capacity, the identification of single genes associated with the wide concept of infertility is problematic. Single-nucleotide polymorphisms (SNPs) can be considered as the most common variant in the human genome since it represents approximately 90% of human genetic variations and may occur once in about every 100–300 base pairs. In parallel, several genome-wide association studies have been performed to assess relative association of SNPs between infertile cases and controls in subgroups of idiopathic infertility using specific Illumina SNP arrays. Insertion-deletion (INDEL) polymorphisms, defined as loses or gains of potential genetic information that may affect introns, exons, or gene promoters, are key to understanding certain scenarios of infertility. Additionally, copy number variations (CNVs) such as Y-chromosome-associated microdeletions have been associated with azoospermia and/or spermatogenic failure. In general, genetic differences cannot be easily ascribed confidently to male infertility, since confounding factors such as the environmental or ethnic influence are essentially impossible to disentangle. Nonetheless, the presence of some of these variations in a given genome may certainly increase the risk of being infertile.

Although the majority of lay people are typically not aware of the details of genetic concepts, one of the main concerns of couples when planning a family is to ensure they have a "genetically balanced" newborn baby. The reality is that the history of family disease is unknown in approximately 80% of a couple hoping to achieve a pregnancy; this phenomenon is all that more significant given the fact that many genetic conditions are present in both parents as recessives. The pros and cons of our available capacity to evaluate this risk is assessed in Chapter 4. Technological advances now allow the identification of different alleles for hundreds of genetic settings to help us calculate the risk of each partner. Subsequently, the possible patterns of inheritance and the possible genetic consequences of producing undesired recessive homozygote individuals of lethal or sublethal condition can be predicted. The potential for matching/mismatching of certain allele combinations can be calculated preconception or prenatal periods. Testing prior to pregnancy provides the best predictive strategy so that unbalanced options for different allele combinations can be avoided or discarded.

Expression of individual genes is not constant for every cell, and genetically identical cells fluctuate in their levels of transcription. However, proteins playing essential roles for cell survival exhibit similar abundances and lower variability. Thus, constitutive genes are those DNA sequences that are transcribed continually as opposed to a facultative gene, which shall be only transcribed when needed. Chapters 5 and 6 review two key groups of DNA sequences that are central for cell survival: telomeres and histone/protamine DNA sequences. Copy number of telomeric DNA sequences is contributed from the oocyte and the sperm, and it is imperative for the maintenance of a minimal and basic copy number associated with each species. This is the only way of providing chromosome stability and integrity to the next generation of cells. Although not profoundly studied in the field of reproduction, telomere biology is important to understanding fertility, since dysfunctional telomeres may provoke undesired aneuploidies and cell death, affecting not only the male and female gametes but also the capacity of the embryo to achieve normal development. Whether telomeres are reflective of newborn longevity, the possibility exists that they are also involved in the susceptibility of common age-related pathologies, and this is still an open question deserving future investigation.

Spermatozoa are unique cells that have the capacity to survive for a relatively long period of time outside the soma from which they were formed prior to syngamy. Protamines are nuclear proteins that replace histones during spermiogenesis and that are essential to condense, stabilize, and provide additional protection to the DNA against external stressors. Human sperm DNA contains two main types of protamines, known as PRM1 and PRM2, and variations in the expected ratio of each one have been associated with different grades of male infertility. Genes involved in the production of such protamines are polymorphic, and some of the SNP variants have been associated with the capacity of these alternatives to bind to the DNA appropriately. Some of these variants and their possible role in infertility have been reviewed in Chapter 6.

A high level of chromatin condensation in the sperm cell was typically considered a restriction for transcriptional activity, but we now know that the sperm may still contribute to the reproductive process via thousands of different RNA molecules. Although most of the functions of these molecules are imprecise and yet to be defined, their possible role in fertility/infertility is gaining interest. Small noncoding RNA populations can be found both at the seminal plasma, some of them linked to prostasomes and also in the spermatozoa. These new molecules are currently under a period of intense characterization, but they are started to be considered as potential molecular biomarkers. The clinical significance of these markers is reviewed in Chapters 7 and 8.

Thoughts of increasing interest in reproduction are targeting the proteome and the metabolome as putative informers of fertility fails. Particularly, Metabolomics could be a low-invasive strategy to understand main causes of fertility fails. These two chapters review the current literature on transcriptomic, proteomic, and metabolomic profiles associated with male factor infertility, emphasizing how such disorders and defects could be associated with infertility pathways. In particular, proteomics and metabolomics extend the coverage of this field to incorporate the areas of implantation and live birth rates in ART. Identification of proteomic and metabolomic profile has allowed the identification of biomarkers for the diagnosis, treatment, and counseling of several infertility-related conditions and may therefore help the couple to achieve a successful pregnancy based on a more objective selection of the best quality gametes and embryos and the most receptive endometrial environment. While promising, science is not definitive, and its clinical practice still remains imperfect; these limitations are also discussed in Chapter 9.

Above the control of genes, RNAs, and proteins, there is a supra level of influence on morphogenesis. Epigenetics is a new concept that attempts to explain how heritable changes in gene function are not directly dependent of changes in the DNA sequence. Current information points out that epigenetic regulation plays crucial role in the control of gene expression. This phenomenon not only has a direct impact in regard to the regulation of the different stages of transition of somatic cell division but also is central to gamete formation in all mammalian species. It is known that this type of regulation is highly linked to a series of environmental effectors that may modify the expected gene expression modulating the action of certain repressor proteins interacting with silencer regions of the DNA. These epigenetic changes may persist through cell divisions, thereby "miming" a certain level of inheritance that passes down the cell line for multiple cycles. Some of these changes not only are related with DNA methylation of developmental and imprinted genes but also include the modification of histones. Chapters 10 and 11 are contributing to some light to this interesting area of research.

Irrespective of obtaining the best quality gametes or being confident of the genome design and even taking care of any possible risk of bad gene regulation, any in vitro-generated embryo still requires a second level of selection at the time of implantation. Today, even using the most modern systems, such as time-lapse cinematography, to assess the development of the embryo, dynamic morphology assessment is the standard reference method for evaluating embryo development and quality. Recent evidence indicates that the combination of time-lapse-based systems combined with proteomic and or metabolomic information may give rise to new scenarios for embryo selection with a higher predictive clinical value to achieving a successful pregnancy. Additionally, recent information about the state of the art in the process of selection against aneuploidy, also known as preimplantation genetic diagnosis (PGD) or preimplantation genetic screening (PGS), is also included in this book. Chapters 12 and 13 summarize the most recent information in these areas.

Although the use of intracytoplasmic sperm injection has revolutionized the human IVF field, there are still repeated occasions when this procedure fails to result in successful fertilization; one possible explanation is the failure of the oocyte to activate, and this defect can be both sperm- and oocyte-dependent. Chapter 14 summarizes the state of the art of the oocyte activation mechanism, placing particularly attention on the importance of calcium transients in this process. The chapter also describes the different sperm- and oocyte-dependent factors that are involved in oocyte activation, which could be linked to fertilization failure, with a special focus on the sperm-activating factor, phospholipase C zeta (PLCζ). The authors emphasize the different diagnostic tools currently available to reveal the oocyte-activating capacity of human spermatozoa and the clinical applications to overcome fertilization failure after ICSI.

Chapter 15 presents the importance of human endometrial physiology, structure, and functionality to achieve a pregnancy. The main focus of the review is to assess the endometrium's receptivity by identifying molecular markers that can be linked to the process. The authors highlight developments in the field of noninvasive omics that can be self-explaining of endometrial receptivity observed in any patient simply though the analysis of endometrial fluids, relieving the need for aggressive and disruptive biopsies.

We have selected two important and singular areas of research to close the book. Chapter 16 covers the importance of producing gametes from somatic cells to be used in those cases of total fail of normal gamete production. Chapter 17 is included because given the possibility of producing massive amounts of information, using some of the techniques included in this book, we need to be aware of how to filter this information to achieve information of clinical value.

Gamete cells derived from a somatic cell is a technical possibility and potential solution for overcoming infertility. In vitro sperm control of gene regulation may give rise to functional gametes using pluripotent stem cells, and in the future, this possibility shall give rise to new opportunities of establishing fertility to patients diagnosed as sterile. The interest in in vitro gamete generation is an expanding area that is progressively giving its assumable results in mouse and in human;

in Chapter 16, the authors have summarized the recent advances in the production of male germ cells in both species. As what usually happens when the rules of biology are challenged, the ethical considerations and security of implementing this technology will need careful consideration. In addition to the state of the art for a technological efficient in vitro production of stem-cell-derived gametes, the authors also highlight the social impact and some potential disadvantages of these next-generation technologies.

Genetic and genomic data are reaching such a high scale of complexity that the biologist now requires the assistance and use of sophisticated and complex computational tools to render observational information in usable form. Thus, the development of -omics high-throughput technologies is undoubtedly increasingly helping us to understand the molecular basis of physiological and pathological processes that cannot be understood from a simplistic assessment of a single-gene change. The last chapter of the book is an overview of the different computational approaches/options for analyzing generated -omics datasets. Databases and the analytic tools currently used in the area of reproductomics are also presented.

The development of new technologies, imposing a positive impact on ART, has yielded a huge panorama of new expectations for biologically infertile couples that became fertile when assisted. Most of these new approaches to assess the main causes of infertility have relieved the incapacity of achieving pregnancy and the psychological issues inherent to this situation. In parallel, all these activities have also promoted a variety of ethical issues that need clear delineation and supervision from ethical committees at the clinics and adaptation of the law to new technological developments to fade this arena of concerns. This book faces some of these challenges waiting that in prompt future what here is collected as forthcoming can be converted in normal praxis.

José A. Horcajadas
Jaime Gosálvez

Introduction: Human Genome Projects: The Omics Starting Point

*Jaime Gosálvez**, *José A. Horcajadas*†

*Department of Biology, University Autónoma of Madrid, Madrid, Spain †Department of Molecular Biology and Biochemical Engineering, Pablo de Olavide University; SINAE, Sevilla, Spain

PREOMICS ERA

In modern times, Gregor Mendel (1822–1884) is recognized as the first author that used an experimental design that was able to describe the existence of "heritable factors" that can be transmitted to the offspring. It was not until 1900 where a reawakening of the Mendelian concept of inheritance was accomplished. Mendel's work remained obscure until 1900, when H. de Vries, C. Correns, and E. von Tschermak revisited his investigations, disseminating the concept of the Mendelian laws of inheritance. When Mendel's work was accepted, K. Pearson in a series of papers entitled *Mathematical Contributions to the Theory of Evolution* established the so-called biometric school for inheritance. There was a certain level of controversy between K. Pearson and W. Bateson, among other supporters of the Mendelian inheritance, because they did not use statistical methodologies to validate their theories. It was statistician and biologist Sir Ronald Aylmer Fisher (1890–1962) the first recognized scientist who combined mathematics, Mendelian genetics, and natural selection. The net result of this contribution later gave rise to a new dimension of Darwinism theory of evolution recognized as the modern synthesis.

By 1915, the basic principles of Mendelian genetics were absorbed by science and were applied to a wide variety of organisms. *Drosophila melanogaster* was a crucial organism to understand how inheritance may operate in nature. T.H. Morgan (1856–1945) and his group developed a conscientious experimental work and established the statistical framework of population genetics, bringing genetic explanations into the study of evolution. H.J. Muller in 1927 while working in the group, which established that exposure to X-rays could cause genetic mutations, published his findings in various articles such as "Artificial Transmutation of the Gene," in which he received the Nobel Prize in Physiology and Medicine in 1946. One of the students in Morgan's group, A Sturtevant, created the first genetic map, which was a landmark event in genetics.

It is interesting to note that during this time, a large debate existed about the nature of the "heritable factors." Scientists first thought that proteins were the required genetic material. Proteins were known to have diverse amino acid sequences that may explain a large part of the observed and expected variability, whereas DNA was considered to be directly related with the structural features because of its poor variability. Whether proteins or DNA are responsible for the "heritable factors," where they are located, and how they are transmitted were mostly answered by T. Boveri, W. Sutton, and T.H. Morgan, when they established that Mendel's heritable factors were most likely carried on chromosomes. T.H. Morgan, in 1933, received the Nobel Prize for Medicine for his work in establishing the chromosomal theory of inheritance. Two years after T.H. Morgan was awarded the Nobel Prize, B. McClintock and H. Creighton demonstrated that crossing over between homologous chromosomes is the cause of recombination.

In parallel to this stimulating time for biology, F. Griffith (1923) detected that DNA carries genes responsible for pathogenicity while assessing bacterial transformation, and later (1928), he discovered that hereditary material from dead bacteria can be incorporated into live bacteria. J.L.A. Brachet is recognized as the first author demonstrating that DNA was found in the chromosomes, while RNA was present in the cytoplasm. J.L.A. Brachet and T.O. Caspersson can be considered as the first authors stating that RNA plays an active role in protein synthesis.

With the arrival of new experimental models such as viruses and bacteria, the understanding of how DNA is the key molecule for genetic information transmission was undeniable, and the discovery of the double helical structure of DNA in 1953 marked the transition to the era of molecular genetics. However, now, it is recognized that before J.D. Watson and F. Crick (1953) proposed the model for the molecular structure of DNA, in 1952, R. Franklin and R. Gosling created an outstandingly clear X-ray diffraction pattern indicating a helical form associated with the DNA molecule. All of these scientific findings established the central dogma of molecular biology, in which DNA is transcribed to RNA and RNA is then translated to proteins. The impact of this dogma was so great and quickly assumed by science that 10 years later, M. Neiman (1964) declared the possibility of using DNA as a substrate for the recording, storage, and retrieval of information. The MNeimONics (Mikhail Neiman OligoNucleotides) era is still waiting to be fully developed, but it provides a proper idea of the revolution associated with disentangling the nature of DNA.

Around the 1950s, two big areas of research were emerging. Firstly, the scientific contributions of J.H. Tjio and A. Levan (1955) determining that the number of chromosomes in humans to be of 46 triggered a large passion for cytogenetics and its clinical implications, which reinforced the previous work consolidated by H.T. Morgan and B. McClintock. In parallel, other scientists such as A. Kornberg and S. Ochoa (1957) showed that the DNA polymerase 1 is required for in vitro synthesis of DNA, in which they were setting the groundwork of a new molecular technology in the laboratory.

In close relationship with modern -omics technologies, we must recognize the in situ hybridization of nucleic acid as the precursor, which allowed for the development of all these areas. The first contributions to the concept of in situ hybridization of nucleic acids emerged from F. Ritossa and S. Spiegelman when they published a paper in 1965, showing the possibility of immobilizing and denaturing DNA on a nitrocellulose filter and later the hybridization of radioactively labeled RNA to single-stranded DNA (DNA-RNA hybridization). The concept of in situ hybridization was later exploited by J. Gall and M.L. Pardue after hybridizing DNA to DNA. The paper "Molecular hybridization of radioactive DNA to the DNA of cytological preparations" was published 1969. It was also in this year where both authors showed, for the first time, the chromosomal localization of satellite DNA associated to the centromeres in mouse chromosomes using a DNA-DNA in situ hybridization protocol.

In 1975, Sir Edwin Mellor Southern was able to detect specific sequences among DNA fragments that were separated by gel electrophoresis. The rationale of the Southern blot, as a molecular technique to localize and characterize DNA-specific DNA molecules, was later exported to protein (Western blot) and RNA (Northern blot) characterization. EM Southern founded a company (Oxford Gene Technology) in Oxford, the United Kingdom, which developed a large part of the DNA microarray technology that today is found in every alley of genetics and is the clear precursor of most -omics technologies.

Behind these advances, there are enormous experimental efforts that made the development of the -omics technologies possible. However, the real value of -omics technologies is not in the nanotechnology, the computing, the sensing, or the imaging. The value of -omics technologies is in the application of the knowledge generated in human research that increased greatly with the information discovered and released by the Human Genome Project.

OMICS ERA

The completion of the Human Genome Project (HGP) in April 2003 yielded a permanent foundation for biological research and launched a new era in biomedicine. The goal of the HGP was to create a reference human genome sequence covering nearly the entire euchromatic genome with an error rate of ~1 in 10,000 bases. In fact, the final sequence covers 99% of the euchromatic genome with fewer than 350 gaps and has an error rate of ~1 in 100,000 bases. Work continues in various labs to close the remaining gaps. Elucidation and interpretation of the human genome is a work in progress at laboratories worldwide. Scientists in the broad community are working to understand its organization and variation and the roles these play in health and disease. One of the most important points is the genetic variability among individuals. The existence of variations, called polymorphisms, makes phenotypical differences among them. Although we cannot say that one individual is better than another of the same species due to genetic and also ethical reasons, we can affirm that one protein is more or less efficient at fulfilling its purpose. These polymorphisms are mutations that normally result from errors during DNA replication or other types of damage to DNA. These mutations can result in many different types of changes in sequences. Mutations in genes can either have no effect, alter the product of a gene, prevent the gene from functioning properly or completely, or even alter its expression. Mutations can also occur in nongenic regions. In that case, they can also have or not have an effect. Therefore, we can say that sequencing of the HGP will go on for decades discovering new genetic variants among human beings.

Genome sequencing is not exclusively conducted in humans. There are an uncountable number of genomes already sequenced, such as small organelles and plasmids, viruses, prokaryotes, and eukaryotes. These efforts support each other because many organisms, even those genealogically far, have homologous genes and proteins with similar functions. This allows for the development of animal models where we can study human disease and disorders.

These ambitious goals require and will continue to demand a variety of new technologies that have made it possible to relatively rapidly construct a first draft of the human genome and to continue to refine that draft. These techniques include the following:

- DNA sequencing
- The employment of restriction fragment length polymorphisms (RFLP)
- Yeast artificial chromosomes (YAC)
- Bacterial artificial chromosomes (BAC)
- The polymerase chain reaction (PCR)

- Quantitative PCR (Q-PCR)
- Electrophoresis
- Microarray

The various aspects of these analyses, mainly used for DNA and RNA analysis, together with other techniques that allow for the analysis of other molecules such as proteins and metabolites were joined under the suffix of "-omics." The English-language neologism omics informally refers to the collective characterization and quantification of pools of biological molecules that translate into the structure, function, and dynamics of an organism or organisms. The related suffix -ome is used to address the objects of study of such fields.

Omics sciences resulted in the assignment of new descriptive terms to familiar concepts, with the common theme that omics approaches attempt to measure all or some biological entity. Although there are many topics that use the suffix -omics, there are basically four omics areas of study that are necessary to identify:

Genomics: It is the study of genomes and the complete collection of genes that they contain. It is the study of the genome that stores the information in a cell to predict what can happen.

Genomics analyzes the DNA, mainly in two different aspects: the sequence (polymorphisms) and the amount (copy number variations). These analyses are mainly performed by microarray (SNPs or CGH) or next-generation sequencing (NGS).

Transcriptomics: It is also known as functional genomics that attempts to analyze patterns of gene expression and to correlate the patterns with the underlying biology. It is basically the study of mRNA changes that would depict what is really happening in a cell. There are a variety of microarray platforms from companies such as Affymetrix, Agilent, NimbleGen, and Illumina that have been developed to accomplish this. The basic idea for each platform is simple: a glass slide or membrane is spotted or arrayed with DNA fragments or oligonucleotides that represent specific gene coding regions. Nowadays, NGS by a method called RNA-Seq is able to analyze the full transcriptomic profile of a biological sample.

Proteomics: It examines the collection of proteins to determine how, when, and where they are expressed. It covers the study of protein molecules that would illustrate the functional roles of molecules in cellular function. Another mechanism is that protein function and activity are regulated or restricted by posttranslational and covalent modifications of protein structure, such as phosphorylation, methylation, and glycosylation, as well as other protein-protein interactions or protein-small molecule interactions. Probably, together, metabolomics is the most complex omics.

Metabolomics: It is a large-scale approach to monitor the compounds involved in cellular processes in a single assay to derive metabolic profiles. It is the study of molecules involved in cellular metabolism that would eventually depict the phenotype of an organism.

With regard to the techniques used in the omics sciences, microarray technology is the most extensively used. The discovery of microarrays for gene expression in 1995 [1] opened the possibility to study the gene expression of thousands of genes in one single experiment. This technology was later applied to DNA analysis. Now, there are microarrays for DNA analysis, including copy number variation (CNV), called comparative genome hybridization (CGH) microarrays, genotyping (SNPs microarrays), and DNA methylation (CpG island microarrays). For RNAm expression, there is gene expression, GE, microarrays. Finally, there are microarrays for protein expression analysis, protein-protein interactions, ligand-DNA interaction analysis, and others that are less known with very specific applications. Next-generation sequencing (NGS) has occupied a large part of the microarray uses applied to genomics and transcriptomics due to the reduction in costs and the advances in the sequencing technology itself. RNA sequencing (RNA-Seq), also called whole-transcriptome shotgun sequencing (WTSS), uses NGS to reveal the presence and quantity of RNA in a biological sample at a given moment. RNA-Seq is used to analyze the continuously changing cellular transcriptome. Transcriptomics also uses multiquantitative PCR (Q-PCR) for gene expression analysis. Proteomics and metabolomics share sophisticated techniques such as high-performance liquid chromatography (HPLC), mass spectrometry (MS), or nuclear magnetic resonance (NMR) that complete the battery of technologies designed for massive studies of biological molecules.

Finally, there is a nonomic component to omics that is bioinformatics. Bioinformatics utilizes the large-scale data sets that are generated by the omics technologies for analyses. These analyses use techniques developed in fields such as computational analysis and statistics that facilitate the understanding of omics data. Common uses of bioinformatics include data normalization, basic and multivariant statistical analysis, correlation and nonlineal relation analysis between methylation and gene expression, identification of candidate genes as biomarkers, sample and gene clustering, single nucleotide polymorphisms (SNPs), meta-analysis, gene ontology, and molecular interactions.

In the last 5–10 years, numerous promising developments have been elucidated using omics science such as the following:

Epigenomics: It is the study of epigenetic modifications across an individual's entire genome, known as the epigenome. An organism has multiple, cell-type-specific epigenomes comprising epigenetic marks such as DNA methylation, histone modification, and specifically positioned nucleosomes. Functionally, epigenetic marks act to regulate gene expression. Therefore, epigenomics allows for the identification of genome sequences that are important for the regulation of gene expression.

Cancers are caused by changes in the genome, the epigenome, or both. Changes in the epigenome can switch on or off genes involved in cell growth or the immune response. These changes can lead to uncontrolled growth, a hallmark of cancer, or to a failure of the immune system to destroy tumors. For example, in a type of brain tumor called glioblastoma, doctors have had some success in treating patients with the drug temozolomide, which kills cancer cells by adding methyl groups to DNA. In some cases, methylation has a welcome secondary effect as it blocks a gene that counteracts temozolomide. Glioblastoma patients whose tumors have such methylated genes are far more likely to respond to temozolomide than those with unmethylated genes.

Interactomics: It is the study of the set of interactions between the biomolecules of a cell, tissue, or organism. It usually refers to the interaction between proteins. The elaboration of the interactome allows for a global vision of the cellular machinery, which allows for the prediction of the repercussions that originate in specific alterations.

It has been observed that the transporter interactome is essential for the acquisition of antimicrobial resistance to antibiotics. The functional interaction of various types of multidrug efflux transporters (MDTs), which remove toxic compounds from the cytoplasm, provides an essential first-line defense mechanism, preventing drugs from reaching lethal concentrations, until a number of stable, more efficient alterations occur that allow for the survival of microbes in the presence of antibiotics.

Cellomics: It refers to the study of cellular phenotype and function. Cellomics is the discipline of quantitative cell analysis using bioimaging methods and informatics. Currently, there are over 40 different application areas that cellomics is used in, including the analysis of 3D cell models, angiogenesis, cell signaling, cancer research, neuroscience research, drug discovery, consumer products safety, and toxicology.

Nutrigenomics: It analyzes how genes and nutrients interact. Nutrigenomics is the study of the effects of food and food constituents on gene expression, and how genetic variations affect the nutritional environment. It focuses on understanding the interaction between nutrients and other dietary bioactives with the genome at the molecular level, to understand how specific nutrients or dietary regimes may affect human health.

Obesity is one of the most widely studied topics in nutrigenomics. Due to genetic variations among individuals, each person could respond to a diet differently. There are studies suggesting genetic factors account for a fair proportion of interindividual body mass index (BMI). Among different types of genetic variation between humans, SNPs are suggested to be the most important marker for the study of nutrigenomics. For example, multiple studies have found an association between SNPs and obesity. One of the most well-known obesity-associated genes is the FTO gene. Among studied individuals, it was found that those with the AA genotype showed higher BMI compared with those with the TT genotype when having a high fat or a low carbohydrate diet. The study of genetic variations in relation with nutrients is called nutrigenetics.

Pharmacogenomics: It is the study of how drug interacts with the genome. As part of this omics, *pharmacogenetics* focuses on the identification of genome variants that influence drug effects, via alterations in pharmacokinetics (absorption, distribution, metabolism, and elimination) or via modulation of pharmacodynamics (modifying a drug's target). It aims to develop rational means to optimize drug therapy, with respect to the patients' genotype, to ensure maximum efficacy with minimal adverse effects.

Azathioprine is an immunosuppressant used to help prevent rejection after organ transplant operations and is also used to treat a variety of inflammatory and autoimmune diseases such as rheumatoid arthritis. In some individuals, azathioprine is not activated in the body properly. As a consequence, it kills developing white blood cells and leaves the individual vulnerable to infection. The conversion of azathioprine into its active form is catalyzed by an enzyme called thiopurine *S*-methyltransferase (TPMT). Some variants of the gene encoding TPMT do not allow individuals to do this conversion. In pharmacogenetics studies, people with rheumatoid arthritis can be tested to find out which variant of the TPMT gene they possess and whether azathioprine will be an effective treatment for them.

Toxicogenomics: The field of toxicogenomics is used to study the structure and output of the genome as it responds to adverse xenobiotic exposure, and it is very closely related to pharmacogenomics. Toxicology has traditionally been evaluated by the dosing of animals to define well-established cytological, physiological, metabolic, and morphological endpoints. The evaluation of the risk to humans cannot be performed in human individuals initially and thus must be derived from studies performed in other species. Typically, rodents are used to identify toxic substances such as carcinogens, reproductive toxins, and neurotoxins. Currently, toxicogenomics applies genomics concepts and technologies to study adverse effects of chemicals. These studies use global gene expression analyses to detect expression changes that influence, predict, or help define drug toxicity. In essence, toxicogenomics combines the tools of traditional toxicology with those of genomics and bioinformatics.

Lipidomics: It is the study of the structure and function of the complete set of lipids (the lipidome) produced in a given cell or organism and their interactions with other lipids, proteins, and metabolites. The lipidome describes the complete lipid profile within a cell, tissue, or organism, and it can be used to monitor temporal changes in response to a certain stimulus or with the onset and evolution of a disease. Lipidomics analysis suggests changes of some potential lipid biomarkers in the development of diseases like diabetes, cardiac dysfunction, and hypertrophy of diabetic cardiomyopathy, which may serve as potential important targets for clinical diagnosis and therapeutic intervention of these diseases in the future.

Secretomics: It describes the global study of proteins that are secreted by a cell, a tissue, or an organism. The term secretome refers to the global group of secreted proteins into the extracellular space by a cell, tissue, organ, or organism at any given time and conditions through known and unknown secretory mechanisms involving constitutive and regulated secretory organelles. As a part of secretomics, we have to highlight *oncosecretomics*, which is the study of alterations in the secretome of cells when they become cancerous. In this part of the secretomics, these analyses have discovered potential new biomarkers in many cancer types, including lung cancer, liver cancer, pancreatic cancer, colorectal cancer, prostate cancer, and breast cancer. Prostate-specific antigen (PSA), the current standard biomarker for prostate cancer, has a low diagnostic specificity (PSA levels cannot always discriminate between aggressive and nonaggressive cancer), and so, a better biomarker is greatly needed. Using secretomics analysis of prostate cell lines, one study was able to discover multiple proteins found in higher levels in the serum of cancer patients than in healthy controls.

Connectomics: It is the study of the neural connections of an organism's nervous system. While the principal focus is the brain, any neural connections could theoretically be mapped by connectomics, including neuromuscular junctions. Connectivity abnormalities have been well established in schizophrenia. Connectomics studies, which map neuronal connections based on functional magnetic resonance imaging, revealed abnormal functional connectivity of the prefrontal cortex in first-episode patients. Functional connectivity is enhanced between the prefrontal cortex and the temporal lobe and reduced between the prefrontal cortex and the parietal lobe, posterior cingulate cortex, thalamus, and striatum of patients. Hypo- and hyperconnectivity in the parietal lobe, occipital lobe, and prominently the frontal and temporal lobes indicate a diffuse functional disconnectivity in schizophrenia.

Metagenomics: It is the study of genetic material recovered directly from environmental samples. It applies a suite of genomics technologies and bioinformatics tools to directly access the genetic content of entire communities of organisms and provides information on the microbial diversity and ecology of a specific environment. Shotgun metagenomics refers to the approach of shearing DNA extracted from the environmental sample and sequencing the small fragments.

Microbiomics: It is the study of all the microorganisms (microbiota) that populate an organism. This could be the microbiota of an environmental sample (e.g., soil or water), a particular body site (e.g., the gut or the mouth), or a particular organism. With metagenomics and microbiomics studies, it has been demonstrated that the human gut has two bacterial divisions, Bacteroidetes and Firmicutes, that constitute over 90% of the known phylogenetic categories that dominate distal gut bacteria. Using the relative gene frequencies found within the gut, these researchers identified 1244 metagenomics clusters that are critically important for the health of the intestinal tract. There are two types of functions in these clusters: housekeeping and those specific to the intestine. The housekeeping gene clusters are required in all bacteria and are often major players in the main metabolic pathways including central carbon metabolism and amino acid synthesis. The gut-specific functions include adhesion to host proteins and the harvesting of sugars from glycolipids. Patients with irritable bowel syndrome were shown to exhibit 25% fewer genes and lower bacterial diversity than individuals not suffering from irritable bowel syndrome indicating that changes in patients' gut biome diversity may be associated with this condition.

Psychogenomics: It is used to describe the process of applying the powerful tools of genomics and proteomics to achieve a better understanding of the biological substrates of normal behavior and of diseases of the brain that manifest themselves as behavioral abnormalities. For example, in hypothesis-driven studies, genes in different brain regions were selectively expressed, down-regulated, or knocked out in animal models of addiction. Recent high-throughput expression-profiling technologies such as microarray and proteomics analyses identified candidate genes and proteins whose expression level changed significantly among different states in drug addiction. Moreover, genetic studies such as animal quantitative trait locus (QTL) studies, genetic linkage studies, and population association studies identified chromosomal regions that may contribute to the vulnerability to addiction.

Glycomics: The term glycomics is derived from the chemical prefix for sweetness or a sugar, glycol. It is an integrated approach to study structure-function relationships between complex carbohydrates or glycans such as glycolipids, glycoproteins, lipopolysaccharides, peptidoglycans, and proteoglycans. Comparative studies of specific carbohydrate chains of glycoproteins can provide useful information for the diagnosis, prognosis, and immunotherapy of tumors. A wide variety of technologies are now being developed to deal with the technically difficult problems of glycan structural analysis and the investigation of the functional roles. Enabling technologies such as high-throughput mass spectroscopy, glycan microarrays, aminoglycoside antibiotic microarrays and glycan sequencing, quantum dots, and gold nanoparticles are currently helping to unravel the complexity resulting from diverse glycans in biological systems.

OMICS IN THE ASSISTED REPRODUCTION ERA

The mentioned omics are now emerging as powerful tools for medical research in different fields. Almost invariably, these advances in omics have been associated with major expectations of transforming not only biological knowledge but also personalized medicine and health. In this sense, in 2012, the term *REPRODUCTOMICS* appeared that gives its name to this book. This name was coined at that time by a group of Spanish researchers [2], and it was defined as the application of omics technologies to the field of reproduction.

Before we enter into the specific chapters of this book, we thought it would be useful to briefly summarize the most important advances in the different fields of reproductomics:

Genomics

In human reproduction, we can classify omics studies in three different areas: preconceptional, preimplantational, and prenatal.

Preconceptional: This can also be subdivided into two different types of studies: adults and gametes. Preconceptional analysis in couples can be performed to analyze the genetic basis of infertility or to detect mutations for monogenic diseases. On one hand, there are specific mutations or copy number variations in the genome that can be related with specific diseases such as azoospermia, premature ovarian failure, recurrent miscarriage due to thrombophilia, or disorders such as atypical ovarian response or polycystic ovarian syndrome. These studies are nowadays carried out by SNP and CGH microarrays and recently by the use of NGS. In many cases, it can anticipate detection of a future problem. On the other hand, the DNA of couples can be analyzed to check for carrier status for thousands of mutations of hundreds of monogenic (Mendelian) diseases. There are more than 7000 monogenic diseases already discovered, but this type of analysis is restricted mainly to recessive and X-linked diseases. After the matching of the results, it is possible to know the reproductive risk of the couple to have an affected baby. In the case that both parents are a carrier of the same disease, a preimplantation genetic diagnosis can be done to assure the transfer of a nonaffected embryo. This analysis is also carried out by SNP microarray and NGS. A comprehensive review about this topic was recently published [3].

Other preconceptional genomics studies analyze the chromosomes of the polar body of the oocyte in order to evaluate the chromosomic composition of the oocyte nucleus. This approach was developed in those countries in which the genetic evaluation of the embryo was not allowed as an alternative method to know more about the genetics of the oocyte. Nowadays, this approach is usually not used, as most countries allow PGD/PGS that investigate the genetics of the growing embryo to detect chromosomal abnormalities, although the proof-of-principle study indicates that the ploidy of the zygote can be predicted with acceptable accuracy by genomics analysis of both polar bodies [4].

Preimplantational: PGS/PGD: Preimplantation genetic testing is a technique used to identify genetic defects in embryos created through in vitro fertilization (IVF) before being transferred to the uterus. Here, we have to distinguish two types of analyses: preimplantation genetic diagnosis (PGD) and preimplantation genetic screening (PGS). PGS is the most commonly used, and it refers to techniques where embryos from presumed chromosomally normal genetic parents are screened for aneuploidy by different genomics techniques. On the other hand, PGD is carried out when one or both genetic parents have a known genetic abnormality. In this case, the embryos are tested for these specific mutations in order to select embryos not affected by one specific disease.

Prenatal

Genomics has also begun to be very common in prenatal analysis. Classically, prenatal testing has been done by using either amniotic fluid or chorionic villus sampling to check the set of chromosomes by cytogenetic techniques. Both of these methods involve collection from inside the uterus and are considered invasive. These methods also increase the risk of miscarriage by around 1% and are therefore only used when other measures show a high risk of disease or abnormality. A few years ago, the NGS techniques have become widely available that allows for the examination of small fragments of fetal DNA that are floating in the mother's blood. From blood, the fetal DNA can be isolated and examined for a range of abnormalities, and it is possible to identify which chromosome each fragment comes from. By looking at the proportion of fragments, it is possible to determine if they come from a person with the standard complement of 46 chromosomes or if there are more or fewer chromosomes. The fragments can also be tested for some specific genetic markers such as some microdeletions.

Transcriptomics

The use of transcriptomics in the reproductive field is useful in determining expression profiles of various cell types, such as the endometrium, the cumulus cells (CCs), the sperm, the trophoblast, and the whole embryo. It could even be utilized to determine the molecular basis of the oocyte maturation process related to aging [5] or the in vitro and in vivo maturation of human oocyte [6].

The endometrium is a very complex tissue. It is very dynamic and can have dramatic physiological changes in response to the absence or presence of hormones. During the past X decades, researchers and clinicians have not developed new tools for endometrial evaluation being the gold standard for clinical evaluation the histological studies and the ultrasound support. With the development of the microarray technology, the capability for monitoring the gene expression of thousands of genes at the same time opened new possibilities. One of the chapters in this book addresses the advances in the transcriptomics of the endometrium in natural, stimulated cycle and pathological situations such as endometriosis or cancer [7].

Transcriptomics analysis has also began to investigate human oocytes and cumulus cells to explain the mechanism involved in oocyte maturation and also provide invaluable molecular markers of gene expression in oocytes. It has correlated the gene expression profile of CCs around the oocyte with its viability to create a specific novel indirect approach to predict embryo quality and potential pregnancy outcome [8].

Other areas of human reproduction, such as andrology, have adapted transcriptomics analysis to further investigate unexplained infertility. For decades, the transcriptional activity in spermatozoa has not been considered due to the absence of ribosomal and other features. There are specific cases of men with normal sperm after basic analysis that are unable of having a full-term pregnancy. First, studies compared global genome expression using microarrays from the sperm of patients whose sperm achieved pregnancy versus those whose sperm failed to do so, in both intrauterine insemination and ICSI cycles. They found a different transcriptomic signature between them, thus supporting the idea that a microarray-based diagnostic tool could be used to predict the possible potential of the sperm to initiate a pregnancy or not [9].

Proteomics and Metabolomics

Proteomics and metabolomics are the two omics closest to the phenotype. They can be used to build a new database for basic knowledge to possibly translate into clinical testing and eventually into medical routine care in this critical branch of health care. These approaches, as in other omics, cover all of the areas of reproduction, gametes, embryos, spent media, serum, placenta, uterine fluids, etc.

In proteomics, the proteomes between diabetic and normal human sperm have been analyzed to look for insights into the effect of diabetes on male reproduction based on the regulation of mitochondria-related proteins. The aim of this study was to identify sources of the reduced fertility of men with type 2 diabetes mellitus. These individuals showed significant reductions in semen volume, sperm concentration, and total sperm count. Proteins that are potentially associated with these sperm defects were identified using proteomics in this study. Comparative analysis of proteomes between diabetic and normal human sperm has led to insights into the effects of diabetes on male reproduction based on the regulation of mitochondria-related proteins [10]. This is only an example, with a very good description of one of the proteomics approaches performed in assisted reproduction and can be found in the paper of Kosteria et al., published in 2017 [11].

In metabolomics, using NMR, researchers have been able to identify a sensitive and specific metabolomic profiling method to distinguish missed abortion in serum with a potential clinical use [12] and also, using the same technique, to identify altered metabolic pathways in the placenta related with preeclampsia [13].

One of the most exciting future applications is the use of metabolomics for noninvasive selection of embryos by analyzing the metabolic profile of the spent media (i.e., the media the embryos have been cultured in for hours or days). The first approaches have analyzed the spent media using proton nuclear magnetic resonance [14] or near-infrared spectroscopy [15].

Lipidomics is one of the sciences in which scientists have focused their investigations. The lipidomic profile of endometrial fluid in women with ovarian endometriosis has been published, and this group found different proteins in the endometrial fluid between the two groups. This technique could potentially be considered as a minimally invasive approach for the diagnosis of endometriosis [16].

Other researchers have found serum lipid biomarkers for preeclampsia using direct-infusion mass spectrometry. These newly developed panels of serum lipidomic biomarkers appear to be able to identify most women at risk for preeclampsia in a given pregnancy at 12–14-week gestation [17]. This has also been found in a different approach in the placenta [18]. Even the profile of lipids in spermatozoa are related with cryoresistance, as it was found that phosphatidylcholines were associated with higher quality of spermatozoa after thawing, especially in a functional capacity, and that lipid semen composition was found to influence the resistance of spermatozoa to cryopreservation and may interfere with motility, membrane integrity, and lipid peroxidation in stallions [19].

Secretomics, which is mainly investigated by proteomics and metabolomics, has focused mainly in uterine fluid and spent media, that is, the media in which the embryo is cultured in. Using protein-array technology, it analyzes the protein profile corresponding to 24h conditioned media of blastocysts that implanted versus those that did not implant. They found that the soluble TNF receptor 1 and IL-10 increased significantly and MSP-a, SCF, CXCL13, TRAILR3, and MIP-1b decreased significantly when the protein profile of the blastocyst culture medium was compared with the control medium. CXCL13 (BLC) and granulocyte-macrophage colony-stimulating factor were also decreased significantly in the implanted blastocyst media compared with that in media from the nonimplanted counterparts with a similar morphology. These differences identified in the protein profile of the culture media in the presence of implanted versus nonimplanted blastocysts could be considered in the future as a new noninvasive approach in the search for new tools to diagnose blastocyst viability [20].

Regarding the uterine secretomics, the two most important papers that have been published to date show part of the molecular dialogue between the endometrium and the embryo. The results obtained in these works, properly validated, could represent a step forward in the development of noninvasive diagnostic tools to assess endometrial receptivity [21,22].

Epigenomics: Assisted reproductive technology involves several steps that subject the gametes and early developing embryos to environmental stress, and this is the primary reason for an increased interest in the putative link between these techniques and imprinting disorders. Although animal studies support a link between assisted reproductive techniques (ARTs) and imprinting disorders, via altered methylation patterns, data in humans are inconsistent. Understanding how epigenetic mechanisms contribute to transgenerational transmission of obesity and metabolic dysfunction is crucial for the development of novel early detection and prevention strategies for programmed metabolic syndrome due to the fact that epigenetic gene modifications may be reversible [23]. Another interesting point is the link between in utero and neonatal exposure to environmental toxicants, such as endocrine-disrupting chemicals (EDCs) and adult female reproductive disorders that is well established in both epidemiological and animal studies [24].

The omics field is greatly changing biomedical research and clinical practice. The simultaneous analysis of thousands of molecules, in cells, tissues, and organs, not only has an impact on our understanding of biological processes but also will improve the prospect of more accurately diagnosing and treating various diseases. The reality of applying omics technologies to unravel disease processes, identify disease biomarkers, and finally make a recommendation for personalized medicine is expected to revolutionize medical practice in all medical fields including assisted reproduction. Therefore, the future of reproductomics continuously integrates these sciences for both research and clinical practice.

References

[1] S. M, D. Shalon, R.W. Davis, P.O. Brown, Quantitative monitoring of gene expression patterns with a complementary DNA microarray, Science 270 (1995) 467–470.

[2] J. Bellver, M. Mundi, F.J. Esteban, S. Mosquera, J. Horcajadas, '-omics' technology and human reproduction: reproductomics, Expert Rev. Obstet. Gynecol. 7 (2012) 493–506.

[3] J. Dungan, Expanded carrier screening: what the reproductive endocrinologist needs to know, Fertil. Steril. 109 (2018) 183–189.

[4] G. J, M. Montag, M.C. Magli, S. Repping, A. Handyside, C. Staessen, J. Harper, A. Schmutzler, J. Collins, V. Goossens, H. van der Ven, K. Vesela, L. Gianaroli, Polar body array CGH for prediction of the status of the corresponding oocyte. Part I: clinical results, Hum. Reprod. 26 (2011) 3173–3180.

[5] M.L. Grøndahl, C. Yding Andersen, J. Bogstad, F.C. Nielsen, H. Meinertz, R. Borup, Gene expression profiles of single human mature oocytes in relation to age, Hum. Reprod. 25 (2010) 957–968.

[6] G.M. Jones, D.S. Cram, B. Song, et al., Gene expression profiling of human oocytes following in vivo or in vitro maturation, Hum. Reprod. 23 (2008) 1138–1144.

[7] J.A. Horcajadas, A. Pellicer, C. Simón, Wide genomic analysis of human endometrial receptivity: new times, new opportunities, Hum. Reprod. Update 13 (2007) 77–86.

[8] S. Assou, D. Haouzi, J. De Vos, S. Hamamah, Human cumulus cells as biomarkers for embryo and pregnancy outcomes, Mol. Hum. Reprod. 16 (2010) 531–538.

[9] N. Garrido, J.A. Martínez-Conejero, J. Jauregui, J.A. Horcajadas, C. Simón, J. Remohí, M. Meseguer, Microarray analysis in sperm from fertile and infertile men without basic sperm analysis abnormalities reveals a significantly different transcriptome, Fertil. Steril. 91 (4 Suppl) (2009) 1307–1310.

[10] T. An, Y.F. Wang, J.X. Liu, Y.Y. Pan, Y.F. Liu, Z.C. He, F.F. Mo, J. Li, L.H. Kang, Y.J. Gu, B.H. Lv, S.H. Gao, G.J. Jiang, Comparative analysis of proteomes between diabetic and normal human sperm: insights into the effects of diabetes on male reproduction based on the regulation of mitochondria-related proteins, Mol. Reprod. Dev. 85 (2018) 7–16.

[11] I. Kosteria, A.K. Anagnostopoulos, C. Kanaka-Gantenbein, G.P. Chrousos, G.T. Tsangaris, The use of proteomics in assisted reproduction, In Vivo 31 (2017) 267–283.

[12] Z. Wu, L. Jin, W. Zheng, C. Zhang, L. Zhang, Y. Chen, J. Guan, H. Fei, NMR-based serum metabolomics study reveals a innovative diagnostic

model for missed abortion, Biochem. Biophys. Res. Commun. 496 (2018) 679–685.

[13] M. Austdal, L.C. Thomsen, L.H. Tangerås, B. Skei, S. Mathew, L. Bjørge, R. Austgulen, T.F. Bathen, A.C. Iversen, Metabolic profiles of placenta in preeclampsia using HR-MAS MRS metabolomics, Placenta 36 (2015) 1455–1462.

[14] E. Seli, L. Botros, D. Sakkas, D.H. Burns, Noninvasive metabolomic profiling of embryo culture media using proton nuclear magnetic resonance correlates with reproductive potential of embryos in women undergoing in vitro fertilization, Fertil. Steril. 90 (2008) 2183–2189.

[15] E. Seli, D. Sakkas, R. Scott, S.C. Kwok, S.M. Rosendahl, D.H. Burns, Noninvasive metabolomic profiling of embryo culture media using Raman and near-infrared spectroscopy correlates with reproductive potential of embryos in women undergoing in vitro fertilization, Fertil. Steril. 88 (2007) 1350–1357.

[16] F. Domínguez, M. Ferrando, P. Díaz-Gimeno, F. Quintana, G. Fernández, I. Castells, C. Simón, Lipidomic profiling of endometrial fluid in women with ovarian endometriosis, Biol. Reprod. 96 (2017) 772–779.

[17] S. Anand, S. Young, M.S. Esplin, B. Peaden, H.D. Tolley, T.F. Porter, M.W. Varner, M.E. D'Alton, B.J. Jackson, S.W. Graves, Detection and confirmation of serum lipid biomarkers for preeclampsia using direct infusion mass spectrometry, J. Lipid Res. 57 (2016) 687–696.

[18] H.A. Korkes, N. Sass, A.F. Moron, N.O. Câmara, T. Bonetti, A.S. Cerdeira, I.D. Da Silva, L. De Oliveira, Lipidomic assessment of plasma and placenta of women with early-onset preeclampsia, PLoS One 9 (10) (2014).

[19] T. Cabrera, C. Ramires-Neto, K.R.A. Belaz, C.P. Freitas-Dell'aqua, D. Zampieri, A. Tata, M.N. Eberlin, M.A. Alvarenga, F.F. Souza, Influence of spermatozoal lipidomic profile on the cryoresistance of frozen spermatozoa from stallions, Theriogenology 108 (2018) 161–166.

[20] F. Domínguez, B. Gadea, F.J. Esteban, J.A. Horcajadas, A. Pellicer, C. Simón, Comparative protein-profile analysis of implanted versus non-implanted human blastocyst, Hum. Reprod. 23 (2008) 1993–2000.

[21] C. Boomsma, C. Heijnen, N. Macklon, Uterine secretomics: a window on the maternal-embryo interface, Fertil. Steril. 99 (2013) 1039–1093.

[22] O. Berlanga, H.B. Bradshaw, F. Vilella-Mitjana, T. Garrido-Gómez, C. Simón, How endometrial secretomics can help in predicting implantation, Placenta 32 (Suppl. 3) (2011) S271–S275.

[23] M. Desai, J.K. Jellyman, M.G. Ross, Epigenomics, gestational programming and risk of metabolic syndrome, Int. J. Obes. 39 (2015) 633–641.

[24] A.M. Zama, M. Uzumcu, Epigenetic effects of endocrine-disrupting chemicals on female reproduction: an ovarian perspective, Front. Neuroendocrinol. 31 (2010) 420–439.

CHAPTER

1

Is the Classic Spermiogram Still Informative? How Did It Develop and Where Is It Going?

William V. Holt

Academic Unit of Reproductive and Developmental Medicine, University of Sheffield, Sheffield, United Kingdom

INTRODUCTION

Although it might seem that human spermatozoa only have to fulfill a single role, namely, to achieve fertilization, this can only occur if a multitude of associated cell functions operate perfectly. If any of the cell systems fail within an individual spermatozoon, it will be prevented from fertilizing the oocyte and supporting subsequent embryonic development. A glance at the complex biochemical diagrams that aim to represent metabolic pathways leads to the thought that there are also multiple functions that could go wrong and that the probability of finding a flawless spermatozoon must indeed be remote. Nevertheless, fertilization events continue to occur despite the drawbacks, and the process is of universal importance for the continuation of life on this planet. It is relevant to this review that the spermatozoa that eventually reach and fuse with oocytes under natural conditions have almost invariably been subjected to stringent selection [1]. If the selection mechanisms that operate in nature are able to discriminate the quality of spermatozoa, it is reasonable to ask whether scientists in a laboratory can do the same thing. To answer this question requires some understanding of the natural mechanisms behind sperm selection. As this reviewer has been charged with assessing the merits of basic semen analysis, the article will focus on whether much relevant information can be gained by using relatively simple approaches.

The technical evaluation of semen quality is also important for any consideration of basic semen analysis. Clearly, if the technical procedures are performed inadequately, it follows that the biological interpretation of results and hence the validity of any fertility-related diagnosis or prediction are undermined. In the United Kingdom, this was recognized more

https://doi.org/10.1016/B978-0-12-812571-7.00002-2

than three decades ago by the British Andrology Society (BAS), which initiated a national quality assurance scheme for clinical andrology laboratories. When the scheme was established, the technical shortcomings within some centers were striking. Single semen samples were reported so differently by diverse laboratories that the same assessments could have identified individuals as being either a semen donor, that is, with a high total sperm count and good progressive motility, or a candidate for infertility treatment. Anecdotal reports persistently indicate that, even today, some laboratories still fail to use a temperature-controlled microscope stage when carrying out subjective sperm motility analyses. In general, the focus on higher standards has paid dividends [2], and consistent and accurate results are now routinely achieved. It is important that these are maintained through technician training, both inhouse and through national organizations, and that standards are overseen through quality assurance schemes.

As a key objective of this review is to revisit "basic" semen analysis and to consider its merits, it will be important here to take account of both the techniques involved in semen assessment and the likely meaning of the results. I will limit my interpretation of "basic semen analysis" to what might be considered as the simplest set of parameters; that is, ejaculate volume, sperm concentration, sperm morphology, sperm viability (or rather plasma membrane integrity), and motility. In fact, none of these parameters can really be regarded as simple, but they were relatively easy to implement at a time when laboratories routinely possessed only a microscope, a warm plate, some pipettes, and some counting chambers.

I also avoid discussing the significance of sperm DNA fragmentation in relation to fertility; however, it is worth pointing out that researchers interested in the development of sperm cryopreservation and storage methods had already recognized in the 1960s that the DNA content of spermatozoa might be affected by ex situ storage [3–5]. At that time, their technology was limited to the use of Feulgen staining, a technique that stains DNA red by the use of Schiff's reagent and spectrophotometric methods that could be used to quantify the amount of DNA in individual cell nuclei. Technologies for more precise measurements of sperm DNA became available in the late 1980s, but their use was limited to a small number of researchers who began to realize the importance of the developing field [6,7].

SPECIFIC ASPECTS OF BASIC SEMEN ANALYSIS

Semen Volume

The measurement of human semen volume is less straightforward than might be expected. Volumetric measurements using pipettes or measuring cylinders have been used for many years [8]; this approach is practical and easy to perform and has previously been recommended for routine use [9]. However, the increased demand for data validation and laboratory accreditation has forced some reevaluation of the optimal methodology for semen volume estimation, and it has been shown that the use of pipettes and cylinders consistently underestimates the real volumes [10]. The degree of underestimation may be 0.2–0.4 mL and occurs because some of the semen is retained within the pipette. The better alternative involves collecting the semen in a container of known weight and estimating the combined weight of the container plus semen, thus permitting the semen weight to be calculated by subtraction. As it is known that the specific gravity of semen is slightly over 1.00 g/mL, the volume can be estimated accurately from the weight. This approach is now recommended by the World Health Organization [11] and supersedes the earlier WHO recommendation [12] to use the volumetric method. An underestimate of 0.2–0.4 mL may seem to be of little significance in terms of patient management, but when the lower reference limit for volume is only 1.5 mL (with 5th and 95th confidence limits of 1.4–1.7 mL) [11], the underestimate could shift a volume assessment of 1.5 mL down to below the reference value. It would also affect the calculation of total sperm number within the ejaculate. Moreover, as pointed out by Matson and colleagues [10], medical laboratories have a responsibility to ensure accuracy and provide an indication of the uncertainties associated with the data they report. The prognostic value of semen volume is nevertheless rather uncertain, as the natural within-subject variation can be as high as 40%. This means that working with values obtained from a single estimate is highly unreliable. This may explain why it has been difficult to identify any correlations between semen volume and pregnancy outcome [13].

The sources of variation in semen volume are not clear, but Schwartz et al. [14] identified abstinence period as a significant positive correlate of semen volume. Variation in semen volume is largely determined by the amount of seminal plasma contained in the ejaculate and, in turn, the amount of fluids that emanated from the accessory sex glands. From a strictly technical point of view, seminal plasma may be seen as little more than a fluid vehicle that facilitates the transfer of spermatozoa into the female reproductive tract. However, extensive research across different mammalian species has shown that seminal plasma contains a complex mixture of proteins whose functions extend far beyond that of a simple fluid. A recent discussion [15] attributed several significant functional roles to seminal plasma, including an influence on sperm capacitation, formation of the oviductal sperm reservoir, and facilitation of sperm-oocyte interactions. As seminal plasma is also a source of antioxidants, it confers protective effects on spermatozoa; however, comparisons of seminal plasma proteins between normal men and infertile patients have demonstrated that infertility can be associated with the presence of inappropriately expressed proteins [16]. While these observations show that seminal plasma analysis may therefore prove useful for the diagnosis of infertility through proteomic analysis, they also demonstrate clearly why a simple estimation of seminal plasma volume may not be a helpful diagnostic indicator of semen quality.

Estimation of Sperm Concentration

Sperm concentration measurements are routinely performed using one of several available designs of counting chambers or disposable slides. The operator usually views these chambers by phase-contrast microscopy and has to count the number of spermatozoa seen within certain parts of a grid. Of the counting chambers themselves, some have inherently higher reproducibility and therefore better precision than others, and it is relevant to consider whether this is important in clinical practice. Christensen and his colleagues [17] compared the accuracy and precision of four types of counting chamber, Makler chambers (Sefi Medical Instruments, Haifa, Israel), Thoma 50 and 100 μm deep chambers (Thoma hemocytometers; Hecht Assistent, Sondheim, Germany), and Bürker-Türk hemocytometers (BT, Brand, Wertheim, Germany). The tests were performed using bull and boar semen and were replicated by including data from more than one technician. Importantly, the counts were calibrated against flow cytometric estimation, which is widely accepted as an independent and objective counting technique. These authors found that the Makler chamber showed highest coefficients of variation (CV; 15%–24%) and consistently underestimated the sperm concentrations by about 25%. This is a significant finding and would lead to serious errors in clinical practice. In the same study, the CVs of other chambers were clustered around 7%–14%. In an earlier study, Mahmoud et al. [18] also tested four different methods for determining sperm concentration and found that the Neubauer hemocytometer produced a CV of about 7%. Where technical differences between individual technicians and between laboratories have been evaluated, the CVs appear to be rather large. A study by Auger et al. [19], which involved allowing 10 teams to estimate sperm concentration for the same sample but using their own techniques, found that the mean interindividual CV was 22.9%. Similarly, a pilot quality control scheme undertaken in the United Kingdom [20] and involving 20 laboratories observed considerable disparities between the concentration values reported, where the highest value was >10-fold higher than the lowest and also found large differences between the intraindividual CV values (four of the returning laboratories having a CV of <10%, nine laboratories reporting between 10% and 20%, and five laboratories >20%). It is interesting to note that Freund et al. [21] had observed these large and significant errors in a systematic study performed more than five decades ago and noted that it would represent a serious problem whenever two or more laboratories wished to cooperate and share research data.

It is clearly desirable to aim for the lowest CV where possible. However, it is worth considering what the CV means in practical terms. Assuming that a particular set of data conform to a statistically normal distribution, individual measurements within the data may fall within three standard deviations (SD) above and below the mean. CV is calculated as $((SD \times 100)/mean)$; therefore, if we consider a semen sample whose actual sperm concentration is $100 \times 10^6/mL$, performing a count using a chamber whose CV is 25% means that the SD is actually $\pm 25 \times 10^6/mL$. The normal distribution could therefore include individual values in the range $25–175 \times 10^6/mL$ (i.e., 3×25 $10^6/mL$ both above and below the mean value). As the current WHO lower reference limit for sperm concentration is $15 \times 10^6/mL$ [11], a count performed using a chamber whose CV = 25% could cause serious diagnostic problems as it would be feasible to obtain individual estimates ranging from around $4 \times 10^6/mL$ to $26 \times 10^6/mL$. The WHO manual and other publications [22] go to great lengths to explain these principles and recommend many practices aimed at improving precision. These include performing replicate counts and using consistent criteria for deciding whether individual spermatozoa should or should not be included in the count. The manual also focuses a great deal of attention on important technical matters such as the best way to introduce the fluid sample into the chamber, making sure that the sperm suspension is well mixed and even ensuring that the microscope optics are correctly set up. It also recognizes the possibility that individual technicians may inadvertently introduce biases and skews when performing sperm counts. This latter problem is not easy to overcome, especially if a technician feels that their competence is being questioned, and therefore, it is a good idea to maintain a program that monitors technician performance over time in an effort to detect such anomalies.

It is worth pointing out that estimating sperm concentration is practiced widely in laboratories dealing with animal andrology. The reasons for checking the concentration of bull, boar, or ram spermatozoa are partly aimed at the identification of poor-quality semen samples, but more often, they are aimed at determining the degree of dilution to be used when preparing semen for artificial insemination or cryopreservation. In an industrial setting, such as a cattle breeding company, where a single semen sample can potentially be diluted and divided into 10,000 semen doses each costing upward of US $10.00, there is a huge economic incentive to maximize the number of doses that can be obtained. This means that the breeding companies need a high degree of accuracy when making their estimations, so that they do not risk their reputation by overestimating the degree of dilution that can be achieved. Although they make extensive use of the Neubauer hemocytometer [23], there is an increasing tendency in agricultural laboratories to move the technology toward the NucleoCounter SP-100 (ChemoMetec A/S Allerod, Denmark). This is an automated system that provides data on sperm concentration in undiluted semen. A semen (50 μL) sample is introduced into a special cassette that incorporates the DNA interacting dye, propidium iodide (PI), and chemicals that disintegrate the sperm plasma membrane. The fluorescent microscope within the system then illuminates the sample, and the sperm nuclei emit red fluorescence; the number of such nuclei is then counted using the imaging system provided. The system can also calculate the proportion of spermatozoa with intact *v* damaged plasma membranes by the expedient of omitting the membrane disintegration step in one sample and only measuring the fluorescence produced by the damaged cells present in the sample. Accuracy of the NucleoCounter is excellent and is regarded as equivalent to the use of flow cytometry, where cells are counted individually as they emerge from the flow cytometer. Hansen et al. [24] showed that reproducibility of the NucleoCounter was 3.1%, which compared well with flow cytometry (2.7%). The same study showed that reproducibility of the Neubauer hemocytometer was 7.1%, which matched almost exactly the value reported by Mahmoud et al. [18]. Surprisingly, although the NucleoCounter has been successfully used for bovine, equine, canine, porcine, and ovine semen [25–32], there are apparently no reports of its use with human semen according to a literature search conducted in June 2017. This seems rather strange as the technique is now known to be reliable, reproducible, and very quick to use; a single sample can be analyzed for the concentration and plasma membrane integrity in <5 min.

In the biological context, a number of studies have attempted to elucidate the relationship between sperm concentration and the likelihood of pregnancy, and it is widely understood that there is a positive, but complex, correlation. What seems a simple rule-of-thumb relationship was reported by Bonde et al. [13] who conducted a prospective investigation of fertility in 430 couples to test the significance of various fertility-related parameters. They showed that a threshold concentration (40×10^6 sperm/mL) marked a significant boundary; below this threshold, there was a positive correlation with pregnancy rate, but above it, there was no particular relationship. On the other hand, Nallella et al. [33] identified sperm concentration as a significant predictor of likely male fertility; their data showed that male fertility was associated with sperm concentrations within or above the range $48.3–120.0\times10^6$/mL. These and other results were reviewed by Lewis [34], who pointed out that most such studies are underpowered because they typically involve <500 individual men. The WHO criteria, where 15×10^6/mL is regarded as the lower threshold of the normal range, represents an authoritative consensus on this topic.

As a final comment, the biological significance of any single ejaculate involves both the semen volume and the sperm concentration. Together, these parameters account for the total number of spermatozoa that would be ejaculated into the female reproductive tract under natural conditions. It is well known from animal studies that the total number of spermatozoa is an important determinant of reproductive success, especially in situations that involve the mating of a single female by two or more males. This situation results in competition between spermatozoa in the female tract. The last four decades have witnessed the development of a body of theory about sperm competition, which formally is said to occur when spermatozoa from more than one male have the opportunity to fertilize eggs from a single female during the same fertile period (for reviews, see [35–37]). In this situation, the male who can produce the most spermatozoa has an advantage over his rivals, and several studies have demonstrated that large testis size, hence a greater sperm production capacity, is a feature of species that exhibit multimale mating systems (see, e.g., [38]). Social dominance is also a determinant in such mating systems, as age, bodyweight, and behavioral differences influence the relative number of spermatozoa contributed by each of the males.

A number of classic experimental studies of sperm competition have, however, established that paternity is still skewed even if confounding factors such as sperm numbers and insemination timing are eliminated [39,40]. Elegant heterospermic insemination (HI) experiments with bull and pig spermatozoa, where equal numbers of spermatozoa from two or more males are mixed and inseminated, have shown that spermatozoa from individual males can be ranked in order of fertilization efficacy [41–44]. Such observations strongly suggest that total sperm numbers cannot alone explain reproductive success and that some aspects of sperm quality per se determine fertilization success. In this context, the term "sperm quality" remains unclear and controversial but implies (i) the existence of a positive correlation between sperm phenotypes and the fitness of the offspring that derives from that particular spermatozoon or (ii) that some spermatozoa simply possess a "fertilization advantage" over others. In 2003 when Foote [45] attempted to identify the value of relatively simple semen parameters in relation to bovine fertility, he concluded that about 40% of the variance could be explained; this left 60% of variance yet to be identified. Given the careful and elaborated reproductive management of experimental animals prior to artificial insemination, this situation is probably much better than could be achieved in humans. More recent research on semen analysis has revealed that investigating more subtle aspects of sperm biology, such as their microRNA content [16,46] and proteomic profile [16,47–49], can now explain a good deal more of this unknown variance source.

Sperm Motility

Ever since the first observations by Antonie van Leeuwenhoek in the 17th century that spermatozoa exhibit motility [50–52], andrologists working in different fields and with a variety of species have reasoned that sperm speed and progression must be a good indicator of likely fertility. The logical basis for this assumption has been based largely on the belief that the spermatozoa need to move swiftly from the site of sperm deposition to the vicinity of the oocyte, where fertilization ultimately occurs. Although this view is now considered to be oversimplified in the physiological context, there is considerable interest in being able to express motility as a quantitative parameter.

The introduction of artificial insemination technology for animal breeding in agriculture stimulated considerable experimental research into finding methods for the short-term storage of semen, and there was thus a need to express sperm motility in quantitative terms. Researchers realized that they were limited to the subjective visualization of spermatozoa through the microscope eyepiece but devised ways to overcome the biases and errors associated with this approach. Typical solutions to this problem involved expressing the vigor of motile spermatozoa using simple scoring methods based on subjective scales of 0–4 or 0–5, where 0 represented total immobility and 4 or 5 represented a very high degree of motility [53–55]. The scoring systems could be used very precisely to describe subtle differences in sperm activity, and some researchers even incorporated intermediate values of ½ and ¼ into their scales. For example, Emmens [54,56,57] showed that when experiments were designed with appropriate methods of randomization and replication, the subjective scores could be analyzed using sophisticated analysis of variance models. A similar scoring system using an A–D scale (where D means immotile) has more recently been developed and standardized for routine use in clinical laboratories [58]. Over the 30–40 years following the introduction of more precise methods for subjectively scoring sperm motility, these approaches have proved to be versatile and valuable. Indeed, the experimental basis of sperm storage for artificial insemination was largely developed using these methods for quantitating sperm motility (for relevant reviews, see [59–62]). A word of caution is appropriate here: the statistical analysis of subjectively scored sperm motility data works well when samples are randomized and coded so that operators do not know their identity. Moreover, the samples should be assessed and scored in duplicate or triplicate. Unfortunately, the literature is still replete with publications that employ subjective assessments but fail to adopt these practices, and the resultant data are consequently unreliable.

Researchers in the 1960s and 1970s were aware of the shortcomings of subjective scoring systems and began to develop novel approaches for the direct determination of sperm speed, track linearity, and trajectory shape. At this time, computerized image analysis was still very much in its infancy and used only for counting objects and measuring their area and shape [63,64]. Nevertheless, some pioneering researchers saw the potential of the available systems for measuring sperm motility and began working with the Quantimet 720 system to measure sperm swimming speeds. This system, which is now regarded as a museum piece (http://www.leica-microsystems.com/science-lab/history/50-years-of-image-analysis/), comprised a built-in phase-contrast microscope connected via an analogue scanning television camera, to a hard-wired computer system. Images were displayed on a TV screen. The operator could adjust image brightness and contrast and, when set up correctly, the system could discriminate between static and moving objects. It was also possible to use a manually operated light-detecting pen to trace sperm trajectories via the television screen and thereby obtain some quantitative data about sperm speed. The system set up by Dr. Hector Dott (Agricultural Research Council Unit of Reproductive Physiology and Biochemistry, Huntingdon Road, Cambridge, the United Kingdom) was used to investigate the motility of rat spermatozoa collected by micropuncture from the epididymis [65,66]. This could be regarded as the first generation of computer-assisted semen analysis (CASA) systems.

While these studies were in progress, other researchers used a more direct and arguably a more accurate albeit also more laborious approach to the measurement of sperm swimming speed. The technique involved examining spermatozoa by dark ground microscopy, where the sperm heads appeared as light objects on a dark background, and taking photographs using long (1–2 s) exposures or multiple exposures of the same field of view [67]. Detailed sperm tracks were easily visible on the photographs and amenable to being measured manually. Aitken and his colleagues undertook some of the most influential studies of human sperm motility at around this time [68–70]. The detailed photographs permitted these researchers to recognize and measure important components of motility, such as not only velocity itself but also linearity of track, the frequency (beat-cross frequency), and distance to which the sperm head swung from one side of the average track to the other (amplitude of lateral head displacement, ALH). These authors compared semen samples from fertile and infertile cohorts of men and showed that the most important diagnostic feature of a semen sample was only apparent when the frequency distributions of sperm velocity were analyzed. The proportion of progressively motile spermatozoa (velocity > 25 μm/s) exhibiting ALH of <10 μm emerged as a significant predictor of success in the zona-free hamster egg test, which was being used widely for a time as a functional test of sperm quality [71].

Also in 1982, I obtained a research grant from the United Kingdom's Medical Research Council aimed at using the newly developed generation of personal computers to develop a semiquantitative method for measuring sperm motility. The system was built around an Apple II computer (which had only 28 KB of memory and no hard disk) linked to an analogue video camera and graphic tablet. The video camera was connected directly to a phase-contrast microscope fitted with a warm stage, and the live video images could be displayed directly on the computer monitor [72]. Using the graphic tablet, it was possible to track individual spermatozoa manually; the accompanying software provided direct readouts of individual sperm swimming speed. The manual tracking was undoubtedly less accurate than the use of time-lapse photography, but with practice, it was possible to measure around 60 sperm tracks in approximately 10 min. Once this equipment was in place, my colleagues and I began to collaborate with the newly established IVF unit at the Hammersmith hospital in London, who provided us with fresh but surplus semen samples for measurement. A research protocol was established whereby the semen samples (a total of 137) were tested using both the zona-free hamster egg test and the sperm-tracking technique. Samples were also checked for sperm concentration and total sperm count. The availability of about 60 tracks per semen sample allowed us to analyze the frequency characteristics of the sperm swimming speeds to check whether these conformed to the normal distribution and also to calculate the proportion (%) of tracks whose velocity was <20 μm/s. The zona-free hamster egg test was used as a basis for classifying each sample as "fertile" or "infertile."

The results [73] mirrored those reported by Aitken's group, even though the techniques used were entirely different. For clarity, the results table is reproduced here (Table 1.1). This study emphasized the discriminatory value of using sperm swimming speed in assays of semen quality and, in particular, also confirmed that there was information to be obtained by analyzing the frequency distributions.

As a follow-up to the initial study, we undertook another study to evaluate relationships between our measurements of sperm velocity and in vivo fertility outcomes following artificial insemination by donor [74]. Frozen semen samples (N = 71), all considered suitable for use as donor samples (sperm concentration > 60 million/mL and motility >40%), were supplied by the Middlesex Hospital, London. Data interpretation involved comparing information from the laboratory tests, which again included the zona-free hamster egg penetration test, with pregnancy rates. Seven of the donor samples that had failed to produce a pregnancy over a range of 3–14 reproductive cycles from different recipients were regarded

TABLE 1.1 Sample Means From 137 Ejaculates (± SEM) Classified as "Fertile" or "Infertile" on the Basis of Zona-Free Hamster Egg Penetration

Parameter	Fertile	Infertile	Statistical Analysis
Mean swimming speed (μm/s)	34.2±0.85	24.2±1.2	$P<0.005$
Sperm concentration (× 10^6)	94.4±7.5	62.1±11.7	NS
Total sperm count (× 10^6)	309±29.0	198±40.5	NS
Proportion of slow swimming sperm (% <20μm/s)	23.8±2.0	57.9±4.6	$P<0.005$

Data summarized from W.V. Holt, H.D.M. Moore, S.G. Hillier, Computer-assisted measurement of sperm swimming speed in human semen: correlation with in vitro fertilization assays, Fertil. Steril. 44 (1985) 112–119.

as an "infertile" group. Twenty-five of the donor samples had produced at least one pregnancy each and were therefore regarded as a "fertile" group. The mean (± SEM) postthaw sperm velocity of the fertile group (65.9±1.8μm/s) was significantly higher than that of the infertile group (50.4±3.2μm/s). An additional parameter reflecting the robustness of the thawed spermatozoa was incorporated into this study by measuring the sperm velocity of each thawed sample twice; the postthaw measurement was repeated after incubating the samples for 3.5h at 37°C. The velocity decline over 3.5h varied between samples, and while some showed declines of >50%, others remained relatively unaffected (<30%). As the fertility data available for this series of samples included success or failure to achieve conception over consecutive reproductive cycles within individual recipients, it was possible to use life table analysis [75] as a way of gaining a more informative insight into the importance of the velocity decline. Samples were classified into two groups, depending on whether the velocity after incubation was still higher than 60% of the original value (Group 1) or had declined to <60% (Group 2). The average postthaw sperm swimming speeds and the zona-free hamster egg penetration results for the two groups were statistically indistinguishable. Nevertheless, life table analysis, which takes account of the time to pregnancy, showed that samples in Group 1 produced a monthly conception rate of nearly 17% over 480cycles (with >65% of patients becoming pregnant within 6cycles), while samples in Group 2 produced a lower ($P=0.024$) monthly conception rate of 11.6% over 362cycles, with 45% of patients becoming pregnant within 6cycles. The velocity decline test represented an effort to determine whether the semen samples could withstand a stressful process. Although the methodology was rather crude, the underlying principle is sound and has been used by others studying human and boar semen [76,77].

It is important to mention that the measurement technique used here would have ignored immotile spermatozoa and therefore the data probably represent a complex response representing all of the processes that govern sperm motility. If this study were to be repeated today, one semen parameter of particular interest would be the measurement of reactive oxygen species (ROS), which are widely recognized as being destructive to spermatozoa [78]. Others would include DNA fragmentation and mitochondrial membrane potential, which both appear to have higher prognostic value than the simple estimation of motility [79]. Nevertheless, the proportion of progressively motile spermatozoa within a semen sample is routinely measured, either subjectively or by the use of a modern CASA system.

THE BIOLOGICAL RELEVANCE OF SPERM MOTILITY

Sperm motility alone, expressed using simple parameters such as average sperm velocity or estimates of the proportion of motile spermatozoa in a sample, is a poor predictor of conception rate, whether estimated in humans or agricultural animals. Experimental approaches to determine the predictive value of sperm motility have to take account of the potential confounding influence of inconsistent numbers of inseminated spermatozoa, where there are positive correlations between sperm numbers and fertility outcomes. Natural experiments involving reproductive skews and sperm competition can, however, provide useful indications about the value of sperm motility and other parameters. Critically important clues about the relevance of sperm motility have come from a series of papers that focused on mouse reproduction (e.g., the *t*-haplotype mouse [80]) and transmission-ratio distortion (TRD). This is an effect whereby genetic mosaicism resulting from meiotic recombination during spermatogenesis leads to the development of genetically distinct sperm subpopulations within a single ejaculate that are either functionally advantaged or disadvantaged with respect to flagellar activity. While spermatozoa with normal flagellar activity are able to cross the uterotubal junction, enter the oviduct, and reach the oocyte, those spermatozoa with abnormal flagellar function are unable to do so. Mechanistically, TRD in the case of the *t*-haplotype occurs because the cell signaling cascades that control flagellar function and motility operate incorrectly. The protein kinases controlling flagellar function in these mice are overexpressed and cause abnormal sperm motility; however, those spermatozoa carrying the *t*-haplotype also possess a *t*-complex responder gene (T_{cr}) that corrects the overexpression of the sperm motility kinase gene (*smok*) and restores normal flagellar action [81]. During meiosis, the T_{cr} cosegregates with the Y-chromosome and causes 95% skewing of the offspring sex ratio in favor of males by promoting unbalanced fertilization success. This extreme example, which results in non-Mendelian inheritance, is paralleled by non-Mendelian transmission of retinoblastoma in humans. Girardet et al. [82] proposed that TRD and sex-ratio distortion among the offspring of males affected with sporadic bilateral retinoblastoma could be explained by the existence of a defectively imprinted gene located on the human X chromosome that produces a subpopulation of spermatozoa that show defective motility.

These data can be viewed as natural sperm competition experiments that provide an intriguing insight into possible mechanisms of sperm selection. They indicate how genetic traits, not overtly reproductive in nature, can be influenced directly by their association with protein kinase-regulated signaling cascades that affect sperm motility. Reproductive skews detected in heterospermic artificial insemination experiments [39,41,83], where spermatozoa from two or more males are mixed in equal proportions, may be partly attributable to similar mechanisms. This suggestion is supported by observations that when porcine sperm populations are activated by bicarbonate [84], a stimulator of adenylate cyclase and hence protein kinase A, heterogeneous responses are seen both within single semen samples and between individual boars [85,86] (Fig. 1.1). Subpopulations of boar spermatozoa respond to bicarbonate in different ways; some are quiescent in the absence of bicarbonate but are rapidly stimulated to maximal progressive motility, some are refractory to stimulation, and others lie somewhere in between.

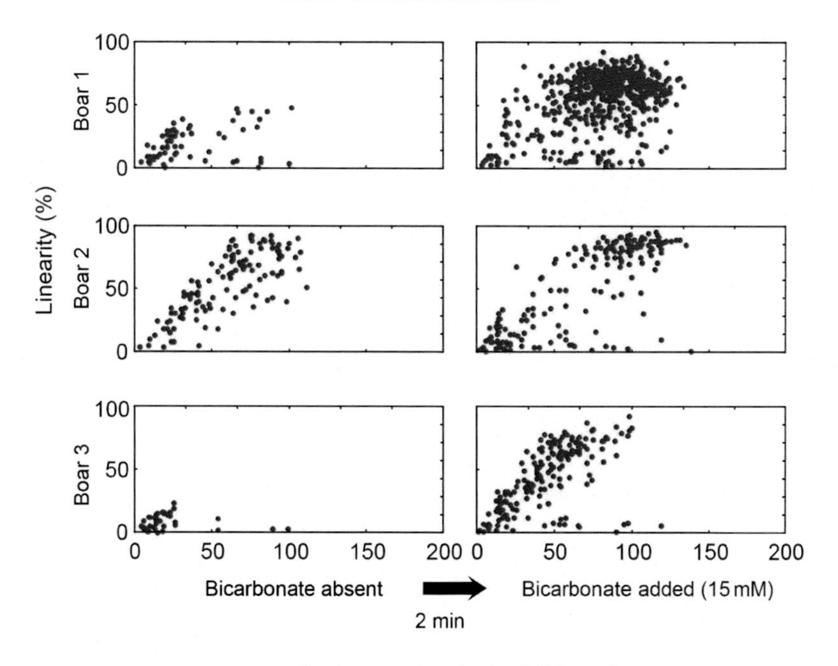

FIG. 1.1 Bicarbonate-induced motility stimulation in boar spermatozoa. Washed spermatozoa from three different boars were incubated at 38°C in a bicarbonate-free HEPES-buffered Tyrode's-based medium. After 10 min incubation, 15 mM bicarbonate/5% CO_2 was added to one-half of the sample, while the other half received 15 mM NaCl (as control treatments). Sperm samples were video-recorded and their motion analyzed using a Hobson Sperm Tracker (Hobson Vision Systems, the United Kingdom). Left-hand panels show that in the absence of bicarbonate, most spermatozoa (points represent individual spermatozoa) moved slowly and nonlinearly (low linearity and average path velocity), but 2 min after bicarbonate addition (right-hand panels), most spermatozoa exhibited significantly increased linearity and velocity. However, while the velocities of most spermatozoa from boar 1 cluster above $50\,\mu m\,s^{-1}$, many of those from boar 2 remain below this threshold. Boar 3 showed the most dramatic bicarbonate response (for more detailed protocols, see [86,128]). Such differential stimulation may be significant in sperm selection mechanisms.

SPERM MORPHOLOGY ANALYSIS

Mammalian spermatozoa from different species come in a surprisingly wide range of shapes and sizes. Some, such as bull, boar, and ram spermatozoa, show a great deal of similarity between species, between individual males, and even between the individual cells. It is therefore relatively easy to spot an abnormally shaped spermatozoon because it may possess a bent or coiled flagellum, an abnormal middle piece, a sperm head crowned with an abnormally shaped acrosome (known in the literature as "knobbed" sperm [87–89]), or a nucleus that contains obvious vacuoles (known as the "crater or diadem" defect [90–92]). A combination of domestication and selection against poor fertility has ensured that spermatozoa from these species are of mainly uniform shape. Morphological examination of semen samples aimed at identifying abnormalities in these agricultural species is therefore relatively straightforward [23], and the spermatozoa that do not stand out as being abnormal are therefore regarded, by default, as normal. Wild mammals with different social mating systems sometimes exhibit a remarkable variety of sperm shapes, even within a single ejaculate [93], a general observation that continues to attract a great deal of attention from researchers with interests in evolutionary biology. Species whose social mating system involves a large element of postejaculatory sperm competition favors the convergence of sperm morphology toward an optimum design best suited for making progress toward the oocyte. In the context of species that do not include sperm competition in their reproductive repertoire and where a high degree of variation is the norm, the "normal" spermatozoon is more difficult to define [94]. In general parlance, "normal" would mean "typical, usual, or ordinary," which is difficult to reconcile with the WHO criteria [11], where the proportion of "normal" spermatozoa in a fertile human ejaculate has to exceed 4%. This figure is, nevertheless, based on detailed observations using strictly defined criteria [95] and the clinical assessment of human spermatozoa has therefore been plagued for many years by the idea that it is possible not only to estimate the proportion of "normal" spermatozoa within an ejaculate but also to make use of the information as a diagnostic correlate of fertility. From a biological point of view, this is unsound as sperm competition is not a major feature of human reproduction [96] and is thus in accordance with the high level of sperm shape heterogeneity seen in human ejaculates. The phenotypic features that define both "normal" and "abnormal" sperm shapes and general appearance in stained preparations are extensively illustrated in the WHO manual [11], and the methodology behind the development of the definition is described. In brief, the authors reasoned that, in vivo, those spermatozoa able to reach the vicinity of the oocyte would represent a fertile population whose phenotype could therefore be considered normal. Prior to the use of such criteria, the subjective assessment of sperm morphology was largely a matter of opinion and led to the production of inconsistent results between different laboratories [97].

A significant study by Auger et al. of human sperm morphology [98] has now clarified the diagnostic significance of human sperm morphology assessment. The morphological data from about 900 fertile men were compared with the data from a similar number of infertile men, and the significance of subtle differences in sperm morphology was compared in considerable detail. Although these authors did not argue that the WHO limit of <4% normal morphology is incorrect, their own data showed that using a value of 23% provided better discrimination between fertile men and infertile men and was also effectively able to distinguish men with testicular cancer who were also included in this study.

It is also worth pointing out that as computer-assisted sperm morphology analysis improves and becomes more reliable, the diagnostic significance of sperm morphology for human semen analysis will also improve.

PLASMA MEMBRANE INTEGRITY

The cytoplasmic components of spermatozoa are all enclosed within a plasma membrane that governs every aspect of sperm function. The plasma membrane is known to be both structurally and biochemically heterogeneous, thus enabling different regions of the membrane to fulfill different functions. Thus, the plasma membrane overlying the acrosome has eventually to fuse with the outer acrosomal membrane when the acrosome reaction occurs, while the plasma membrane overlying the flagellum has to be more robust but able to interact sensitively with the surrounding environment. As with all cells, if the sperm plasma membrane becomes damaged, it can no longer fulfill its myriad functions. Assessing sperm function has, therefore, been seen for many years as a key component of semen quality testing. The usual outcome of such tests is an estimate of the relative proportions of spermatozoa with damaged and undamaged plasma membranes. This is often regarded as a test of "sperm viability" or "percentage of live sperm," even though the use of the terms "live" and "dead" is not appropriate in this context. The simplest tests for sperm plasma membrane integrity rely on the exclusion of dyes such as eosin, trypan blue, Chicago sky blue, and other histological stains from the cell because they cannot traverse the membrane. Thus, when the semen sample is briefly incubated with these dyes and then examined by bright field microscopy, the "live" cells are seen to be unstained, while the damaged cells have taken up the red (eosin) or blue dyes [99–102]. Eosin is frequently used in combination with nigrosine, a stain that provides a dark background when viewing stained smears on the microscope, and thus, use of "eosin/nigrosine" method [103] has long been the most widely used method for sperm plasma membrane evaluation in human clinical medicine and veterinary science, even though it can sometimes be quite tricky to decide whether a particular cell is or is not stained pink.

The development of fluorescent stains in the 1970–80s brought about a major revolution in the methodology available for assessing sperm plasma membrane integrity [104–107]. These included the use of techniques based on the cellular uptake of nonfluorescent esterified fluorescein or carboxyfluorescein, which became fluorescent when the intracellular esterases released the free fluorescent dyes. These dyes caused the entire cell to fluoresce, a property that could be exploited for the exploration of dynamic processes, such as cryopreservation and cooling, where the induction sperm plasma membrane damage is temperature-dependent [108,109]. An alternative suite of fluorescent dyes, SYBR-14 and propidium iodide, was also developed for the estimation of sperm plasma membrane integrity. SYBR-14 enters all of the cells regardless of membrane integrity and stains them green. Propidium iodide, which stains nuclei red, only enters those spermatozoa whose plasma membrane is damaged; it is thus the fluorescent analogue of a dye exclusion test. It is easy to estimate the proportion of plasma membrane-intact spermatozoa, either by direct observation and counting or by more sophisticated methods such flow cytometry [110,111]. A comparison of methods for estimating bovine sperm plasma membrane integrity [102] found high levels of correlation (correlation coefficients of approximately 0.9) between the eosin/nigrosine and trypan blue methods and results with the fluorescent stains. It is worth mentioning that these fluorescent dyes are used for sperm plasma membrane integrity estimation by the NucleoCounter SP-100, mentioned previously in relation to sperm counting. The assay procedure takes about 2 min [23], and the results have been validated against flow cytometry in several mammalian species [26,32].

A somewhat different approach to the evaluation of sperm plasma membrane integrity was based on work carried out in the 1960s and 1970s by Drevius [112,113]. Drevius realized that if spermatozoa were exposed to hypotonic conditions, they would experience an osmotically driven influx of water through the plasma membrane and the sperm tail would therefore become swollen and distorted. The degree of swelling was correlated with the osmolarity of the surrounding medium but only as long as the sperm plasma membrane remained intact. These observations were exploited by Jeyendran et al. [114] and Revell and Mrode [115], who developed an osmotic resistance test (ORT; also known as the hypoosmotic swelling test,

HOS) for the evaluation of fresh and cryopreserved semen. Spermatozoa that could swell in the hypotonic environment were regarded as plasma membrane intact. A fertility trial conducted for the assessment of cryopreserved bull semen resulted in a correlation ($r=0.79$) between the 49-day nonreturn rate (an indicator of conception in cattle) and the ORT test. De Castro et al. [116] used the HOS test for the evaluation of prevasectomy semen samples, and numerous authors have attempted, with varying degrees of success, to use the HOS test as a method of choosing suitably viable spermatozoa for use in intracytoplasmic sperm injection (ICSI) [117–120].

CONCLUSIONS

The researcher who is attempting to develop an optimal method of sperm storage or cryopreservation needs the ability to find out whether the new technique of interest is better than the old one. In this scenario, it is not necessary to predict fertility, although the results should be robust and as precise as possible. The underlying assumption that fertility should be improved if the laboratory tests also show improved responses is therefore sufficiently useful for the interpretation of the experimental outcomes. It is important to point out that considerable progress has been achieved since the 1930s and 1940s. During this period, artificial insemination technology grew in global importance, and the worldwide dairy industry was revolutionized by the success of semen cryopreservation technology.

The agricultural industry has, nevertheless, a long-standing interest in predicting the relative fertility of individual male animals, principally bulls, boars, and rams, but this is still rather an imprecise science despite the availability of increasingly sophisticated methods for semen analysis. This uncertainty is reflected in the application of semen assessment methods that aim to diagnose fertility and infertility in humans. Here, the patient attending a clinic wants to know whether his semen is defective or potentially fertile; that is, he is asking for a very personalized diagnosis or prediction. At present, it is clear that despite our best efforts, the use of a limited set of sperm function tests can reliably diagnose only the most serious of problems such as the complete absence of spermatozoa in the ejaculate or the absence of the acrosome. It seems to this reviewer that, in clinical practice, sometimes, the results of the spermiogram are largely ignored and the patient is immediately directed toward ICSI or IVF. Perhaps, a more thoughtful use of the spermiogram is still required.

More subtle reasons for male infertility are now known to include the inability of spermatozoa to stimulate oocyte activation and embryonic development [121–123]. It is also becoming clear that the nature of RNA in spermatozoa is more important than was previously thought [124–127], and indeed, it is fair to remark that until 20–30 years ago, reports of RNA in spermatozoa were viewed with a good deal of skepticism. We should therefore be warned that the ability to make reliable predictions about fertility is probably still some way in the future.

References

[1] W.V. Holt, A. Fazeli, The oviduct as a complex mediator of mammalian sperm function and selection, Mol. Reprod. Dev. 77 (2010) 934–943.
[2] L. Bjorndahl, C.L. Barratt, L.R. Fraser, U. Kvist, D. Mortimer, ESHRE basic semen analysis courses 1995–1999: immediate beneficial effects of standardized training, Hum. Reprod. 17 (2002) 1299–1305.

[3] I.I. Ivanov, N.V. Korban, V.I. Sharobaiko, Mechanism of the reduction in DNA content in sperm cells preserved in vitro, Bull. Exp. Biol. Med. USSR 67 (1969) 383–385.
[4] Y. Morita, H. Mitsuya, K. Miyake, Study on DNA content of individual human spermatozoa, Int. Urol. Nephrol. 4 (1972) 423–433.
[5] D.R. Ackerman, U.A. Sod-Moriah, DNA content of human spermatozoa after storage at low temperatures, J. Reprod. Fert. 17 (1968) 1–7.
[6] D. Royere, S. Hamamah, J.C. Nicolle, C. Barthelemy, J. Lansac, Freezing and thawing alter chromatin stability of ejaculated human spermatozoa: fluorescence acridine orange staining and Feulgen-DNA cytophotometric studies., Gamete Res., 21 (1988) 51–57.
[7] J. Auger, M. Mesbah, C. Huber, J.P. Dadoune, Aniline blue staining as a marker of sperm chromatin defects associated with different semen characteristics discriminates between proven fertile and suspected infertile men, Int. J. Androl. 13 (1990) 452–462.
[8] J. MacLeod, The male factor in fertility and infertility; an analysis of ejaculate volume in 800 fertile men and in 600 men in infertile marriage, Fertil. Steril. 1 (1950) 347–361.
[9] M. Freund, R. Peterson, Semen evaluation and fertility, in: E. Hafez (Ed.), Human Semen and Fertility Regulation in Men, CV Mosby Co, St Louis, USA, 1976, pp. 344–354.
[10] P.L. Matson, K. Myssonski, S. Yovich, L. Morrison, J. Irving, H.W. Bakos, The density of human semen and the validation of weight as an indicator of volume: a multicentre study, Reprod. Biol. 10 (2010) 141–153.
[11] WHO Laboratory Manual for the Examination and Processing of Human Semen, World Health Organization, Geneva, Switzerland, 2010.
[12] WHO Laboratory Manual for the Examination of Human Semen and Sperm-Cervical Mucus Interaction, Cambridge University Press, 1999.
[13] J.P.E. Bonde, E. Ernst, T.K. Jensen, N.H.I. Hjollund, H. Kolstad, T. Scheike, A. Giwercman, N.E. Skakkebæk, T.B. Henriksen, J. Olsen, Relation between semen quality and fertility: a population-based study of 430 first-pregnancy planners, Lancet 352 (1998) 1172–1177.
[14] D. Schwartz, A. Laplanche, P. Jouannet, G. David, Within-subject variability of human semen in regard to sperm count, volume, total number of spermatozoa and length of abstinence, J. Reprod.Fert. 57 (1979) 391–395.
[15] I. Caballero, I. Parrilla, C. Almiñana, D. Del Olmo, J. Roca, E. Martínez, J. Vázquez, Seminal plasma proteins as modulators of the sperm function and their application in sperm biotechnologies, Reprod. Domest. Anim. 47 (2012) 12–21.
[16] A. Agarwal, A. Ayaz, L. Samanta, R. Sharma, M. Assidi, A.M. Abuzenadah, E. Sabanegh, Comparative proteomic network signatures in seminal plasma of infertile men as a function of reactive oxygen species, Clin. Proteom 12 (2015) 23.
[17] P. Christensen, H. Stryhn, C. Hansen, Discrepancies in the determination of sperm concentration using Burker-Turk, Thoma and Makler counting chambers, Theriogenology 63 (2005) 992–1003.
[18] A.M. Mahmoud, B. Depoorter, N. Piens, F.H. Comhaire, The performance of 10 different methods for the estimation of sperm concentration, Fertil. Steril. 68 (1997) 340–345.
[19] J. Auger, F. Eustache, B. Ducot, T. Blandin, M. Daudin, I. Diaz, S.E. Matribi, B. Gony, L. Keskes, M. Kolbezen, A. Lamarte, J. Lornage, N. Nomal, G. Pitaval, O. Simon, I. Virant-Klun, A. Spira, P. Jouannet, Intra- and inter-individual variability in human sperm concentration, motility and vitality assessment during a workshop involving ten laboratories, Hum. Reprod. 15 (2000) 2360–2368.
[20] P.L. Matson, Andrology: external quality assessment for semen analysis and sperm antibody detection: results of a pilot scheme, Hum. Reprod. 10 (1995) 620–625.
[21] M. Freund, B. Carol, Factors affecting haemocytometer counts of sperm concentration in human semen, J. Reprod.Fert. 8 (1964) 149–155.
[22] L. Bjorndahl, C.L. Barratt, D. Mortimer, P. Jouannet, 'How to count sperm properly': checklist for acceptability of studies based on human semen analysis, Hum. Reprod. 31 (2016) 227–232.
[23] S.P. Lorton, Evaluation of Semen in the Andrology Laboratory, Animal Andrology: Theories and Applications, CAB International, Wallingford, UK, 2014, pp. 100–143.
[24] C. Hansen, T. Vermeiden, J.P. Vermeiden, C. Simmet, B.C. Day, H. Feitsma, Comparison of FACSCount AF system, improved Neubauer hemocytometer, corning 254 photometer, SpermVision, UltiMate and NucleoCounter SP-100 for determination of sperm concentration of boar semen, Theriogenology 66 (2006) 2188–2194.
[25] M. Anzar, T. Kroetsch, M.M. Buhr, Comparison of different methods for assessment of sperm concentration and membrane integrity with bull semen, J. Androl. 30 (2009) 661–668.

[26] J.M. Morrell, A. Johannisson, L. Juntilla, K. Rytty, L. Backgren, A.M. Dalin, H. Rodriguez-Martinez, Stallion sperm viability, as measured by the Nucleocounter SP-100, is affected by extender and enhanced by single layer centrifugation, Vet. Med. Int. 2010 (2010) 659862.
[27] J.M. Morrell, H. Rodriguez-Martinez, A. Johannisson, Single layer centrifugation of stallion spermatozoa improves sperm quality compared with sperm washing, Reprod. BioMed. Online 21 (2010) 429–436.
[28] A. Camus, S. Camugli, C. Leveque, E. Schmitt, C. Staub, Is photometry an accurate and reliable method to assess boar semen concentration? Theriogenology 75 (2011) 577–583.
[29] M.K. Hoogewijs, S.P. de Vliegher, J.L. Govaere, C. de Schauwer, A. de Kruif, A. van Soom, Influence of counting chamber type on CASA outcomes of equine semen analysis, Equine Vet. J. 44 (2012) 542–549.
[30] L.F. Brito, A multilaboratory study on the variability of bovine semen analysis, Theriogenology 85 (2016) 254–266.
[31] I. Ibanescu, C. Leiding, S.G. Ciornei, P. Rosca, I. Sfartz, D. Drugociu, Differences in CASA output according to the chamber type when analyzing frozen-thawed bull sperm, Anim. Reprod. Sci. 166 (2016) 72–79.
[32] L. Daub, A. Geyer, S. Reese, J. Braun, C. Otzdorff, Sperm membrane integrity in fresh and frozen-thawed canine semen samples: a comparison of vital stains with the NucleoCounter SP-100, Theriogenology 86 (2016) 651–656.
[33] K.P. Nallella, R.K. Sharma, N. Aziz, A. Agarwal, Significance of sperm characteristics in the evaluation of male infertility, Fertil. Steril. 85 (2006) 629–634.
[34] S.E. Lewis, Is sperm evaluation useful in predicting human fertility? Reproduction 134 (2007) 31–40.
[35] G.A. Parker, Sperm competition and its evolutionary consequences in the insects, Biol. Rev. Camb. Philos. Soc. 45 (1970) 525–567.
[36] G.A. Parker, M.A. Ball, P. Stockley, M.J.G. Gage, Sperm competition games: a prospective analysis of risk assessment, Proc. R. Soc. Lond. Ser. B Biol. Sci. 264 (1997) 1793–1802.
[37] G.A. Parker, Sperm competition and the evolution of ejaculates: toward a theory base, in: T.R. Birkhead, A.P. Møller (Eds.), Sperm Competition and Sexual Selection, Academic Press, London, 1998, pp. 3–54.
[38] A.H. Harcourt, P.H. Harvey, S.G. Larson, R.V. Short, Testis weight, body weight and breeding system in primates, Nature 293 (1981) 55–57.
[39] J.M. Robl, P.J. Dziuk, Comparison of heterospermic and homospermic inseminations as measures of male fertility, J. Exp. Zool. 245 (1988) 97–101.
[40] P.J. Dziuk, Factors that influence the proportion of offspring sired by a male following heterospermic insemination, Anim. Reprod. Sci. 43 (1996) 65–88.
[41] D.L. Stewart, R.L. Spooner, G.H. Bennett, R.A. Beatty, J.L. Hancock, A second experiment with heterospermic insemination in cattle, J. Reprod. Fert. 36 (1974) 107–116.
[42] R.A. Beatty, D.L. Stewart, R.L. Spooner, J.L. Hancock, Evaluation by the heterospermic insemination technique of the differential effect of freezing at −196°C on the fertility of individual bull semen, J. Reprod. Fert. 47 (1976) 377–379.
[43] J.J. Parrish, R.H. Foote, Fertility differences among male rabbits determined by heterospermic insemination of fluorochrome-labeled spermatozoa, Biol. Reprod. 33 (1985) 940–949.
[44] T. Berger, Proportion of males with lower fertility spermatozoa estimated from heterospermic insemination, Theriogenology 43 (1995) 769–775.
[45] R.H. Foote, Fertility estimation: a review of past experience and future prospects, Anim. Reprod. Sci. 75 (2003) 119–139.
[46] A. Salas-Huetos, J. Blanco, F. Vidal, J.M. Mercader, N. Garrido, E. Anton, New insights into the expression profile and function of micro-ribonucleic acid in human spermatozoa, Fertil. Steril. 102 (2014) 213–222 (e214).
[47] R. Oliva, S. de Mateo, J.M. Estanyol, Sperm cell proteomics, Proteomics 9 (2009) 1004–1017.
[48] I.A. Brewis, B.M. Gadella, Sperm surface proteomics: from protein lists to biological function, Mol. Hum. Reprod. 16 (2010) 68–79.
[49] A.-J. Xin, L. Cheng, H. Diao, P. Wang, Y.-H. Gu, B. Wu, Y.-C. Wu, G.-W. Chen, S.-M. Zhou, S.-J. Guo, H.-J. Shi, S.-C. Tao, Comprehensive profiling of accessible surface glycans of mammalian sperm using a lectin microarray, Clin. Proteom. 11 (2014) 10.
[50] M. Karamanou, E. Poulakou-Rebelakou, M. Tzetis, G. Androutsos, Anton van Leeuwenhoek (1632–1723): father of micromorphology and discoverer of spermatozoa, Rev. Argent. Microbiol. 42 (2010) 311–314.
[51] J. Kremer, The significance of Antoni van Leeuwenhoek for the early development of andrology, Andrologia 11 (1979) 243–249.

[52] J. Kremer, Antoni van Leeuwenhoek, originator of spermatology, Ned. Tijdschr. Geneeskd. 121 (1977) 1929–1934.

[53] H.A. Lardy, P.H. Phillips, The effect of certain inhibitors and activators on sperm metabolism, J. Biol. Chem. 138 (1941) 195–202.

[54] C.W. Emmens, The effect of variations in osmotic pressure and electrolyte concentration on the motility of rabbit spermatozoa at different hydrogen-ion concentrations, J. Physiol. 107 (1948) 129–140.

[55] C.W. Emmens, G.I. Swyer, Observations on the motility of rabbit spermatozoa in dilute suspension, J. Gen. Physiol. 32 (1948) 121–138.

[56] C.W. Emmens, A.W. Blackshaw, The low temperature storage of ram, bull and rabbit spermatozoa, Aust. Vet. J. 26 (1950) 226–228.

[57] A.W. Blackshaw, C.W. Emmens, The interaction of pH, osmotic pressure and electrolyte concentration on the motility of ram, bull and human spermatozoa, J. Physiol. 114 (1951) 16–26.

[58] C.-H. Yeung, T.G. Cooper, E. Nieschlag, A technique for standardization and quality control of subjective sperm motility assessments in semen analysis, Fertil. Steril. 67 (1997) 1156–1158.

[59] W.M. Maxwell, S. Salamon, Liquid storage of ram semen: a review, Reprod. Fert. Dev. 5 (1993) 613–638.

[60] W.M. Maxwell, A.J. Landers, G. Evans, Survival and fertility of ram spermatozoa frozen in pellets, straws and minitubes, Theriogenology 43 (1995) 1201–1210.

[61] S. Salamon, W.M. Maxwell, Storage of ram semen, Anim. Reprod. Sci. 62 (2000) 77–111.

[62] L.A. Johnson, K.F. Weitze, P. Fiser, W.M.C. Maxwell, Storage of boar semen, Anim. Reprod. Sci. 62 (2000) 143–172.

[63] W. Sawicki, J. Rowinski, J. Abramczuk, Image analysis of chromatin in cells of preimplantation mouse embryos, J. Cell Biol. 63 (1974) 227–233.

[64] O.A. Husain, Television scanning of cervical smears: the Quantimet, J. Clin. Pathol. 22 (1969) 743–744.

[65] B.T. Hinton, H.M. Dott, B.P. Setchell, Measurement of the motility of rat spermatozoa collected by micropuncture from the testis and from different regions along the epididymis, J. Reprod. Fert. 55 (1979) 167–172.

[66] D.F. Katz, H.M. Dott, Methods of measuring swimming speed of spermatozoa, J. Reprod. Fert. 45 (1975) 263–272.

[67] S.G. Revell, P.D. Wood, A photographic method for the measurement of motility of bull spermatozoa, J. Reprod. Fert. 54 (1978) 123–126.

[68] R.J. Aitken, F.S. Best, D.W. Richardson, O. Djahanbakhch, M.M. Lees, The correlates of fertilizing capacity in normal fertile men, Fertil. Steril. 38 (1982) 68–76.

[69] R.J. Aitken, F.S. Best, D.W. Richardson, O. Djahanbakhch, D. Mortimer, A.A. Templeton, M.M. Lees, An analysis of sperm function in cases of unexplained infertility: conventional criteria, movement characteristics, and fertilizing capacity, Fertil. Steril. 38 (1982) 212–221.

[70] R.J. Aitken, F.S.M. Best, D.W. Richardson, O. Djahanbakhch, M.M. Lees, The correlates of fertilizing capacity in normal fertile men, Fertil. Steril. 38 (1982) 68–76.

[71] R.J. Aitken, R.A. Elton, Significance of Poisson distribution theory in analysing the interaction between human spermatozoa and zona-free hamster oocytes, J. Reprod. Fert. 72 (1984) 311–321.

[72] W.V. Holt, H.D.M. Moore, A. Parry, S. Hillier, A computerized technique for the measurement of sperm cell velocity: correlation of the results with *in vitro* fertilization assays, in: W. Thompson, R.F. Harrison, J. Bonnar (Eds.), The Male Factor in Human Infertility Diagnosis and Treatment, MTP Press, Lancaster, 1983, pp. 3–7.

[73] W.V. Holt, H.D.M. Moore, S.G. Hillier, Computer-assisted measurement of sperm swimming speed in human semen: correlation with in vitro fertilization assays, Fertil. Steril. 44 (1985) 112–119.

[74] W.V. Holt, F. Shenfield, T. Leonard, T.D. Hartman, R.D. North, H.D.M. Moore, The value of sperm swimming speed measurements in assessing the fertility of human frozen semen, Hum. Reprod. 4 (1989) 292–297.

[75] D.W. Cramer, A.M. Walker, I. Schiff, Statistical methods in evaluating the outcome of infertility treatments, Fertil. Steril. 32 (1979) 80–86.

[76] P.S. Fiser, C. Hansen, L. Underhill, G.J. Marcus, New thermal stress test to assess the viability of cryopreserved boar sperm, Cryobiology 28 (1991) 454–459.

[77] B.A. Kolb, V.C. Acosta, R.S. Jeyendran, Accessing the fertilizing potential of cryopreserved sperm by its ability to maintain quality in a glycerol-free medium, Arch. Androl. 43 (1999) 221–225.

[78] H.D. Guthrie, G.R. Welch, Effects of reactive oxygen species on sperm function, Theriogenology 78 (2012) 1700–1708.

[79] S. Malić Vončina, B. Golob, A. Ihan, A.N. Kopitar, M. Kolbezen, B. Zorn, Sperm DNA fragmentation and mitochondrial membrane potential combined are better for predicting natural conception than standard sperm parameters, Fertil. Steril., 105 (2016) 637–644.e631.

[80] P. Olds-Clarke, L.R. Johnson, T-haplotypes in the mouse compromise sperm flagellar function, Dev. Biol. 155 (1993) 14–25.

[81] B.G. Herrmann, B. Koschorz, K. Wertz, K.J. McLaughlin, A. Kispert, A protein kinase encoded by the t-complex responder gene causes non-Mendelian inheritance, Nature 402 (1999) 141–146.

[82] A. Girardet, M.S. McPeek, E.P. Leeflang, F. Munier, N. Arnheim, M. Claustres, F. Pellestor, Meiotic segregation analysis of RB1 alleles in retinoblastoma pedigrees by use of single-sperm typing, Am. J. Hum. Genet. 66 (2000) 167–175.

[83] L.D. Nelson, B.W. Pickett, G.E. Seidel Jr., Effect of heterospermic insemination on fertility of cattle, J. Anim. Sci. 40 (1975) 1124–1129.

[84] Y. Tajima, N. Okamura, Y. Sugita, The activating effects of bicarbonate on sperm motility and respiration at ejaculation, Biochim. Biophys. Acta 924 (1987) 519–529.

[85] R.A.P. Harrison, W.V. Holt, Modelling boar sperm signalling systems, in: L.A. Johnson, H.D. Guthrie (Eds.), Boar Semen Preservation. IV, Allen Press Inc., Lawrence, Kansas, 2000, pp. 13–19.

[86] W.V. Holt, R.A. Harrison, Bicarbonate stimulation of boar sperm motility via a protein kinase A-dependent pathway: between-cell and between-ejaculate differences are not due to deficiencies in protein kinase A activation, J. Androl. 23 (2002) 557–565.

[87] A.D. Barth, The knobbed acrosome defect in beef bulls, Can. Vet. J. 27 (1986) 379–384.

[88] D.G. Cran, H.M. Dott, The ultrastructure of knobbed bull spermatozoa, J. Reprod. Fert. 47 (1976) 407–408.

[89] T. Bonadonna, A particular form of the head (knobbed head) of the bovine sperms shown by electron microscopy, Riv. Biol. 47 (1955) 297–301.

[90] S. Esser, A. Landaeta-Hernandez, J.P. Hernandez-Fonseca, J. Verlander, G. Stunic, F. Escalona, N. Madrid-Bury, A. Quintero-Moreno, H. Bollwein, M. Kaske, P.J. Chenoweth, Diadem/crater defect in spermatozoa of a Brahman bull: seminal traits, microscopic findings and IVF fertility. Genetic predisposition? Mol. Reprod. Dev. 77 (2010) 1000.

[91] R.E. Larsen, P.J. Chenoweth, Diadem/crater defects in spermatozoa from two related Angus bulls, Mol. Reprod. Dev. 25 (1990) 87–96.

[92] E. Heath, R.S. Ott, Diadem/crater defect in spermatozoa of a bull, Vet. Rec. 110 (1982) 5–6.

[93] S. Pitnick, D.J. Hosken, T.R. Birkhead, Sperm morphological diversity, in: T.R. Birkhead, D.J. Hosken, S. Pitnick (Eds.), Sperm Biology: An Evolutionary Perspective, Academic Press, London, 2009, pp. 69–149.

[94] G. van der Horst, L. Maree, Sperm form and function in the absence of sperm competition, Mol. Reprod. Dev. 81 (2014) 204–216.

[95] J.H. Check, H.G. Adelson, B.R. Schubert, A. Bollendorf, Evaluation of sperm morphology using Kruger's strict criteria, Arch. Androl. 28 (1992) 15–17.

[96] R.D. Martin, The evolution of human reproduction: a primatological perspective, Am. J. Phys. Anthropol. (Suppl. 45) (2007) 59–84.

[97] N. Jorgensen, J. Auger, A. Giwercman, D.S. Irvine, T.K. Jensen, P. Jouannet, N. Keiding, C. Le Bon, E. MacDonald, A.M. Pekuri, T. Scheike, M. Simonsen, J. Suominen, N.E. Skakkeboek, Semen analysis performed by different laboratory teams: an intervariation study, Int. J. Androl. 20 (1997) 201–208.

[98] J. Auger, P. Jouannet, F. Eustache, Another look at human sperm morphology, Hum. Reprod. 31 (2016) 10–23.

[99] E.F. Graham, K.L. Schmehl, B.K. Evensen, D.S. Nelson, Viability assays for frozen semen, Cryobiology 15 (1978) 242–244.

[100] G. Kútvölgyi, J. Stefler, A. Kovács, Viability and acrosome staining of stallion spermatozoa by Chicago sky blue and Giemsa, Biotech. Histochem. 81 (2006) 109–117.

[101] T. Somfai, S.Z. Bodó, S.Z. Nagy, E. Gócza, J. Iváncsics, A. Kovács, Simultaneous evaluation of viability and acrosome integrity of mouse spermatozoa using light microscopy, Biotech. Histochem. 77 (2002) 117–120.

[102] L.F. Brito, A.D. Barth, S. Bilodeau-Goeseels, P.L. Panich, J.P. Kastelic, Comparison of methods to evaluate the plasmalemma of bovine sperm and their relationship with *in vitro* fertilization rate, Theriogenology 60 (2003) 1539–1551.

[103] J.L. Hancock, The morphology of boar spermatozoa, J. R. Microscop. Soc. 76 (1956) 84–97.

[104] S.A. Ericsson, D.L. Garner, D. Redelman, K. Ahmad, Assessment of the viability and fertilizing potential of cryopreserved bovine spermatozoa using dual fluorescent staining and two-flow cytometric systems, Gamete Res. 22 (1989) 355–368.

[105] R. Fichorova, D.J. Anderson, Use of sperm viability and acrosomal status assays in combination with immunofluorescence technique to ascertain surface expression of sperm antigens, J. Reprod. Immunol. 20 (1991) 1–13.
[106] S. Sukardi, M.R. Curry, P.F. Watson, Simultaneous detection of the acrosomal status and viability of incubated ram spermatozoa using fluorescent markers, Anim. Reprod. Sci. 46 (1997) 89–96.
[107] R. Fierro, M.C. Bene, B. Foliguet, G.C. Faure, G. Grignon, Evaluation of human sperm acrosome reaction and viability by flow cytometry, Ital. J. Anat. Embryol. 103 (1998) 75–84.
[108] W.V. Holt, R.D. North, Effects of temperature and restoration of osmotic equilibrium during thawing on the induction of plasma-membrane damage in cryopreserved ram spermatozoa, Biol. Reprod. 51 (1994) 414–424.
[109] W.V. Holt, M.F. Head, R.D. North, Freeze-induced membrane damage in ram spermatozoa is manifested after thawing—observations with experimental cryomicroscopy, Biol. Reprod. 46 (1992) 1086–1094.
[110] J.K. Graham, E. Kunze, R.H. Hammerstedt, Analysis of sperm cell viability, acrosomal integrity, and mitochondrial-function using flow-cytometry, Biol. Reprod. 43 (1990) 55–64.
[111] D.P. Evenson, Z. Darzynkiewcz, M.R. Melamed, Simultaneous measurement by flow cytometry of sperm cell viability and mitochondrial membrane potential related to cell motility, J. Histochem. Cytochem. 30 (1982).
[112] L.O. Drevius, Bull spermatozoa as osmometers, J. Reprod. Fert. 28 (1972) 29–39.
[113] L.O. Drevius, H. Eriksson, Osmotic swelling of mammalian spermatozoa, Exp. Cell Res. 42 (1966) 136–156.
[114] R.S. Jeyendran, H.H. Van der Ven, M. Perez-Pelaez, B.G. Crabo, L.J. Zaneveld, Development of an assay to assess the functional integrity of the human sperm membrane and its relationship to other semen characteristics, J. Reprod. Fert. 70 (1984) 219–228.
[115] S.G. Revell, R.A. Mrode, An osmotic resistance test for bovine demen, Anim. Reprod. Sci. 36 (1994) 77–86.
[116] M. de Castro, R.S. Jeyendran, L.J. Zaneveld, Hypo-osmotic swelling test: analysis of prevasectomy ejaculates, Arch. Androl. 24 (1990) 11–16.
[117] A. Bollendorf, J.H. Check, D. Kramer, The majority of males with subnormal hypoosmotic test scores have normal vitality, Clin. Exp. Obstet. Gynecol. 39 (2012) 25–26.
[118] V. Mangoli, R. Mangoli, S. Dandekar, K. Suri, S. Desai, Selection of viable spermatozoa from testicular biopsies: a comparative study between pentoxifylline and hypoosmotic swelling test, Fertil. Steril. 95 (2011) 631–634.
[119] H.N. Sallam, F. Ezzeldin, A. Sallam, A.F. Agameya, A. Farrag, Sperm velocity and morphology, female characteristics, and the hypo-osmotic swelling test as predictors of fertilization potential: experience from the IVF model, Int. J Fertil. Womens. Med. 48 (2003) 88–95.
[120] J.H. Check, D. Katsoff, M.L. Check, J.K. Choe, K. Swenson, In vitro fertilization with intracytoplasmic sperm injection is an effective therapy for male factor infertility related to subnormal hypo-osmotic swelling test scores, J. Androl. 22 (2001) 261–265.
[121] I. Gat, R. Orvieto, "This is where it all started"—the pivotal role of PLCzeta within the sophisticated process of mammalian reproduction: a systemic review, Basic Clin. Androl. 27 (2017) 9, https://doi.org/10.1186/s12610-017-0054-y (eCollection 2017).
[122] K. Swann, F.A. Lai, The sperm phospholipase C-zeta and Ca^{2+} signalling at fertilization in mammals, Biochem. Soc. Trans. 44 (2016) 267–272.
[123] S.N. Amdani, M. Yeste, C. Jones, K. Coward, Phospholipase C zeta (PLCzeta) and male infertility: clinical update and topical developments, Adv. Biol. Regul. 61 (2016) 58–67.
[124] N. Garrido, S. Garcia-Herrero, M. Meseguer, Assessment of sperm using mRNA microarray technology, Fertil. Steril. 99 (2013) 1008–1022.
[125] N. Garrido, J.A. Martinez-Conejero, J. Jauregui, J.A. Horcajadas, C. Simon, J. Remohi, M. Meseguer, Microarray analysis in sperm from fertile and infertile men without basic sperm analysis abnormalities reveals a significantly different transcriptome, Fertil. Steril. 91 (2009) 1307–1310.
[126] D. Miller, G.C. Ostermeier, Towards a better understanding of RNA carriage by ejaculate spermatozoa, Hum. Reprod. Update 12 (2006) 757–767.
[127] G.C. Ostermeier, D.J. Dix, D. Miller, P. Khatri, S.A. Krawetz, Spermatozoal RNA profiles of normal fertile men, Lancet 360 (2002) 772–777.
[128] T. Abaigar, W.V. Holt, R.A.P. Harrison, G. del Barrio, Sperm subpopulations in boar (*Sus scrofa*) and gazelle (*Gazella dama mhorr*) semen as revealed by pattern analysis of computer-assisted motility assessments, Biol. Reprod. 60 (1999) 32–41.

CHAPTER

2

Genetic Variations and Male Infertility

Sezgin Gunes[*,†], *Ashok Agarwal*[*], *Aysenur Ersoy*[*,‡], *Ralf Henkel*[*,§]

[*]American Center for Reproductive Medicine, Cleveland Clinic, Cleveland, OH, United States [†]Department of Medical Biology, Ondokuz Mayis University, Samsun, Turkey [‡]Capa Medical School, Istanbul University, Istanbul, Turkey [§]Department of Medical Bioscience, University of the Western Cape, Bellville, South Africa

INTRODUCTION

Infertility is a complex disorder with multiple genetic and environmental causes influencing approximately 15% of couples around the world and affecting about 7% of men [1]. Furthermore, male infertility is a heterogeneous disorder that can arise from chromosomal aberrations, mutations in a single or more than one gene, or disturbance in the interaction of multiple genes by environmental factors. The genetic factors causing male infertility include cytogenetic abnormalities, Y chromosome deletions, and monogenic disorders and constitute about 30% of all infertility cases. Roughly 2300 genes play a role in spermatogenesis, and a proper coordination of these genes is essential for normal sperm production [2]. Owing to this complexity and since most cases do not conform to any classic monogenic pattern of inheritance, it is a challenge to explore the effects of polymorphisms on male infertility.

Alterations of the DNA sequences such as deletions, duplications, mutations, polymorphisms, or variations can result in spermatogenic failure [3]. Although polymorphisms may not cause diseases directly, they can affect the susceptibility and severity of the disease [4]. Despite some specific mutations that have been identified, other factors including genetic variations responsible for sperm defects remain to be defined. The causes of approximately 30% of male factor infertility cases are still unexplained [5]. Significant progress has been achieved in the management of infertility, and new technologies including *in vitro* fertilization (IVF) and intracytoplasmic sperm injection (ICSI) have been developed to treat infertile couples. As a result, particularly due to the use of ICSI, lesions in the sperm genome could

https://doi.org/10.1016/B978-0-12-812571-7.00003-4

potentially be transmitted unrepaired to the offspring. Therefore, comprehensive analysis of infertile men that can help to understand the underlying causes of the infertility and its mechanism is eagerly needed.

DNA sequence changes in a locus with minor allele frequency >1% in the general population are called polymorphism. Polymorphisms can be classified into three subgroups: single-nucleotide polymorphisms (SNPs), insertion-deletion (INDEL) polymorphisms, and copy-number variation (CNV) polymorphisms (Table 2.1). CNVs are structural variations in a DNA region ranging from a kilobase (kb) to several megabases (Mbs), occurring as a result of translocations, inversions, and insertions/deletions with two or multiple number of alleles [4].

In this chapter, we aim to summarize the current data on the genetic variations that might be involved in male infertility in various populations and discuss the limitations of these studies. Many studies have indicated associations between polymorphisms or genetic variations and male infertility. Our aim is to understand the effects of these genetic variants on spermatogenic failure.

TABLE 2.1 DNA Variations, General Features, Data Sources, and Methods for DNA Polymorphism Analysis

Polymorphisms	Features	Data Sources	Methods
SNP	• Alterations in a single nucleotide in a population • Simple and most common type • Usually two alleles • 1:1000 base pairs • Increase susceptibility to disease	• OMIM • JSNP • The Human Gene Mutation Database • dbSNP • Entrez Gene • GenBank • HapMap • Ensemble • SNPedia	• SNP arrays
Insertion-deletion (INDEL)	• Result of deletions/insertions of DNA units • 1–10,000 bp • Two to multiple number of alleles • *Microsatellite*: DNA units of 2–4 nucleotides are repeated one to few dozen times (TGTG) therefore called short tandem repeat polymorphism (STRP) • *Mini satellites*: DNA units of 10–100 base pairs repeated 100–1000 times therefore called variable number tandem repeats (VNTR)	• OMIM • JSNP • The Human Gene Mutation Database • dbSNP • Entrez Gene • GenBank • HapMap • Ensemble • SNPedia	• MLPA • PAGE
CNV	• Number of copies varying from 1 kb to 2 Mb that can account for two to multiple number of alleles	• OMIM • Toronto Database • Genomic variants • DECIPHER • ECARUCA	• Chromosomal microarray • Array comparative genomic hybridization

Abbreviations: *OMIM*, Online Mendelian Inheritance in Man; *JSNP*, Japanese SNP; *DECIPHER*, Database of Chromosomal Imbalance and Phenotype in Humans using Ensemble Resources; *ECARUCA*, European Cytogeneticists Association Register of Unbalanced Chromosome; *MLPA*, multiplex ligation-dependent probe amplification; *PAGE*, polyacrylamide gel electrophoresis.

SINGLE NUCLEOTIDE POLYMORPHISMS

The most common or natural variants in the human genome are SNPs, representing single-base substitutions within the genome and varying between individuals within populations. SNPs make up approximately 90% of human genetic variations and are occurring once in about every 100–300 base pairs (bps). They are the major source of genetic heterogeneity [4]. 1000 Genomes Project consortium scientists identified 38 million SNPs in their cohort [6] (Table 2.1).

MTHFR Gene Polymorphisms

Methylenetetrahydrofolate reductase (MTHFR) is one of the regulatory enzymes involved in folate metabolism, DNA replication, and methylation of both DNA and protein. MTHFR is a precursor of *S*-adenosylmethionine that functions as methyl donor for DNA and amino acid methylation in spermatogenesis [7]. Alterations in methylation patterns affect several cellular processes including gene expression [8]. The expression of the *MTHFR* gene is higher in the testicles than in other organs in the adult male mouse [9]. Moreover, inactivation of the *MTHFR* gene that occurs due to deficiency in folate results in hyperhomocysteinemia and infertility in male mice. This condition supports the role of *MTHFR* in spermatogenesis. Folate metabolism is important for spermatogenesis because of its association with DNA synthesis and methylation [10–12]. Thus, hyperhomocysteinemia is considered as a male infertility risk factor. To date, research regarding the association between *MTHFR* gene polymorphisms and male infertility has been extensively conducted (Table 2.2).

Recently, the association between *MTHFR* gene polymorphisms and male infertility has been investigated in multiple case-control studies in different populations [7,14,28,64]. One of the most frequently studied *MTHFR* gene polymorphisms is c.677C>T (rs1801133), which causes a substitution of alanine with valine and reduces the enzymatic activity of MTHFR. Mfady and colleagues investigated two variants (c.677C>T and c.1298C>A) of the *MTHFR* gene and one variant of the methionine synthase reductase (*MTRR*) gene (c.66A>G) [16]. The authors concluded that *MTHFR* 677TT homozygosity is associated with infertility in Jordanian and Indian men [24,27] and Korean [26,28], Italian [65], and Brazilian [66] populations. On the other hand, no link was identified between *MTHFR* 677TT homozygosity and male infertility in French [11], Dutch [67], and Caucasian [18] populations. A recent study showed an interaction between *MTHFR* gene c.677C>T polymorphisms and male infertility after a 5-year follow-up. The *MTHFR* 677TT genotype significantly increases the risk of infertility compared with the 677CC genotype or heterozygous genotype. This finding is consistent with a 1-year follow-up study of idiopathic male infertility conducted by Paracchini and colleagues [65].

MTHFR gene c.1298C>A is a missense polymorphism that reduces the activity of *MTHFR*. A meta-analysis of seven studies with different ethnic and geographic origin demonstrated a significant association with azoospermia but not with oligozoospermia. However, there was no association between *MTHFR* gene c.1298C>A polymorphism and male infertility in Jordanian men. A similar situation was reported in Caucasian [18], Brazilian [66], and Asian populations [18], but not in Indian men [27]. The association between *MTRR* c.66A>G polymorphism and male infertility was investigated in a study, but the difference was not

TABLE 2.2 Studies Investigating the Relationship Between SNPs and Male Infertility

Gene/Polymorphism	Effect	Patients (*n*)/ Control (*n*)	OR (95% CI) or P	Ethnicity/ Country	Reference
MTHFR					
C677T (rs1801133)					
	Decrease in spermatogenesis	1843/1791	1.39 (1.15–1.69)	Caucasian and Asian	[7]
	TT genotype is associated with increased risk of infertility during 5-year follow-up	215 Infertile men	10.24 (1.26–83.46)	Russia	[13]
	High association with combination of polymorphisms 677C>T and 1298 A>C	215 Infertile men	11.82 (1.42–98.67)	Russia	[13]
	Risk of azoospermia by the use of fixed effects model	275/349	1.24 (1.09–1.40)	Russia	[13]
	T allele is associated with a higher risk of infertility	637/364	1.12 (0.24–5.25)	North Indian	[14]
	T allele is associated with a higher risk of infertility	118/132	1.84 (1.11–3.04)	Iranian	[15]
	No association	150/150	2.23 (0.89–5.59)	Jordanian	[16]
	No association with azoospermia, severe oligozoospermia, and oligozoospermia	344/450	0.78 (0.60–1.02)	Moroccan	[17]
	Associated with the increased risk of male infertility in Asian population	2217/2312	1.57 (1.05–2.37)	Caucasian and Asian	[18]
	Increase the risk of male infertility in Asian population Increase in the risk of azoospermia	2275/1958	1.79 (1.08–2.96) 1.45 (1.18–1.79)	Caucasian and Asian	[19]
	No association	206/212	1.36 (0.83–2.22)	Indian	[20]
	Associated with azoospermia and severe oligozoospermia	133/173	1.97 (1.15–3.37) 3.66 (2.34–5.73)	Brazilian	[21]
	TC and TT genotypes are associated with male infertility	164/328	1.60 (1.21–2.75) 2.68 (1.84–3.44)	Iranian	[22]
	T allele associated with a higher risk of infertility and OAT	522/315	1.58(1.17–2.13) 1.90 (1.15–3.15)	Indian	[23]
	No association	179/200	1.65 (0.87–3.15)	Indian	[24]
	T allele associated with a higher risk of infertility	355/252	1.58 (1.24–2.02)	China	[25]
	TT genotype was associated with male infertility	360/325	1.60 (1.00–2.56)	Korean	[26]

	TC and TT genotypes were associated with male infertility	151/200	1.68 (1.01–2.794) 1.93 (1.17–3.17)	Indian	[27]
	T allele associated with a higher risk of infertility	396/373	1.29 (1.06–1.59)	Korean	[28]
A1298C (rs1801131)					
	No association	275/349	0.86 (0.48–1.57)	Russia	[13]
	No association	215 Infertile men	1.14 (0.71–1.834)	China	[29]
	No association	150/150	1.178 (0.52–2.64)	Jordanian	[16]
	No association	215 Infertile men	0.80 (0.23–2.73)	China	[29]
	AA genotype associated with varicocele	107/109		Turkish	[30]
	No relation	2217/2312	1.30 (0.87–1.95)	Caucasian and Asian	[18]
	Associated with increased risk of severe oligozoospermia	344/450	3.372 (1.27–8.24)	Moroccan	[17]
	C allele is associated with susceptibility to infertility AA genotype associated with azoospermia	1633/1735	1.12 (1.00–1.26) 1.66 (1.01–2.73)	Asian and Caucasian	[31]
	No association	164/328	0.94 (0.84–1.74)	Iranian	[22]
	No association with azoospermia and oligozoospermia	133/173	0.72 (0.47–1.10) 1.08 (0.73–1.57)	Brazilian	[21]
	CC genotype was associated with male infertility	151/149	3.45 (1.01–11.79)	Indian	[32]
	No association	179/200	0.79 (0.32–1.94)	Indian	[24]
	No relation	360/325	1.23 (0.60–2.54)	Korean	[26]
	No association	179/200	0.79 (0.32–1.94)	Indian	[24]
	No association	396/373	1.18 (0.92–1.52)	Korean	[28]
G1793A					
	No relation	164/328	0.98 (0.84–1.17)	Iranian	[22]
MTRR					
A66G (rs1801394)					
	No association	150/150	1.432 (0.71–2.88)	Jordanian	[16]
	GG genotype was associated with male infertility	360/325	1.78 (1.01–3.17)	Korean	[26]
	No association with azoospermia and oligozoospermia	133/173	1.48 (1.01–2.16)	Brazilian	[21]

Continued

TABLE 2.2 Studies Investigating the Relationship Between SNPs and Male Infertility—cont'd

Gene/Polymorphism	Effect	Patients (*n*)/ Control (*n*)	OR (95% CI) or P	Ethnicity/ Country	Reference
DAZ and *DAZL*					
T54A	Susceptibility to oligozoospermia and azoospermia			Chinese	[7]
T54A	No association	234/131	$P=0.67$	Japanese	[33]
T54A	Significant association with infertility in Asians (AG vs AA)	2556/1997	4.36 (2.21–8.63) (codominant) 4.584 (2.32–9.06) (dominant)	Asians and Caucasians	[34]
T54A	No association			Caucasian	[35]
T54A	No association	93/63		Italian	[36]
T54A	No association	165 idiopathic infertile/200		Indian	
-792G>A, -669A>C, and -309T>C	-792G>A polymorphism associated with infertility -792G>A polymorphism associated with sperm concentration -792G>A polymorphism associated with sperm motility	337/203	$P=0.0005$ $P=0.0025$ $P=1.5\times10\ (-7)$	Taiwanese	[37]
T12A	Associated with azoospermia or oligozoospermia	234/131	$P=0.67$	Japanese	[33]
T12A	Significant association with infertility (AG vs AA)	2556/1997	1.26 (1.00–1.59)	Asians and Caucasians	[34]
T12A	No association				[7]
BRDT					
rs3088232	Associated with infertility in Albanian	60 Azoospermic/76 oligozoospermic/161	1.65 (1.11–2.47)	Macedonian and Albanian	[38]
TERT and *TEP1*					
TERT (rs2736100)	Inversely associated with male infertility risk	580/580	0.66 (0.47–0.92)	Chinese	[39]
TEP1 (rs1713449)	TT genotype associated with increased risk of male infertility	580/580	1.39 (1.20–1.62)	Chinese	[39]
CYP1A1					
CYP1A1*2A	Increased risk of male infertility in Asians	1060/1225		Asians and Caucasians	[40]

CYP1A1*2A and CYP1A1*2C	No association with CYP1A1*2A Association with CYP1A1*2C polymorphisms	192/226	$P=0.019$	Chinese	[41]
3801T>C	T allele and TT genotype is associated with idiopathic male infertility	1060/1225	1.36 (1.01–1.83) 2.18 (1.15–4.12)	Caucasian and Asian	[42]
CYP1A1*2C	CYP1A1*2C is associated with infertility in smokers	203/227	1.91 (1.01–3.59)		[43]
CYP1A1*2A	No association	150/200		Iran	[44]
CYP1A1*2A	CC genotype is associated with increased risk of male infertility	206/230	6.08 (1.91–25.27)	Indian	[45]
FSHR					
Thr307Ala (rs6165) and Asn680Ser (rs6166)	Asn680Ser with total testes volume No association Asn680Ser between the Baltic cohort and Estonian male infertility group	684 Oligozoospermic men/54 azoospermic men/1052 Baltic male	$P<0.000066$ $P>0.70$		[46]
A2039G rs6166	No association	1231		Germany	[47]
Ala307Thr and Asn680Ser	No association			Brazilian	[48]
G-29A, Thr 307Ala, and Asn680Ser	No significant association between the three polymorphisms and risk of male infertility	1644/1748		Chinese	[12]
-29 and codon 680	No association with -29 polymorphism Associated with fertility but not serum FSH levels	270/240	$P<0.05$	Turkish	[49]
Codon 307 and codon 680	The heterozygous genotype Thr/Ala-Ser/Asn was significantly increased in infertile patients	352/145	$P<0.05$	Japanese	[50]
Thr307 and Asn680	Thr307/Asn680 allelic variant was associated with increased sperm motility	200/250		Greek	[51]
-29 and codons 307 and 680	No influence on FSH levels in both normal and infertile males	364/281		Chinese	[52]
ESR1 and *ESR2*					
ERα (PvuII and *Xba*I) and ERβ (*Alu*I and *Rsa*I)	No association with none of the polymorphic sites	187/216	$P>0.05$	Brazilian	[53]
ER-α gene and ER-β gene (rs1256049 and rs4986938 and rs605059)	*ER-α* gene (rs1801132 and rs2228480) and *ER-β* gene (rs1256049 and rs4986938) are associated with sperm concentration	467/210	$P<0.01$	Taiwan	[54]

Continued

TABLE 2.2 Studies Investigating the Relationship Between SNPs and Male Infertility—cont'd

Gene/Polymorphism	Effect	Patients (*n*)/ Control (*n*)	OR (95% CI) or P	Ethnicity/ Country	Reference
ER-α gene (ESR1) *Pvu*II and XbaI and ER-β gene (ESR2) RsaI and Alul polymorphisms	ER-α Pvull TC, ER-α XbaI AG, and ER-β Alul GG genotypes have a protective effect for infertility ER-β RsaI AG and ER-β Alul AG genotypes are associated with increased risk for infertility	164/164	0.56 (0.26–0.80) 0.51 (0.31–0.84) 0.48 (0.26–0.84) 2.32(1.61–3.22) 2.76 (1.64–3.66)	Iran	[22]
ER-α gene (rs2077647, rs1801132, rs2228480) and ER-β gene (rs4986938 and rs1256049)	ER-α gene (rs180113) and ER-β gene (rs1256049) associated with spermatogenic defects	183/120	<0.0001 <0.0001	Taiwanese	[55]
Pvull (T397C) ESR1 g.938T>C ESR2 *39A>G	Excess of homozygous genotype in infertile men	104/95	<006	Spanish	[56]
(TA)n repeat polymorphism	Inversely correlated TA repeat number and sperm concentration	191/156	$P<0.05$	Italian	[57]
Exon 4 codon 325	Relevant to spermatogenetic failure in patients with idiopathic azoospermia		$P<0.01–0.001$	Japanese	[58]
G1082A	Increased risk among Caucasians with GA versus GG	Meta-analysis	2.263 (1.07–4.78)		[59]
G1730A	Reduced risk in GA versus AA among azoospermia or severe oligozoospermia	Meta-analysis	0.687 (0.50–0.94)		[59]
PRM1 and *PRM2*					
R34S (rs35576928)	Arg > Ser change is associated with severely defective spermatogenesis	309/377	0.0079	Chinese	[60]
PRM1 (C321A) and PRM2 (C248T)	No association	96/100	$P=0.805$	Iran	[61]
PRM1 (c.54G>A, c.102G>T, c.230A>C) *PRM2* (c.298G>C and c.373C>A)	An association between common haplotype formed by *PRM1* and *PRM2* and fertility	88/83/77		German	[62]
Direct sequencing of *PRM1* and *PRM2*		226/270		Japanese	[63]

statistically significant. In contrast to a study performed in Korean men, no association was found between *MTRR* c.66A>G polymorphism and male infertility in French and Brazilian populations. Discrepancies in the studies of *MTHFR* gene polymorphisms in different populations might be associated with environmental factors such as the lack of folate in diet [68].

DAZL and *DAZ* Genes Polymorphisms

The deleted-in-azoospermia (DAZ) gene family is a multicopy gene family with two autosomal homologues, *BOLL* and *DAZL*, and one Y chromosomal homologue *DAZ*. The deleted-in-azoospermia-like (*DAZL*) gene is essential for the germ-cell lineage development [69,70] and expressed in all stages of spermatogenic development [71]. Decreased *DAZ/DAZL* ratio can cause the lack of synapsis during meiosis because the *DAZ* family encodes proteins that have RNA-binding motifs that are necessary for meiosis and are engaged in regulating protein expression regulation in the germ line [7]. Two nonsynonymous SNPs were identified: The first in exon 2 at position 260 causes an amino acid substitution (T12A), and the second in exon 3 at position 386 (rs12918346) causes an amino acid change T54A. SNP 260 variation was not associated with male infertility except in Taiwanese populations [7,37]. Similarly, SNP 386 was significantly associated with oligozoospermia and azoospermia in Chinese men, whereas the same association cannot be confirmed in Caucasian [35,36] and Japanese men [33] (Table 2.2). These conflicting findings of these studies demonstrate how important the ethnic and geographic background is in genetic variations; the resulting phenotypes may be different among different populations.

BRDT Gene Polymorphisms

Bromodomain testis-specific protein (BRDT), encoded by *BRDT* gene, is essential for normal spermatogenesis and plays a role in regulation of transcription [72]. *BRDT* gene polymorphisms were reported in infertile men in some studies [2]. A study performed on males from Albanian and Macedonian origins concluded that *BRDT* variant rs3088232 is strongly associated with azoospermia in Macedonian men [38] (Table 2.2). However, no association was found between low levels of *BRDT* expression or *BRDT* gene mutations and impaired fertility in men.

TERT and *TEP1* Genes

Telomeres are repeated DNA sequences that are associated with proteins at the end of linear eukaryotic chromosomes and are essential for genomic stability and integrity by protecting the chromosomal termini against degradation, end-to-end fusion, and irregular recombination. These sequences are shortened during DNA replication and oxidative stress. Telomerase is an enzyme that extends the telomeric repeats of chromosomes, maintains genomic integrity and stability of chromosome, and is active especially in male germ cells, stem cells, and tumor cells [73,74]. Yan and colleagues investigated eight SNPs that have potential function in telomerase reverse transcriptase (TERT) and telomerase-associated protein 1 (TEP1) in infertile men in Han Chinese ethnicity. They have found that the GG genotype of TERT polymorphism (rs27366100) is associated with a reduced risk for male infertility compared with common the TT genotype [39]. In addition, males with TT genotype of *TEP1* polymorphism (rs1713449) had an increased risk of infertility compared with CC genotype. Lastly, the risk of male infertility was twofold higher in subjects with risk genotypes of both loci, and the results of this study draw a more reliable picture because subjects were racially homogenous (Han Chinese) and well matched (BMI, habits, etc.) [39] (Table 2.2).

Cytochrome P4501A1

Cytochrome P4501A1 (*CYP1A1*) gene is located on chromosome 15 (15q22-q24) and is crucial in metabolizing endogenous metabolites such as steroids, prostaglandins, and fatty acids and exogenous metabolites like drugs and chemicals that might be toxic to the male reproductive system [75]. Several studies yielded inconclusive results regarding the association between *CYP1A1* gene polymorphism and infertility [41–44]. *CYP1A1**2A and *CYP1A1**2C are the two common and functionally significant SNPs. Recently published meta-analysis results indicated an increased risk for idiopathic male infertility in individuals with *CYP1A1**2A genotypes among Asians but not in Caucasians [40] (Table 2.2). Discrepancies in findings of these reports might be related to ethnic and geographic origin of the study population.

FSHR Gene Polymorphisms

Follicle-stimulating hormone (FSH) receptor (*FSHR*) is a member of a G-protein-coupled receptor family and plays a role in stimulating spermatogenesis [76]. It is crucial for the normal functioning of Sertoli cells [77]. Some SNPs are defined in exon 10 of *FSHR* gene. Simoni and colleagues indicated that there is no significant difference between infertile and normal men in terms of distribution of the SNPs [78], whereas Ahda and colleagues have claimed that exon 10 SNPs combined with -29 SNP are responsible for severe spermatogenic impairment [79]. In addition, their results indicated that there is no relation between A2039, G29A, and A919G SNPs of *FSHR* gene and serum FSH levels in a German population [79]. A study on a Turkish population consisting of 270 infertile and 240 fertile men showed no relationship among *FSHR* gene polymorphism, FSH serum levels, and effect on spermatogenesis [49] (Table 2.2). The results of these investigations on *FSHR* polymorphisms and male infertility are contradictory.

ESR1 and *ESR2* Gene Polymorphisms

Estrogen receptors (ER) are encoded by *ESR1* (*ERα*) (MIM 133430) and *ESR2* (*ERβ*) (MIM 601663) genes and located on 6q25 and 14q23-24, respectively [76]. A study in ERα knockout mice indicated infertility in males [80]. Human *ER* genes contain several polymorphic sites. Galan and colleagues reported an association between *ESR1* gene c.938C>T polymorphism and infertility in a Spanish population [56]. In addition, a significant association was reported between polymorphism at exon 4 of *ER* genes encoding hormone-binding domain and idiopathic infertility [58]. Similarly, the results of a small case–control study indicated an association between *ER* gene and infertility [81]. An association study performed on *ESR2* SNPs, c.1082G>A and c.1730G>A, found a threefold higher frequency of the heterozygous *Rsa*I AG genotype in infertile men than in controls. Cai and colleagues performed a meta-analysis, and their results suggested that polymorphisms in the genes of ERs (*ESR1* and *ESR2*) may have differential roles in the predisposition to male infertility depending on the ethnicity [59] (Table 2.2). Well-designed studies with larger sample sizes and diverse ethnic backgrounds should be conducted for further confirmation of the findings.

PRM1 and *PRM2* Genes Polymorphisms

During human spermiogenesis, histone proteins are first substituted by transition nuclear proteins (TNPs), which are then replaced with protamines. This replacement of protamines causes chromatin condensation and compaction of the sperm nucleus [82]. Grassetti and colleagues performed a study in an Italian population and concluded that highly alleviated protamine expression is related to premeiotic arrest [83]. In a study on the *TNP1* gene, SNPs in regions c.55G>A, c.54A>C, and c.-222A>G were investigated with no significant difference in terms of prevalence being detected between infertile men compared with fertile controls [82]. However, in a study performed in Chinese men with nonobstructive azoospermia (NOA) and severe oligozoospermia, protamine 1 (*PRM1*) (rs35576928) variant was found to be associated with severely defective spermatogenesis and that the dominant type of the SNP acts protectively against the disease. The same study investigated four variants of the *PRM1* gene and three variants of protamine 2 (*PRM2*) gene. However, the authors did not find any significant difference between cases and controls in terms of expression [84]. A study conducted by Tanaka and colleagues that involved 496 Japanese men investigated four SNPs of *PRM1* gene and three SNPs of *PRM2* gene. No difference in prevalence between fertile and infertile men was observed for both gene variations [63]. SNPs in the *PRM1* and *PRM2* genes in men mainly with azoospermia showed no association with infertility either [68]. Recent large-scale meta-analysis studies demonstrated that SNPs have a small effect on the etiology of infertility (Table 2.2). Despite the many association studies of common variations, the underlying mechanism for most of the infertility cases remains unknown.

INSERTION DELETION POLYMORPHISMS

INDELs are small deletions or insertions of DNA sequences ranging from 1 to 10,000 bp in length and the human genome harbors at least 1.6–2.5 million INDEL polymorphisms. About 36% of these INDEL polymorphisms are located within promoters, exons, and introns, and therefore, they might affect gene function [85]. Several small INDEL polymorphisms have been identified in individual genomes and populations [86], and a few studies demonstrated their associations with male infertility in some populations [1,87] (Table 2.3).

POLG Gene Polymorphisms

Human mitochondrial DNA polymerase gamma (POLG) is a nuclear enzyme. The *POLG* gene codes for the only DNA polymerase for mitochondrial DNA (mtDNA) replication and repair [47], and impairment of this gene causes mitochondrial dysfunction [68]. The *POLG* gene contains a polymorphic CAG repeat region, $(CAG)_{10}$CAACAGCAG, coding for 13 glutamines located in the first exon. In most ethnicities, up to 10 CAG repeats in the *POLG* gene

TABLE 2.3 Studies Investigating the Relationship Between INDEL Polymorphisms and Male Infertility

INDEL Polymorphisms					
AR					
CAG repeats	15–18 CAG repeats and 22–28 CAG repeats are associated with testis volume and Sertoli cell function in men	110/61	$P<0.0005$	Italy	[88]
CAG repeats	Long CAG repeats ($n=24$ and 25) are associated with severe oligozoospermia and azoospermia. Short GGN repeats ($n<23$) are associated with defective spermatogenesis	125/96	2.45 (1.33–4.52) 5.40 (2.61–11.13) 2.20 (1.12–4.34)	China	[89]
CAG repeats	Long AR alleles (≤2 repeat) with the increased sperm motility in both normozoospermic and oligozoospermic patients	200 Oligozoospermic/250 normozoospermic	$P<0.003$	Greek	[51]
CAG repeats	Longer AR CAG repeat lengths are associated with reduced male fertility	3027/2722	SMD 0.19 (0.09–0.29)	Caucasian and Asian	[90]
CAG repeats	Impaired sperm production (>23 repeats a risk factor)	217/131		Singaporean	[91]
POLG					
CAG repeats (not-10/not-10 CAG repeats)	Spermatogenic failure (oligozoospermia and oligoasthenozoospermia)	124/60	$P<0.05$	Indian	[92]
	Associated with heterozygous common and the homozygous noncommon genotypes in infertile patients Significant association with azoospermia and oligoasthenoteratozoospermia	216/123	$P=1.3\times10\ (-3)$ $P=4.2\times10\ (-7)$	Tunisian	[93]
	No association with asthenozoospermia and oligoasthenozoospermia No association with infertility in meta-analysis	150/126	0.94 (0.60–1.48)	China	[94]
	Significant difference between normozoospermic men and nonnormozoospermic men	93 Normozoospermic men/192 nonnormozoospermic men	$P<0.025$	New Zealand	[95]

TABLE 2.3 Studies Investigating the Relationship Between INDEL Polymorphisms and Male Infertility—cont'd

INDEL Polymorphisms					
	No relationship between the polymorphic CAG repeat in the *POLG* gene and male infertility	433 Infertile/91		French	[96]
	No association, considered common allele	195/190		Italian	[97]
	No association, considered common allele	509/241		Indian	[98]
GSTT1 and *GSTM1*					
	Double-null genotype is associated with infertility	150/200	3.75 (2.42–6.45)	Iran	[44]
ESR1 and *ESR2*					
(TA)n repeat polymorphism	Inversely correlated TA repeat number and sperm concentration	191/156	$P<0.05$	Italian	[57]

are considered as common alleles [92]. Krausz and colleagues performed a study on 385 Italian men and concluded that there is no association between *POLG* CAG repeat polymorphism and idiopathic male infertility [97]. This finding is in line with studies performed on populations from New Zealand [95] and France [96] and also confirmed that there is no association between idiopathic male infertility and *POLG* CAG polymorphism. Furthermore, two studies reported an association between *POLG* CAG polymorphism and populations from South India [92] and Tunisia [93]. Recently, a meta-analysis including 13 case-control studies reported no association between male infertility and *POLG* repeat polymorphism [99].

AR Gene Polymorphisms

Androgens are essential hormones for both the development of the male phenotype and spermatogenesis. Massart and colleagues suggested that mutations or polymorphisms in genes involved in this mechanism may be responsible for idiopathic male infertility [100]. The androgen receptor (*AR*) gene (OMIM: 313700) is located on the long arm of the X chromosome. The amino terminal transactivation domain of the *AR* gene bears two polymorphic repeat sequences, polyglutamine (CAG) and polyglycine (GGC), in the exon 1 [76]. Expansion of CAG repeats in the *AR* gene results in an adult onset of neuromuscular disorder, spinal and bulbar muscular atrophy (SBMA), with azoospermia or oligozoospermia and testicular atrophy [101]. Some studies have indicated that there is an association between impaired sperm production and long CAG repeats [76]. In 5 out of 11 studies, it was reported that in Singaporean, North American, Japanese, and Australian infertile men, having >23 allele repeats serves as an inherent risk factor for infertility, whereas these data were not validated in European studies [91]. On the other hand, Castro-Nallar and colleagues reported a significant association between 21 CAG repeats and an increased risk of SCO syndrome and GGN 24 repeat allele and cryptorchidism and spermatogenic failure [102]. Large population studies show that CAG repeat length is not a risk factor for male infertility. Although the impact of genetic variations of male infertility has only recently begun to be investigated, numerous genes have already been shown to be related to male infertility (Table 2.3).

COPY NUMBER VARIATIONS

CNVs are widespread genomic copy-number gains and losses ranging in size from 1 kb to Mb that are present in variable copy numbers [103]. Deletions, duplications, insertions, and translocations could result in CNVs. These variations can lead to microdeletion syndromes, sporadic single-gene disorders, and common complex traits including infertility and disease susceptibility [104]. Recent studies indicated that CNVs take part in human genetic variation [103] and rare and de novo CNVs are potential risk factors for complex human genetic diseases depending on penetrance and expressivity [105]. CNVs can be classified in two subgroups based on large-scale association studies. The first class of CNVs named as common copy-number polymorphisms (CNPs) is multicopy variations with a population frequency higher than 1% and ranged from 0 to 30 copies per genome [106]. The second class of CNVs is low-copy variations ranging from 1 to 3 copies per genome with a population frequency lower than 1% [103]. Several mechanisms are proposed for pathogenicity of CNVs. They could change the copy number of dosage-specific gene or contiguous genes or can alter the expression of dosage-specific genes by position effect. In addition, CNVs can unmask the recessive mutations or functional polymorphisms of the remaining allele (Table I).

Several CNV studies were performed to assess the relative association of CNVs between infertile cases and controls over the last several years [107–109]. The first whole-genome CNV study was performed on 89 infertile patients with severe oligozoospermia and 37 infertile men with Sertoli-cell-only syndrome (SCOS) and 100 controls using array comparative genomic hybridization (CGH). Ten recurring patient-specific CNVs in severe oligozoospermia, three recurring patient-specific CNVs in the SCOS group, and one in both patient groups in more than one patient were found. None of these CNVs were observed in normozoospermic men. Additionally, a high frequency of CNVs on the X chromosome in patients with SCOS and negative association between the number of CNV and sperm counts in normozoospermic men were found. As a result, four autosomal and eight X chromosomal genes, most of which are expressed in Sertoli or germ cells with unknown functions, were identified [107]. Indeed, X chromosomal CNVs in 96 idiopathic infertile men and 103 normozoospermic men using a high-resolution CGH array were analyzed. A total of 73 CNVs in 102 samples were identified, Thirty-two of these CNVs were inside the genes or transcriptional units, and 31 of them were patient-specific. The frequency of X chromosomal deletions was higher in infertile men compared with normozoospermic controls. When the study was extended to analyze these patient-specific CNVs by including a total of 359 patients and 370 normozoospermic controls, it was confirmed that 12 X chromosome deletions have a role in spermatogenesis.

In the second study from Belgium and the Netherlands, a 244K oligonucleotide array was used to screen CNVs in 9 patients with complete meiotic arrest and 20 controls. The authors identified 10 promising candidate regions expressed in the testis and then analyzed

130 additional controls to verify their results to increase statistical power. The authors found eight potential maturation arrest-specific CNVs [110]. The largest whole-genome CNV and male infertility study was performed on 323 infertile cases and 1136 controls using SNP microarray. The authors reported a high number of CNVs in infertile men, especially in sex chromosomes. In addition, recurrent deletion in doublesex and mab-3-related transcription factor 1 (*DMRT1*) gene that has a role in sex determination and differentiation of the testis was identified in 5 azoospermic patients of 1306 (0.38%) men with idiopathic spermatogenic impairment in contrast with none in 7754 controls. Indeed, *DMRT1* mutations of 171 patients with idiopathic cryptozoospermia or nonobstructive azoospermia and 215 normozoospermic controls were investigated. Four putative mutations in a total of six cases (3.5%) compared with normozoospermic controls (0.9%) were reported [111]. Regarding male infertility, although some patient-specific CNVs were identified, there are some limitations of these studies. However, the underlying mechanism of CNVs in infertility has not been explained yet. Several mechanisms are proposed for the role of CNVs in male infertility. CNVs could change the copy number of dosage-specific genes or contiguous genes or can alter the expression of dosage-specific genes by position effect. In addition, CNVs can unmask recessive mutations or functional polymorphisms of the remaining allele.

Y Chromosome Microdeletions

Microdeletions on the Y chromosome are the second most common genetic cause of male infertility. The Y chromosome contains highly homologous repeated DNA sequences in an inverted orientation, roughly in eight palindromic sequences. Deletions of the genes in these intervening sequences as a result of faulty recombination result in microdeletions. These microdeletions are too small to be observed in standard karyotype analysis [5,71,112,113]. Microdeletions that are observed on the long arm of the Y chromosome are associated with spermatogenic failure and are the outcome of nonallelic homologous recombination [71]. Approximately 1 in 4000 men in the general population has Y chromosomal microdeletions, and the frequency is significantly higher in infertile men [114]. The frequency of Y chromosomal microdeletions in nonobstructive azoospermia and severe oligozoospermia is about 10% and 5%, respectively [3].

Microdeletions in the azoospermia factor (AZF) genes cause azoospermia and spermatogenic failure. On the other hand, AZF microdeletions have no effect on phenotype or health of patients carrying them other than spermatogenic failure. Almost all AZF microdeletions on the Y chromosome develop *de novo* and are inherited to the offspring. The AZF region is located on the long arm of the Y chromosome (Yq11) and contains many genes critical for male germ-cell development, arranged in detached intervals of the AZFa, P5/proximal P1 (AZFb), b2/b4 (AZFc), and partially overlapped P5/distal P1 and p4/distal P1 (AZFbc) regions. All these three AZF regions are constitutional domains for proper spermatogenesis [115].

The majority of deletions occurring are in the AZFc region (~80%), followed by AZFa with approximately 0.5%–4% and AZFb regions with about 1%–5% deletions [1,116]. Deletions of both AZFa and AZFb regions occur with a frequency of 1%–3% [87].

The AZFa region is located at the proximal portion of interval 5 and includes single-copy *USP9Y* and *DDX3Y* (*DBY*) genes [117,118]. Complete deletions of this region remove both genes and result in SCOS and complete absence of sperm in microsurgical testicular sperm extraction (micro-TESE) [68,116,119]. However, loss of the *USP9Y* gene alone results in maturation arrest in spermatogenesis [118].

The AZFb region is located at the distal portion of interval 5 where it coincides with the AZFc region and the proximal portion of interval 6. Major genes of the AZFb region are *RBMY* and *PRY* that encode testis-specific splicing factors and involved in apoptosis [120,121]. Deletions in the AZFb region cause maturation arrest of spermatogenesis at the primary spermatocyte stage [122].

The AZFc region is located at interval 6 and composed of 12 genes. There are four copies of deleted azoospermia (*DAZ*), two copies of *CDY1*, and three copies of *BPY2* genes in this region. AZFc region is constituted by amplicons that show high homology and is prone to rearrangements in the AZFc region such as deletions, duplications, and inversions [112,123]. Repping and colleagues established AZFc partial deletions (subdeletions). The three main groups of Y chromosome AZFc partial deletions are b1/b3, b2/b3, and gr/gr [87] (Fig. 2.1). gr/gr is the most common type of subdeletion and leads to loss of half of the AZFc content. These deletions have functions in diverse spermatogenic statuses from normozoospermia to azoospermia by decreasing the copy number in AZFc regions. Several studies demonstrated a possible association between male infertility and gr/gr subdeletion. However, various studies did not observe any significant relation between gr/gr subdeletions and male infertility. Nevertheless, a significant relationship between gr/gr deletion and infertility in cases from Holland, Spain, and Italy has been reported. Conversely, the same authors reported no relationship between case studies from eight ethnic groups in East Asia. Likewise, there was no relationship between gr/gr deletions and infertility in studies from Israel, China, Malaysia, Egypt, Turkey, and Chile. In a meta-analysis (6388 cases and 6011 controls), Stouffs and colleagues reported the frequency of gr/gr deletions as 4.69% in normozoospermic controls and 6.86% in cases with spermatogenic impairment. Their results demonstrated that gr/gr deletions were significantly higher in oligozoospermic (not azoospermic) cases

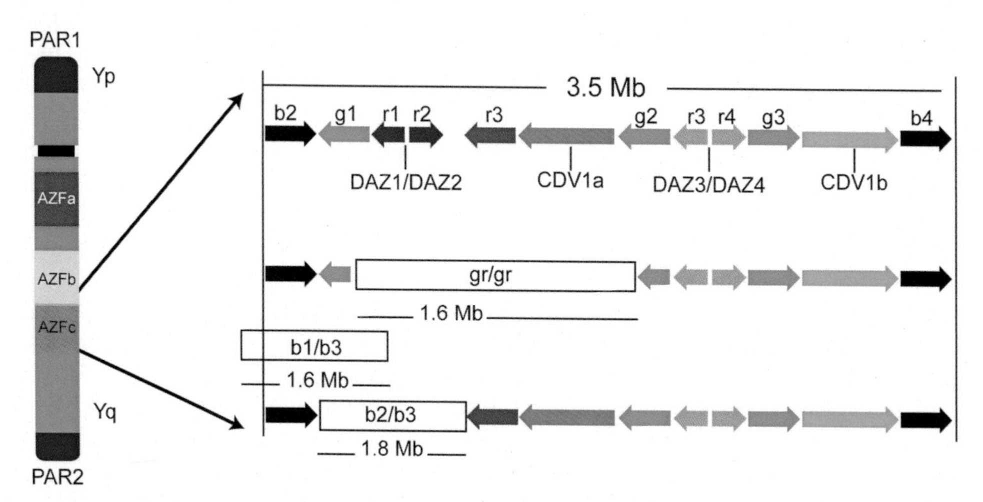

FIG. 2.1 AZFc partial deletions on Y chromosome. *Modified from C.C. Beyaz, S. Gunes, K. Onem, T. Kulac, R. Asci, Partial deletions of Y-chromosome in infertile men with non-obstructive azoospermia and oligoasthenoteratozoospermia in a Turkish population, In Vivo 31 (2017) 365–371.*

($P < 0.001$). Yq chromosome gr/gr phenotypes are heterogeneous, and their effect depends on environmental factors, penetrance of the gene, additional microdeletions on AZF regions, germinal mosaicism, and gene-gene interactions [1]. The frequency of deletion genotypes may be influenced by geographic and ethnic origins of the population [5].

Two copies of *DAZ* genes and one copy of *CDY1* gene are deleted in gr/gr and b2/b3 deletions. The results of gr/gr deletion association studies indicate ethnic and geographic variability. Ye and colleagues performed a study to search for AZF subdeletions on the Yi ethnic minority in China that is the seventh largest ethnicity in China. They have detected gr/gr deletions in 7.6% of cases and 8.5% of controls and b2/b3 deletions in 6.3% of cases and 8.5% of controls [124]. These results are consistent with East Asian studies, but are not in agreement with Giacchini's study on an Italian population [125]. Another study performed on 689 men (473 with spermatogenic failure) by Repping and colleagues concluded that gr/gr deletion frequency was higher in men with spermatogenic failure [87]. Besides gr/gr subdeletions of the AZFc region, the b2/b3 deletion is considered the most common reason for spermatogenic failure that causes the complete loss of the AZFc region. However, ethnic differences, methodology, and study design variations cause discrepancies in the evaluation of significance of gr/gr deletions [5].

GENOME-WIDE ASSOCIATION STUDIES

Recently, several genome-wide association studies (GWAS) were performed to assess relative association of SNPs between infertile cases and controls in subgroups of idiopathic infertility cases [2,126]. The first study was performed in 92 azoospermic and severe oligozoospermic men using Illumina 370K SNP array. The authors demonstrated a significant association between 21 SNPs and spermatogenic impairment. The second GWAS was using the Affymetrix Human SNP Array 6.0 to screen 488 patients with testicular dysgenesis syndrome (TDS) and 439 controls. The results of the most significant SNPs were replicated in a cohort of 436 men with TDS and 235 controls. The authors identified a relationship between SNPs in *TGFBR3* and *BMP7* genes and several TDS phenotypes, more particularly cryptorchidism and testicular cancer.

Two Han Chinese studies reported three stages of GWAS results using Affymetrix 6.0 microarrays in nonobstructive azoospermic males. GWAS of 2927 individuals with NOA and 5734 controls was performed, and significant associations between NOA risk and common variants near *PRMT6* (rs12097821 at 1p13.3, odds ratio (OR) = 1.25), *PEX10* (rs2477686 at 1p36.32, OR = 1.39), and *SOX5* (rs10842262 at 12p12.1, OR = 1.23) have been reported. The results of the study demonstrated that genetic variants at 1p13.3, 1p36.32, and 12p12.1 may have a role in the etiology of NOA in Han Chinese men [127].

Zhao and colleagues have genotyped 802 NOA men and 1863 controls. In two independent cohorts, 37 regions were genotyped, 818 azoospermia cases and 1755 controls from northern China and 606 azoospermia cases and 958 controls from central and southern China. The study revealed that four SNPs at human leukocyte antigen (HLA) locus were associated with NOA [128]. In a recent GWAS, 269 Hutterite men have been genotyped for two fertility traits (family size and birth rate) using three Affymetrix SNP arrays. Forty-one promising SNPs were validated with 123 ethnically diverse men from Chicago, and nine of these selected

SNPs were found to be associated with poor semen parameters. Although some GWAS have been performed with subgroups of infertile men in different ethnicities, significant and replicative evidence for association has been difficult to obtain with inconsistent findings across the studies [129].

LIMITATIONS OF GENETIC ASSOCIATION STUDIES

Genetic variations are not direct causes of male infertility, but these variations might increase the susceptibility to spermatogenic failure and infertility. Association or linkage studies are a challenge for investigating impaired spermatogenesis since male infertility is a heterogeneous condition and can result from aberrations of many different genes and environmental conditions or both. Histological patterns of spermatogenesis are varied and can be associated with specific phenotype, but they are not well characterized yet. The genetic variations have become a powerful tool in investigation of most multifactorial disorders, but yet, there are some limitations in this promising area of research for male infertility [7,76].

Results of association studies regarding male infertility often could not be repeated or reproduced. Definition of cases and controls is important since certain polymorphisms may be related with specific phenotypes. Therefore, only the analysis of this subgroup of infertile patients gives chance to identify their clinical significance. The most appropriate controls are fertile men with normal spermatogenesis. Inconsistent results might arise from phenotypic heterogeneity of the condition (oligozoospermia, azoospermia, etc.); selection of controls or engagement of different genes can result in the same type of infertility phenotype [7].

On the other hand, there are some other confounding factors in the etiology of the disease such as environmental factors including occupational factors, lifestyle, epigenetic factors, and gene-gene and/or gene-environment interactions. Sample size is another major step in designing a study as small studies may often result in unrealistic assumptions. The proper sample size for both infertile males and controls should be calculated by power analysis. However, SNPs related to the idiopathic male infertility subgroup make up a very small percentage of the cases.

CONCLUSION

Karyotype abnormalities, microdeletions on the Y chromosome, and *CFTR* gene mutations are well-known genetic causes of infertility in azoospermic or severely oligozoospermic men. Recently, some CNVs have been described to be associated with severe oligozoospermia or SCOS or both. In addition, some autosomal deletions, rare X-linked CNVs, DNA repair mechanism defects, and some SNPs have been found to be associated with male factor infertility [5].

Genetic abnormalities are one of the major causes of male infertility that can result in spermatogenic arrest, azoospermia, oligozoospermia, etc. Causes for abnormal spermatogenesis remain elusive almost half of the cases. Instead of holding a single SNP responsible for complex diseases including male infertility, which involves the contribution of many genes and their interactions, it is more rational to follow a more realistic approach. However, investigation of the genetic background in male infertility is a challenge in many ways. Traditional linkage analyses may not be feasible because infertility is a complex disease that is influenced by numerous variants. Since heterogeneity of genes is highly prevalent in sample populations, larger sample sizes are required for adequate statistical power. Additionally, genetic association studies to infertility could also be affected by many factors such as sample selection and ethnical differences.

Conflicting results may stem from the complexity and heterogeneity of male infertility etiology; differences in the selection of population characteristics, sample size, geographic and ethnic origins, including interaction with environmental factors; and gene-gene and gene-environment interactions.

References

[1] N. Rives, Y chromosome microdeletions and alterations of spermatogenesis, patient approach and genetic counseling, Ann. Endocrinol. (Paris) 75 (2014) 112–114.

[2] K.I. Aston, Genetic susceptibility to male infertility: news from genome-wide association studies, Andrology 2 (2014) 315–321.

[3] J. Hotaling, D.T. Carrell, Clinical genetic testing for male factor infertility: current applications and future directions, Andrology 2 (2014) 339–350.

[4] P.E. Thomas, R. Klinger, L.I. Furlong, M. Hofmann-Apitius, C.M. Friedrich, Challenges in the association of human single nucleotide polymorphism mentions with unique database identifiers, BMC Bioinformatics 12 (2011) S4.

[5] C. Krausz, L. Hoefsloot, M. Simoni, F. Tuttelmann, EAA/EMQN best practice guidelines for molecular diagnosis of Y-chromosomal microdeletions: state-of-the-art 2013, Andrology 2 (2014) 5–19.

[6] M. Nikpay, A. Goel, H.H. Won, L.M. Hall, C. Willenborg, S. Kanoni, D. Saleheen, T. Kyriakou, C.P. Nelson, J.C. Hopewell, et al., A comprehensive 1,000 genomes-based genome-wide association meta-analysis of coronary artery disease, Nat. Genet. 47 (10) (2015) 1121–1130.

[7] F. Tuttelmann, E. Rajpert-De Meyts, E. Nieschlag, M. Simoni, Gene polymorphisms and male infertility—a meta-analysis and literature review, Reprod. BioMed. Online 15 (2007) 643–658.

[8] W. Chen, Y. Li, P. Xu, G. Yang, G. Wang, K. Ma, Q. Li, X. Li, B. Cao, A comparative study on three-dimensional digital model of knee joint anterior cruciate ligament between normal adult and corpse based on magnetic resonance imaging reconstruction, Zhongguo Xiu Fu Chong Jian Wai Ke Za Zhi 25 (2011) 1314–1318.

[9] T.L. Kelly, O.R. Neaga, B.C. Schwahn, R. Rozen, J.M. Trasler, Infertility in 5,10-methylenetetrahydrofolate reductase (MTHFR)-deficient male mice is partially alleviated by lifetime dietary betaine supplementation, Biol. Reprod. 72 (2005) 667–677.

[10] T. Forges, H. Pellanda, C. Diligent, P. Monnier, J.L. Gueant, Do folates have an impact on fertility? Gynecol. Obstet. Fertil. 36 (2008) 930–939.

[11] C. Ravel, S. Chantot-Bastaraud, C. Chalmey, L. Barreiro, I. Aknin-Seifer, J. Pfeffer, I. Berthaut, E.E. Mathieu, J. Mandelbaum, J.P. Siffroi, et al., Lack of association between genetic polymorphisms in enzymes associated with folate metabolism and unexplained reduced sperm counts, PLoS One 4 (2009) e6540.

[12] W. Wu, O. Shen, Y. Qin, X. Niu, C. Lu, Y. Xia, L. Song, S. Wang, X. Wang, Idiopathic male infertility is strongly associated with aberrant promoter methylation of methylenetetrahydrofolate reductase (MTHFR), PLoS One 5 (2010) e13884.

[13] A.S. Weiner, U.A. Boyarskikh, E.N. Voronina, A.E. Tupikin, O.V. Korolkova, I.V. Morozov, M.L. Filipenko, Polymorphisms in folate-metabolizing genes and risk of idiopathic male infertility: a study on a Russian population and a meta-analysis, Fertil. Steril. 101 (2014) 87–94 (e83).

[14] H. Naqvi, S.R. Hussain, M.K. Ahmad, F. Mahdi, S.P. Jaiswar, S.N. Shankhwar, A.A. Mahdi, Role of 677C→T polymorphism a single substitution in methylenetetrahydrofolate reductase (MTHFR) gene in North Indian infertile men, Mol. Biol. Rep. 41 (2014) 573–579.

[15] M. Karimian, A.H. Colagar, Association of C677T transition of the human methylenetetrahydrofolate reductase (MTHFR) gene with male infertility, Reprod. Fertil. Dev. 28 (2016) 785–794.

[16] D.S. Mfady, M.F. Sadiq, O.F. Khabour, A.S. Fararjeh, A. Abu-Awad, Y. Khader, Associations of variants in MTHFR and MTRR genes with male infertility in the Jordanian population, Gene 536 (2014) 40–44.

[17] A. Eloualid, O. Abidi, M. Charif, B. El Houate, H. Benrahma, N. Louanjli, E. Chadli, M. Ajjemami, A. Barakat, A. Bashamboo, et al., Association of the MTHFR A1298C variant with unexplained severe male infertility, PLoS One 7 (2012) e34111.

[18] B. Wei, Z. Xu, J. Ruan, M. Zhu, K. Jin, D. Zhou, Z. Xu, Q. Hu, Q. Wang, Z. Wang, MTHFR 677C>T and 1298A>C polymorphisms and male infertility risk: a meta-analysis, Mol. Biol. Rep. 39 (2012) 1997–2002.

[19] W. Wu, O. Shen, Y. Qin, J. Lu, X. Niu, Z. Zhou, C. Lu, Y. Xia, S. Wang, X. Wang, Methylenetetrahydrofolate reductase C677T polymorphism and the risk of male infertility: a meta-analysis, Int. J. Androl. 35 (2012) 18–24.

[20] G.T. Vani, N. Mukesh, P. Rama Devi, P. Usha Rani, P.P. Reddy, Methylenetetrahydrofolate reductase C677T polymorphism is not associated with male infertility in a South Indian population, Andrologia 44 (2012) 252–259.

[21] M.M. Gava, E.A. Kayaki, B. Bianco, J.S. Teles, D.M. Christofolini, A.C. Pompeo, S. Glina, C.P. Barbosa, Polymorphisms in folate-related enzyme genes in idiopathic infertile Brazilian men, Reprod. Sci. 18 (12) (2011) 1267–1272.

[22] M.R. Safarinejad, N. Shafiei, S. Safarinejad, Relationship between genetic polymorphisms of methylenetetrahydrofolate reductase (C677T, A1298C, and G1793A) as risk factors for idiopathic male infertility, Reprod. Sci. 18 (2011) 304–315.

[23] N. Gupta, S. Gupta, M. Dama, A. David, G. Khanna, A. Khanna, S. Rajender, Strong association of 677 C>T substitution in the MTHFR gene with male infertility—a study on an indian population and a meta-analysis, PLoS One 6 (7) (2011) e22277.

[24] V.S. Dhillon, M. Shahid, S.A. Husain, Associations of MTHFR DNMT3b 4977 bp deletion in mtDNA and GSTM1 deletion, and aberrant CpG island hypermethylation of GSTM1 in non-obstructive infertility in Indian men, Mol. Hum. Reprod. 13 (2007) 213–222.

[25] Z.C. A, Y. Yang, S.Z. Zhang, N. Li, W. Zhang, Single nucleotide polymorphism C677T in the methylenetetrahydrofolate reductase gene might be a genetic risk factor for infertility for Chinese men with azoospermia or severe oligozoospermia, Asian J. Androl. 9 (2007) 57–62.

[26] H.C. Lee, Y.M. Jeong, S.H. Lee, K.Y. Cha, S.H. Song, N.K. Kim, K.W. Lee, S. Lee, Association study of four polymorphisms in three folate-related enzyme genes with non-obstructive male infertility, Hum. Reprod. 21 (2006) 3162–3170.

[27] K. Singh, S.K. Singh, R. Sah, I. Singh, R. Raman, Mutation C677T in the methylenetetrahydrofolate reductase gene is associated with male infertility in an Indian population, Int. J. Androl. 28 (2005) 115–119.

[28] J.H. Park, H.C. Lee, Y.M. Jeong, T.G. Chung, H.J. Kim, N.K. Kim, S.H. Lee, S. Lee, MTHFR C677T polymorphism associates with unexplained infertile male factors, J. Assist. Reprod. Genet. 22 (2005) 361–368.

[29] S.S. Li, J. Li, Z. Xiao, A.G. Ren, L. Jin, Prospective study of MTHFR genetic polymorphisms as a possible etiology of male infertility, Genet. Mol. Res. 13 (2014).

[30] V.B. Ucar, B. Nami, H. Acar, M. Kilinc, Is methylenetetrahydrofolate reductase (MTHFR) gene A1298C polymorphism related with varicocele risk? Andrologia 47 (2015) 42–46.

[31] O. Shen, R. Liu, W. Wu, L. Yu, X. Wang, Association of the methylenetetrahydrofolate reductase gene A1298C polymorphism with male infertility: a meta-analysis, Ann. Hum. Genet. 76 (2012) 25–32.

[32] K. Singh, S.K. Singh, R. Raman, MTHFR A1298C polymorphism and idiopathic male infertility, J. Postgrad. Med. 56 (2010) 267–269.

[33] X.J. Yang, T. Shinka, S. Nozawa, H.T. Yan, M. Yoshiike, M. Umeno, Y. Sato, G. Chen, T. Iwamoto, Y. Nakahori, Survey of the two polymorphisms in DAZL, an autosomal candidate for the azoospermic factor, in Japanese infertile men and implications for male infertility, Mol. Hum. Reprod. 11 (2005) 513–515.

[34] S. Zhang, Q. Tang, W. Wu, B. Yuan, C. Lu, Y. Xia, H. Ding, L. Hu, D. Chen, J. Sha, et al., Association between DAZL polymorphisms and susceptibility to male infertility: systematic review with meta-analysis and trial sequential analysis, Sci. Rep. 4 (2014) 4642.

[35] L. Becherini, E. Guarducci, S. Degl'Innocenti, M. Rotondi, G. Forti, C. Krausz, DAZL polymorphisms and susceptibility to spermatogenic failure: an example of remarkable ethnic differences, Int. J. Androl. 27 (2004) 375–381.

[36] L. Bartoloni, C. Cazzadore, A. Ferlin, A. Garolla, C. Foresta, Lack of the T54A polymorphism of the DAZL gene in infertile Italian patients, Mol. Hum. Reprod. 10 (2004) 613–615.

[37] Y.N. Teng, Y.P. Chang, J.T. Tseng, P.H. Kuo, I.W. Lee, M.S. Lee, P.L. Kuo, A single-nucleotide polymorphism of the DAZL gene promoter confers susceptibility to spermatogenic failure in the Taiwanese Han, Hum. Reprod. 27 (2012) 2857–2865.

[38] T. Plaseski, P. Noveski, Z. Popeska, G.D. Efremov, D. Plaseska-Karanfilska, Association study of single-nucleotide polymorphisms in FASLG, JMJDIA, LOC203413, TEX15, BRDT, OR2W3, INSR, and TAS2R38 genes with male infertility, J. Androl. 33 (2012) 675–683.

[39] L. Yan, S. Wu, S. Zhang, G. Ji, A. Gu, Genetic variants in telomerase reverse transcriptase (TERT) and telomerase-associated protein 1 (TEP1) and the risk of male infertility, Gene 534 (2014) 139–143.

[40] J. Fang, S. Wang, H. Wang, S. Zhang, S. Su, Z. Song, Y. Deng, J. Qian, J. Gu, B. Liu, et al., The cytochrome P4501A1 gene polymorphisms and idiopathic male infertility risk: a meta-analysis, Gene 535 (2014) 93–96.

[41] N. Lu, B. Wu, Y. Xia, W. Wang, A. Gu, J. Liang, C. Lu, L. Song, S. Wang, Y. Peng, et al., Polymorphisms in CYP1A1 gene are associated with male infertility in a Chinese population, Int. J. Androl. 31 (2008) 527–533.

[42] H. Luo, H. Li, N. Yao, L. Hu, T. He, Association between 3801T>C polymorphism of CYP1A1 and idiopathic male infertility risk: a systematic review and meta-analysis, PLoS One 9 (2014) e86649.

[43] S.L. Yarosh, E.V. Kokhtenko, N.I. Starodubova, M.I. Churnosov, A.V. Polonikov, Smoking status modifies the relation between CYP1A1*2C gene polymorphism and idiopathic male infertility: the importance of gene-environment interaction analysis for genetic studies of the disease, Reprod. Sci. 20 (2013) 1302–1307.

[44] Z. Salehi, L. Gholizadeh, H. Vaziri, A.H. Madani, Analysis of GSTM1, GSTT1, and CYP1A1 in idiopathic male infertility, Reprod. Sci. 19 (2012) 81–85.

[45] G.T. Vani, N. Mukesh, B. Siva Prasad, P. Rama Devi, M. Hema Prasad, P. Usha Rani, P. Pardhanandana Reddy, Association of CYP1A1*2A polymorphism with male infertility in Indian population, Clin. Chim. Acta 410 (2009) 43–47.

[46] M. Grigorova, M. Punab, O. Poolamets, S. Sober, V. Vihljajev, B. Zilaitiene, J. Erenpreiss, V. Matulevicius, I. Tsarev, M. Laan, Study in 1790 Baltic men: FSHR Asn680Ser polymorphism affects total testes volume, Andrology 1 (2013) 293–300.

[47] F. Tuttelmann, M. Laan, M. Grigorova, M. Punab, S. Sober, J. Gromoll, Combined effects of the variants FSHB -211G>T and FSHR 2039A>G on male reproductive parameters, J. Clin. Endocrinol. Metab. 97 (2012) 3639–3647.

[48] M. Ghirelli-Filho, C. Peluso, D.M. Christofolini, M.M. Gava, S. Glina, C.P. Barbosa, B. Bianco, Variants in follicle-stimulating hormone receptor gene in infertile Brazilian men and the correlation to FSH serum levels and sperm count, Reprod. Sci. 19 (2012) 733–739.

[49] M. Balkan, A. Gedik, H. Akkoc, O. Izci Ay, M.E. Erdal, H. Isi, T. Budak, FSHR single nucleotide polymorphism frequencies in proven fathers and infertile men in Southeast Turkey, J. Biomed. Biotechnol. 2010 (2010) 640318.

[50] C. Shimoda, E. Koh, K. Yamamoto, F. Matsui, K. Sugimoto, H.S. Sin, Y. Maeda, J. Kanaya, A. Yoshida, M. Namiki, Single nucleotide polymorphism analysis of the follicle-stimulating hormone (FSH) receptor in Japanese with male infertility: identification of codon combination with heterozygous variations of the two discrete FSH receptor gene, Endocr. J. 56 (2009) 859–865.

[51] L. Lazaros, N. Xita, A. Takenaka, N. Sofikitis, G. Makrydimas, T. Stefos, I. Kosmas, K. Zikopoulos, E. Hatzi, I. Georgiou, Semen quality is influenced by androgen receptor and aromatase gene synergism, Hum. Reprod. 27 (2013) 3385–3392.

[52] Y. Chen, L.N. Li, C.Q. Yang, Z.P. Hao, H.K. Sun, Y. Li, Countermeasures for priority control of toxic VOC pollution, Huan Jing Ke Xue 32 (2011) 3469–3475.

[53] B. Bianco, C. Peluso, M.M. Gava, M. Ghirelli-Filho, M.V. Lipay, M.A. Lipay, D.M. Christofolini, C.P. Barbosa, Polymorphisms of estrogen receptors alpha and beta in idiopathic, infertile Brazilian men: a case-control study, Mol. Reprod. Dev. 78 (2011) 665–672.

[54] I.W. Lee, P.H. Kuo, M.T. Su, L.C. Kuan, C.C. Hsu, P.L. Kuo, Quantitative trait analysis suggests polymorphisms of estrogen-related genes regulate human sperm concentrations and motility, Hum. Reprod. 26 (2011) 1585–1596.

[55] M.T. Su, C.H. Chen, P.H. Kuo, C.C. Hsu, I.W. Lee, H.A. Pan, Y.T. Chen, P.L. Kuo, Polymorphisms of estrogen-related genes jointly confer susceptibility to human spermatogenic defect, Fertil. Steril. 93 (2010) 141–149.

[56] J.J. Galan, B. Buch, N. Cruz, A. Segura, F.J. Moron, L. Bassas, L. Martinez-Pineiro, L.M. Real, A. Ruiz, Multilocus analyses of estrogen-related genes reveal involvement of the ESR1 gene in male infertility and the polygenic nature of the pathology, Fertil. Steril. 84 (2005) 910–918.

[57] E. Guarducci, F. Nuti, L. Becherini, M. Rotondi, G. Balercia, G. Forti, C. Krausz, Estrogen receptor alpha promoter polymorphism: stronger estrogen action is coupled with lower sperm count, Hum. Reprod. 21 (2006) 994–1001.

[58] Y. Suzuki, I. Sasagawa, K. Itoh, J. Ashida, K. Muroya, T. Ogata, Estrogen receptor alpha gene polymorphism is associated with idiopathic azoospermia, Fertil. Steril. 78 (2002) 1341–1343.
[59] Y. Cai, T. Liu, H. Li, C. Xiong, Meta-analysis of the association of oestrogen receptor-beta gene RsaI (G/A) and AluI (A/G) polymorphisms with male infertility, Andrologia 47 (2015) 257–265.
[60] W. Wu, Z. Hu, Y. Qin, J. Dong, J. Dai, C. Lu, W. Zhang, H. Shen, Y. Xia, X. Wang, Seminal plasma microRNAs: potential biomarkers for spermatogenesis status, Mol. Hum. Reprod. 18 (2012) 489–497.
[61] E. Siasi, A. Aleyasin, J. Mowla, H. Sahebkashaf, Association study of six SNPs in PRM1, PRM2 and TNP2 genes in Iranian infertile men with idiopathic azoospermia, Iran J. Reprod. Med. 10 (2012) 329–336.
[62] F. Tuttelmann, P. Krenkova, S. Romer, A.R. Nestorovic, M. Ljujic, A. Stambergova, M. Macek Jr., M. Macek Sr., E. Nieschlag, G.J. Romoll, et al., A common haplotype of protamine 1 and 2 genes is associated with higher sperm counts, Int. J. Androl. 33 (2010) e240–e248.
[63] H. Tanaka, Y. Miyagawa, A. Tsujimura, K. Matsumiya, A. Okuyama, Y. Nishimune, Single nucleotide polymorphisms in the protamine-1 and -2 genes of fertile and infertile human male populations, Mol. Hum. Reprod. 9 (2003) 69–73.
[64] A. Eloualid, H. Rhaissi, A. Reguig, S. Bounaceur, B. El Houate, O. Abidi, M. Charif, N. Louanjli, E. Chadli, A. Barakat, et al., Association of spermatogenic failure with the b2/b3 partial AZFc deletion, PLoS One 7 (2012) e34902.
[65] V. Paracchini, S. Garte, E. Taioli, MTHFR C677T polymorphism, GSTM1 deletion and male infertility: a possible suggestion of a gene-gene interaction? Biomarkers 11 (2006) 53–60.
[66] M.M. Gava, O. Chagas Ede, B. Bianco, D.M. Christofolini, A.C. Pompeo, S. Glina, C.P. Barbosa, Methylenetetrahydrofolate reductase polymorphisms are related to male infertility in Brazilian men, Genet. Test Mol. Biomarkers 15 (2011) 153–157.
[67] I.M. Ebisch, W.L. van Heerde, C.M.N. Thomas van der Put, W.Y. Wong, R.P. Steegers-Theunissen, C677T methylenetetrahydrofolate reductase polymorphism interferes with the effects of folic acid and zinc sulfate on sperm concentration, Fertil. Steril. 80 (2003) 1190–1194.
[68] F. Nuti, C. Krausz, Gene polymorphisms/mutations relevant to abnormal spermatogenesis, Reprod. BioMed. Online 16 (2008) 504–513.
[69] M. Ruggiu, R. Speed, M. Taggart, S.J. McKay, F. Kilanowski, P. Saunders, J. Dorin, H.J. Cooke, The mouse Dazla gene encodes a cytoplasmic protein essential for gametogenesis, Nature 389 (1997) 73–77.
[70] M. Ruggiu, P.T. Saunders, H.J. Cooke, Dynamic subcellular distribution of the DAZL protein is confined to primate male germ cells, J. Androl. 21 (2000) 470–477.
[71] J.M. Hotaling, Genetics of male infertility, Urol. Clin. North Am. 41 (2014) 1–17.
[72] S. Barda, L. Yogev, G. Paz, H. Yavetz, O. Lehavi, R. Hauser, T. Doniger, H. Breitbart, S.E. Kleiman, BRDT gene sequence in human testicular pathologies and the implication of its single nucleotide polymorphism (rs3088232) on fertility, Andrology 2 (2014) 641–647.
[73] E.H. Blackburn, C.W. Greider, J.W. Szostak, Telomeres and telomerase: the path from maize, Tetrahymena and yeast to human cancer and aging, Nat. Med. 12 (2006) 1133–1138.
[74] D.L. Keefe, L. Liu, Telomeres and reproductive aging, Reprod. Fertil. Dev. 21 (2009) 10–14.
[75] M.E. McManus, W.M. Burgess, M.E. Veronese, A. Huggett, L.C. Quattrochi, R.H. Tukey, Metabolism of 2-acetylaminofluorene and benzo(a)pyrene and activation of food-derived heterocyclic amine mutagens by human cytochromes P-450, Cancer Res. 50 (1990) 3367–3376.
[76] C. Krausz, C. Giachini, Genetic risk factors in male infertility, Arch. Androl. 53 (3) (2007) 125–133.
[77] L. Casarini, E. Pignatti, M. Simoni, Effects of polymorphisms in gonadotropin and gonadotropin receptor genes on reproductive function, Rev. Endocr. Metab. Disord. 12 (2011) 303–321.
[78] M. Simoni, G.F. Weinbauer, J. Gromoll, E. Nieschlag, Role of FSH in male gonadal function, Ann. Endocrinol. (Paris) 60 (1999) 102–106.
[79] Y. Ahda, J. Gromoll, A. Wunsch, K. Asatiani, M. Zitzmann, E. Nieschlag, M. Simoni, Follicle-stimulating hormone receptor gene haplotype distribution in normozoospermic and azoospermic men, J. Androl. 26 (2005) 494–499.
[80] E.M. Eddy, T.F. Washburn, D.O. Bunch, E.H. Goulding, B.C. Gladen, D.B. Lubahn, K.S. Korach, Targeted disruption of the estrogen receptor gene in male mice causes alteration of spermatogenesis and infertility, Endocrinology 137 (1996) 4796–4805.
[81] A. Kukuvitis, I. Georgiou, I. Bouba, A. Tsirka, C.H. Giannouli, C. Yapijakis, B. Tarlatzis, J. Bontis, D. Lolis, N. Sofikitis, et al., Association of oestrogen receptor alpha polymorphisms and androgen receptor CAG trinucleotide repeats with male infertility: a study in 109 Greek infertile men, Int. J. Androl. 25 (2002) 149–152.

[82] Y. Miyagawa, H. Nishimura, A. Tsujimura, Y. Matsuoka, K. Matsumiya, A. Okuyama, Y. Nishimune, H. Tanaka, Single-nucleotide polymorphisms and mutation analyses of the TNP1 and TNP2 genes of fertile and infertile human male populations, J. Androl. 26 (2005) 779–786.

[83] D. Grassetti, D. Paoli, M. Gallo, A. D'Ambrosio, F. Lombardo, A. Lenzi, L. Gandini, Protamine-1 and -2 polymorphisms and gene expression in male infertility: an Italian study, J. Endocrinol. Investig. 35 (2012) 882–888.

[84] X.J. He, J. Ruan, W.D. Du, G. Chen, Y. Zhou, S. Xu, X.B. Zuo, Y.X. Cao, X.J. Zhang, PRM1 variant rs35576928 (Arg>Ser) is associated with defective spermatogenesis in the Chinese Han population, Reprod. BioMed. Online 25 (2012) 627–634.

[85] R.E. Mills, C.T. Luttig, C.E. Larkins, A. Beauchamp, C. Tsui, W.S. Pittard, S.E. Devine, An initial map of insertion and deletion (INDEL) variation in the human genome, Genome Res. 16 (2006) 1182–1190.

[86] J.M. Mullaney, R.E. Mills, W.S. Pittard, S.E. Devine, Small insertions and deletions (INDELs) in human genomes, Hum. Mol. Genet. 19 (2010) R131–136.

[87] S. Repping, H. Skaletsky, L. Brown, S.K. van Daalen, C.M. Korver, T. Pyntikova, T. Kuroda-Kawaguchi, J.W. de Vries, R.D. Oates, S. Silber, et al., Polymorphism for a 1.6-Mb deletion of the human Y chromosome persists through balance between recurrent mutation and haploid selection, Nat. Genet. 35 (2003) 247–251.

[88] V.A. Giagulli, M.D. Carbone, G. De Pergola, E. Guastamacchia, F. Resta, B. Licchelli, C. Sabba, V. Triggiani, Could androgen receptor gene CAG tract polymorphism affect spermatogenesis in men with idiopathic infertility? J. Assist. Reprod. Genet. 31 (2014) 689–697.

[89] T.T. Han, J. Ran, X.P. Ding, L.J. Li, L.Y. Zhang, Y.P. Zhang, S.S. Nie, L. Chen, Cytogenetic and molecular analysis of infertile Chinese men: karyotypic abnormalities, Y-chromosome microdeletions, and CAG and GGN repeat polymorphisms in the androgen receptor gene, Genet. Mol. Res. 12 (2014) 2215–2226.

[90] C.A. Davis-Dao, E.D. Tuazon, R.Z. Sokol, V.K. Cortessis, Male infertility and variation in CAG repeat length in the androgen receptor gene: a meta-analysis, J. Clin. Endocrinol. Metab. 92 (2007) 4319–4326.

[91] K. Asatiani, S. von Eckardstein, M. Simoni, J. Gromoll, E. Nieschlag, CAG repeat length in the androgen receptor gene affects the risk of male infertility, Int. J. Androl. 26 (2000) 255–261.

[92] J. Poongothai, No CAG repeat expansion of polymerase gamma is associated with male infertility in Tamil Nadu, South India, Indian J. Hum. Genet. 19 (2013) 320–324.

[93] S. Baklouti-Gargouri, M. Ghorbel, N. Chakroun, A. Sellami, F. Fakhfakh, L. Ammar-Keskes, The CAG repeat polymorphism of mitochondrial polymerase gamma (POLG) is associated with male infertility in Tunisia, Andrologia 44 (2012) 68–73.

[94] S.Y. Liu, C.J. Zhang, H.Y. Peng, Y.F. Yao, L. Shi, J.B. Chen, K.Q. Lin, L. Yu, X.Q. Huang, H. Sun, et al., CAG-repeat variant in the polymerase gamma gene and male infertility in the Chinese population: a meta-analysis, Asian J. Androl. 13 (2011) 298–304.

[95] T.P. Harris, K.P. Gomas, F. Weir, A.J. Holyoake, P. McHugh, M. Wu, Y. Sin, I.L. Sin, F.Y. Sin, Molecular analysis of polymerase gamma gene and mitochondrial polymorphism in fertile and subfertile men, Int. J. Androl. 29 (2006) 421–433.

[96] I.E. Aknin-Seifer, R.L. Touraine, H. Lejeune, C. Jimenez, J. Chouteau, J.P. Siffroi, K. McElreavey, T. Bienvenu, C. Patrat, R. Levy, Is the CAG repeat of mitochondrial DNA polymerase gamma (POLG) associated with male infertility? A multi-centre French study, Hum. Reprod. 20 (2005) 736–740.

[97] C. Krausz, E. Guarducci, L. Becherini, S. Degl'Innocenti, L. Gerace, G. Balercia, G. Forti, The clinical significance of the POLG gene polymorphism in male infertility, J. Clin. Endocrinol. Metab. 89 (2004) 4292–4297.

[98] D.S. Rani, S.J. Carlus, J. Poongothai, A. Jyothi, K. Pavani, N.J. Gupta, A.G. Reddy, M.M. Rajan, K. Rao, B. Chakravarty, et al., CAG repeat variation in the mtDNA polymerase gamma is not associated with oligoasthenozoospermia, Int. J. Androl. 32 (2009) 647–655.

[99] J. Zhang, W. Jiang, Q. Zhou, M. Ni, S. Liu, P. Zhu, Q. Wu, W. Li, M. Zhang, X. Xia, CAG-repeat polymorphisms in the polymerase gamma gene and male infertility: a meta-analysis, Andrologia 48 (2016) 882–889.

[100] A. Massart, W. Lissens, H. Tournaye, K. Stouffs, Genetic causes of spermatogenic failure, Asian J. Androl. 14 (2012) 40–48.

[101] A.R. La Spada, E.M. Wilson, D.B. Lubahn, A.E. Harding, K.H. Fischbeck, Androgen receptor gene mutations in X-linked spinal and bulbar muscular atrophy, Nature 352 (1991) 77–79.

[102] E. Castro-Nallar, K. Bacallao, A. Parada-Bustamante, M.C. Lardone, P.V. Lopez, M. Madariaga, R. Valdevenito, A. Piottante, M. Ebensperger, A. Castro, Androgen receptor gene CAG and GGN repeat polymorphisms in Chilean men with primary severe spermatogenic failure, J. Androl. 31 (2010) 552–559.

[103] F.M. Mikhail, Copy number variations and human genetic disease, Curr. Opin. Pediatr. 26 (2014) 646–652.

[104] P. Stankiewicz, J.R. Lupski, Structural variation in the human genome and its role in disease, Annu. Rev. Med. 61 (2010) 437–455.

[105] C.L. Usher, S.A. McCarroll, Complex and multi-allelic copy number variation in human disease, Brief. Funct. Genomics 14 (2015) 329–338.

[106] P.H. Sudmant, J.O. Kitzman, F. Antonacci, C. Alkan, M. Malig, A. Tsalenko, N. Sampas, L. Bruhn, J. Shendure, P. Genomes, et al., Diversity of human copy number variation and multicopy genes, Science 330 (2010) 641–646.

[107] F. Tuttelmann, M. Simoni, S. Kliesch, S. Ledig, B. Dworniczak, P. Wieacker, A. Ropke, Copy number variants in patients with severe oligozoospermia and Sertoli-cell-only syndrome, PLoS One 6 (2011) e19426.

[108] Y. Dong, Y. Pan, R. Wang, Z. Zhang, Q. Xi, R.Z. Liu, Copy number variations in spermatogenic failure patients with chromosomal abnormalities and unexplained azoospermia, Genet. Mol. Res. 14 (2015) 16041–16049.

[109] S. Eggers, K.D. DeBoer, J. van den Bergen, L. Gordon, S.J. White, D. Jamsai, R.I. McLachlan, A.H. Sinclair, M.K. O'Bryan, Copy number variation associated with meiotic arrest in idiopathic male infertility, Fertil. Steril. 103 (2015) 214–219.

[110] K. Stouffs, D. Vandermaelen, A. Massart, B. Menten, S. Vergult, H. Tournaye, W. Lissens, Array comparative genomic hybridization in male infertility, Hum. Reprod. 27 (2012) 921–929.

[111] A.C. Tewes, S. Ledig, F. Tuttelmann, S. Kliesch, P. Wieacker, DMRT1 mutations are rarely associated with male infertility, Fertil. Steril. 102 (2014) 816–820.

[112] T. Kuroda-Kawaguchi, H. Skaletsky, L.G. Brown, P.J. Minx, H.S. Cordum, R.H. Waterston, R.K. Wilson, R.S. Silber, S. Oates Rozen, et al., The AZFc region of the Y chromosome features massive palindromes and uniform recurrent deletions in infertile men, Nat. Genet. 29 (2001) 279–286.

[113] K.L. O'Flynn O'Brien, A.C. Varghese, A. Agarwal, The genetic causes of male factor infertility: a review, Fertil. Steril. 93 (2010) 1–12.

[114] C. Krausz, C. Chianese, Genetic testing and counselling for male infertility, Curr. Opin. Endocrinol. Diabetes Obes. 21 (2014) 244–250.

[115] P.H. Vogt, A. Edelmann, S. Kirsch, O. Henegariu, P. Hirschmann, F. Kiesewetter, F.M. Kohn, W.B. Schill, S. Farah, C. Ramos, et al., Human Y chromosome azoospermia factors (AZF) mapped to different subregions in Yq11, Hum. Mol. Genet. 5 (1996) 933–943.

[116] A. Ferlin, B. Arredi, E. Speltra, C. Cazzadore, R. Selice, A. Garolla, A. Lenzi, C. Foresta, Molecular and clinical characterization of Y chromosome microdeletions in infertile men: a 10-year experience in Italy, J. Clin. Endocrinol. Metabol. 92 (2007) 762–770.

[117] C. Sun, H. Skaletsky, B. Birren, K. Devon, Z. Tang, S. Silber, R. Oates, D.C. Page, An azoospermic man with a de novo point mutation in the Y-chromosomal gene USP9Y, Nat. Genet. 23 (1999) 429–432.

[118] C.V. Hopps, A. Mielnik, M. Goldstein, G.D. Palermo, Z. Rosenwaks, P.N. Schlegel, Detection of sperm in men with Y chromosome microdeletions of the AZFa, AZFb and AZFc regions, Hum. Reprod. 18 (2003) 1660–1665.

[119] P. Navarro-Costa, C.E. Plancha, J. Goncalves, Genetic dissection of the AZF regions of the human Y chromosome: thriller or filler for male (in)fertility? J. Biomed. Biotechnol. 2010 (2010) 936569.

[120] R. Lavery, M. Glennon, J. Houghton, A. Nolan, D. Egan, M. Maher, Investigation of DAZ and RBMY1 gene expression in human testis by quantitative real-time PCR, Arch. Androl. 53 (2007) 71–73.

[121] X.W. Feng, X.M. Zhou, W.X. Qu, Y. Li, S.Y. Li, L. Zhao, Retrospective analysis of related factors for patients with weaning difficulties in medical intensive care unit, Zhonghua Yi Xue Za Zhi 91 (2011) 2688–2691.

[122] P.H. Vogt, Azoospermia factor (AZF) in Yq11: towards a molecular understanding of its function for human male fertility and spermatogenesis, Reprod. BioMed. Online 10 (2005) 81–93.

[123] H. Skaletsky, T. Kuroda-Kawaguchi, P.J. Minx, H.S. Cordum, L. Hillier, L.G. Brown, S. Repping, T. Pyntikova, J. Ali, T. Bieri, et al., The male-specific region of the human Y chromosome is a mosaic of discrete sequence classes, Nature 423 (2003) 825–837.

[124] J.J. Ye, L. Ma, L.J. Yang, J.H. Wang, Y.L. Wang, H. Guo, N. Gong, W.H. Nie, S.H. Zhao, Partial AZFc duplications not deletions are associated with male infertility in the Yi population of Yunnan Province, China, J. Zhejiang Univ. Sci. B 14 (2013) 807–815.

[125] C. Giachini, I. Laface, E. Guarducci, G. Balercia, G. Forti, C. Krausz, Partial AZFc deletions and duplications: clinical correlates in the Italian population, Hum. Genet. 124 (2008) 399–410.

[126] K.I. Aston, C. Krausz, I. Laface, E. Ruiz-Castane, D.T. Carrell, Evaluation of 172 candidate polymorphisms for association with oligozoospermia or azoospermia in a large cohort of men of European descent, Hum. Reprod. 25 (2010) 1383–1397.

[127] Z. Hu, Y. Xia, X. Guo, J. Dai, H. Li, H. Hu, Y. Jiang, F. Lu, Y. Wu, X. Yang, et al., A genome-wide association study in Chinese men identifies three risk loci for non-obstructive azoospermia, Nat. Genet. 44 (2011) 183–186.
[128] H. Zhao, J. Xu, H. Zhang, J. Sun, Y. Sun, Z. Wang, J. Liu, Q. Ding, S. Lu, R. Shi, et al., A genome-wide association study reveals that variants within the HLA region are associated with risk for nonobstructive azoospermia, Am. J. Hum. Genet. 90 (2012) 900–906.
[129] Y. Sato, A. Tajima, K. Tsunematsu, S. Nozawa, M. Yoshiike, E. Koh, J. Kanaya, M. Namiki, K. Matsumiya, A. Tsujimura, et al., An association study of four candidate loci for human male fertility traits with male infertility, Hum. Reprod. 30 (2015) 1510–1514.

Further Reading

[130] C.C. Beyaz, S. Gunes, K. Onem, T. Kulac, R. Asci, Partial deletions of Y-chromosome in infertile men with non-obstructive azoospermia and oligoasthenoteratozoospermia in a Turkish population, In Vivo 31 (2017) 365–371.

CHAPTER

3

New Genetic Point Mutations in Male Infertility

Xavier Vendrell

Sistemas Genómicos Ltd, Valencia, Spain

INTRODUCTION

Human reproductive incompetence is either expressed as infertility or subfertility and is an imbricated trait with several etiological causes and heterogeneous phenotypic presentation. Nowadays, in developed countries, close to 15% of couples have problems to reproduce, which represents a serious global disability that has been recognized by the World Human Organization in the WHO/World Bank World report on disability [1]. In fact, the diagnosis and treatment of the failure of the reproductive system is a milestone in private and public health care programs. There are several "nonbiological" causes of a decreased birth rate in modern societies such as the delay in the active search for the first conception, new social models, environmental pollution, and chemical exposition. The analysis of these factors is out of the objective of this chapter. However, it is important to note that it could mask the biological basis of the disorder.

Traditionally, it is assumed that the infertility causes are shared between males and females and, in particular cases, it is extremely difficult to determine the exact contribution to the reproductive failure from each partner. In this scenario, it is typically assumed that the so-called male factor is responsible for close to half of the clinical causes of this phenotype. The etiology of male infertility has been thoroughly studied in the past decades. International specialist groups assume several causes: congenital or acquired urogenital abnormalities, malignancies, urogenital tract infections, increased scrotal temperature (e.g., because of varicocele), endocrine disturbances, immunological factors and genetic abnormalities (see a deep review in Ref. [2]). In 30%–40% of cases, any cause associated to male infertility is found after a complete diagnostic work-up. This idiopathic situation is assumed to be caused by several factors and a lot of research is directed to explain part of this inconclusive diagnosis. Under the definition of "idiopathic," there are a few proposed causes such as: endocrine disruption because of environmental pollution, reactive oxygen species, or genetic and epigenetic variations. It is highly relevant to elucidate the genetic basis of the idiopathic male factor infertility.

Reproductomics
https://doi.org/10.1016/B978-0-12-812571-7.00004-6

Nowadays, the advances of assisted reproductive techniques allow men with suboptimal reproductive prognosis to conceive a viable zygote. In these circumstances, the lack of knowledge about the inheritance of this trait could have an impact on future generations, through passing pathological traits to offspring. Therefore, an accurate genetic diagnosis is crucial in the frame of reproductive genetic counseling concerning the effectiveness and safety of treatments.

From the genetic point of view, the human spermatogenesis depends on the synchronized action of thousands of genes whose transcripts are expressed mainly in the germ tissues. The role of a few known genetic regions' variations on sperm's function is clear (e.g., AZF regions on Y chromosome, sex chromosomal aneuploidies, or chromosomal rearrangements). On the other hand, there are an increasing number of studies that suggest the implication of new genes. These studies will be focused upon in the present chapter. In the recent years, the efforts of researchers have been to identify candidate genes involved in spermatogenesis. Global genome-based approaches have yielded close to 3000 genes related with spermatogenesis or germ cell function [3,4]. Recently, new technological approaches are available and offer novel ways to conduct the experiments. Comparative genome hybridization (CGH)-based microarrays, single-nucleotide polymorphisms (SNPs)-based microarrays, exome sequencing or whole-genome sequencing are used to interrogate the genomic constitution of an organism, yielding different levels of resolutions (reviewed in Refs. [5,6]). The massive data analysis is revealing new knowledge and the findings show a sophisticated network of complex molecular pathways. However, the causative effects of new genes on reproductive failure remain to be proven. Furthermore, to identify the genes related to spermatogenesis is not the only challenge. The discovery of thousands of small noncoding RNA acting on the mRNA stability as well as translation and protein modification offers new data to understand this highly regulated process. In this sense, the importance of epigenetics (reviewed in Ref. [7]) and the role of micro-RNA in the regulation of spermatogenesis have been highlighted recently [8,9], and both of these issues will be the aim of other chapters.

The objective of this chapter is to review the current knowledge concerning the genetic role on male reproductive failure, emphasizing new discoveries in relation to new point variants in the genome sequence of specific nuclear genes. In some cases, there is a clear correlation between new specific variants and infertile and subfertile traits. On the contrary, there are cases where the discovery of new changes in the genome and its relationship with male factor are purely speculative.

GENES AND MALE FACTOR INFERTILITY

The fertility function in men is a highly coordinated process where many pathways are involved and are genetically regulated. These include the development of the urogenital system; differentiation of the spermatogonial stem cells; formation of the spermatid acrosome and flagellum; acquisition of motility in the ejaculated spermatozoa; searching mechanism to oocyte-corona-cumulus complex arrival; chemical, enzymatic, and mechanical equipment to penetrate the membranes when joining with the female machinery and finally completing the fertilization and activation of genomes. In parallel, the endocrine regulation of gonadal function and the physiological mechanism of erection or ejaculation is genetically directed.

From the genetic point of view, several genetic causes, with different degrees of certainty, center around the analysis of infertile males. The chromosomal abnormalities explain close to 5% of infertility in males (see review [10]) and increase to 15% in cases of azoospermy [11]. Altered somatic karyotypes, including aneuploidy (e.g., Klinefelter syndrome) or chromosomal translocations (reciprocal and robertsonian translocations), are clearly related with infertility or subfertility in male populations (reviewed in Refs. [10,12]). In addition, the aneuploidy of sperm chromosomes has been deeply related with subfertility (reviewed in Ref. [13]). Furthermore, Copy Number Variations (CNVs) have been described in association with fertility impairment. CNVs are complete gains or losses of sequences distributed along the genome, conferring variability. In male infertility, the only CNV clearly associated with spermatogenic failure is AZF microdeletions on the Y chromosome [6,14–16]. However, new candidate CNVs have been revised recently in autosomes and the X chromosome with strong correlation with spermatogenic failure [16,17].

It is difficult to study the reproductive function as a single process in order to find new causative genome variations. From the clinical research point of view, the study of the etiopathogenesis of male infertility obligates authors to separate the investigations in detached biological pathways (deeply reviewed in Ref. [18]). The analysis of genes related with distinct routes causing different forms of infertility has been focused upon by researchers in the past years [3]. Several strategies have been proposed using different tools (reviewed in Ref. [5]), which include Sanger sequencing for candidate genes, genome-whole association (GWA)-microarray's based analysis, whole exome or genome sequencing based on next generation sequencing (NGS) or NGS's resequencing studies. Conclusively, a plethora of genes expressed in the male germ line, and other tissues nondirectly related with testicular function, have been reported. Following Matzuk and Lamb's proposal [18], we can differentiate three big groups of phenotypes in abnormal male reproductive function. These are: (i) male sexual differentiation disorders and gonadal development impairment, (ii) spermatogenesis alterations causing defects on spermatic functions, and (iii) systemic disorders affecting fertility. The last category is out of the scope of this chapter. There are a large number of systemic disorders (single-gene, congenital or nongenetic disorders) that present infertility or subfertility as a secondary trait and are included in a syndromic presentation. Some classical examples include autoimmune diseases, progressive neurological disorders (e.g., specific types of Charcot-Marie-Tooth disease or Duchenne's muscular dystrophy), systemic cilia-proteins affectation (Kartagener syndrome), hematological disorders (e.g., sickle-cell anemia or thalassemias), urogenital abnormalities (e.g., Alport syndrome), anatomical malformations, infections, and toxic exposure [19]. In these cases, the diagnostic work-up is based on the main symptoms of the disorder and not for the disrupted fertility. The clinical diagnosis as well as the patient's management and treatment options have special clinical protocols, with genetic counseling being highly recommended in all cases.

Fig. 3.1 represents the distribution of genes (per chromosome) that has been related with these phenotypes, in order to have an overall view. Following the view of Matzuk and Lamb [18], genes have been separated into two big groups depending on the main traits associated to the gene function. In addition, genes have been associated with phenotypic presentation, with a total of 16 possible phenotypes being reported (see legend). The list of genes and phenotypes has been updated with data from recent publications. Several authors have described the implication of these genes with different levels of evidence. For instance, case-control

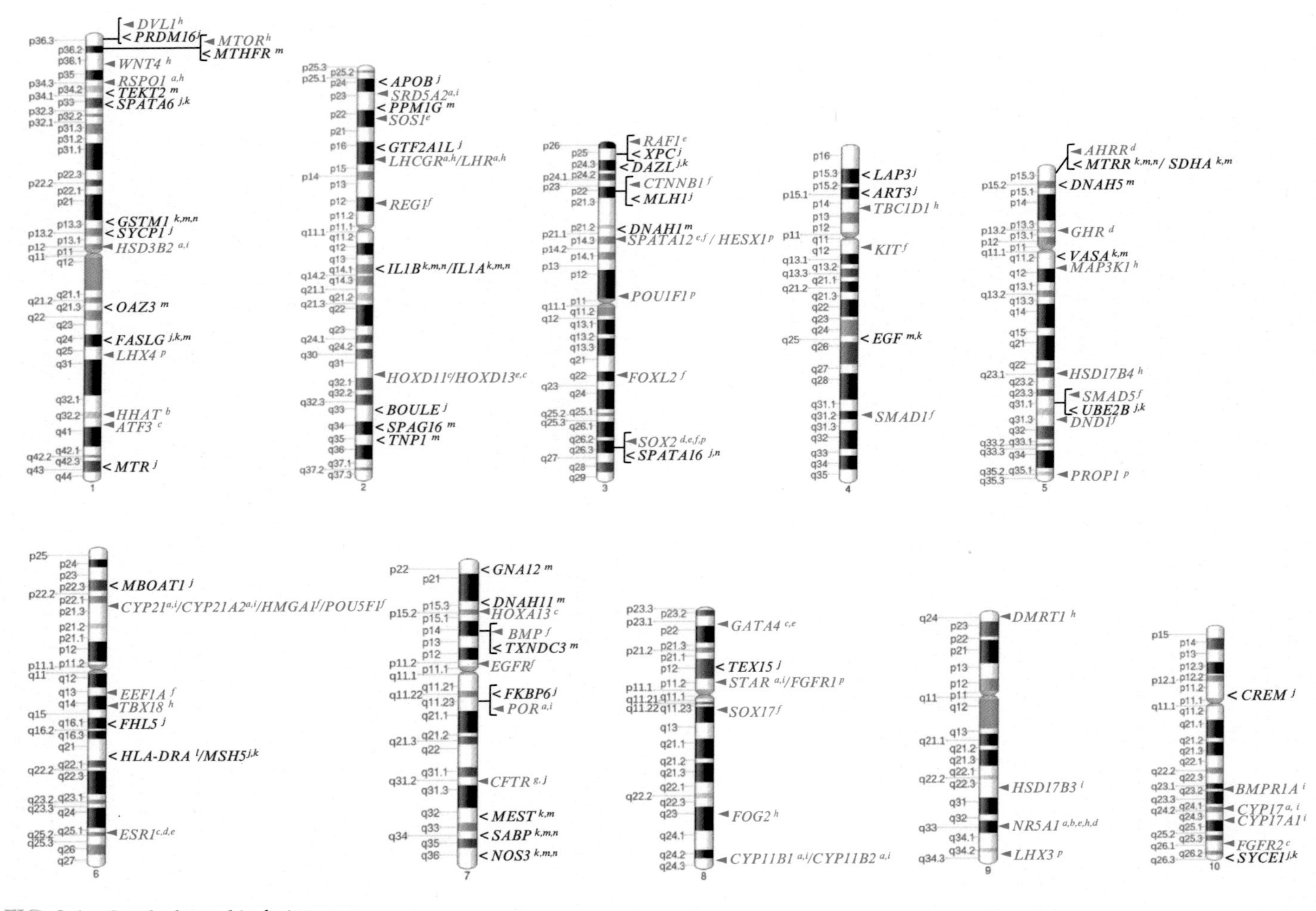

FIG. 3.1 See the legend in facing page.

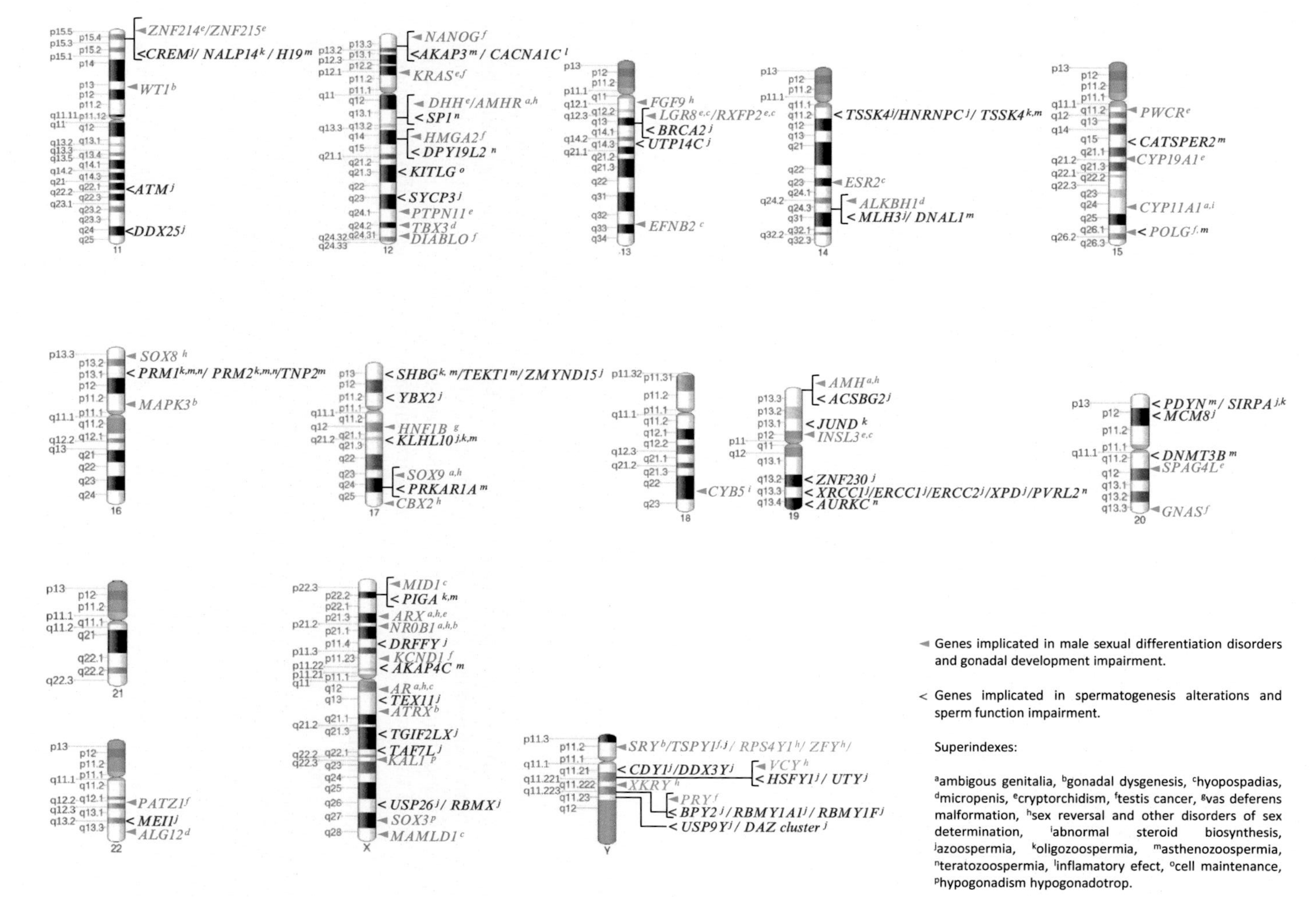

FIG. 3.1, CONT'D Distribution of genes implicated in male infertility. The genes have been located in their chromosome cytobands. The figure represents ideograms of human chromosomes showing a G-banded pattern at resolution of 850 bands per haploid set (from NCBI's Genome Decoration Page, https://www.ncbi.nlm.nih.gov/genome/tools/gdp). The genes are super indexed with letters that represent the reported associated phenotype (see legend). Two groups have been differentiated: genes related with sex male differentiation and gonadal development impairment (*in blue*) and genes related with spermatogenesis alteration and sperm function deficiency (*in red*). Information concerning phenotypes has been updated with data from [6,10,12,16–18,20–29].

studies with candidate genes have found more or less consistency in function of population size, with single-case studies observing an uncertain inheritance pattern. Even GWAs studies have reported genes directly related with complete testicular failure expressed as azoospermia. Furthermore, SNPs-based studies have suggested genetic susceptibility to male factor. In particular cases, the relationship between the genetic variant and the phenotype has been reported as a direct cause-effect relationship, which is responsible for the phenotype. However, in most of the cases, a collaborative action with other genes has been suggested (oligogenic model). Likewise, in particular cases, the authors propose the interaction of the gene products with other biomolecules, such as micro-RNA or epigenetic regulators, concluding that etiopathology remains to be elucidated. On the other hand, in many cases, functional studies are pending and the biological effect of the protein-coding genes is not fully understood. Finally, a very interesting issue that has been described in many studies is the crucial role of the ethnic background in order to explain incomplete penetrance and phenotype variability. The following subsections intend to update the knowledge related to genetic causes of male factor infertility, emphasizing new point variants recently described. The description of new discoveries has been separated by previously described phenotypes, i.e., male sexual differentiation disorders and spermatogenesis alterations.

Male Sexual Differentiation Disorders and Gonadal Development Impairment

First of all, it is important to emphasize that eutherian mammals have a special early development, characterized by similar steps in both sexes until the fetal stage. At the initial fetal stages, males and females have two distinct pathways. The common primordial organization is driven by specific programs that, in males, conclude with the development of the testis (deeply reviewed in Ref. [30]). The primary sex determination in human males is genetically regulated by a large number of genes acting synergistically and antagonistically (see review [20]). The understanding of the genetic network responsible for the development of the testis has been the focus of developmental researchers for decades. Svingen and Koopman [30] reported on three special characteristics in the differentiation of male germinal tissue, which confer high complexity. These include: (1) the cell lineages implicated are bipotential, meaning the differentiation (into ovaries or testis) depends on the signals received; (2) the differentiation of the Sertoli cells regulates the differentiation of the rest of the cell types implicated in the formation of the testis, and (3) the influence of somatic factors on male germ cells. Eventually, all this orchestration fails, resulting in what we know as Disorders of sex Determination (DSD). The DSD has been defined as "congenital conditions in which the development of chromosomal, gonadal, or anatomical sex is atypical" [31]. It is estimated than around 50% of these patients have causative mutations in genes involved in differentiation. However, the molecular diagnosis is still only achieved in around 20% of cases, excluding cases of steroidogenic block, which is detected biochemically [31].

From the genetic point of view, point mutations and small deletions have been described in a few crucial genes in the early development (reviewed in Refs. [21,22]). In mammals, the Y chromosome plays a pivotal role in male sex determination and is essential for normal sperm production (reviewed in Refs. [23,32,33]). It is well known that alterations in the Y-chromosome genes affect male fertility. One of the clearly implicated genes in sexual differentiation in males is the Y-linked testis-determining gene *SRY*. The *SRY*'s protein is the testis-determining factor, which initiates male sex determination by directing the development of supporting cell precursors (pre-Sertoli cells). Close to 15% of gonadal dysgenesis cases result from a mutation involving *SRY*. In many cases, the reported variants in *SRY* are de novo. However, there are cases with pathogenic variants inherited from a fertile father, which suggests incomplete penetrance or the concomitant action of other variants [34]. Another related Y-chromosome gene is the *TSPY* gene. This gene is involved in spermatic differentiation and proliferation, tumor suppression as well as cell cycle regulation, and it has been described in relation to sudden infant death with dysgenesis [23].

Other variants related with sexual developmental disorders have been reported in autosomal genes regulated by the Y-chromosome's genes. Particularly, DSD has been reported to be associated to many genes, such as *SOX9, GATA4, FOG2, NR5A1, WT, DHH, CBX2, ATRX, MAP3K1*, and *FGF9* (reviewed in Ref. [22]). The synergistic action of the *NR5A* gene and the *SRY* gene coactivating the action of the *SOX9* gene is highly relevant. The key role of the *SOX9* gene on sex differentiation has been highlighted in classical haploinsufficiency studies [35]. The product of the *SOX9K* gene acts during chondrocyte differentiation and, with steroidogenic factor 1, regulates the transcription of the anti-Muellerian hormone (AMH) gene. Deficiencies of this product lead to sex reversal. In this sense, it is important to highlight cases of testicular dysgenesis linked to heterozygous missense mutations in the *NR5A1* gene. In these cases, men have normal development of external genitalia [36]. The *NR5A1* gene product is a transcriptional activator. It appears to be essential for sexual differentiation and formation of the primary steroidogenic tissues as well as also regulating the AMH/Muellerian balance. Specifically, the point mutation p.R103Q in the *NR5A1* gene has been associated to impaired activation of steroidogenic genes. Additionally, the specific point change p.R92W is the only monogenic variant described that has recurrently been associated with nonsyndromic errors in primary sex determination [37]. This point variant shows incomplete penetrance and its physiopathological function is unclear. However, a reduced ability to interact with β-catenin to synergistically activate the reporter gene has been reported [37,38]. These findings stress the importance of this variant on upregulation of the *SOX9* gene expression (clearly explained in Ref. [20]). Finally, the new p.L230R point variant in the *NR5A1* gene has been described to be related to nonclassical DSD's phenotypes [39]. This fact emphasizes the importance of overlapping phenotypes. Authors have proposed the need to clearly distinguish between strict gonadal dysgenesis and androgen biosynthesis defects, thus allowing for a new classification of DSD based on new molecular findings.

Point variants have been also identified in the cofactors *GATA4* and *FOG2* genes, which cooperatively interact with the NR5A1 protein for testis determination and differentiation [22]. The *GATA4* gene participates in the regulation of other genes involved in embryogenesis and in myocardial differentiation and function, and it is also necessary for normal testicular development. In turn, the *FOG2* gene's protein interacts with GATA-family products regulating hematopoiesis and cardiogenesis in mammals. It has also been demonstrated that this protein can both activate and downregulate the expression of GATA-target genes. In particular, the heterozygous variant p.G221R in the *GATA4* gene disrupts the synergistic activation of the AMH promoter, thus losing the ability to bind to the FOG2 protein [40]. However, the single-nucleotide variant p.S402R in the *FOG2* gene exhibits a truncated protein, which lacks the ability to bind to the GATA4 protein. Other two point mutations in the *FOG2* gene (e.g., the homozygous point variant p.M544I and de novo heterozygous variant p.R260Q) have revealed its importance in this pathway. The combination of both of these variants eliminates FOG2's interaction with the GATA4 protein [41]. Furthermore, another gene that plays a key role in the differentiation of male gonads is the *DHH* gene and mutations in this gene have been related with dysgenetic testis. *DHH* encodes a signaling molecule that plays an important role in regulating morphogenesis. The point homozygous mutation p.R124Q has been related with impairment of Sertolli cell-Leydig cell interaction in early testis formation [42]. Additionally, other cases of gonadal dysgenesis have been related with the variant p.G287V in homozygosis in the *HHAT* gene [43], which is directly associated with the proteasome mediated degradation pathway.

Furthermore, the key role of genes implicated with MAPKs (mitogen-activated protein kinases) in the initial stages of sex determination in mammals is well known. The MAPKs are key proteins that are evolutionarily conserved and associated with the signaling transduction pathway. Bashamboo and McElreavey [22] reviewed the role of six candidate pathogenic variants in the *MAP3K1* gene: p.G616R, p.L189P, p.L189R, c.634-8T>A, p.P153L, and c.2180-2A>G. The MAP3K1 gene codes to a serine/threonine kinase, which is part of some signal transduction cascades. Mutations are related with cases of sex reversal and gonadal dysgenesis. However, a direct relationship with a phenotype has not yet been established. Finally, other genes like *DMRT1*, *SOXs*, *RSPO1*, *WNT4*, and FOXL2 have been correlated with initial stages of sex differentiation. However, the information related with point mutations is diffuse and fragmented, and even though it has been deeply studied [21] the physiopathologic information related to specific variants must be confirmed.

Unfortunately, classical genetic techniques for diagnosis or linkage studies in specific families of genes are tedious approaches; however, they are needed in order to conclusively identify novel pathogenic variants in this complicated process. In addition, another serious limitation consists of the lack of clear models for research. For example, the murine model to study the mechanism of sex determination and development in men is insufficient and, in some aspects, not fully appropriate, as the differences with the reproductive mouse systems are too large. In this context, the "omics" approaches are encouraging. Particularly, Next-Generation Sequencing (NGS) approaches are guiding the research to the identification of new genes or loci associated with these phenotypes, as previously stated. NGS strategies include a targeted approach on candidate or causative genes, sequencing of all the coding regions (exome) or whole-genome sequencing. These are very powerful approaches, both in diagnosis and clinical research. The application of exome sequencing in severe forms of DSD has been reported in close to 20% of patients where there was a misdiagnosis or atypical clinical presentation [30]. In addition, NGS applications have diagnosed close to 40% of patients with a wide phenotype spectrum. New genes such as *TBC1D1*, *TBX18*, *HSD17B4*, *MBOAT*, *MTOR*, and *DVL1* have been postulated to have associations with syndromic forms of DSD. Nowadays, the challenge remains in determining the causative effect of the new variants. Researchers are centering their efforts on functional assays in order to determine the biological consequences of the new variants on the gene product.

Other conditions associated with incomplete male virilization and compromised fertility are defects in the development of the genital tract. In this case, we are talking about hormonal, gonadal, or reproductive organ defects. It has been previously described alterations in genes responsible for the biosynthesis of steroid hormones, specifically the *AR* gene of androgen receptor, and in the signal pathway of AMH: the *AMH*, *BMPR1A*, and *SMAD* genes (transcription factor) (see review [18]). Furthermore, it is well established that mutations in the genes coding for enzymes and proteins supporting the early biosynthesis pathways cause classic androgen biosynthesis defects [44]. Mutations have been described in the *StAR, CYP11A1, HSD3B2, CYP17A1, HSD17B3, and SRD5A2* genes as well as the *POR and CYB5* genes, which are cofactors for the enzymatic reaction regulated by the CYP17 protein. Finally, from the genetic point of view, it is especially relevant to mention congenital abnormalities caused by abnormal genitourinary tract development. These malformations cause serious fertility impairment and the genetic etiology has been proven in some cases. In particular, these cases include cryptorchidism, hypospadias, and congenital absence (uni/bilateral) of the vas deferens (CBAVD). Extremely rare mutations have been described in the *INSL3* and *RXFP2* genes, which are implicated in the signal transduction pathways in tissues, which command the abdominal phase of testis descent [18]. However, a large body of evidence shows an association between CBAVD and the cystic fibrosis transmembrane conductance regulator (*CFTR*) gene. It is well known that there is a clear correlation between the phenotype and the combination of mutations F508del (c.1521_1523delCTT) and p.R117H (c.350G>A) with the polymorphism 5T, localized at the splice acceptor site of exon 8. In addition, the penetrance of this conformation might be determined by the length of the adjacent TG repeats (TG)12_13 [45].

Spermatogenesis Alterations and Sperm Function Impairment

Reduced sperm production is one of the more usual clinical traits detected during the clinical male examination of infertile couples. The affectation grade varies from low sperm counting (oligozoospermia) to complete absence of spermatozoa in the ejaculated seminal fluid (azoospermia). From the spermatogenesis alteration point of view, azoospermia is the most severe form of male infertility and has been studied for a long time. Approximately, 1% of men are azoospermic [46]. Azoospermia and low sperm counts might have a multifactorial etiology; but typically, it has been associated with a Sertolli-cell only phenotype, meiotic or maturation arrest, hypospermatogenesis (low cellularity) or normal spermatogenesis but an obstruction of the efferent vias. The biological pathways involved in these pathologies are highly regulated genetically. For example, defects in genes related with endocrine function, cellular proliferation and differentiation, DNA repair, apoptosis, recombination, chromatin remodeling, motility, cell-to-cell interactions, and mitosis and meiosis regulation could all possibly be implicated [47]. Mitchell and coworkers [17] differentiated between pretesticular, testicular, and posttesticular defects in order to organize the study of the genetic etiology of this imbricated network. Under this definition, pretesticular damage can be caused principally by endocrine abnormalities (low levels of sex steroids or gonadotropins). The cause might be congenital (e.g., single-gene defects), acquired (e.g., disorders of the hypothalamic-pituitary axis), or iatrogenic (e.g., environmental factors). Secondly, posttesticular causes could be caused by ejaculatory disturbances or obstructions, which impede the transport of spermatozoa. The genetic variants related with these disturbances have been addressed in previous sections. Thirdly, the causes of inefficient spermatogenesis at the testicular level could be separated into three broad groups: (i) Y-chromosome deletions, (ii) germ cell aplasia (including Sertolli cell-only syndrome), or (iii) spermatogenesis arrest. Other causes of drastic reduction in sperm counts at the testicular level include traumas, neoplasia, infections, or inflammatory diseases. These processes have been deeply studied in humans and animal models. However, clear evidence does not really exist at the genetic level. Nowadays, abnormal karyotype and Y-linked CNVs remains as the only recurrent genetic cause of nonobstructive azoospermia (NOA). The correlation between chromosomal translocations, aneuploidy of sex chromosomes (e.g., 47,XXY), chromosome Y microdeletions affecting the AZF region, and hypoespermatogenesis is being intensively studied (see reviews [6,10,15,18]).

Historically, the Y chromosome has been the center for the study of male infertility. Beyond the study of the AZF region and its CNVs (deeply reviewed in Ref. [48]), a few variants have been correlated with infertility phenotypes in more than 20 genes (reviewed in Ref. [23]). Recently, two reviews [24,49] collected the discoveries of new massive technologies in this field. In particular, it is important to highlight the role of variants in genes outside the AZF regions (e.g., the *ZFY*, *SRY*, *RPS4Y1*, and *TG1FLY* genes). However, the clinical association of the new variants has not been clearly established. Furthermore, Krausz and Casamonti [24] reviewed the role of the human protein-coding Y genes *TSPY*, *VCY*, *XKRY*, *CDY*, *HSFY*, *RBMY1*, *PRY*, *BPY2*, and *DAZ* as well as their autosomal or X-linked homologues in the infertility phenotype. The contribution of NGS technologies and the massive data analysis has revealed the variability of genotype-phenotype correlation concerning infertility phenotypes and the importance of noncoding transcripts from male-specific regions of Y chromosome (MYS) DNA. This fact suggests the possible regulatory function of these genic products [49]. Due to large population stratification and geographic variability of these genome regions, the establishment of a clear correlation between variants and particular phenotypes is challenging.

Alternatively, the new approaches in genomics have revealed novel mutations related with infertile patients and severe reduced counting in autosomal genes. In particular, Miyamoto and colleagues [25] and most recently Mitchell and colleagues [17] have reviewed the role of new point mutations in the autosomal genes *SYCP3*, *SYCE1*, *KLHL10*, *AURKC*, and *SPATA6*. have described the high impact of the genes coding for synaptonemal complex proteins such as the *SYCP3* and *SYCE1* genes on human fertility [26]. It has been suggested that variants c.643delA, IVS7-16_19del, c.657T>C, c.548T>C, and 666A>G in the *SYCP3* gene, and c.613C>T, c.197-2A>G in the *SYCE1* gene, correlate with meiotic arrest. These genes codify to a DNA-binding protein in the synaptonemal complex involved in the key steps of germ cell meiosis. Variants in these genes cause meiotic arrest, sperm aneuploidy, or synaptic failure [26]. In regards to the *KLHL10* gene, its product is a substrate-specific adapter of a CUL3-based E3 ubiquitin-protein ligase complex, which mediates the ubiquitination and subsequent proteasomal degradation of target proteins during spermatogenesis. Single-nucleotide missense variants p.A313T and p.Q216P were detected in heterozygosis in severe oligozoospermic males. However, a new analysis performed in control men and revision of data in the Exome Aggregation Consortium database (ExAC; url: http://exac.broadinstitute.org) now exclude these variants. Another point mutation has been described in homozygosis in the *AURKC* gene, which is expressed in the testis and has been implicated in cytokinesis, meiosis, and mitosis. The single-nucleotide variant p.L49W has been related with abnormal spermatogenesis, as well as polyploid and morphologically abnormal spermatozoa. Other mutations in the *SPATA16* gene, involved in the formation of the sperm acrosome, have been reported in patients with impaired spermatogenesis showing diverse phenotypes (e.g., asthenozoospermia, see later).

A relevant point in the nonobstructive azoospermia has been the analysis of the X chromosome and autosomal genes in consanguineous families. Recent data (reviewed in

Ref. [17]) indicate the relevance of the *TEX11* gene. This gene is expressed in the testis and is a regulator of crossover during meiosis, involved in initiation and/or maintenance of chromosome synapsis, as well as the formation of crossovers. The variants p.Asp435Leufs*10, c.1838-1G>A, c.792+1G>A, c.1837+1G>C, and p.Thr218_Lys296del have been found in azoospermic patients, whereas they have not been found in controls in the ExAC database. On the contrary, the study of recessive forms of NOA has generated an enormous interest. Recessive causative genes of NOA have been described in autosomes. In this context nonsense, splice and missense variants have been reported in azoospermic patients. Examples include p.Leu50fs*30 in the *ZMYND15* gene (chr. 17), p.Arg611* in the *TAFA4B* gene (chr.18); c.198-2A>G in the *SYCE1* gene (chr. 10), c.1954-1G in the *MCM8* gene (chr. 20), p.Y710* in the *TEX15* gene (chr. 8), and p.Pro455Ala in the *NPAS2* gene (chr. 2) (data obtained from review of Ref. [17]). The confirmation of these data is crucial and functional studies are necessary. Nevertheless, complete exome and genome approaches have increased new discoveries for these phenotypes exponentially. It is important to note that heterozygous variants, found in particular patients, have been refuted when population frequencies from thousands of complete exomes and genomes have been published. This fact has occurred in variants found in genes such as *USP26, NANOS1, MTHFR, GSTM1, FSHB,* or *NR5A1* [17]. For this reason, "candidate genes" sequencing strategies have been reconsidered and new massive strategies are strongly recommended. A clear example has been proposed by Quaynor and coworkers [27] who described 18 new genes, which are probably implicated with hypogonadotropic hypogonadism and Kallmann syndrome, by using targeted NGS techniques.

Finally, other frequent sperm abnormalities, beyond sperm quantity, are astenozoospermia and teratozoospermia. Genetic causes of these motility and morphological sperm disorders are still unknown. Deletions in the *AKAP3* and *AKAP4* genes were described as possibly being responsible for sperm motility affectation due to fibrous sheath dysplasia (reviewed in Ref. [18]). Ben Khelifa and coworkers [50] coined the acronym MMAF (Multiple Morphological Abnormalities of sperm Flagella), in order to identify the nonsyndromic sperm motility affectation and pathogenic variants in the *DNAHI* gene, which encodes a protein in the axonemal dynein cluster related with microtubule motor activity. Recently, three point variants in the *DNAHI* gene have been associated with MMAF phenotype: c.8626-1G>A, Val1287Gly, and Asp1293Asn [28] and thus concluding that this gene is one of the main genes involved in the phenotype. In regards to morphological abnormalities, it is difficult to detect a "pure" phenotype. Usually, several abnormalities of sperm coexist in the same sample with varying grades of severity (e.g., round or abnormal heads or multiflagellation). In addition, abnormal sperm shapes could appear jointly with reduced sperm count or limited motility. In this scenario with a multiplicity of phenotypes, it is difficult to establish a clear phenotype-genotype correlation. However, a monomorphic teratozoospermia with round heads and acrosome-lacked spermatozoa exists. This is the globozoospermia, which is a quasipure phenotype. Three genes have been described as responsible for the phenotype: *SPATA16, PRKCA1,* and *DPY19L2* (reviewed in Ref. [29]). The homozygous point variant c.848G>A in the *SPATA16* gene has been clinically associated with the abnormal phenotype. The codified protein is expressed in the Golgi apparatus and is involved in the formation of the sperm acrosome, which explains its potential role in spermatogenesis and sperm-egg fusion.

Single-Nucleotide Polymorphism

A critical aspect that has been focused upon in the study of male factor infertility researcher is the probable concomitant function of specific polymorphic changes on the infertile phenotype. The role of these genome variants has been associated with a "genetic susceptibility" or "genetic predisposition" to infertility. In this sense, it is interesting to review the role of specific single-nucleotide polymorphisms (SNPs) in particular genes implicated in meiosis, spermatogenic function and regulation. Krausz and coworkers [16] reviewed a large set of metaanalysis, case-control, and GWAs studies performed on more than 370,000 SNPs in key genes related with cell maintenance (general cell functions), apoptotic process, DNA repair, response to reactive oxygen species, etc. in regards to expression in germ cells. The intensive comparison of studies concluded that only a few number of SNPs have been clearly correlated with infertility phenotypes. An interesting point is the crucial role of ethnography in the phenotypic expression. Concretely, a positive association has been reported for five genes: *MTHFR*, *ESR1/ESR2*, *NOS3*, *DAZL*, *MSH5*, *SIRPA/SIRPG*, and *HLA-DRA*. The point polymorphic variants A222V (c.677C > T) and E429A (c.1298C > A) in the *MTHFR* gene play a role in processing amino acids in folate metabolism, and have been associated with azoospermia and oligoastenoteratozoospermia. Variable penetrance has been reported and related to folate intake. This information could be relevant in relation with the response to folate supplementation in certain patients. Other SNPs in the estrogen receptor genes *ESR1* and *ESR*, which are highly expressed in testicular germ cells, are c.453-397T > C, c.453-351A > G in the *ESR1* gene, and c.984G > A in the *ESR2* gene. Once again, its modulatory effect has been reported depending on geographic background. In the same sense, clinical relevancy has been conferred to SNPs c.-786C > T and 4a4b (27 base pairs tandem repeats) in the *NOS3* gene, which are responsible for nitric oxide (NO) production. The missense point variant c.160A > G (p.Thr54Ala) in the *DAZ* gene codifies RNA-binding proteins that are key in spermatogenesis, and predispose carriers to spermatogenic failure. Concordant results between studies have been reported for the c.85C > T (p.Pro29Ser) SNP in the *MSH5* gene, which has been implicated in DNA repair and apoptosis. A significant susceptibility for oligozoospermia and NOA has been reported for c.*273G > T in the *SIRPA* gene and c.*223T > G in the *SIRPG* gene, with both of them playing a key role in intracellular signaling during synaptogenesis and in synaptic function. Finally, three point polymorphic variants in the *HLA-DRA* gene participating in the immune system show significant association to male infertility. Concretely, c.82 + 926A > C, c.*63G > A, and c.724T > G have been defined as susceptibility factors for autoimmune diseases. Probably these nucleotide changes may have a role in inflammatory responses of patients with antecedents of urogenital inflammation.

The continuous updating and publication of new SNPs, sequence variations, uncertain significance changes in DNA, etc. exceed the review capability of this chapter. New metaanalysis and case-control studies investigating the probable biological functions of specific SNPs are being published constantly. In the future, the sharing of massive databases in the context of big data platforms will offer a whole and systemic perspective of the biological pathways involved in the etiopathogenesis of idiopathic male factor. In this new scenario, possible gene interactions will probably help to elucidate the role of genome variability and will expand the diagnosis potential.

FINAL COMMENTS

The reproductive male failure is an extremely complex trait that includes overlapping phenotypes. This circumstance complicates the study of its etiology. The reproductive function in males is a highly coordinated process. Thousands of genes are working together beginning in the first stages of embryo development. In addition, the spermatogenic function is a continuous and prolonged process (with duration close to 3 months) and, for this reason, it is highly susceptible to environmental, genetic, or epigenetic changes. Genetically speaking, hundreds of studies have been published during the past years in order to look for candidate genes. Several studies propose new genes or new mutations in key genes. However, a great limitation of the majority of studies is the inclusion of specific patients in particular families, and that the extrapolation to other phenotypically similar patients is not clear. Moreover, the ethnicity and geographical background is crucial in these phenotypes, and the sequence-based phylogenetic analysis has revealed complex mechanisms of selection (positive/negative selection) of particular traits. In addition, in some cases, a clear clinical correlation failed to be proved (obviously, the "infertility" trait implies the impossibility of natural transmission of the phenotype). Other milestones in order to elucidate the clear genetic contribution to the idiopathic male fertility disorders are reduced penetrance, locus heterogeneity, and multigenic (digenic or trigenic) inheritance detected in particular features. In this imbricated scenario, the contribution of new massive technologies is highly encouraging. The exome or whole genome sequencing approaches are allowing for the discovery of thousands of single-nucleotide variants. In the majority of cases, these variants have been described previously as disease-causing variants or frequent nonpathogenic population polymorphisms. However, the new found variants should be described as new polymorphisms, new pathogenic variants or accidental sequencing errors. This is the real challenge of these new approaches. For this purpose, it is extremely important to filter and cure the data concerning these new findings, depending on factors such as inheritance pattern of a particular trait in our experimental population (considering de novo appearance), family history, variable expressivity, sample size, and other confounding factors (e.g., eventual iatrogenic or environmental factors). In this sense, general databases should be used with caution because they may be "contaminated" by nonclearly defined variants. The next widespread use of these techniques for clinical diagnosis or screening might accomplish strict criteria concerning accurate sequence assembly, annotating of the new variants, implementing of comprehensive collection of functionally validated datasets and new precise function predictor algorithms. Most likely, these premises will help with the systemic understanding of the male reproductive process.

References

[1] WHO/World Bank, World Report on Disability. http://www.who.int/disabilities/world_report/2011/report/en/.

[2] A. Jungwirth, T. Diemer, Z. Kopa, C. Krausz, H. Tournaye, B. Kelly, R. Pal, European Association of Urology Guidelines, http://uroweb.org/guideline/male-infertility/, 2016.

[3] R. Hochstenbach, J.H. Hackstein, The comparative genetics of human spermatogenesis: clues from flies and other model organisms, Results Probl. Cell Differ. 28 (2000) 271–298.

[4] N. Schultz, F.K. Hamra, D.L. Garbers, A multitude of genes expressed solely in meiotic or postmeiotic spermatogenic cells offers a myriad of contraceptive targets, Proc. Natl. Acad. Sci. USA 100 (21) (2003) 12201–12206.

[5] K.I. Aston, Genetic susceptibility to male infertility: news from genome-wide association studies, Andrology 2 (3) (2014) 315–321.
[6] S.H. Song, K. Chiba, R. Ramasamy, D.J. Lamb, Recent advances in the genetics of testicular failure, Asian J. Androl. 18 (3) (2016) 350–355.
[7] S. Gunes, M.A. Arslan, G.N.T. Hekim, R. Asci, The role of epigenetics in idiopathic male infertility, J. Assist. Reprod. Genet. 33 (5) (2016) 553–569.
[8] X. Chen, X. Li, J. Guo, P. Zhang, Z. Wenxian, The roles of microRNAs in regulation of mammalian spermatogenesis, J. Anim. Sci. Biotechnol. 8 (2017) 35.
[9] A. Salas-Huetos, J. Blanco, F. Vidal, M. Grossmann, M.C. Pons, N. Garrido, E. Anton, Spermatozoa from normozoospermic fertile and infertile individuals convey a distinct miRNA cargo, Andrology 4 (6) (2016) 1028–1036.
[10] K. O'Brien, A. Varghese, A. Agarwal, The genetic causes of male factor infertility: a review, Fertil. Steril. 93 (1) (2010) 1–12.
[11] A. Ferlin, F. Raicu, V. Gatta, D. Zuccarello, G. Palka, C. Foresta, Male infertility: role of genetic background, Reprod. BioMed. Online 14 (6) (2007) 734–745.
[12] F.T. Neto, P.V. Bach, B.B. Najari, P.S. Li, M. Goldstein, Genetics of male infertility, Curr Urol Rep 17 (10) (2016) 70.
[13] X. Vendrell, M. Ferrer, E. García-Mengual, P. Muñoz, J.C. Triviño, C. Calatayud, V.Y. Rawe, M. Ruiz-Jorro, Correlation between aneuploidy, apoptotic markers and DNA fragmentation in spermatozoa from normozoospermic patients, Reprod. BioMed. Online 28 (4) (2014) 492–502.
[14] P.H. Vogt, A. Edelmann, S. Kirsch, O. Henegariu, P. Hirschmann, F. Kiesewetter, F.M. Köhn, W.B. Schill, S. Farah, C. Ramos, M. Hartmann, W. Hartschuh, D. Meschede, H.M. Behre, A. Castel, E. Nieschlag, W. Weidner, H.J. Gröne, A. Jung, W. Engel, G. Haidl, Human Y chromosome azoospermia factors (AZF) mapped to different subregions in Yq11, Hum. Mol. Genet. 5 (7) (1996) 933–943.
[15] C. Krausz, L. Hoefsloot, M. Simoni, F. Tüttelmann, European Academy of Andrology, European Molecular Genetics Quality Network, EAA/EMQN best practice guidelines for molecular diagnosis of Y-chromosomal microdeletions: state-of-the-art 2013, Andrology 2 (1) (2014) 5–19.
[16] C. Krausz, A.R. Escamilla, C. Chianese, Genetics of male infertility: from research to clinic, Reproduction 150 (5) (2015) R159–R174.
[17] M.J. Mitchell, C. Metzler-Guillemain, A. Toure, C. Coutton, C. Arnoult, P.F. Ray, Single gene defects leading to sperm quantitative anomalies, Clin. Genet. 91 (2) (2017) 208–216.
[18] M.M. Matzuk, D.J. Lamb, The biology of infertility: research advances and clinical challenges, Nat. Med. 14 (11) (2008) 1197–1213.
[19] A. Mahmoud, F. Comhaire, Systemic causes of male infertility, in: W.-B. Schill, F. Comhaire, T.B. Hargreave (Eds.), Andrology for the Clinician, Springer-Verlag, Berlin; Heidelberg, 2006, pp. 57–63.
[20] A. Bashamboo, C. Eozenou, S. Rojo, K. McElreavey, Anomalies in human sex determination provide unique insights into the complex genetic interactions of early gonad development, Clin. Genet. 91 (2) (2017) 143–156.
[21] A. Bashamboo, K. McElreavey, Gene mutations associated with anomalies of human gonad formation, Sex Dev. 7 (1–3) (2013) 126–146.
[22] A. Bashamboo, K. McElreavey, Human sex-determination and disorders of sex-development (DSD), Semin. Cell Dev. Biol. 45 (2015) 77–83.
[23] J. Dhanoa, C. Mukhopadhyay, J. Arora, Y-chromosomal genes affecting male fertility: a review, Vet. World 9 (7) (2016) 783–791.
[24] C. Krausz, E. Casamonti, Spermatogenic failure and the Y chromosome, Hum. Genet. 136 (5) (2017) 637–655.
[25] T. Miyamoto, G. Minase, K. Okabe, H. Ueda, K. Sengoku, Male infertility and its genetic causes, J. Obstet. Gynaecol. Res. 41 (10) (2015) 1501–1505.
[26] A. Geisinger, R. Benavente, Mutations in genes coding for synaptonemal complex proteins and their impact on human fertility, Cytogenet. Genome Res. 150 (2) (2016) 77–85.
[27] S.D. Quaynor, M.E. Bosley, C.G. Duckworth, K.R. Porter, S.H. Kim, H.G. Kim, L.P. Chorich, M.E. Sullivan, J.H. Choi, R.S. Cameron, L.C. Layman, Targeted next generation sequencing approach identifies eighteen new candidate genes in normosmic hypogonadotropic hypogonadism and Kallmann syndrome, Mol. Cell. Endocrinol. 437 (2016) 86–96.
[28] A. Amiri-Yekta, C. Coutton, Z.E. Kherraf, T. Karaouzène, P. Le Tanno, M.H. Sanati, M. Sabbaghian, N. Almadani, M.A. Sadighi Gilani, S.H. Hosseini, S. Bahrami, A. Daneshipour, M. Bini, C. Arnoult, R. Colombo, H. Gourabi, P.F. Ray, Whole-exome sequencing of familial cases of multiple morphological abnormalities of the sperm flagella (MMAF) reveals new DNAH1 mutations, Hum. Reprod. 31 (12) (2016) 2872–2880.

[29] H. Ghédir, S. Ibala-Romdhane, O. Okutman, G. Viot, A. Saad, S. Viville, Identification of a new DPY19L2 mutation and a better definition of DPY19L2 deletion breakpoints leading to globozoospermia, Mol. Hum. Reprod. 22 (1) (2016) 35–45.

[30] T. Svingen, P. Koopman, Building the mammalian testis: origins, differentiation, and assembly of the component cell populations, Genes Dev. 27 (22) (2013) 2409–2426.

[31] I.A. Hughes, C. Houk, S.F. Ahmed, P.A. Lee, Lawson Wilkins Pediatric Endocrine Society/European Society for Paediatric Endocrinology Consensus Group, Consensus statement on management of intersex disorders, J. Pediatr. Urol. 2 (3) (2006) 148–162.

[32] J.F. Hughes, S. Rozen, Genomics and genetics of human and primate y chromosomes, Annu. Rev. Genomics Hum. Genet. 13 (2012) 83–108.

[33] P. Navarro-Costa, Sex, rebellion and decadence: the scandalous evolutionary history of the human Y chromosome, Biochim. Biophys. Acta 1822 (12) (2012) 1851–1863.

[34] N.B. Phillips, J. Racca, Y.S. Chen, R. Singh, A. Jancso-Radek, J.T. Radek, N.P. Wickramasinghe, E. Haas, M.A. Weiss, Mammalian testis-determining factor SRY and the enigma of inherited human sex reversal: frustrated induced fit in a bent protein-DNA complex, J. Biol. Chem. 286 (42) (2011) 36787–36807.

[35] T. Wagner, J. Wirth, J. Meyer, B. Zabel, M. Held, J. Zimmer, J. Pasantes, F.D. Bricarelli, J. Keutel, E. Hustert, U. Wolf, N. Tommerup, W. Schempp, G. Scherer, Autosomal sex reversal and campomelic dysplasia are caused by mutations in and around the SRY-related gene SOX9, Cell 79 (6) (1994) 1111–1120.

[36] A. Bashamboo, B. Ferraz-de-Souza, D. Lourenço, L. Lin, N.J. Sebire, D. Montjean, J. Bignon-Topalovic, J. Mandelbaum, J.P. Siffroi, S. Christin-Maitre, U. Radhakrishna, H. Rouba, C. Ravel, J. Seeler, J.C. Achermann, K. McElreavey, Human male infertility associated with mutations in NR5A1 encoding steroidogenic factor 1, Am. J. Hum. Genet. 87 (4) (2010) 505–512.

[37] A. Bashamboo, P.A. Donohoue, E. Vilain, S. Rojo, P. Calvel, S.N. Seneviratne, F. Buonocore, H. Barseghyan, N. Bingham, J.A. Rosenfeld, S.N. Mulukutla, M. Jain, L. Burrage, S. Dhar, A. Balasubramanyam, B. Lee, Members of UDN, M.C. Dumargne, C. Eozenou, J.P. Suntharalingham, K. de Silva, L. Lin, J. Bignon-Topalovic, F. Poulat, C.F. Lagos, K. McElreavey, J.C. Achermann, A recurrent p.Arg92Trp variant in steroidogenic factor-1 (NR5A1) can act as a molecular switch in human sex development, Hum. Mol. Genet. 25 (16) (2016) 3446–3453.

[38] M. Igarashi, K. Takasawa, A. Hakoda, J. Kanno, S. Takada, M. Miyado, T. Baba, K.I. Morohashi, T. Tajima, K. Hata, K. Nakabayashi, Y. Matsubara, R. Sekido, T. Ogata, K. Kashimada, M. Fukami, Identical NR5A1 missense mutations in two unrelated 46,XX individuals with testicular tissues, Hum. Mutat. 38 (1) (2017) 39–42.

[39] R. Werner, I. Mönig, J. August, C. Freiberg, R. Lünstedt, B. Reiz, L. Wünsch, P.M. Holterhus, A. Kulle, U. Döhnert, S.A. Wudy, A. Richter-Unruh, C. Thorns, O. Hiort, Novel insights into 46,XY disorders of sex development due to NR5A1 gene mutation, Sex Dev. 9 (5) (2015) 260–268.

[40] D. Lourenço, R. Brauner, M. Rybczynska, C. Nihoul-Fékété, K. McElreavey, A. Bashamboo, Loss-of-function mutation in GATA4 causes anomalies of human testicular development, Proc. Natl. Acad. Sci. USA 108 (4) (2011) 1597–1602.

[41] A. Bashamboo, R. Brauner, J. Bignon-Topalovic, S. Lortat-Jacob, V. Karageorgou, D. Lourenco, A. Guffanti, K. McElreavey, Mutations in the FOG2/ZFPM2 gene are associated with anomalies of human testis determination, Hum. Mol. Genet. 23 (14) (2014) 3657–3665.

[42] R. Werner, H. Merz, W. Birnbaum, L. Marshall, T. Schröder, B. Reiz, J.M. Kavran, T. Bäumer, P. Capetian, O. Hiort, 46,XY gonadal dysgenesis due to a homozygous mutation in desert hedgehog (DHH) identified by exome sequencing, J. Clin. Endocrinol. Metab. 100 (7) (2015) E1022–E1029.

[43] P. Callier, P. Calvel, A. Matevossian, P. Makrythanasis, P. Bernard, H. Kurosaka, A. Vannier, C. Thauvin-Robinet, C. Borel, S. Mazaud-Guittot, A. Rolland, C. Desdoits-Lethimonier, M. Guipponi, C. Zimmermann, I. Stévant, F. Kuhne, B. Conne, F. Santoni, S. Lambert, F. Huet, F. Mugneret, J. Jaruzelska, L. Faivre, D. Wilhelm, B. Jégou, P.A. Trainor, M.D. Resh, S.E. Antonarakis, S. Nef, Loss of function mutation in the palmitoyl-transferase HHAT leads to syndromic 46,XY disorder of sex development by impeding Hedgehog protein palmitoylation and signaling, PLoS Genet. 10 (5) (2014) e1004340.

[44] C.E. Flück, A.V. Pandey, Steroidogenesis of the testis new genes and pathways, Ann. Endocrinol. (Paris) 75 (2) (2014) 40–47.

[45] J. Yu, Z. Chen, Y. Ni, Z. Li, CFTR mutations in men with congenital bilateral absence of the vas deferens (CBAVD): a systemic review and meta-analysis, Hum. Reprod. 27 (1) (2012) 25–35.

[46] E.H. Stephen, A. Chandra, Declining estimates of infertility in the United States: 1982–2002, Fertil. Steril. 86 (2006) 516–523.

[47] M.M. Matzuk, D.J. Lamb, Genetic dissection of mammalian fertility pathways, Nat. Cell Biol. 4 (Suppl (10)) (2002) s41–s49.

[48] P. Navarro-Costa, C.E. Plancha, J. Gonçalves, Genetic dissection of the AZF regions of the human Y chromosome: thriller or filler for male (in)fertility? J Biomed Biotechnol 2010 (2010) 1–18.

[49] M.A. Jobling, C. Tyler-Smith, Human Y-chromosome variation in the genome-sequencing era, Nat. Rev. Genet. 18 (8) (2017) 485–497.

[50] M. Ben Khelifa, C. Coutton, R. Zouari, T. Karaouzène, J. Rendu, M. Bidart, S. Yassine, V. Pierre, J. Delaroche, S. Hennebicq, D. Grunwald, D. Escalier, K. Pernet-Gallay, P.S. Jouk, N. Thierry-Mieg, A. Touré, C. Arnoult, P.F. Ray, Mutations in DNAH1, which encodes an inner arm heavy chain dynein, lead to male infertility from multiple morphological abnormalities of the sperm flagella, Am. J. Hum. Genet. 94 (1) (2014) 95–104.

CHAPTER

4

Carrier Screening for Inherited Genetic Disorders: A Review of Current Practices

*Neha Kumar**, *Sally Ann Rodríguez*†

*Genetic Counselor, New York, NY, United States †Certified Genetic Counselor, New York, NY, United States

INTRODUCTION TO GENETICS

Our genetic information is encoded in deoxyribonucleic acid (DNA), which is composed of a four-letter alphabet made up of adenine (A), cytosine (C), guanine (G), and thymine (T). Stretches of DNA are sectioned into genes, which provide the necessary instructions for a cell to produce specific proteins. These proteins impact many functions, such as our body's ability to grow and repair itself, our appearance (e.g., hair and eye color), and our physical and cognitive development, and they can even lead to certain genetic diseases. The entire complement of our genes is referred to as our genome. Our genes are made up of introns and exons, with exons being the segments of the gene that are the most important for creating proteins. The human exome refers to all of the exons in all of our genes, which accounts for <2% of our entire genome.

DNA is tightly wound and stored within the nucleus of each cell in the form of chromosomes (Fig. 4.1). Humans typically have 23 pairs of chromosomes for a total of 46 chromosomes within each cell. Chromosome pairs 1–22 are referred to as autosomes. For the autosomes, both members of each chromosome pair contain the same genes, so that every cell has two copies of each gene. The 23rd chromosome pair is the sex chromosomes. Females have two X chromosomes, whereas males have one X and one Y chromosome. Humans inherit one set of chromosomes (i.e., one set of genes) from their mother and one from their father.

GENETIC DISEASE AND INHERITANCE

Changes in the sequence of letters in our DNA can affect the production of the specific protein that the gene normally encodes, and these changes are what make us unique. Some changes, called mutations, can cause genetic disease. These mutations can either be passed

Reproductomics
https://doi.org/10.1016/B978-0-12-812571-7.00005-8

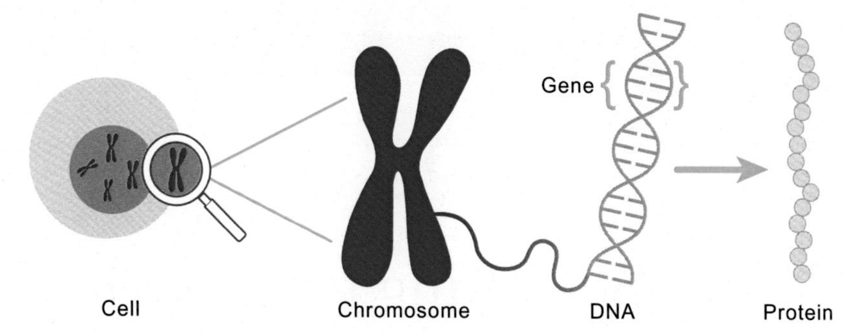

FIG. 4.1 DNA storage. DNA is wound and compacted into structures called chromosomes, which allow the DNA to be stored within our cells. The DNA provides a code from which the body can build proteins.

down to a child from a parent or they can occur randomly in the DNA of egg or sperm cells (referred to as de novo mutations).

Genetic conditions can be inherited in several different ways. This is based on the location of the gene within the genome, how many working copies of a gene are required to maintain health, and from which parent a mutation is inherited. The main modes of disease inheritance are as follows: autosomal dominant, autosomal recessive, X-linked dominant, X-linked recessive, mitochondrial, and multifactorial:

Autosomal dominant inheritance: Autosomal dominant (AD) conditions occur in genes where both copies of the gene must be functional to maintain health. Consequently, a mutation in one of the two copies of these genes is sufficient to cause disease. As one mutated copy is enough to confer disease, any child of a person affected with an AD genetic condition has a 50% chance to be affected as well (Fig. 4.2). However, AD conditions can often be de novo and are not inherited from either parent. In this case, the recurrence risk for future siblings of an affected child is very low. However, the reproductive risk for an individual with a de novo AD condition is still 50%.

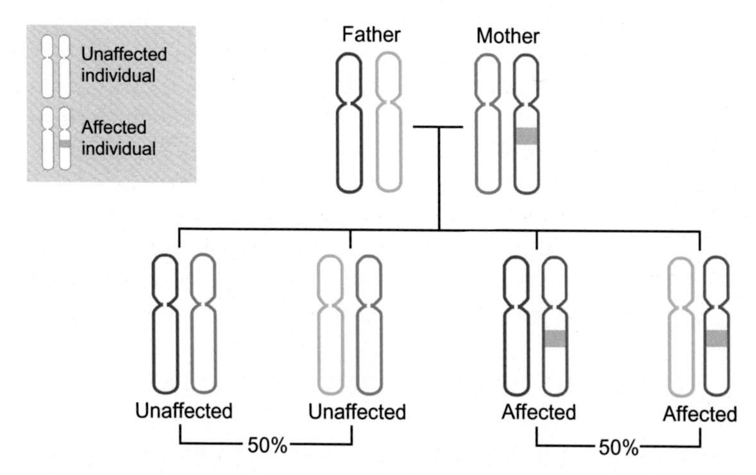

FIG. 4.2 Autosomal dominant inheritance. For an individual to be affected with an autosomal dominant condition, he/she must have one nonfunctional copy of the disease-causing gene. Affected individuals have a 50% risk of passing the mutation to their offspring.

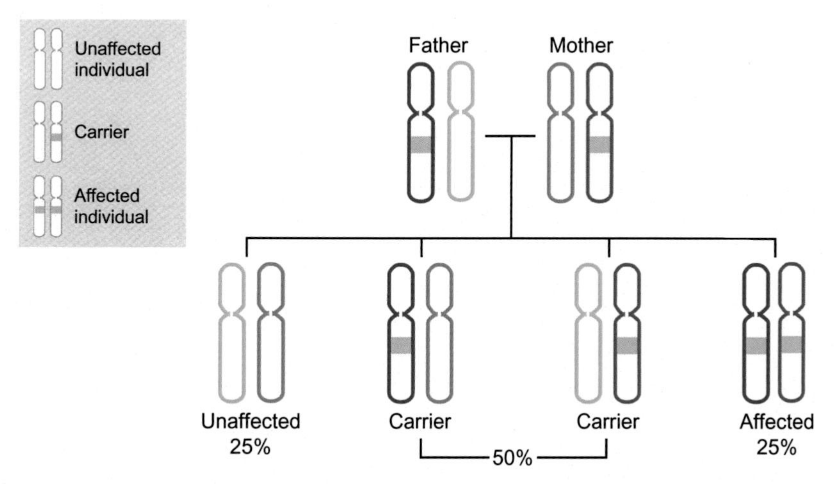

FIG. 4.3 Autosomal recessive inheritance. For an individual to be affected with an autosomal recessive condition, he/she must inherit two nonfunctional copies of the disease-causing gene. Thus, couples that are both carriers for the same autosomal recessive disorder have a 25% chance to have a child affected by the disorder. The couple has a 50% chance to have a child that is not affected, but who will be a carrier of the disorder.

Autosomal recessive inheritance: In autosomal recessive (AR) genetic conditions, both copies of a particular gene do not work properly. AR conditions must be inherited from both parents for a child to be affected. When a child has an AR condition, both parents are considered to be obligate carriers of the condition. De novo mutations are possible but are rare. Unlike with AD conditions, carriers of AR conditions typically have no symptoms. Furthermore, they are only at risk to have an affected child if both partners are carriers for a mutation within the same gene. For this reason, most individuals with an AR condition have no family history of the disease. For a couple where both partners are carriers for an AR condition, each child has a 25% chance of being affected with the condition (Fig. 4.3).

X-linked inheritance: The last pair of chromosomes is the sex chromosomes. These chromosomes determine whether an individual is physically male or female. Females have two X chromosomes, whereas males have one X chromosome and one Y chromosome. X-linked conditions are caused by mutations in genes on the X chromosome. Therefore, due to the fact that women have two X chromosomes and men have only one X chromosome, X-linked conditions are expressed in different proportions in males and females:

X-linked recessive conditions typically occur when there are no functional copies of a particular X-linked gene. X-linked recessive conditions are much more common in males than in females, as males only have one copy of the X chromosome (hemizygous). Typically, males inherit X-linked recessive conditions from their carrier mothers. Every male child of an X-linked recessive carrier mother has a 50% chance of being affected. Female children of X-linked recessive carrier mothers are typically not at risk to be affected, but have a 50% chance of being carriers (Fig. 4.4). Females who are affected with an X-linked recessive genetic condition are expected *either* to have a carrier mother and an affected father *or* to have skewed X-inactivation. X-inactivation

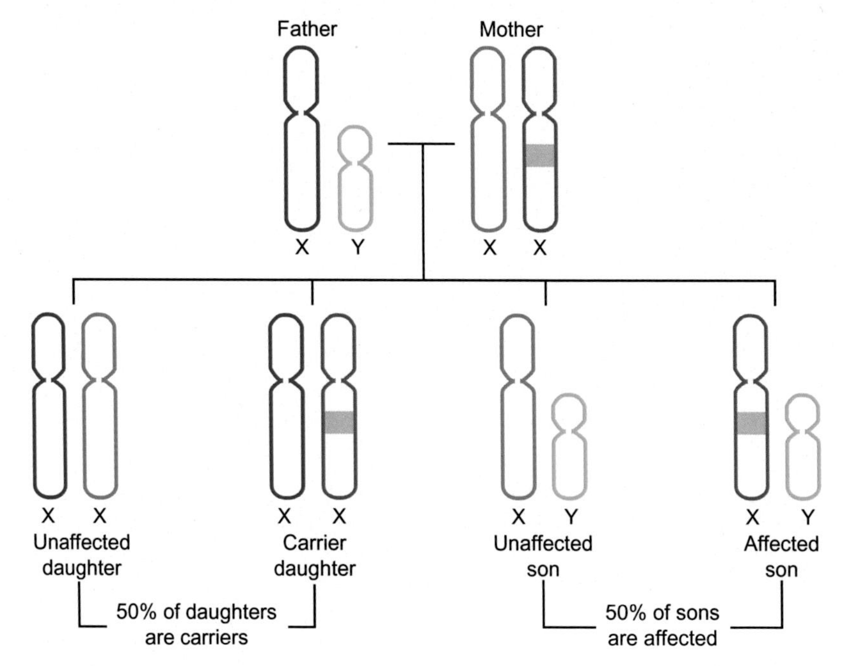

FIG. 4.4 X-linked recessive inheritance. X-linked recessive conditions tend to be passed down through the mother. Female carriers of an X-linked recessive condition have a 50% risk of having an affected son because he inherits only one X chromosome. Daughters are not typically affected, but have a 50% chance of being a carrier.

occurs early in embryonic development, causing each cell to have only one active X chromosome (which could be the chromosome with or without the mutation). Therefore, the exhibition of symptoms in a female carrier of an X-linked recessive condition is dependent upon the X-inactivation ratio in specific tissues.

X-linked dominant conditions occur when one nonworking copy of a particular X-linked gene is sufficient to cause disease. X-linked dominant conditions are traditionally only seen in females, as they are typically lethal in males. However, there are rare exceptions to this rule.

Mitochondrial inheritance: The mitochondrion is the organelle that produces energy for the cell. Mitochondria have their own genome, which includes 13 genes. Mutations in any of these genes can cause mitochondrial genetic conditions. Mitochondrial conditions are only inherited from mothers, as all of the mitochondria an individual has are descended from the mitochondria in the egg. Often, an individual's mitochondria do not all have the same genotype, which is called heteroplasmy. The proportion of wild-type to mutated mitochondrial DNA determines whether an individual expresses symptoms of a mitochondrial genetic condition.

Multifactorial inheritance: The majority of common health conditions, such as diabetes and autoimmune diseases, are the result of multifactorial inheritance. This means that a combination of many genetic and environmental factors determines whether an individual is affected with a particular condition. It is often difficult to determine what these factors are and to differentiate them from each other. Individuals with a family history of these types of diseases will often have an increased risk compared with the general population to develop the same (or a similar) condition, and empirical risks are sometimes available if familial studies have been performed.

INTRODUCTION TO CARRIER SCREENING

Carrier screening is the process of testing individuals before or during pregnancy to determine their risk of having a child with a specific heritable genetic condition. Everyone is a carrier for something, and in fact, the average person is thought to carry 3–5 disease-causing mutations. Carrier screening is most commonly offered for autosomal recessive conditions, where both partners would need to be identified as carriers in order to be at a high risk (25%) of having an affected child. Carrier screening can also be offered to women for X-linked recessive conditions, where carrier females are at high risk (50%) of having affected sons. Learning about any potential risks to future children prior to or during pregnancy can empower individuals with more reproductive options, such as preimplantation genetic diagnosis (PGD) or prenatal diagnosis. As the majority (80%) of children born with a genetic condition have no family history of that disease, carrier screening may be the only way to help identify couples with a high risk of having a child with a genetic disorder.

Carrier screening can be performed concurrently (testing both partners simultaneously) or sequentially (testing one partner, typically the female, first and then determining whether or not testing is necessary for the second partner). Sequential testing allows for the selection of the most appropriate testing methodology for the second partner. Furthermore, by screening one partner at a time, costs may be reduced if testing is not needed for the second partner. On the other hand, concurrent screening can reduce testing turnaround time, which can be valuable if testing is being performed during pregnancy or if there are other time-limiting factors. Concurrent testing may also help reduce patient anxiety. Both of these testing protocols are valid, and the choice is often dependent on a patient's unique situation. Genetic counseling can help determine which methodology is most appropriate for any particular scenario.

GUIDELINES FOR CARRIER SCREENING

Carrier screening guidelines vary significantly across the world and are typically limited. In the United States, the American Congress of Obstetricians and Gynecologists (ACOG) and the American College of Medical Genetics and Genomics (ACMG) currently recommend carrier screening for specific genetic disorders in ethnic populations where the prevalence is high, as outlined in Table 4.1. However, screening based on ethnicity alone is problematic, as the genetic pool is homogenizing. Recent technological advancements have made it possible to screen for multiple conditions with a single test. This is referred to as expanded carrier screening (ECS), which takes a more panethnic approach and can screen for upward of 300 diseases.

TABLE 4.1 Carrier Screening Recommendations

	Ethnic Population	
Disease	**ACOG Recommendation**	**ACMG Recommendation**
Alpha thalassemia	All	–
Beta thalassemia (including sickle cell disease)	All	–
Bloom syndrome	Ashkenazi Jewish[a]	Ashkenazi Jewish
Canavan disease	Ashkenazi Jewish	Ashkenazi Jewish
Cystic fibrosis	All	All
Familial dysautonomia	Ashkenazi Jewish	Ashkenazi Jewish
Fanconi anemia, type C	Ashkenazi Jewish[a]	Ashkenazi Jewish
Fragile X syndrome	All[b]	All[b]
Gaucher disease	Ashkenazi Jewish[a]	Ashkenazi Jewish
Mucolipidosis, type IV	Ashkenazi Jewish[a]	Ashkenazi Jewish
Niemann-Pick disease, type A	Ashkenazi Jewish[a]	Ashkenazi Jewish
Spinal muscular atrophy	All	All
Tay-Sachs disease	Ashkenazi Jewish	Ashkenazi Jewish

ACOG and ACMG recommend preconception and/or prenatal screening for these diseases, based on ethnicity as indicated.
[a] *If requested by the patient.*
[b] *If indicated by medical or family history.*

In response to these advances, a number of professional societies, including ACOG, ACMG, and the European Society of Human Genetics (ESHG), have published statements regarding responsible development and implementation of ECS panels. The societies acknowledge that the paradigm of ECS allows for all individuals, regardless of race or ethnicity, to be offered screening for the same set of conditions. Each statement emphasizes that conditions included on ECS panels should have a significant impact on reproductive decision-making or medical management of a child and the importance of informed consent prior to screening and posttest genetic counseling.

CARRIER SCREENING TECHNOLOGIES

Carrier screening is typically carried out via one of two available genetic testing methodologies: genotyping or sequencing. Genotyping technologies target specific known pathogenic mutations to determine if the tested individual carries those mutations. Due to advancements in genetic testing technology, an individual can now be genotyped for thousands of mutations on a single testing platform (i.e., a microarray), allowing for very cost-effective carrier screening. A limitation of this technology is that it can only determine an individual's genotype at a particular, targeted location; thus, carriers of rare or novel mutations, which would not be included in the test design, would not be identified.

Alternatively, sequencing technologies will determine the exact sequence of an entire exon, gene, or gene region. Thus, sequencing technologies will identify not only known pathogenic mutations but also novel variants that are located throughout the sequenced region. While sequencing is a well-established technology, recent advancements have led to the development of next-generation sequencing (NGS), which allows for the sequencing of hundreds of genes on a single platform. Unlike traditional sequencing methodologies, NGS-based sequencing platforms typically target only the exons (or coding regions) of the tested genes, which makes it more cost-effective but reduces the sensitivity of the test. However, as most disease-causing mutations are located within the exons of a gene, the reduction in sensitivity is typically small.

Furthermore, as sequencing determines the sequence of a given region and does not simply target known pathogenic mutations, all identified variants undergo a process called variant curation to determine their pathogenicity. Variant curation involves reviewing (1) population allele-frequency databases, (2) disease- and locus-specific databases, (3) computational predictions of variant effect on protein function, (4) sequence conservation metrics, and (5) reports in the primary literature to classify variants as benign, pathogenic, or unknown. Those that fall in the "unknown" category are referred to as variants of uncertain significance (VUS). These prove to be particularly problematic in the preconception and prenatal setting; therefore, US professional societies, such as ACOG and ACMG, recommend that VUS not be reported in the context of carrier screening.

While most carrier screening can be performed via genotyping and/or sequencing, there are some additional testing technologies that can be utilized to determine carrier status for specific genetic conditions. These supplementary tests may provide a more complete picture of an individual's reproductive risk.

CBC and Hemoglobin Electrophoresis

Alpha thalassemia, beta thalassemia, and sickle cell disease, collectively called hemoglobinopathies, are blood disorders with variable presentation (ranging from chronic but treatable disease to a neonatal lethal phenotype) depending on the number and type of mutations inherited. Hemoglobinopathies are most common in individuals of Asian, African, and Mediterranean descent and can be tested for via molecular (genetic) methods but have historically been screened for using two simple blood tests: complete blood count (CBC) and quantitative hemoglobin electrophoresis. On a CBC, both alpha and beta thalassemia carriers will typically present with low mean corpuscular volume (MCV), low mean corpuscular hemoglobin (MCH), and low hemoglobin (Hb), with values as outlined in Table 4.2. A quantitative hemoglobin electrophoresis will also identify beta thalassemia carriers, as they typically present with decreased HbA and increased HbA_2. It will also detect other hemoglobinopathies, such as Hb S (sickle cell trait). In many cases, a combination of CBC/hemoglobin electrophoresis and molecular methods is necessary to assess reproductive risk.

HEXA (Tay-Sachs) Enzyme Analysis

Tay-Sachs disease (TSD) is a fatal neurological disorder with early childhood onset that is commonly seen in individuals of Ashkenazi Jewish and French Canadian descent. The disease

TABLE 4.2 Red Blood Cell Indexes in Alpha and Beta Thalassemia

	Normal		Alpha Thalassemia				Beta Thalassemia Trait (β-Thal Minor)	
	Male	Female	Silent Carrier		Trait (–/α or -α/-α)			
MCV (fl)	89.1±5.01	87.6±5.5	81.2±6.9		71.6±4.1		<79	
MCH (pg)	30.9±1.9	30.2±2.1	26.2±2.3		22.9±1.3		<27	
Hb (g/dL)	15.9±1.0	14.0±0.9	Male	Female	Male	Female	Male	Female
			14.3±1.4	12.6±1.2	13.9±1.7	12.0±1.0	11.5–15.3	9.1–14

is caused by mutations in the *HEXA* gene. Carrier screening for TSD has traditionally been performed via HEXA enzyme analysis, in which an individual's blood is analyzed to determine the amount of HEXA enzymatic activity, thus identifying them as a carrier or noncarrier for TSD. Tay-Sachs enzyme (TSE) analysis is considered the gold standard approach to carrier screening for TSD due to its comprehensive nature, although the recent availability of cost-effective NGS-based molecular methods may soon provide an equally comprehensive alternative. While TSE analysis has a high carrier detection across all ethnicities, it is important to note that there are two pseudodeficiency alleles that exist and are not disease-causing but will lead to a "carrier" result on TSE analysis—that is, a false-positive result. Therefore, the combination of TSE analysis and molecular methods that target disease-causing mutations and pseudodeficiency alleles provides the most comprehensive approach to carrier screening for TSD.

Fragile X Syndrome—CGG and AGG Analysis

Fragile X syndrome is the most common cause of inherited intellectual disability, and it is caused by an expansion of a CGG repeat region within the *FMR1* gene. Generally, large repeat regions are difficult to assess via genotyping or sequencing methods; therefore, PCR analysis is the recommended technology for assessing CGG repeat length to determine fragile X syndrome carrier status. Females with a CGG repeat length in the premutation range (55–200 CGG repeats) are at risk of having an affected child, and this risk increases as the CGG repeat length increases. However, recent studies have shown that AGG interruptions within a CGG repeat region can confer stability to the DNA and therefore reduce the risk of expansion in the next generation. Thus, AGG testing can further refine the risk for expansion. AGG interruption analysis is also performed via PCR analysis.

Spinal Muscular Atrophy (SMA) Dosage Analysis

Spinal muscular atrophy (SMA) is a variable but typically severe disorder that affects muscle movement due to the loss of motor neurons. The disease is caused by mutations in the *SMN1* gene, with the vast majority of affected individuals (~95%) harboring a deletion of exon 7 in both copies of the *SMN1* gene. SMA dosage or copy number analysis has been the longstanding gold standard approach to carrier screening for this disease. This method determines carrier status by assessing the number of functional *SMN1* gene copies in an individual's DNA. This analysis can be accomplished by various methodologies, including qPCR (quantitative

PCR), ddPCR (digital droplet PCR), and MLPA (multiplex ligation-dependent probe amplification). Recent advancements indicate that carrier screening for SMA may also be possible via NGS. Though considered the gold standard, dosage analysis for SMA still has important limitations. While a test result indicating the presence of 2 *SMN1* copies is considered normal, there is a residual carrier risk due to the possibility of 2 *SMN1* gene copies on a single chromosome—a phenomenon that is present in approximately 5%–8% of the general population and is referred to as silent carrier status. Furthermore, copy number analysis cannot detect any point mutations, which are generally rare but can account for 2%–5% of affected individuals (who carry one deletion and one point mutation). Lastly, approximately 2% of affected individuals inherit a de novo (or sporadic) deletion from a parent's gamete (egg or sperm).

CARRIER RATES, DETECTION RATES, AND RESIDUAL RISKS

A carrier rate is the frequency with which individuals are found to be carriers for a particular genetic disease within a given population. High carrier rates can be due to a few mutations with a high frequency in a given population or collectively due to hundreds of mutations with small frequencies within a given population. For example, Tay-Sachs disease is predominantly caused by two specific mutations in individuals of Ashkenazi Jewish descent. Mutations like these, which are common and account for a large percentage of a specific disease within a single population, are referred to as founder mutations. On the other hand, phenylketonuria (PKU) is common in the European population, but this high carrier rate is due to hundreds of different mutations. Mutations that are individually rare and tend to be family-specific are referred to as private mutations.

In order to determine how effective a particular carrier screen will be for an individual, it is necessary to assess the screen's detection rate for each disease. The detection rate is determined by the percentage of the disease accounted for by the mutations that can be identified by the screen. For example, if a test can detect mutations that collectively account for 75% of a particular genetic disease within a specific ethnic population, the test can be said to have a 75% detection rate for individuals of that ethnic group. For genotyping-based carrier screens, detection rates for each disease will typically vary across ethnic groups, as different mutations will be more or less common in each ethnic group. For sequencing-based carrier screens, detection rates are typically higher than those of a genotyping-based screen and are also more uniform across ethnicities.

As genotyping-based carrier screens typically only test for the most common mutations and sequencing-based carrier screens typically only target certain gene regions, having a negative result for a disease on a carrier screen only reduces, but does not eliminate, an individual's chance of being a carrier. This remaining risk after a negative test result is referred to as a residual carrier risk. This residual carrier risk is calculated using Bayes' theorem, which is a form of conditional probability. To illustrate, 1 in 25 individuals of European descent is a carrier for cystic fibrosis. If a European individual has a screening test for cystic fibrosis that has a 92% detection rate for the European population, that individual's residual risk of being a carrier after a negative test result is 1 in 301. Residual carrier risks can then be used to calculate a couple's residual reproductive risk, which is the risk to have an affected child after carrier screening. See Fig. 4.5 for a detailed description of how to calculate these risks.

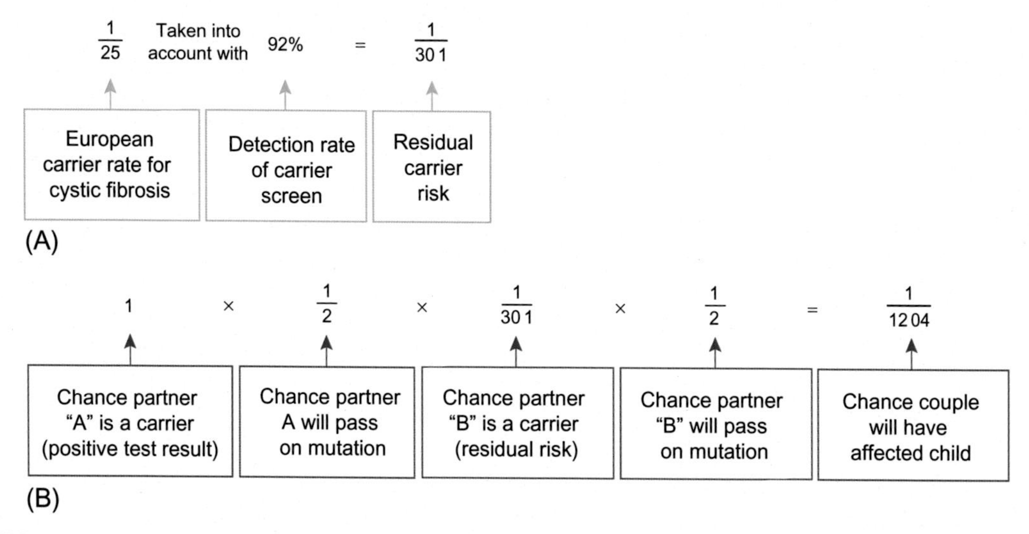

FIG. 4.5 Calculating residual risks. In this example, a couple of European descent was screened for cystic fibrosis (CF). Partner A was found to be a carrier and Partner B was not found to be a carrier. (A) A residual carrier risk is the risk of being a carrier after a negative test result, taking into account the population-based carrier rate for CF and the detection rate for that screen. (B) A residual reproductive risk is the risk of passing on a genetic condition after carrier screening. This risk is calculated by taking into account the carrier status of each partner and the likelihood of passing on the mutation, as well as the inheritance pattern of the condition.

REPRODUCTIVE OPTIONS FOR HIGH-RISK COUPLES

Couples determined to be at increased risk of passing on an inherited condition to their offspring have several options available to them:

Preimplantation genetic diagnosis (PGD): PGD is an option that can be pursued prior to pregnancy. During PGD, embryos created through in vitro fertilization (IVF) are screened for a particular genetic disease, and only unaffected embryos are transferred to the woman's uterus. PGD enables a couple to have an unaffected biological child while avoiding possible termination; however, PGD requires IVF, which can be costly. PGD test preparation can also be time-consuming and require testing of additional family members.

Prenatal diagnosis: For couples who do not wish to pursue PGD or are already pregnant, prenatal testing via chorionic villus sampling (CVS) or amniocentesis is an available option:

Chorionic villus sampling (CVS): The chorionic villi are part of the placenta and are therefore derived from the fetus. As such, the genetic information contained within the chorionic villi is typically the same as that of the fetus. In rare cases, the placenta may not exactly match the fetus' DNA—this is referred to as confined placental mosaicism, which can lead to inaccurate genetic testing results. During a chorionic villus sampling (CVS), a needle is inserted either transabdominally or transcervically (depending on the position of the placenta), and a small sample of the chorionic villi is taken (Fig. 4.6). A CVS procedure can typically be performed between weeks 10 and 13 of pregnancy. Due to the invasive nature of the test, there is a small risk of miscarriage associated with a CVS procedure.

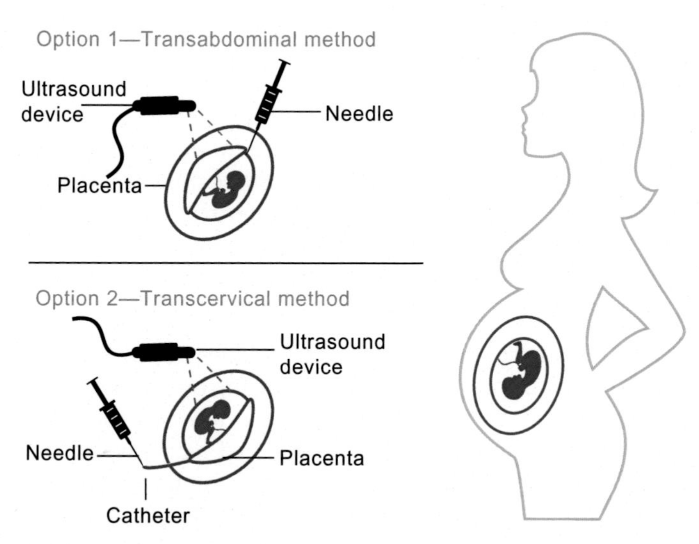

FIG. 4.6 CVS. Depending on the location of the placenta, a thin needle is inserted either transabdominally or transcervically, and a small sample of the chorionic villi is removed, allowing for genetic testing to be performed.

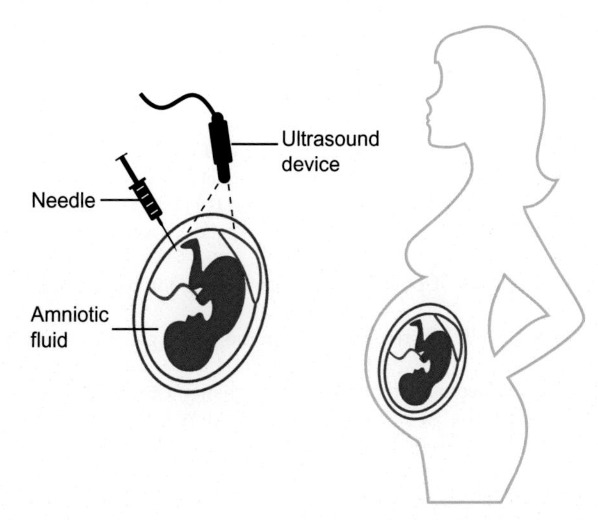

FIG. 4.7 Amniocentesis. Using the guidance of an ultrasound, a thin needle is inserted into the abdomen, and a small amount of amniotic fluid is withdrawn. Cells derived from the fetus are contained within the fluid, allowing for genetic testing to be performed.

Amniocentesis: During development, the fetus sheds skin and other cells containing genetic material into the amniotic fluid. During an amniocentesis, a needle, under the guidance of an ultrasound, is inserted into the abdomen, and a small amount of amniotic fluid is extracted (Fig. 4.7). Since this test is being performed on fetal cells, there is less risk of discordant results between the amniocentesis and the actual fetal outcome when compared with CVS. An amniocentesis can be performed after the 15th week of pregnancy and also carries a small risk of miscarriage, although this risk is lower than that associated with a CVS procedure.

CVS and amniocentesis procedures are typically performed by a maternal-fetal medicine (MFM) specialist, and the samples obtained can be used for genetic diagnosis of the fetus for the disease of interest when there is a high risk. It should be noted that DNA testing typically has a longer turnaround time when compared with other testing more routinely performed on CVS and amniocentesis samples (such as karyotypes, which are used to identify chromosomal abnormalities). Prenatal diagnosis can help couples make a decision about termination or help them to prepare for raising a child with a genetic condition.

Gamete donors: The use of an egg or sperm donor that does not carry the relevant condition is another option available to couples at high risk of passing on a genetic disease. The selected donor should have undergone comprehensive molecular analysis for the condition in question, and egg donors should also be screened for X-linked recessive conditions, particularly fragile X syndrome. Like PGD, the use of gamete donors can be time-consuming and costly, but this option avoids termination and enables a couple to have a child biologically related to one parent.

High-risk couples also have the option to pursue adoption, to continue with family planning without further testing, or to live child-free. Genetic counselors can help high-risk couples fully understand the options available to them.

Further Reading

[1] ACOG Committee Opinion No. 432: Spinal muscular atrophy, Obstet. Gynecol. 113 (5) (2009) 1194–1196.
[2] ACOG Committee Opinion No. 442: Preconception and prenatal carrier screening for genetic diseases in individuals of Eastern European Jewish descent, Obstet. Gynecol. 114 (4) (2009) 950–953.
[3] ACOG Committee Opinion No. 469: Carrier screening for fragile X syndrome, Obstet. Gynecol. 116 (4) (2010) 1008–1010.
[4] ACOG Committee Opinion No. 486: Update on carrier screening for cystic fibrosis, Obstet Gynecol. 117 (4) (2011) 1028–1031.
[5] ACOG Committee Opinion No. 690: Carrier screening in the age of genomic medicine, Obstet. Gynecol. 129 (2017) e35–e40.
[6] ACOG Practice Bulletin No. 78: Hemoglobinopathies in pregnancy, Obstet. Gynecol. 109 (1) (2007) 229–237.
[7] E. Armenti, E. Cameron, I.B. Carlsson, A. Jordan, N. Kumar, K. McWilliams, S. Munné, N. Paolino, S. Rodríguez, S. Sehnert, E. Shehan, S. Yarnall, Genetics in Reproductive Medicine: A Handbook for Clinicians, CooperSurgical, Inc., Livingston, NJ, 2016.
[8] J.G. Edwards, G. Feldman, J. Goldberg, et al., Expanded carrier screening in reproductive medicine—points to consider, Obstet. Gynecol. 125 (3) (2015) 653–662.
[9] E.E. Eichler, J.J.A. Holden, B.W. Popovich, et al., Length of uninterrupted CGG repeats determines instability in the FMR1 gene, Nat. Genet. 8 (1) (1994) 88–94.
[10] L. Henneman, P. Borry, D. Chokoshvili, et al., Responsible implementation of expanded carrier screening, Eur. J. Hum. Genet. 24 (6) (2016) e1–e12.
[11] W.W. Grody, B.H. Thompson, A.R. Gregg, L.H. Bean, K.G. Monaghan, A. Schneider, R.V. Lebo, ACMG position statement on prenatal/preconception expanded carrier screening, Genet. Med. 15 (6) (2013) 482–483.
[12] S.J. Gross, B.A. Pletcher, K.G. Monaghan, For the Professional Practice and Guidelines Committee, Carrier screening in individuals of Ashkenazi Jewish descent, Genet. Med. 10 (1) (2008) 54–56.
[13] G.A. Lazarin, I.S. Haque, S. Nazareth, et al., An empirical estimate of carrier frequencies for 400+ causal Mendelian variants: results from an ethnically diverse clinical sample of 23,453 individuals, Genet. Med. 15 (3) (2013) 178–186.

[14] K.G. Monaghan, E. Lyon, E.B. Spector, ACMG Standards and Guidelines for fragile X testing: a revision to the disease-specific supplements to the Standards and Guidelines for Clinical Genetics Laboratories of the American College of Medical Genetics and Genomics, Genet. Med. 15 (7) (2013) 575–586.

[15] S.L. Nolin, F.A. Lewis, L.L. Ye, et al., Familial transmission of the FMR1 CGG repeat, Am. J. Hum. Genet. 59 (6) (1996) 1252–1261.

[16] S.L. Nolin, W.T. Brown, A. Glicksman, et al., Expansion of the fragile X CGG repeat in females with premutation or intermediate alleles, Am. J. Hum. Genet. 72 (2) (2003) 454–464.

[17] S.L. Nolin, S. Sah, A. Glicksman, et al., Fragile X AGG analysis provides new risk predictions for 45-69 repeat alleles, Am. J. Hum. Genet. A 161A (4) (2013) 771–778.

[18] S.L. Nolin, A. Glicksman, et al., Fragile X full mutation expansions are inhibited by one or more AGG interruptions in premutation carriers, Genet. Med. 17 (5) (2015) 358–364.

[19] R. Origa, P. Moi, Alpha-thalassemia, in: R.A. Pagon, M.P. Adam, H.H. Ardinger, et al. (Eds.), GeneReviews® [Internet], University of Washington, Seattle, Seattle, WA, 1993–2016. Available from: https://www.ncbi.nlm.nih.gov/books/NBK1435/. 2005 November 1 [Updated 2016 December 29].

[20] R. Origa, Beta-thalassemia, in: R.A. Pagon, M.P. Adam, H.H. Ardinger, et al. (Eds.), GeneReviews® [Internet], University of Washington, Seattle, Seattle, WA, 1993–2016. Available from: https://www.ncbi.nlm.nih.gov/books/NBK1426/. 2000 September 28 [Updated 2015 May 14].

[21] T.W. Prior, Spinal muscular atrophy, in: R.A. Pagon, M.P. Adam, H.H. Ardinger, et al. (Eds.), GeneReviews® [Internet], University of Washington, Seattle, Seattle, WA, 1993–2016. Available from: https://www.ncbi.nlm.nih.gov/books/NBK1352/. 2000 February 24 [Updated 2016 December 22].

[22] S.A. Scott, L. Edelmann, L. Liu, M. Luo, R.J. Desnick, R. Kornreich, Experience with carrier screening and prenatal diagnosis for sixteen Ashkenazi Jewish genetic diseases, Hum. Mutat. 31 (11) (2010) 1240–1250.

[23] S. Sherman, B.A. Pletcher, D.A. Driscoll, Fragile X syndrome: diagnostic and carrier testing, Genet. Med. 7 (8) (2005) 584–587.

[24] B.S. Srinivasan, E.A. Evans, J. Flannick, A.S. Patterson, C.C. Chang, T. Pham, S. Young, A. Kaushal, J. Lee, J.L. Jacobson, P. Patrizio, A universal carrier test for the long tail of Mendelian disease, Reprod. BioMed. Online 21 (4) (2010 Oct) 537–551.

[25] E.A. Sugarman, N. Nagan, H. Zhu, et al., Panethnic carrier screening and prenatal diagnosis for spinal muscular atrophy: clinical laboratory analysis of >72,400 specimens, Eur. J. Hum. Genet. 20 (1) (2012) 27–32.

[26] W. Thomas, Carrier screening for spinal muscular atrophy, Genet. Med. 10 (11) (2008) 840–842.

[27] M.S. Watson, G.R. Cutting, R.J. Desnick, D.A. Driscoll, K. Klinger, M. Mennuti, G.E. Palomaki, B.W. Popovich, V.M. Pratt, E.M. Rohlfs, C.M. Strom, C.S. Richards, D.R. Witt, W.W. Grody, Cystic fibrosis population carrier screening: 2004 revision of American College of Medical Genetics mutation panel, Genet. Med. 6 (5) (2004) 387–391.

[28] N. Zhong, W. Ju, J. Pietrofesa, D. Wang, C. Dobkin, W.T. Brown, Fragile X "gray zone" alleles: AGG patterns, expansion risks, and associated haplotypes, Am. J. Med. Genet. 64 (1996) 261–265.

CHAPTER

5

Telomeres in Germ Line and Early Embryo: An Overview

José Luís Fernández[*,†], *Iria González-Vasconcellos*[*,‡]

[*]Unidad de Genética—Complejo Hospitalario Universitario A Coruña-INIBIC, A Coruña, Spain [†]Laboratorio de Genética Molecular y Radiobiología, Centro Oncológico de Galicia, A Coruña, Spain [‡]Institute of Radiation Biology, Helmholtz Zentrum München, Deutsches Forschungszentrum für Gesundheit und Umwelt, Neuherberg, Germany

HISTORICAL LANDMARKS

The terminal ends of the chromosomes were recognized as differentiated areas in the late 1930s. The pioneer work of Hermann Muller in the fruit fly (*Drosophila melanogaster*) in 1938 and Barbara McClintock in corn (*Zea mays*) in 1940 showed that natural chromosome ends do not tend to rejoin unlike induced interstitial chromosome breaks [1,2]. A particular structure should therefore be capping the end of the chromosomes stabilizing and preventing end-to-end fusions. These structures were designated by Muller as telomeres, a word derived from the Greek word telos (end) and meros (part).

Later, DNA was demonstrated to be the genetic material capable of semiconservative replication by DNA polymerases. There was however a problem concerning DNA replication that was discovered by James Watson when studying the discontinuous replication of the lagging strand. DNA polymerases must be primed by short RNA sequences. After replication, the internal RNA primers are degraded and replaced. However, the RNA primer located at the distal 5′ end of the new replicated lagging strand cannot be replaced after degradation, resulting in progressive shortening of the DNA strands with each round of replication [3]. Telomeres were proposed to be the structures that deal with this end-replication problem.

Leonard Hayflick showed in 1961 that human fibroblasts cultured in vitro have a limited life span, only achieving 30–50 population doublings, appointing the so-called Hayflick limit [4]. Proliferation rates progressively decline during the life span of a cell up to the point where they stop dividing, arresting the cycle in G1 to finally die. This phenomenon is called replicative senescence (Fig. 5.1). Alexey Olovnikov suggested that the progressive telomere

Reproductomics
https://doi.org/10.1016/B978-0-12-812571-7.00006-X

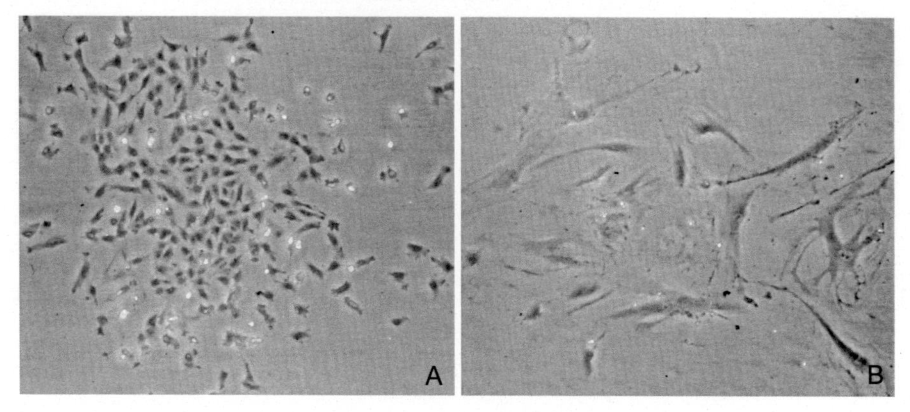

FIG. 5.1 Human amniocytes. (A) Early passage, young cells. (B) Late passage, senescent cells.

shortening due to the end-replication problem is a biological clock that determines the potential number of replications of a cell [5].

The DNA sequence of telomeres was first discovered by Elizabeth Blackburn within a project led by Joseph Gall, in *Tetrahymena thermophila*, a ciliate living in freshwater ponds. *Tetrahymena* produces about 200–300 small chromosome fragments and near 10,000 minichromosomes of 21 kb containing amplified ribosomal RNA coding genes, within their polyploid macronucleus. All of them are linear and stable due to telomere addition to both ends. These telomeric sequences were identified as tandem 50–70 repeats of the hexanucleotide CCCCAA [6]. In collaboration with Jack Szostak, Blackburn elegantly confirmed that these sequences were functional since their terminal assembly conferred stability to artificial yeast linear plasmids [7,8]. Subsequent work by Elizabeth Blackburn and her graduate student Carol Greider, using extracts from *Tetrahymena* and primers constituted by telomeric DNA sequences, allowed the discovery of the telomerase, responsible of de novo addition of tandem terminal TTGGGG repetitions to the primers [9]. These findings shed light over the cellular mechanism used by certain cells to solve the end-replication problem.

Studies of telomere structure and function were rapidly expanded to the vast majority of organisms and cell types, and it was evident that the physiology of telomere was strongly linked to genome integrity and stability and to aging and cancer.

STRUCTURE OF TELOMERES

Telomeres are highly specialized chromatin, that is, nucleoprotein, structures assembled at the distal ends of eukaryotic chromosomes (Fig. 5.2A). In humans, as in vertebrates, telomeres rely on a highly conserved structure composed of approximately 5–15 kb of tandem repeats of a G-rich sequence in the leading strand (5′-TTAGGG-3′) [10,11]. The G-rich strand has a C-rich complementary strand (5′-CCCTAA-3′) but protrudes at the terminus constituting a 3′ single-stranded overhang of around 150–200 pb long. This G-rich overhang folds back invading the upstream double-stranded region and pairing with the complementary C-rich strand. This constitutes the telomeric loop (T-loop) important for the protection of chromosome ends and its capping function.

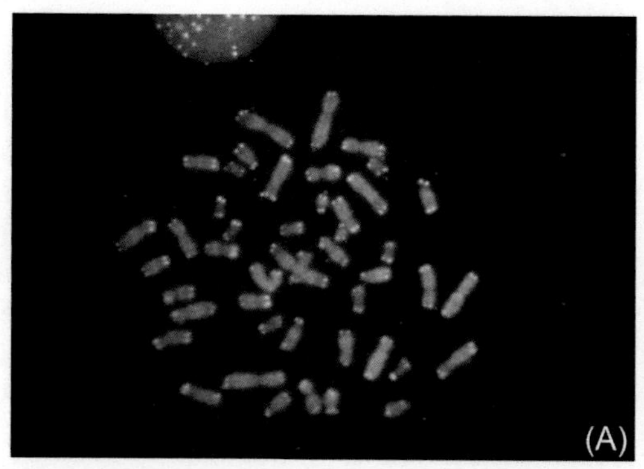

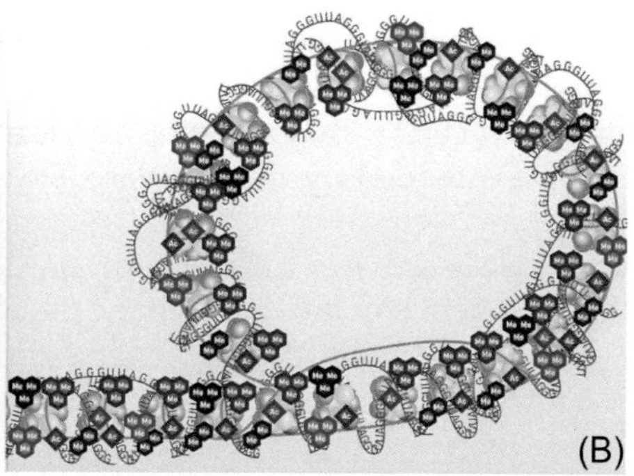

FIG. 5.2 (A) Metaphase human lymphocytes processed by telomeric FISH. The PNA-Cy3 labeling of heterogeneous telomeric signals in the distal end of the chromosomes can be observed in red. The chromosomes counterstained with DAPI can be observed in blue. (B) Telomere structural representation. The two DNA strands forming the telomeric T-loop (orange) are packed in arrays of nucleosomes (represented as octameres of colored histones) showing different histone modifications, di- or trimethylations (me), and acetylations (ac). The TERRA long noncoding telomeric RNA molecules are represented in pink. The RNA sequences are intercalated within the telomeric structure aiding condensation.

Multimeric protein complexes colocalize with the telomere sequences to constitute the shelterin complex or telosome that is essential for the organization and stability of telomeric chromatin and for the regulation of telomere length [12]. The shelterin complex in humans consists of six proteins, TRF1, TRF2, RAP1, TIN2, TPP1, and POT1, which seems to work as a functional unit because disruption of one component can affect the activities of the remaining components, resulting in abnormal capping. Moreover, telomeres contain the classical histone octamers, that is, nucleosomes, enriched with specific epigenetic modifications mainly underacetylated lysine residues and several hypermethylations including H3K9me3 and H4K20me3 that suggest a compacted heterochromatic state (Fig. 5.2B). Changes in the histone epigenetic markers can lead to changes in telomere homeostasis [12,13].

Recently, a structural long noncoding RNA called telomere repeat-containing RNA (TERRA) has been reported to be transcribed by RNA polymerase II from the subtelomeric region of most telomeres [14–17]. TERRA transcripts contain a telomeric repeat track and sequences arising from subtelomeric regions and may associate with their complementary telomeric DNA strand forming DNA/RNA hybrids named R-loops [18,19] (Fig. 5.2B). TERRA also shows a tight relationship with components of the shelterin complex such as TRF1 and TRF2 and different members of the hnRNP family. Moreover, it interacts with the histone methyltransferase SUV39H1 that promotes H3K9 methylation and with the H3K9me3 residues from telomeric nucleosomes [11,20]. All these findings suggest that TERRA may be an anchor molecule that recruits proteins and specific enzymatic activities to telomeres, participating in the constitution of a local packed heterochromatin conformation and proper telomere capping [13,17,21].

Telomere Physiology

Telomere length is restored by telomerase during S phase of the cell cycle. Telomerase is a specialized reverse transcriptase that extends the 3′ end of the G-rich parental leading strand. It contains an RNA subunit, a telomerase RNA component (TERC) with an almost perfect tandem repeat complementary to the telomere sequence. This RNA component acts as a template to prime the DNA synthesis by the catalytic protein subunit telomerase reverse transcriptase (TERT). By extending the leading strand, the complementary lagging strand can then be extended by the standard DNA polymerase, but leaving the G-rich overhang [20,22,23].

Telomerase is active in human germ line, oogonia, and spermatogonia, but is not active or insufficiently active in most of somatic tissues. Immortalized cell lines and around 85%–90% of cancer cells maintain their telomeres through the activation of telomerase, whereas the rest use a recombination-based mechanism, the alternative lengthening of telomeres (ALT) [24,25].

In the absence of telomerase or alternative lengthening mechanisms, telomeres progressively shorten after each DNA replication cycle, due to the inability of the DNA polymerase to use the 3′ end as a template for DNA replication. When telomeric sequences become critically shortened or the higher-order chromatin architecture is compromised, telomeres lose their capping function and trigger an ATM/ATR-mediated DNA-damage response (DDR). This leads to replicative senescence that operates as a tumor suppressor mechanism; therefore, cells stop dividing or die through apoptosis [10,22].

Cells harboring mutations in certain tumor suppressor genes such as *p53* and/or *Rb1* can overcome the barrier of replicative senescence resulting in dysfunctional telomeres that may activate the nonhomologous end-joining (NHEJ) pathway of double-strand DNA break repair, leading to end-to-end chromosome fusions. Resulting dicentric chromosomes can be mechanically pulled between the two poles of the mitotic spindle forming anaphase bridges that may break during the next division producing new broken ends, thus initiating breakage-fusion-bridge cycles in following mitoses, progressively increasing genome instability, characteristic of cancer cells [26,27].

Telomeres in Aging and Age-Related Disease

Same-age individuals show a wide variation in telomere length, attributed to both genetic and environmental factors [28]. On the one hand, telomere size seems genetically influenced, heritability being estimated between 36 and 84% [29]. But on the other hand, evidence exists that chronic environmental stressors and even psychosocial adversities, mainly in childhood, may accelerate telomere shortening [30]. This influence could be mediated by accelerated tissue proliferation and/or oxidative stress. Increased reactive oxygen species (ROS) damage DNA, the guanine sequences being especially sensitive. Telomeres are G-rich sequences so prone to oxidative damage and excision (telomere shortening) [31].

Although no clear relationship was evident between replicative senescence in vitro and organism aging, it is obvious that telomere length decreases with age and is being considered a valuable marker of biological age. Several studies, including one with >64,000 subjects [32], suggest that short telomeres measured in leukocytes or saliva correlate with decreased life span [32,33]. Most of these studies have not been performed throughout the life of those individuals by analyzing repeated samples over time, so it is difficult to separate influences at the individual or population levels. Some longitudinal surveys indicate that telomere loss is greater during early life [34]. Heidinger et al. [35] monitored telomere length in red cells from the bird zebra finch (*Taeniopygia guttata*) and observed that telomere length in early life, that is, at 25 days, toward the end of the growth period, was a very strong predictor of realized longevity.

Moreover, diverse metaanalyses correlate decreased telomere length with many age-related diseases. For example, individuals with short telomeres in blood leukocytes are at 80% increased risk of concurrent cardiovascular disease and a 40% of developing this disease in the future [36]. This risk is also evident with diabetes [37] and several cancers [38,39]. Genome-wide association cohort studies indicate that common sequence variants in seven genes involved in telomere regulation increased the risk for cardiovascular, pulmonary, and Alzheimer's diseases [40,41]. Regarding human reproduction, a study including 25 couples with recurrent pregnancy loss also revealed shorter leukocyte telomere length of both male and female partners when compared with 20 fertile proved couples [42].

Telomeres in Male Germ Line

Whereas in somatic cells telomere length decreases with aging, germ-line stem cells ensure telomere maintenance throughout the spermatogenic maturation continuously during the adult life of the male. This is achieved through high telomerase activity. Using a *Tert* reporter mice strain, it was confirmed that the undifferentiated spermatogonia is the specific lineage responsible for the germ-line stem cell activity, showing high telomerase expression [25]. This self-renewal compartment corresponds to the immortal cell line from metazoans, which maintains species' highest telomeres. In fact, telomere erosion through inactivation of telomerase in model organisms results in testicular atrophy and depletion of germ cells, as well as infertility [43,44].

Undifferentiated spermatogonia give rise to differentiated spermatogonia that undergo several transit-amplifying mitoses to differentiate into meiotic spermatocytes. Telomerase activity significantly decreases during these more differentiated stages and disappears in spermatozoa [45]. Furthermore, the long noncoding telomeric RNA TERRA forms part of the telomeric structure during mammalian meiosis. TERRA transcripts increase from spermatogonia to spermatocyte II and then decreases during spermiogenesis [46].

Telomeres play a crucial role in maintaining genome integrity during meiosis. To produce the haploid gamete, a ploidy reduction must be achieved by the meiotic division. To this purpose, homologous chromosomes must correctly pair during the first meiotic prophase of the primary spermatocyte to constitute the bivalents. The accurate movements of the chromosomes are driven by the movement of the telomeres along the nuclear envelope, known as rapid prophase chromosome movements (RPMs), forming the so-called bouquet organization in zygotene [47] (Fig. 5.3A). The movements of telomeres and chromosomes allow high-fidelity pairing of the homologous chromosomes and facilitate the recombination process and formation of chiasmata that physically link homologous chromosomes, being critical for the adequate reductional segregation at the first meiotic division [48].

Leptotene telomeres attach to the nucleoplasmic inner face of the nuclear membrane (Fig. 5.3A), associating with protein complexes that link the cytoskeleton and the nucleoskeleton and enabling cytoplasmic motors like dynein to move the telomeres along the nuclear envelope. Sad1/Unc-84 (SUN)-domain proteins are located in the inner membrane of the nuclear envelope, connected to nucleoskeletal structures like the nuclear lamina.

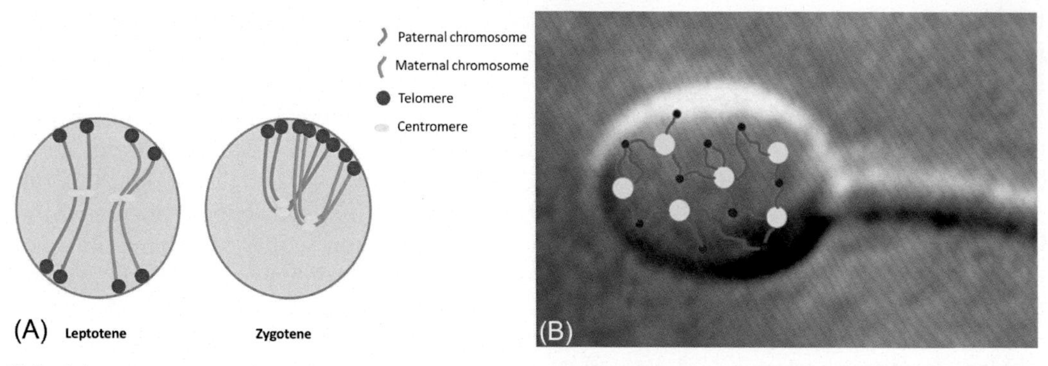

FIG. 5.3 (A) Representation of the bouquet formation in the prophase of the first meiotic division in germ cells. Telomeres cluster in one side of the nucleus helping the pairing of homologous chromosomes. (B) Model of chromosome organization within the sperm nucleus according to Ioannou et al. [56]. Clustered centromeres form chromocenters along the nucleus (yellow), telomere clusters scatter throughout the nucleus (red), and both clusters are connected with the p and q arms of the chromosomes (blue).

The Klarsicht/ANC-1/Syne homology (KASH)-domain proteins located in the outer face of the nuclear membrane interact with cytoskeletal structures like microtubules or actin microfilaments. SUN- and KASH-domain proteins interact at the perinuclear space constituting the complexes that connect the nucleoskeleton and the cytoskeleton (LINC, the linker between nucleoskeleton and cytoskeleton) [49]. The attachment of telomeres to the nuclear envelope possibly requires changes or modification in telomere proteins and meiosis-specific factors, like cohesin CMC1B [50] and CCDC79 [51]. In this respect, the cyclin-dependent kinase CDK2 has an essential role in meiosis possibly mediating the attachment of telomeres to the nuclear envelope and driving chromosomal movements [52]. E-type cyclins also seem to play a relevant role in regulating telomere integrity during the meiotic prophase, as their deficiency led to abnormal telomeres in mouse spermatocytes, altered shelterin protein complexes, and induced telomere associations and fusions, as well as detachment from the nuclear envelope and chromosome rearrangements [53].

Secondary spermatocytes undergo the second meiotic division giving rise to the haploid spermatids that differentiate to mature spermatozoa throughout spermiogenesis. At the spermatid elongation stage, most of the nucleosomes are removed, so histones are replaced by transition proteins and finally by protamines. This results in the unique and strong chromatin packaging of sperm DNA [54]. The nucleus occupies most of the sperm head with its particular shape. Individual chromosomes organize into the sperm nucleus following a particular arrangement where telomeres play a remarkable role. FISH studies hybridizing centromeric, telomeric, and locus-specific probes on moderately decondensed human sperm nuclei suggested a "hairpin-loop" organization. Centromeric sequences would cluster in the interior forming a chromocenter, whereas telomeres from both chromosome arms attach to the nuclear membrane in the periphery, so the chromosome would have a loop-like appearance [55].

Recent 2-D and 3-D radial and longitudinal studies confirm that telomeres form clusters of dimers or tetramers, but not necessarily sitting at the periphery. Their organization would be segmental, with many of them forming a belt in the mid part of the human sperm nucleus [56]. Moreover, the centromeres form many clusters with a more segmental organization throughout the nucleus and not restricted to the interior (Fig. 5.3B). Whole chromosomes seem to be ordered in a polar-like organization along the head-tail axis [57]. This architecture may be important for the ordered exodus of paternal chromosomes, adequate DNA repair, chromatin remodeling, and pronuclear formation, after fertilization.

As indicated previously, late-generation telomerase null mice show a practically total depletion of germ cells. This is a consequence of excessive telomere shortening that seriously compromises meiotic progression, so spermatocyte arrest at meiotic prophase and apoptosis is triggered [43,44]. An association between unusually short sperm telomeres and oligospermia has been recently described [58,59]. Thus, a germ cell telomere surveillance mechanism would prevent the transmission of dysfunctional telomeres that might lead to chromosomal abnormalities. The third generation of telomerase null mice begin to show critically short telomeres, but spermatozoa may still be obtained in the epididymis. Remarkably, many of these sperm contain fragmented DNA that should occur after the spermatid elongation stage [60]. Telomere dysfunction may result in aneuploid gametes, and DNA fragmentation is more frequent in aneuploid sperm [61]. These data suggest the existence of a genomic surveillance mechanism that operates during or after spermiogenesis [62], inducing DNA fragmentation to genetically inactivate sperm with a defective genome. This checkpoint and that of the first meiotic prophase would ensure an adequate chromosome constitution of the gamete.

Human sperm telomeric size is substantially variable within and between individuals, usually ranging from <9 to over 17 kb [63]. Variation of sperm telomeric size could be due to differences in telomerase activity, effect of oxidative stress, or mutational events. The mature and highly motile sperm population selected after swim-up or after density-gradient centrifugation exhibit longer telomeres [64,65]. Some sperm may contain severely truncated telomeres, even shorter than those from senescent cells [63]. These sperm perhaps could contribute to aneuploidy in the early embryo and affect embryo implantation and development.

Remarkably, several studies point that telomere length in sperm appears to increase with age and this may correlate with a longer leukocyte telomere size in the offspring of older fathers [66,67]. Furthermore, sperm telomere length is more heterogeneous with aging, perhaps as a consequence of the ALT activation during spermatogenesis [68]. The increased sperm telomere size occurring with paternal aging may elongate the life span of older male's offspring. A large longitudinal study analyzing multigenerational samples from Cebu, Philippines, concluded that telomeres measured in blood leukocytes were longer in individuals whose paternal grandfathers were older at their father's birth [69]. Thus, paternal age at the moment of reproduction influences telomere length, and this influence is transmitted across at least two generations. A higher testicular telomerase activity with advanced age or a selective replicative advantage of germ cells with longer telomeres may be possible explanations of the increase of telomere size with age in sperm. Nonetheless, a longitudinal follow-up of telomere size of sperm from the same individuals would be necessary to accurately confirm these findings [70]. Although a paternal age effect on telomere length of offspring seems convincing, a maternal age effect is not clear [69]. Taking into account the presumed influence of paternal telomere length in offspring life span, it is of great interest to investigate the longevity of individuals conceived by assisted reproduction using sperm with longer telomeres selected when using swim-up or density-gradient centrifugation. But possibly during natural intercourse, there may be a natural selection within the female genital tract, so highly motile sperm, with longer telomeres, reach the oocyte.

In a group of normozoospermic samples from men within a limited age range, telomere size from sperm was positively correlated with motility, vitality, and protamination and negatively associated with sperm DNA fragmentation [71]. There are some reports available indicating that individuals with idiopathic infertility show compromised telomere homeostasis in spermatocytes [72] and shorter telomeres in sperm [73]. As cited before, oligospermic sperm samples show particularly small telomeres [59]. Moreover, certain genetic variants of TERT and of telomerase-associated protein 1 (TEP1) also seem to be associated with male infertility [74].

Telomeres in Female Germ Line

Spermatogonia from men produce sperm throughout their adult life. However, oogonia are not present in adult women's ovaries. A fixed cohort of nearly 7 million oocytes is formed during early fetal life, and this pool is progressively depleted by apoptotic death. Around 1 million remain at birth; 500,000 at puberty; and near 1000 at menopause [75,76]. These oocytes remain arrested at the dictyate stage of the prophase of the first meiotic division, even during decades, until ovulation. The later the oocyte ovulation, the higher the duration of prophase I arrest and the higher the oocyte aging and segregation errors. As a consequence, oocytes are more prone than spermatocytes to develop aneuploidy during meiosis. In fact, most of human aneuploid embryos are of maternal meiotic origin, and their prevalence increases with age, especially from their mid-30s and accelerating through their 40s.

Chromosomal nondisjunction seems responsible for aneuploidy in oocytes from younger women, whereas premature separation and missegregation of sister centromeres are the main mechanism of aneuploidy in oocytes from older women [77–79]. Thus, the majority of MI errors related with oocyte aging are single chromatid losses or gains. This phenomenon seems to be caused by deterioration and depletion of cohesins that maintain sister centromeres attached together [80]. Dysfunction of the spindle meiotic apparatus, deficient chiasmata between homologous chromosomes, and less efficient spindle assembly checkpoint are also invoked as factors involved in the age-related oocyte aneuploidy [81].

Telomere length in oocytes is shorter than in sperm and somatic cells [82]. The size of telomeres starts decreasing during fetal oogenesis and further decreases with aging during the meiotic arrest, possibly by oxidative stress and DNA damage. Telomere dysfunction is an important element that can originate aneuploid oocytes leading to infertility and miscarriage [48].

Late-generation null mice for telomerase develop telomere dysfunction due to telomere erosion. Oocytes from these mice show abnormal spindles and misaligned lagging chromosomes [83]. Unlike in spermatocytes, oocytes with misaligned chromosomes do not undergo apoptosis or do not arrest and progress from MI to MII, demonstrating a less efficient checkpoint control and promoting higher aneuploidy levels. Moreover, meiotic synapsis and recombination are also impaired [84]. These same alterations had been described in spermatocytes, where telomere loss alters chromosome pairing at the leptotene stage, thus compromising adequate chromosome segregation. It is not clear if the telomere dysfunction can influence the premature separation of sister chromatids.

Polar bodies mirror the genetic pool of oocytes. Thus, determination of telomere length in polar bodies provides indirect information of the telomeric length in the oocytes. A remarkable study using single-cell SNP microarray to assess human 24-chromosome aneuploidy and quantifying telomere DNA demonstrated that aneuploid polar bodies have less telomeric DNA than euploid polar bodies from the same IVF patient and cycle [85].

Granulosa cells are quite relevant for oocyte maturation, so telomere length and telomerase activity in granulosa cells have been proposed as biomarkers of oocyte and embryo quality. Longer telomere size is believed to indicate better quality granulosa cells that would support an adequate maturation of the oocyte and good-quality embryos when fertilized [86,87].

Telomeres in Early Embryo

Telomeres from the parental and maternal gametes are integrated in the embryo and play a crucial role during its development. Telomerase expression is undetectable in mice zygote and cleavage stages, but it is highly active in the blastocyst stage [82]. Nevertheless, telomeres significantly lengthen during the first or second cycles. This effect has been remarkable in telomerase knockout mice and must be dependent on a local recombination mechanism or ALT [82,88]. Using transgenic mice with tagged telomeres, it was demonstrated that telomeres were not copied from one chromosome to its homologous chromosome in the preimplantation embryo [89]. This suggests that telomere lengthening should occur by using the telomere sequence from its own chromosome as a template for ALT. Perhaps the telomere sequences from sperm could guide this lengthening, thus explaining the preferential paternal influence of telomere size in the offspring [88]. Nevertheless, ALT was also found in parthenogenetically activated eggs without sperm input.

A study measuring telomeres in human preimplantation embryos by quantitative fluorescence in situ hybridization (qFISH) indicated that telomere length decreased in the cleavage stage, being even shorter than in oocytes and then increasing at blastocyst stage [90]. However, these results could be possibly affected by the variable efficiency of in situ hybridization. The accessibility of the telomeric probe could be influenced by the different chromatin stages in the different pronuclei or embryo stages [48].

Aneuploidy is often found in human embryos, being that the leading cause of genetic miscarriage. Abnormal chromosome number in the embryo can be derived from aneuploid gametes, and telomere dysfunction could play a role in their origin during meiosis, as described before. Moreover, initially, euploid preimplantation embryos are frequently prone to develop aneuploidies during the first mitotic cycles, resulting in chromosomal mosaicism. Even a chaotic pattern with diverse aneuploidies in different cells from the same embryo can be detected. This is a consequence of mitotic nondisjunction, anaphase lagging, or even breakage-fusion-bridge cycles [91]. Although initial studies using FISH reported a very high frequency of aneuploid embryos, the SNP microarray technique established a range of 25%–30% [85]. The particular vulnerability of early mitoses could rely on dysfunction of checkpoint maternal factors that control cell cycle [92] or impaired function of the sperm-derived centrosome [93]. Additionally, telomeres could also be involved in embryo aneuploidy.

The study of Treff et al. [85] assessing the 24-chromosome aneuploidy and the amount of telomere DNA evidenced less telomere DNA sequences in human aneuploid blastomeres than in euploid ones from the same embryo. This association disappeared at the blastocyst stage due to telomere normalization. Mania et al. point to a similar result [94]. In their report, the telomere FISH signal was quantified followed by ploidy determination in the same cell.

The aforementioned showed a shorter mean telomere size in aneuploid versus euploid cells in embryos that did not achieve the blastocyst stage, but telomere size being very similar in blastocysts at day 5 after fertilization.

In the same report [94], they also observed that embryo cells obtained from women of advanced age and history of repeated miscarriage tended to have substantially shorter telomeres. On the other hand, a significant direct correlation has been found between the sperm telomere length and the quality of the embryo obtained in IVF with transplantable rates, but not with pregnancy rates [95].

Despite the presumed telomerase-dependent reset and restoration of telomeres at the morula-blastocyst transition initially demonstrated in cattle and mice cloned embryos whose nuclei where transferred from fibroblasts [96], it remains intriguing that initial telomere sizes in the preimplantation embryo could influence the rest of the development, perhaps even after birth [97]. Furthermore, how species-specific telomere size range is properly established or how the particular genetic determinant or inheritance influences this setting and modulates the future dynamic behavior of telomeres constitute challenges yet to be investigated.

CONCLUDING CONSIDERATIONS

Transmission of the genome to the next generation may be considered a form of immortality at the species level. In eukaryotes with sexual reproduction, male sperm and female oocyte ensure the proper continuity of a stable genome. Accurate telomeric chromatin must be adequately maintained in the precursors of germ cells to preserve this continuity, analogous to the state of immortalized cell lines. Thus, the knowledge of telomere architecture and its dynamic regulation is of outstanding relevance in the reproductive field and in the pathology of reproduction genetics. Moreover, telomeres transmitted through the germ line strongly influence somatic cells, mainly regarding aging and common age-related diseases.

Acknowledgments

We are grateful to Prof. Mike Kjelland for checking the English in the manuscript.

References

[1] H.J. Muller, The remaking of chromosomes, Collect. Net. 13 (1938) 181–198.
[2] B. McClintock, The stability of broken ends of chromosomes in *Zea mays*, Genetics 26 (2) (1941) 234–282.
[3] J.D. Watson, Origin of concatemeric T7 DNA, Nat. New Biol. 239 (94) (1972) 197–201.
[4] L. Hayflick, P.S. Moorhead, The serial cultivation of human diploid cell strains, Exp. Cell Res. 25 (1961) 585–621.
[5] A.M. Olovnikov, Principle of marginotomy in template synthesis of polynucleotides, Dokl. Akad. Nauk. SSSR 201 (6) (1971) 1496–1499.
[6] M.C. Yao, E. Blackburn, J. Gall, Tandemly repeated C-C-C-C-A-A hexanucleotide of *Tetrahymena* rDNA is present elsewhere in the genome and may be related to the alteration of the somatic genome, J. Cell Biol. 90 (2) (1981) 515–520.
[7] J.W. Szostak, E.H. Blackburn, Cloning yeast telomeres on linear plasmid vectors, Cell 29 (1) (1982) 245–255.
[8] J. Shampay, J.W. Szostak, E.H. Blackburn, DNA sequences of telomeres maintained in yeast, Nature 310 (5973) (1984) 154–157.
[9] C.W. Greider, E.H. Blackburn, Identification of a specific telomere terminal transferase activity in *Tetrahymena* extracts, Cell 43 (2 Pt 1) (1985) 405–413.

[10] S. Schoeftner, M.A. Blasco, A "higher order" of telomere regulation: telomere heterochromatin and telomeric RNAs, EMBO J. 28 (2009) 2323–2336.
[11] A. Galati, E. Micheli, S. Cacchione, Chromatin structure in telomere dynamics, Front. Oncol. 3 (2013) 46.
[12] S.M. Bailey, J.P. Murnane, Telomeres, chromosome instability and cancer, Nucleic Acids Res. 34 (2006) 2408–2417.
[13] E. Cusanelli, P. Chartrand, Telomeric noncoding RNA: telomeric repeat-containing RNA in telomere biology, Wiley Interdiscip. Rev. RNA 5 (3) (2014) 407–419.
[14] B.O. Farnung, E. Giulotto, C.M. Azzalin, Promoting transcription of chromosome ends, Transcription 1 (3) (2010) 140–143.
[15] E. Cusanelli, P. Chartrand, Telomeric repeat-containing RNA TERRA: a noncoding RNA connecting telomere biology to genome integrity, Front. Genet. 6 (2015) 143.
[16] I. López de Silanes, O. Graña, M.L. De Bonis, O. Dominguez, D.G. Pisano, M.A. Blasco, Identification of TERRA locus unveils a telomere protection role through association to nearly all chromosomes, Nat. Commun. 5 (2014) 4723.
[17] C.M. Azzalin, P. Reichenbach, L. Khoriauli, E. Giulotto, J. Lingner, Telomeric repeat containing RNA and RNA surveillance factors at mammalian chromosome ends, Science 318 (2007) 798–801.
[18] B.O. Farnung, C.M. Brun, R. Arora, L.E. Lorenzi, C.M. Azzalin, Telomerase efficiently elongates highly transcribing telomeres in human cancer cells, PLoS One 7 (4) (2012) e35714.
[19] B. Balk, A. Maicher, M. Dees, J. Klermund, S. Luke-Glaser, K. Bender, B. Luke, Telomeric RNA-DNA hybrids affect telomere-length dynamics and senescence, Nat. Struct. Mol. Biol. 20 (10) (2013) 1199–1205.
[20] S. Schoeftner, M.A. Blasco, A "higher order" of telomere regulation: telomere heterochromatin and telomeric RNAs, EMBO J. 28 (16) (2009) 2323–2336.
[21] Z. Crees, J. Girard, Z. Rios, G.M. Botting, K. Harrington, C. Shearrow, L. Wojdyla, A.L. Stone, S.B. Uppada, J.T. Devito, N. Puri, Oligonucleotides and G-quadruplex stabilizers: targeting telomeres and telomerase in cancer therapy, Curr. Pharm. Des. (2014).
[22] M.A. Blasco, The epigenetic regulation of mammalian telomeres, Nat. Rev. Genet. 8 (4) (2007) 299–309.
[23] C.A. Armstrong, K. Tomita, Fundamental mechanisms of telomerase action in yeasts and mammals: understanding telomeres and telomerase in cancer cells, Open Biol. 7 (3) (2017) 160338.
[24] W.E. Wright, M.A. Piatyszek, W.E. Rainey, W. Byrd, J.W. Shay, Telomerase activity in human germ line and embryonic tissues and cells, Dev. Genet. 18 (2) (1996) 173–179.
[25] M.F. Pech, A. Garbuzov, K. Hasegawa, M. Sukhwani, R.J. Zhang, B.A. Benayoun, S.A. Brockman, S. Lin, A. Brunet, K.E. Orwig, S.E. Artandi, High telomerase is a hallmark of undifferentiated spermatogonia and is required for maintenance of male germ line stem cells, Genes Dev. 29 (23) (2015) 2420–2434.
[26] I. Gonzalez-Vasconcellos, N. Anastasov, B. Sanli-Bonazzi, O. Klymenko, M.J. Atkinson, M. Rosemann, Rb1 haploinsufficiency promotes telomere attrition and radiation-induced genomic instability, Cancer Res. 73 (14) (2013) 4247–4255.
[27] C. Frias, J. Pampalona, A. Genesca, L. Tusell, Telomere dysfunction and genome instability, Front. Biosci. Landmark Ed. 17 (2012) 2181–2196.
[28] P. Monaghan, Telomeres and life histories: the long and the short of it, Ann. N. Y. Acad. Sci. 1206 (1) (2010) 130–142.
[29] A. Aviv, Genetics of leukocyte telomere length and its role in atherosclerosis, Mutat. Res. 730 (1–2) (2012) 68–74.
[30] E. Puterman, A. Gemmill, D. Karasek, D. Weir, N.E. Adler, A.A. Prather, E.S. Epel, Lifespan adversity and later adulthood telomere length in the nationally representative US Health and Retirement Study, Proc. Natl. Acad. Sci. U. S. A. 113 (42) (2016) E6335–E6342.
[31] J.M.J. Houben, H.J.J. Moonen, F.J. van Schooten, G.J. Hageman, Telomere length assessment: biomarker of chronic oxidative stress? Free Radic. Biol. Med. 44 (3) (2008) 235–246.
[32] L. Rode, B.G. Nordestgaard, S.E. Bojesen, Peripheral blood leukocyte telomere length and mortality among 64,637 individuals from the general population, J. Natl. Cancer Inst. 107 (6) (2015) djv074.
[33] J. Deelen, M. Beekman, V. Codd, S. Trompet, L. Broer, S. Hägg, K. Fischer, P.E. Thijssen, H.E.D. Suchiman, I. Postmus, A.G. Uitterlinden, A. Hofman, A.J.M. de Craen, A. Metspalu, N.L. Pedersen, C.M. van Duijn, J.W. Jukema, J.J. Houwing-Duistermaat, N.J. Samani, P.E. Slagboom, Leukocyte telomere length associates with prospective mortality independent of immune-related parameters and known genetic markers, Int. J. Epidemiol. 43 (3) (2014) 878–886.
[34] S.L. Zeichner, P. Palumbo, Y. Feng, X. Xiao, D. Gee, J. Sleasman, M. Goodenow, R. Biggar, D. Dimitrov, Rapid telomere shortening in children, Blood 93 (9) (1999) 2824–2830.

[35] B.J. Heidinger, J.D. Blount, W. Boner, K. Griffiths, N.B. Metcalfe, P. Monaghan, Telomere length in early life predicts lifespan, Proc. Natl. Acad. Sci. 109 (5) (2012) 1743–1748.
[36] P.C. Haycock, E.E. Heydon, S. Kaptoge, A.S. Butterworth, A. Thompson, P. Willeit, Leucocyte telomere length and risk of cardiovascular disease: systematic review and meta-analysis, BMJ 349 (3) (2014) g4227.
[37] J. Zhao, K. Miao, H. Wang, H. Ding, D.W. Wang, Association between telomere length and type 2 diabetes mellitus: a meta-analysis, PLoS One 8 (11) (2013) e79993.
[38] I.M. Wentzensen, L. Mirabello, R.M. Pfeiffer, S.A. Savage, The association of telomere length and cancer: a meta-analysis, Cancer Epidemiol. Biomark. Prev. 20 (6) (2011) 1238–1250.
[39] H. Ma, Z. Zhou, S. Wei, Z. Liu, K.A. Pooley, A.M. Dunning, U. Svenson, G. Roos, H.D. Hosgood, M. Shen, Q. Wei, Shortened telomere length is associated with increased risk of cancer: a meta-analysis, PLoS One 6 (6) (2011) e20466.
[40] V. Codd, C.P. Nelson, E. Albrecht, M. Mangino, J. Deelen, J.L. Buxton, J.J. Hottenga, K. Fischer, T. Esko, I. Surakka, L. Broer, D.R. Nyholt, I. Mateo Leach, P. Salo, S. Hägg, M.K. Matthews, J. Palmen, G.D. Norata, P.F. O'Reilly, D. Saleheen, N. Amin, A.J. Balmforth, M. Beekman, R.A. de Boer, S. Böhringer, P.S. Braund, P.R. Burton, A.J.M. de Craen, M. Denniff, Y. Dong, K. Douroudis, E. Dubinina, J.G. Eriksson, K. Garlaschelli, D. Guo, A.-L. Hartikainen, A.K. Henders, J.J. Houwing-Duistermaat, L. Kananen, L.C. Karssen, J. Kettunen, N. Klopp, V. Lagou, E.M. van Leeuwen, P.A. Madden, R. Mägi, P.K.E. Magnusson, S. Männistö, M.I. McCarthy, S.E. Medland, E. Mihailov, G.W. Montgomery, B.A. Oostra, A. Palotie, A. Peters, H. Pollard, A. Pouta, I. Prokopenko, S. Ripatti, V. Salomaa, H.E.D. Suchiman, A.M. Valdes, N. Verweij, A. Viñuela, X. Wang, H.-E. Wichmann, E. Widen, G. Willemsen, M.J. Wright, K. Xia, X. Xiao, D.J. van Veldhuisen, A.L. Catapano, M.D. Tobin, A.S. Hall, A.I.F. Blakemore, W.H. van Gilst, H. Zhu, Cardi: CARDIoGRAM consortium, J. Erdmann, M.P. Reilly, S. Kathiresan, H. Schunkert, P.J. Talmud, N.L. Pedersen, M. Perola, W. Ouwehand, J. Kaprio, N.G. Martin, C.M. van Duijn, I. Hovatta, C. Gieger, A. Metspalu, D.I. Boomsma, M.-R. Jarvelin, P.E. Slagboom, J.R. Thompson, T.D. Spector, P. van der Harst, N.J. Samani, Identification of seven loci affecting mean telomere length and their association with disease, Nat. Genet. 45 (4) (2013) 422–427. 427e1–2.
[41] Y. Zhan, C. Song, R. Karlsson, A. Tillander, C.A. Reynolds, N.L. Pedersen, S. Hägg, Telomere length shortening and Alzheimer disease—a Mendelian randomization study, JAMA Neurol. 72 (10) (2015) 1202–1203.
[42] J. Thilagavathi, S.S. Mishra, M. Kumar, K. Vemprala, D. Deka, V. Dhadwal, R. Dada, Analysis of telomere length in couples experiencing idiopathic recurrent pregnancy loss, J. Assist. Reprod. Genet. 30 (6) (2013) 793–798.
[43] E. Herrera, E. Samper, J. Martín-Caballero, J.M. Flores, H.W. Lee, M.A. Blasco, Disease states associated with telomerase deficiency appear earlier in mice with short telomeres, EMBO J. 18 (11) (1999) 2950–2960.
[44] M.T. Hemann, K.L. Rudolph, M.A. Strong, R.A. DePinho, L. Chin, C.W. Greider, Telomere dysfunction triggers developmentally regulated germ cell apoptosis, Mol. Biol. Cell 12 (7) (2001) 2023–2030.
[45] M.V. Achi, N. Ravindranath, M. Dym, Telomere length in male germ cells is inversely correlated with telomerase activity, Biol. Reprod. 63 (2) (2000) 591–598.
[46] R. Reig-Viader, M. Vila-Cejudo, V. Vitelli, R. Buscà, M. Sabaté, E. Giulotto, M.G. Caldés, A. Ruiz-Herrera, Telomeric repeat-containing RNA (TERRA) and telomerase are components of telomeres during mammalian Gametogenesis1, Biol. Reprod. 90 (5) (2014) 103.
[47] C.-Y. Lee, M.N. Conrad, M.E. Dresser, Meiotic chromosome pairing is promoted by telomere-led chromosome movements independent of bouquet formation, PLoS Genet. 8 (5) (2012) e1002730.
[48] K.H. Kalmbach, D.M. Fontes Antunes, R.C. Dracxler, T.W. Knier, M.L. Seth-Smith, F. Wang, L. Liu, D.L. Keefe, Telomeres and human reproduction, Fertil. Steril. 99 (1) (2013) 23–29.
[49] H. Shibuya, Y. Watanabe, The meiosis-specific modification of mammalian telomeres, Cell Cycle 13 (13) (2014) 2024–2028.
[50] C. Adelfalk, J. Janschek, E. Revenkova, C. Blei, B. Liebe, E. Göb, M. Alsheimer, R. Benavente, E. de Boer, I. Novak, C. Höög, H. Scherthan, R. Jessberger, Cohesin SMC1β protects telomeres in meiocytes, J. Cell Biol. 187 (2) (2009) 185–199.
[51] K. Daniel, D. Tränkner, L. Wojtasz, H. Shibuya, Y. Watanabe, M. Alsheimer, A. Tóth, Mouse CCDC79 (TERB1) is a meiosis-specific telomere associated protein, BMC Cell Biol. 15 (1) (2014) 17.
[52] A. Viera, M. Alsheimer, R. Gomez, I. Berenguer, S. Ortega, C.E. Symonds, D. Santamaria, R. Benavente, J.A. Suja, CDK2 regulates nuclear envelope protein dynamics and telomere attachment in mouse meiotic prophase, J. Cell Sci. 128 (1) (2015) 88–99.
[53] M. Manterola, P. Sicinski, D.J. Wolgemuth, E-type cyclins modulate telomere integrity in mammalian male meiosis, Chromosoma 125 (2) (2016) 253–264.

[54] W.S. Ward, D.S. Coffey, DNA packaging and organization in mammalian spermatozoa: comparison with somatic cells, Biol. Reprod. 44 (4) (1991) 569–574.
[55] L. Solov'eva, M. Svetlova, D. Bodinski, A.O. Zalensky, Nature of telomere dimers and chromosome looping in human spermatozoa, Chromosom. Res. 12 (8) (2004) 817–823.
[56] D. Ioannou, N.M. Millan, E. Jordan, H.G. Tempest, A new model of sperm nuclear architecture following assessment of the organization of centromeres and telomeres in three-dimensions, Sci. Rep. 7 (2017) 41585.
[57] N.M. Millan, P. Lau, M. Hann, D. Ioannou, D. Hoffman, M. Barrionuevo, W. Maxson, S. Ory, H.G. Tempest, Hierarchical radial and polar organisation of chromosomes in human sperm, Chromosom. Res. 20 (7) (2012) 875–887.
[58] A. Ferlin, E. Rampazzo, M.S. Rocca, S. Keppel, A.C. Frigo, A. De Rossi, C. Foresta, In young men sperm telomere length is related to sperm number and parental age, Hum. Reprod. 28 (12) (2013) 3370–3376.
[59] F. Cariati, S. Jaroudi, S. Alfarawati, A. Raberi, C. Alviggi, R. Pivonello, D. Wells, Investigation of sperm telomere length as a potential marker of paternal genome integrity and semen quality, Reprod. BioMed. 33 (3) (2016) 404–411.
[60] S. Rodriguez, V. Goyanes, E. Segrelles, M. Blasco, J. Gosalvez, J. Fernandez, Critically short telomeres are associated with sperm DNA fragmentation, Fertil. Steril. 84 (4) (2005) 843–845.
[61] L. Muriel, V. Goyanes, E. Segrelles, J. Gosalvez, J.G. Alvarez, J.L. Fernandez, Increased aneuploidy rate in sperm with fragmented DNA as determined by the sperm chromatin dispersion (SCD) test and FISH analysis, J. Androl. 28 (1) (2006) 38–49.
[62] A.J. Ross, K.G. Waymire, J.E. Moss, A.F. Parlow, M.K. Skinner, L.D. Russell, G.R. MacGregor, Testicular degeneration in Bclw-deficient mice, Nat. Genet. 18 (3) (1998) 251–256.
[63] D.M. Baird, B. Britt-Compton, J. Rowson, N.N. Amso, L. Gregory, D. Kipling, Telomere instability in the male germ line, Hum. Mol. Genet. 15 (1) (2005) 45–51.
[64] R. Santiso, M. Tamayo, J. Gosálvez, M. Meseguer, N. Garrido, J.L. Fernández, Swim-up procedure selects spermatozoa with longer telomere length, Mutat. Res. Mol. Mech. Mutagen. 688 (1–2) (2010) 88–90.
[65] Q. Yang, N. Zhang, F. Zhao, W. Zhao, S. Dai, J. Liu, I. Bukhari, H. Xin, W. Niu, Y. Sun, Processing of semen by density gradient centrifugation selects spermatozoa with longer telomeres for assisted reproduction techniques, Reprod. BioMed. 31 (1) (2015) 44–50.
[66] M. Kimura, L.F. Cherkas, B.S. Kato, S. Demissie, J.B. Hjelmborg, M. Brimacombe, A. Cupples, J.L. Hunkin, J.P. Gardner, X. Lu, X. Cao, M. Sastrasinh, M.A. Province, S.C. Hunt, K. Christensen, D. Levy, T.D. Spector, A. Aviv, Offspring's leukocyte telomere length, paternal age, and telomere elongation in sperm, PLoS Genet. 4 (2) (2008) e37.
[67] O.T. Njajou, R.M. Cawthon, C.M. Damcott, S.-H. Wu, S. Ott, M.J. Garant, E.H. Blackburn, B.D. Mitchell, A.R. Shuldiner, W.-C. Hsueh, Telomere length is paternally inherited and is associated with parental lifespan, Proc. Natl. Acad. Sci. U. S. A. 104 (29) (2007) 12135–12139.
[68] D.M.F. Antunes, K.H. Kalmbach, F. Wang, R.C. Dracxler, M.L. Seth-Smith, Y. Kramer, J. Buldo-Licciardi, F.B. Kohlrausch, D.L. Keefe, A single-cell assay for telomere DNA content shows increasing telomere length heterogeneity, as well as increasing mean telomere length in human spermatozoa with advancing age, J. Assist. Reprod. Genet. 32 (11) (2015) 1685–1690.
[69] D.T.A. Eisenberg, M.G. Hayes, C.W. Kuzawa, Delayed paternal age of reproduction in humans is associated with longer telomeres across two generations of descendants, Proc. Natl. Acad. Sci. U. S. A. 109 (26) (2012) 10251–10256.
[70] R. Stindl, The paradox of longer sperm telomeres in older men's testes: a birth-cohort effect caused by transgenerational telomere erosion in the female germ line, Mol. Cytogenet. 9 (2016) 12.
[71] M.S. Rocca, E. Speltra, M. Menegazzo, A. Garolla, C. Foresta, A. Ferlin, Sperm telomere length as a parameter of sperm quality in normozoospermic men, Hum. Reprod. 31 (6) (2016) 1158–1163.
[72] R. Reig-Viader, L. Capilla, M. Vila-Cejudo, F. Garcia, B. Anguita, M. Garcia-Caldés, A. Ruiz-Herrera, Telomere homeostasis is compromised in spermatocytes from patients with idiopathic infertility, Fertil. Steril. 102 (3) (2014) 728–738.e1.
[73] J. Thilagavathi, S. Venkatesh, R. Dada, Telomere length in reproduction, Andrologia 45 (5) (2013) 289–304.
[74] L. Yan, S. Wu, S. Zhang, G. Ji, A. Gu, Genetic variants in telomerase reverse transcriptase (TERT) and telomerase-associated protein 1 (TEP1) and the risk of male infertility, Gene 534 (2) (2014) 139–143.
[75] F.J. Broekmans, E.A.H. Knauff, E.R. te Velde, N.S. Macklon, B.C. Fauser, Female reproductive ageing: current knowledge and future trends, Trends Endocrinol. Metab. 18 (2) (2007) 58–65.

[76] M. Dorland, R.J. van Kooij, E.R. te Velde, General ageing and ovarian ageing, Maturitas 30 (2) (1998) 113–118.
[77] R.R. Angell, Predivision in human oocytes at meiosis I: a mechanism for trisomy formation in man, Hum. Genet. 86 (4) (1991) 383–387.
[78] F. Pellestor, B. Andréo, T. Anahory, S. Hamamah, The occurrence of aneuploidy in human: lessons from the cytogenetic studies of human oocytes, Eur. J. Med. Genet. 49 (2) (2006) 103–116.
[79] F. Pellestor, B. Andréo, F. Arnal, C. Humeau, J. Demaille, Maternal aging and chromosomal abnormalities: new data drawn from in vitro unfertilized human oocytes, Hum. Genet. 112 (2) (2003) 195–203.
[80] M. Tsutsumi, R. Fujiwara, H. Nishizawa, M. Ito, H. Kogo, H. Inagaki, T. Ohye, T. Kato, T. Fujii, H. Kurahashi, Age-related decrease of meiotic Cohesins in human oocytes, PLoS One 9 (5) (2014) e96710.
[81] M. Herbert, D. Kalleas, D. Cooney, M. Lamb, L. Lister, Meiosis and maternal aging: Insights from Aneuploid oocytes and trisomy births, Cold Spring Harb. Perspect. Biol. 7 (4) (2015) a017970.
[82] L. Liu, S.M. Bailey, M. Okuka, P. Muñoz, C. Li, L. Zhou, C. Wu, E. Czerwiec, L. Sandler, A. Seyfang, M.A. Blasco, D.L. Keefe, Telomere lengthening early in development, Nat. Cell Biol. 9 (12) (2007) 1436–1441.
[83] L. Liu, M.A. Blasco, D.L. Keefe, Requirement of functional telomeres for metaphase chromosome alignments and integrity of meiotic spindles, EMBO Rep. 3 (3) (2002) 230–234.
[84] L. Liu, S. Franco, B. Spyropoulos, P.B. Moens, M.A. Blasco, D.L. Keefe, Irregular telomeres impair meiotic synapsis and recombination in mice, Proc. Natl. Acad. Sci. 101 (17) (2004) 6496–6501.
[85] N.R. Treff, J. Su, D. Taylor, R.T. Scott, Telomere DNA deficiency is associated with development of human embryonic aneuploidy, PLoS Genet. 7 (6) (2011) e1002161.
[86] E.-H. Cheng, S.-U. Chen, T.-H. Lee, Y.-P. Pai, L.-S. Huang, C.-C. Huang, M.-S. Lee, Evaluation of telomere length in cumulus cells as a potential biomarker of oocyte and embryo quality, Hum. Reprod. 28 (4) (2013) 929–936.
[87] W. Wang, H. Chen, R. Li, N. Ouyang, J. Chen, L. Huang, M. Mai, N. Zhang, Q. Zhang, D. Yang, Telomerase activity is more significant for predicting the outcome of IVF treatment than telomere length in granulosa cells, Reproduction 147 (5) (2014) 649–657.
[88] C. de Frutos, A.P. López-Cardona, N. Fonseca Balvís, R. Laguna-Barraza, D. Rizos, A. Gutierrez-Adán, P. Bermejo-Álvarez, Spermatozoa telomeres determine telomere length in early embryos and offspring, Reproduction 151 (1) (2016) 1–7.
[89] A.A. Neumann, C.M. Watson, J.R. Noble, H.A. Pickett, P.P.L. Tam, R.R. Reddel, Alternative lengthening of telomeres in normal mammalian somatic cells, Genes Dev. 27 (1) (2013) 18–23.
[90] S. Turner, H.P. Wong, J. Rai, G.M. Hartshorne, Telomere lengths in human oocytes, cleavage stage embryos and blastocysts, Mol. Hum. Reprod. 16 (9) (2010) 685–694.
[91] E. Vanneste, T. Voet, C. Le Caignec, M. Ampe, P. Konings, C. Melotte, S. Debrock, M. Amyere, M. Vikkula, F. Schuit, J.-P. Fryns, G. Verbeke, T. D'Hooghe, Y. Moreau, J.R. Vermeesch, Chromosome instability is common in human cleavage-stage embryos, Nat. Med. 15 (5) (2009) 577–583.
[92] L. Voullaire, L. Wilton, J. McBain, T. Callaghan, R. Williamson, Chromosome abnormalities identified by comparative genomic hybridization in embryos from women with repeated implantation failure, Mol. Hum. Reprod. 8 (11) (2002) 1035–1041.
[93] L. Rodrigo, V. Peinado, E. Mateu, J. Remohí, A. Pellicer, C. Simón, M. Gil-Salom, C. Rubio, Impact of different patterns of sperm chromosomal abnormalities on the chromosomal constitution of preimplantation embryos, Fertil. Steril. 94 (4) (2010) 1380–1386.
[94] A. Mania, A. Mantzouratou, J.D.A. Delhanty, G. Baio, P. Serhal, S.B. Sengupta, Telomere length in human blastocysts, Reprod. BioMed. Online 28 (5) (2014) 624–637.
[95] Q. Yang, F. Zhao, S. Dai, N. Zhang, W. Zhao, R. Bai, Y. Sun, Sperm telomere length is positively associated with the quality of early embryonic development, Hum. Reprod. 30 (8) (2015) 1876–1881.
[96] S. Schaetzlein, A. Lucas-Hahn, E. Lemme, W.A. Kues, M. Dorsch, M.P. Manns, H. Niemann, K.L. Rudolph, Telomere length is reset during early mammalian embryogenesis, Proc. Natl. Acad. Sci. U. S. A. 101 (21) (2004) 8034–8038.
[97] J.L. Simpson, D. Wells, Telomere length and aneuploidy: clinical and biological insights into human preimplantation embryos, Reprod. BioMed. Online 28 (5) (2014) 531–532.

CHAPTER

6

Human Protamine Genes' Polymorphisms as a Possible Cause Underlying Male Infertility

Anaís García Rodríguez, Rosa Roy Barcelona
Department of Biology, Autonomous University of Madrid, Madrid, Spain

INTRODUCTION

Protamines are small basic proteins that replace histones in the sperm DNA of vertebrates, thus constituting the most common protein in the sperm head. Protamines were first discovered in 1874 by Friedrich Miescher in the spermatozoa of the salmon [1], and their nature was first established by Albrecht Kossel in 1896 showing that somatic cells and spermatozoon differ in their protein composition [2]. During decades, they were thought to be present exclusively in the sperm of fish. By 1926, they had already been found in almost 20 species of fish including carp, herring, sturgeon, and sardine [3]; later, they were found in all *Vertebrata* subphylum including amphibians, reptiles, birds, and mammals. Moreover, proteins similar to the vertebrate's protamines have also been sequenced in insects and algae [4]. In all cases, the primary common feature is a high content of arginine organized into clusters.

It is well known that a correct protamination has a capital importance for the spermatozoa DNA condensation and protection as well as for the ultimate spermatozoa's head shape and its aerodynamic behavior. However, little information is known about the reasons leading to changes in the protamine's levels or structure that in turn lead to a defective protamination. It is known that the protamine genes are very polymorphic, and it has been proposed that some of those single-nucleotide polymorphisms (SNPs) may lead to changes in the expression of the genes or in the capacity of the proteins to bind to the DNA. The aim of this chapter is to compile the available information regarding the genetic causes that may be involved in a defective protamination.

Reproductomics
https://doi.org/10.1016/B978-0-12-812571-7.00007-1

PROTAMINE STRUCTURE AND FUNCTIONS

As previously said, protamines are small basic proteins that replace histones in the sperm DNA in order to get a higher folding level [5]. In 1999, Ausió carried out an elegant study in which he compared the sequences of protamines from different phylogenetic groups in order to offer a common definition of protamine based on their amino acid composition. He concluded that protamines are arginine-rich (Arg content ≥30 mol%) highly basic proteins (His + Lys + Arg = 45–80 mol% and Ser + Thr + Gly = 10–25 mol%) of relatively small molecular mass [6].

The protamines' main function is to protect the DNA. Protamines possess two main characteristics that are essential for this function: (i) They contain a high number of positively charged amino acids (arginines) that produce a tight binding to the DNA (negative charge) and (ii) they contain a significant and variable number of cysteine residues that form disulfide bridges inter- and intraprotamines increasing both folding rate and stability on the resulting chromatin. This allows the spermatozoa to get smaller and more aerodynamic nucleus and head that facilitates its movement and at the same time increases the protection of the DNA against external agents [7]. Protamines are also thought to have other relevant functions [8]: (i) They compete with transcriptional factors leaving the sperm DNA free from epigenetic information, (ii) they participate in the paternal imprinting, (iii) they act as a checkpoint for spermiogenesis, and (iv) they may act in the fertilized oocyte during pronuclei formation.

Two types of protamines have been described in mammals: protamine 1 (PRM1) and protamine 2 (PRM2). However, while all mammals contain PRM1 in their sperm DNA, only some species have PRM2 [9]. Moreover, there are important differences in the structure of PRM1 between the three groups of extant mammals. PRM1 in placental mammals are the shortest ones (around 50 amino acids), and they show a content in arginine residues of approximately the 50%. In contrast, PRM1 in marsupials and monotremes are longer (around 60 amino acids) and do not contain cysteine residues with the exception of *Planigale maculata sinualis*. Marsupials and monotremes are also characterized because the content in arginine residues is slightly superior to the content in placental mammals [10,11].

In humans, both PRM1 and PRM2 are expressed and can be found in the mature sperm DNA. PRM1 is composed of 50 amino acids (6.8 kDa) and contains 6 cysteines and 24 arginines, while PRM2 is composed of 57 amino acids (13 kDa) and contains 5 cysteines and 32 arginines.

PROTAMINE GENES ORGANIZATION

In mammals, PRM1 and PRM2 genes are coded together in a loop domain of about 28 kb along with the transition protein 2 gene and a small sequence called *gene4* whose function has still not been elucidated. This organization allows a coordinated expression of both genes during spermiogenesis. However, it must be highlighted that whereas PRM1 is expressed in all mammals, PRM2 is not. In fact, although PRM2 gene is coded in all mammals, it is only expressed in some particular species (primates and most rodents, lagomorphs, and perissodactyls) [12]. Interestingly, it has been demonstrated that the way by which the expression of

PRM2 is suppressed differs depending on the species. For example, while PRM2 expression in bull is suppressed due to several mutations within the gene inactivating gene expression, in rat, it is due to both a transcriptional and translational suppression.

In humans, the DNA loop containing both protamine genes is coded in chromosome 16p13.3 and flanked by matrix attachment regions (MARs). Both protamine genes contain two exons and one intron, the intron in PRM2 gene being slightly bigger than the one in PRM1. Moreover, in both cases, the gene promoter contains two common regulatory elements: a TATA box and a cAMP response element (CRE). However, the synthesis of both protamines is different. While protamine 1 is synthesized directly as a mature protein, protamine 2 is synthesized as an immature precursor that is afterward proteolyzed to produce the different members of the PRM2 family that differ only in the N-terminal extension of 1–4 residues (see Fig. 6.1).

The process of sperm histone-protamine substitution is perfectly organized [13]. In late pachytene spermatocytes, the DNA cluster containing the protamine genes is potentiated into a transcriptionally active state, and afterward, it is transcribed in round spermatids. The resulting mRNAs are stored in translationally repressed ribonucleoproteins until the early stages of spermiogenesis. When somatic histones in elongating spermatids are hyperacetylated, nucleosomes are disassembled, and around 80% of the somatic histones are replaced by transition nuclear proteins (TNPs). Protamines are then synthetized, quickly phosphorylated, and assembled with the DNA molecule while replacing TNPs. After they are bound to the DNA, protamines are dephosphorylated with the exception of some specific residues. Finally, during the last steps of spermiogenesis and the transit through the epididymis, intra- and intermolecular disulfide bridges are formed allowing a further condensation and stabilization of the DNA molecule.

Probably, the most exciting thing about this process is how the transcription and translation of the protamine genes are temporary separated. When histone replacement begins, the sperm DNA enters into a transcriptionally silenced state, and for that reason, the protamine genes cannot be transcribed at the moment in which they are needed. Consequently, both transcription and translation of the protamine genes are precisely organized [14]. During late

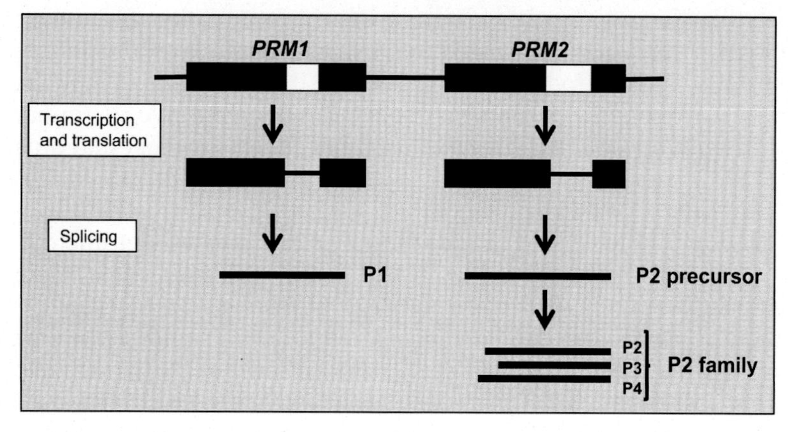

FIG. 6.1 Transcription, translation, and processing of human protamine genes. *Adapted from R Oliva, Protamines and male infertility, Hum. Reprod. Update. 12 (2006) 417–435.*

pachytene, the gene cluster is potentiated via a matrix attachment system, thus entering into a transcriptionally active state [15]. Afterward, transcription is initiated in round spermatids via both TATA box and CRE systems. Once the mRNAs are completely elongated, they undergo some processing in the nucleus, including polyadenylation, which contributes to translation suppression. Finally, translation of the mRNAs is repressed by two mechanisms: (i) binding of translation repressors both to the poly-A and to the 3′-UTR and (ii) physical separation from the translation machinery by storage in messenger ribonucleoprotein particles (mRNP) near the sperm nucleus [16].

PRM1/PRM2 RATIO

As we have already mentioned, in humans, there are two functional protamines, PRM1 and PRM2. These two protamines are represented in the sperm DNA approximately in the same quantity, thus leaving a PRM1/PRM2 ratio around 1. Different studies have concluded that a normal PRM1/PRM2 ratio can vary from 0.8 to 1.2 while ratios outstanding those levels are considered abnormal ratios and are detrimental for fertility.

Several studies have correlated abnormal PRM1/PRM2 ratios with infertility [17]. If fewer protamines are represented in the sperm chromatin, a smaller number of disulfide bridges will be held between them. This unbalanced disulfide bridge formation will leave the DNA less folded in mature sperm. Moreover, PRM1 has more cysteine residues than PRM2 (6 vs. 5, respectively), and consequently, changes in the quantity of them present in the DNA will correlate with changes in the number of disulfide bridges formed.

Abnormal PRM1/PRM2 ratios have detrimental consequences over sperm functionality and over fertility. On the one hand, as the DNA is poorly folded, it is more exposed to external agents, and this increases sperm DNA fragmentation [18–20]. But also, changes in the PRM1/PRM2 ratio correlate with abnormal sperm head shapes, thus reducing sperm motility and increasing sperm abnormal morphology accounts [21,22]. This situation specially affects the first steps of fertilization including capacitation, acrosomal reaction, membrane fusion, and penetration in the oocyte [23]. Consequently, it has been shown that abnormal PRM1/PRM2 ratios reduce fertilization results when applying in vitro fecundation. In contrast, the effect is less important when applying ICSI procedures; in this assisted reproduction technique, one spermatozoon is artificially introduced into the oocyte, thus overlapping with the early stages of fertilization and avoiding the negative consequences of a defective sperm-oocyte recognition [24]. It has been proposed that it may be useful to develop an easy and quick method to assess PRM1/PRM2 ratio in clinics in order to derive patients with abnormal PRM1/PRM2 ratios to ICSI procedures instead of IVF ones.

Although abnormal PRM1/PRM2 ratios in both directions are detrimental, it has been shown that the effect of an abnormal PRM1/PRM2 ratio is bigger when the ratio is under normal values than when it is over normal values. The unbalanced elevated PRM1/PRM2 ratio is also more common than the unbalanced reduced ratio, and it is normally due to a PRM2 underexpression concomitant with an increased level of PRM2 precursors [25]. This demonstrates that PRM1 has a more central structural role than PRM2 and explains why the first one appears in all mammals and its amino acid sequence is more conserved than the second one. Moreover, while there are some cases described of patients lacking PRM2 in their sperm DNA and even getting normal fertilization with the help of ICSI procedures, there are no cases described of patients lacking PRM1 [23]. Also, it has been shown that for PRM1, not only the quantity transcribed but also the moment of the transcription is essential, as premature transcription of PRM1 causes precocious nuclear condensation and arrests spermatid differentiation as demonstrated in mice [26].

PROTAMINE 1 GENE POLYMORPHISMS

More than 20 polymorphisms have been identified in the gene of PRM1 (see Table 6.1). Although more than a half of them are located in the coding region, only six lead to a nonsynonymous change in the amino acid sequence.

TABLE 6.1 Overview of PRM1 SNPs Identified in Several Studies

Polymorphism	Amino Acid	Region	Effect	Author
bc-248a		5′ promoter		Jodar, 2010
c-191a		5′ promoter		
g-107c		5′ promoter		Ravel, 2010
a133g	14	Exon 1		Tanaka, 2003
c49t	17	Exon 1	Arg→Cys	Jodar, 2010
g54a	18	Exon 1		Ravel, 2010
g65a	22	Exon 1	Ser→Asn	Imken, 2009
c160a	23	Exon 1		Tanaka, 2003
g93c	31	Exon 1	Gln→His	Ravel, 2010
g197t	34	Exon 1	Arg→Ser	Iguchi, 2006
g113t	38	Exon 2	Arg→Met	Jodar, 2010
g119a	40	Exon 2	Cys→Tyr	Ravel, 2010
g320a	46	Exon 2		Tanaka, 2003
c321a	47	Exon 2		
a139c	47	Exon 2		Ravel, 2010
g152a		Intron		Jodar, 2010
a431g		3′ noncodifying		Tanaka, 2003
g*51c		3′ noncodifying		Ravel, 2010
g*54c		3′ noncodifying		

To enable an easiest understanding of the information, the polymorphisms' nomenclature from the original papers has been maintained.

Polymorphisms in the Promoter Region

As previously mentioned, both protamine genes have regular promoter regions including TATA box and CRE sites. In this region of the PRM1 gene, two SNPs with consequences over fertility have been described.

SNP c-191a has been widely studied [27–29]. In 2008, Gazquez et al. concluded that the recessive AA variant was related with male infertility as it was associated with increased numbers of morphologically altered spermatozoa and with increased PRM1/PRM2 ratios [30]. These observations are explained because this SNP is located in a DNAse protection footprint that also contains binding sites for transcription factors BNC and SIP1. Consequently, this polymorphism may alter PRM1 gene expression and at the same time render the sequence of the gene less protected against DNase activity. In a recent study, SNP c-191a has been related with severe oligozoospermia [31].

Another SNP localized in the PRM1 gene promoter that is thought to have consequences over fertility is SNP g-107c. This SNP is localized inside a region of the PRM1 promoter of 77 nucleotides (from −77 to −150 positions) that is known to be specifically needed for transcription in the spermatid stage. Moreover, the SNP g-107c creates a binding site for HNF3 transcription factor and can thus affect the gene expression and PRM1/PRM2 ratio [29]. It has also been related with severe oligozoospermia [28]. Curiously, a study made by Kichine et al. in sub-Saharan population stated that in that population, this SNP seems to be a common SNP not related with infertility [32].

Polymorphisms in the Coding Region

At least 13 SNPs have been identified in the coding region of the PRM1 gene. Most of them (12) are located in the exonic regions, seven in exon 1 and five in exon 2, while only one is located in the intron. Among the first ones, only half of them actually lead to a change in the corresponding amino acid (Fig. 6.2A).

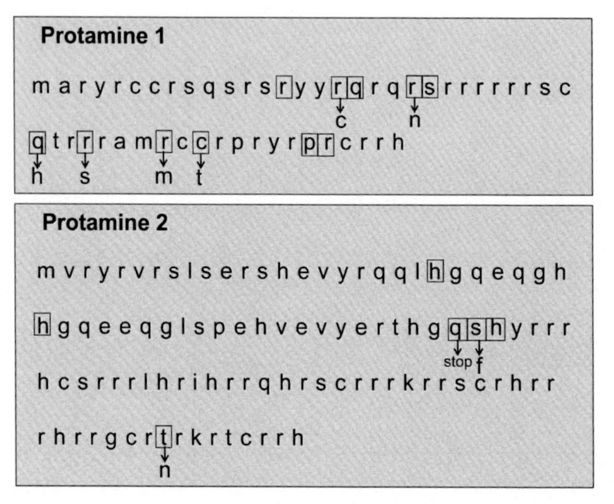

FIG. 6.2 Schematic overview of SNPs localized in the exonic regions of the protamine genes.

SNP g197t is located in the first exon and leads to a substitution in amino acid 34 from an arginine to a serine just in the middle of an arginine cluster; this is the most frequent mutation identified in protamine 1. This change creates a new phosphorylation site for enzyme SR protein-specific kinase 1, which is one of the enzymes implicated in the phosphorylation suffered by PRM1 prior to its assembly into the DNA. Consequently, the g197t SNP for PRM1 could derive into a change in the phosphorylation pattern of PRM1 altering both protamine-to-DNA and protamine-to-protamine interactions. Moreover, the change from an arginine to a serine in amino acid 34 also disrupts a beta-sheet structure changing protein conformation [33]. This SNP may have important consequences over chromatin compaction and has indeed been related with increased levels of DNA fragmentation and azoospermia [34].

Another SNP located in the codifying region of PRM1 that has been related with severe oligozoospermia is SNP g65a [28]. This SNP is located in exon 1 and leads to a substitution of a serine to an asparagine.

Polymorphisms in the 3′ Noncodifying Region

Among the different SNPs that have been identified within the 3′UTR region of the PRM1 gene, SNP g51c appears to be the most interesting one. This SNP alters the binding sites of two germ cell-specific binding factors from the Y-box gene family, MSY2 and MSY4 [29]. Both MSY2 and MSY4 are part of the ribonucleoprotein particles that bind to the 3′UTR region of the PRM1 mRNA in round spermatids to prevent mRNA translation until spermiogenesis begins. Thus, alterations in the binding sites of those factors may lead to a failure in PRM1 mRNA translational control.

PROTAMINE 2 GENE POLYMORPHISMS

In the case of protamine 2, 22 polymorphisms have been identified (see Table 6.2), but only three of them correlate with a nonsynonymous change in the amino acid sequence.

Polymorphisms in the Promoter Region

An important number of SNPs have been identified in the promoter region of the PRM2 gene; however, none of them has shown important effects over gene expression or over fertility [27]. Interestingly, one of those SNPs, c-67t, was thought to have important effects over the expression of the gene as it alters motive BS2, a common transcription factor-binding motive; none definitive conclusion can be derived since this mutation has only been described in one patient with azoospermia [28].

Polymorphisms in the Coding Region

Several SNPs have been identified in the codifying regions of the PRM2 gene, especially in exon 1 (five SNPs) and in the intron (eight SNPs) (Fig. 6.2). Among the SNPs located in the exons of the gene, two of them lead to an amino acid change, while one leads to a STOP

TABLE 6.2 Overview of PRM2 SNPs Identified in Several Studies

Polymorphism	Amino Acid	Region	Effect	Author
t-512g		5′ promoter		Jodar, 2010
g-392a		5′ promoter		
t-389c		5′ promoter		
g-371c		5′ promoter		
c-321t		5′ promoter		
g-226a		5′ promoter		
c-123g		5′ promoter		
c-67t		5′ promoter		Imken, 2009
t66c	22	Exon 1		Jodar, 2010
c87t	29	Exon 1		Imken, 2009
c248t	50	Exon 1	Gln → STOP	Tanaka, 2003
c153t	51	Exon 1	Ser → Phe	He, 2012
c201t	52	Exon 1		Jodar, 2010
c443a	94	Exon 2	Thr → Asn	Venkatesh, 2011
g398c		Intron		Tanaka, 2003
a473c		Intron		
g298c		Intron		Jodar, 2010
c281t		Intron		
c290t		Intron		
c373a		Intron		
c406t		Intron		
c278t		Intron		Aoki, 2006

To enable an easiest understanding of the information, the polymorphisms' nomenclature from the original papers has been maintained.

codon. None of them have been deeply studied as their incidence in population is extremely low [34–36].

Regarding the SNPs located in the intron, they are thought to lead to changes in the splicing sites. Again, no significant relation has been found between them and infertility.

Polymorphisms in the 3′ Noncodifying Region

To our knowledge, there are still no polymorphisms identified in the 3′ noncodifying region of the PRM2 gene.

SINGLE NUCLEOTIDE POLYMORPHISM GENOTYPING METHODS

A SNP is a position in the DNA in which two different bases can occur. Nowadays, really efficient methods are available for genetic mutation genotyping. This has led to an exponential increase in the knowledge about the genetic causes behind some of the main pathologies affecting high percentages of the population. However, it must be taken into account that most genetic differences or polymorphisms found are not related with any disease or health advantage. On the one hand, polymorphisms in coding regions may lead to nonsense mutations due to the redundancy of the genetic code. That means that a change in a nucleotide base does not necessary have to cause a change in the corresponding amino acid of the coded protein. Even if a polymorphism causes a change in the corresponding amino acid (missense mutation) or in the sites for transcription factor binding, the consequence may not be really advantageous or deleterious as most diseases display a polygenic background (see Table 6.3).

Moreover, it is also important to note that genetic mechanisms are so complex that an important number of the polymorphisms identified appear in extremely low percentages of the population. There are even polymorphisms that have been only described in one person. Consequently, for this type of polymorphisms, any possible study to correlate their presence with a disease is almost impossible as statistically significant results will not be achieved.

In general, a mutational association study usually covers two steps: In the first step, SNP genotyping is carried on the gene or genes of interest in order to try to find the existing SNPs for that gene in the population. In the second step, statistical analysis is made to find statistically significant associations between the described SNPs and the presence of a disease. This association study can be made either for individual SNPs or at a more complex level to find specific combinations of various SNP genotypes (haplotype) with the studied disease.

Although there are many different commercially available platforms for massive SNP genotyping, the mechanisms by which they work can be classified basically in four options: enzymatic cleavage-based, PCR-based, hybridization-based, or ligation-based protocols [37,38].

To our knowledge, there are still no commercially available platforms specifically designed for protamine genotyping. The SNPs present in the protamine's genes that we have compiled in the previous sections of this chapter have all been identified in experimental settings. As

TABLE 6.3 Types of SNPs With Analysis Interest

Type of SNP	Biological Effect	Relation With Disease	Analysis Utility
Missense SNP in coding region	Changes protein structure and/or function	Can cause monogenic disorder	Diagnosis
		Can alter drug metabolism	Pharmacogenetics
		Can contribute to common polygenic diseases	To assess risk for a disease
SNP in regulatory region	May modify transcription factor binding	Can influence risk of common disease	To assess risk for a disease
SNP in noncoding region	No known effect	No impact on phenotype	Populations and evolutionary genetics

the genes codifying for the protamines are quite small, the genotyping mechanism employed in most studies has been an amplification of the genes by regular PCR followed by sequenciation and comparison of the obtained sequences between participants in order to identify polymorphic differences. There exist however a few studies in which specific already-known SNPs have been studied using the enzymatic-based procedure restriction fragment length polymorphism (RFLP).

Enzymatic Cleavage-Based Techniques

- Restriction enzyme-based: the RFLP is the term used to name the collection of DNA fragments of different sizes that are obtained after digestion of the DNA with a restriction enzyme and that differ between individuals. The explanation for this situation is that a punctual change in the DNA sequence or polymorphisms can eliminate or create a cleavage site for a restriction enzyme. The RFLP can be applied into an easy method for genotyping specific SNPs. The method comprises three steps: (i) The target gene is amplified by PCR obtaining a not too long DNA fragment containing the site in which the SNP is located, (ii) the DNA fragment is digested with a restriction enzyme whose cleavage site meets the site of the SNP and is eliminated or created depending on the SNP variant, and (iii) the resulting fragments are separated in an electrophoresis gel and the band pattern interpreted (see Fig. 6.3A).
- Invasive cleavage approach: more complex technique in which you use two types of probes. One is an allele-specific probe that binds the DNA near the SNP position with a 5′ region noncomplementary to the DNA sequence (flap). The other is an upstream invader sequence-specific probe whose 3′ end binds exactly to the SNP position but is not complementary. If the allele-specific probe matches the SNP variant, the 3-D structure formed is recognized by an endonuclease that ligates both probes releasing the flap [39]. The releasing of the flap can be monitored by mass spectrometry or by fluorescence methods (see Fig. 6.3B).

PCR-Based Techniques

Two main approaches are possible when using PCR techniques for polymorphism genotyping:

- Sequencing approach: in this case, the target gene is amplified by PCR and afterward subjected to complete sequencing with one of the multiple available techniques (conventional Sanger sequencing, next-generation sequencing, pyrosequencing, mass spectrometry, etc.). Once the target sequence is obtained, you can you can identify the polymorphic site in it.
- Allele-specific PRC: in this option, you use specific forward PCR primers whose 3′ end matches up with the SNP position so that only if the allele is the one complementary to the primer, the amplification will take place. There are different ways to detect the formation of the product, from a simple electrophoretic gel to complex systems using target-specific probes that fluoresce only when the DNA polymerase cleaves them when the PCR reaction takes place (see Fig. 6.4A). Another option is to use a fluorescent dye that intercalates with the double-stranded PCR product.

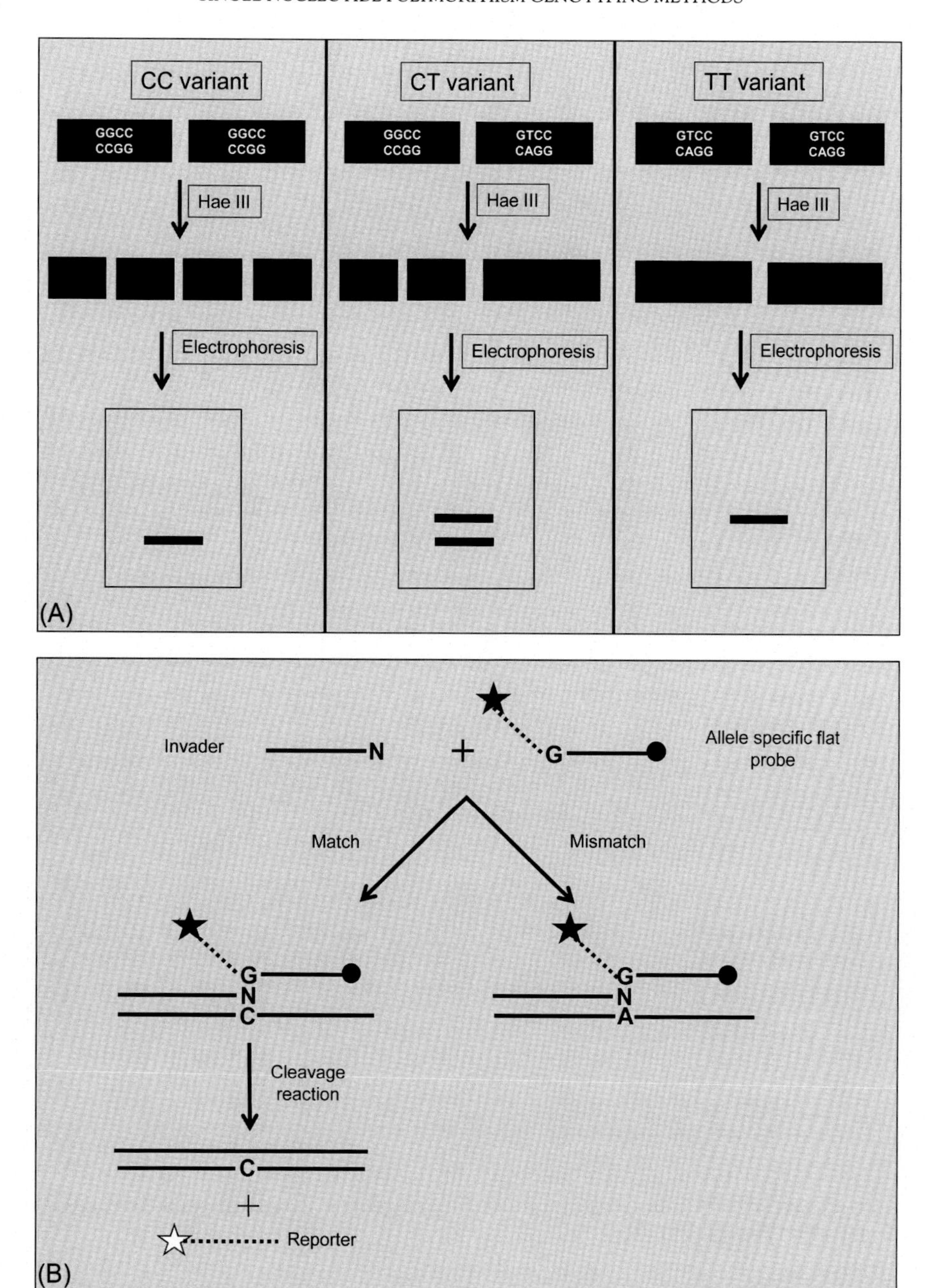

FIG. 6.3 Enzymatic cleavage genotyping techniques. (A) RFLP technique. (B) Invasive cleavage. Black star: inactivated fluorophore. White star: active fluorophore. Black circle: quencher.

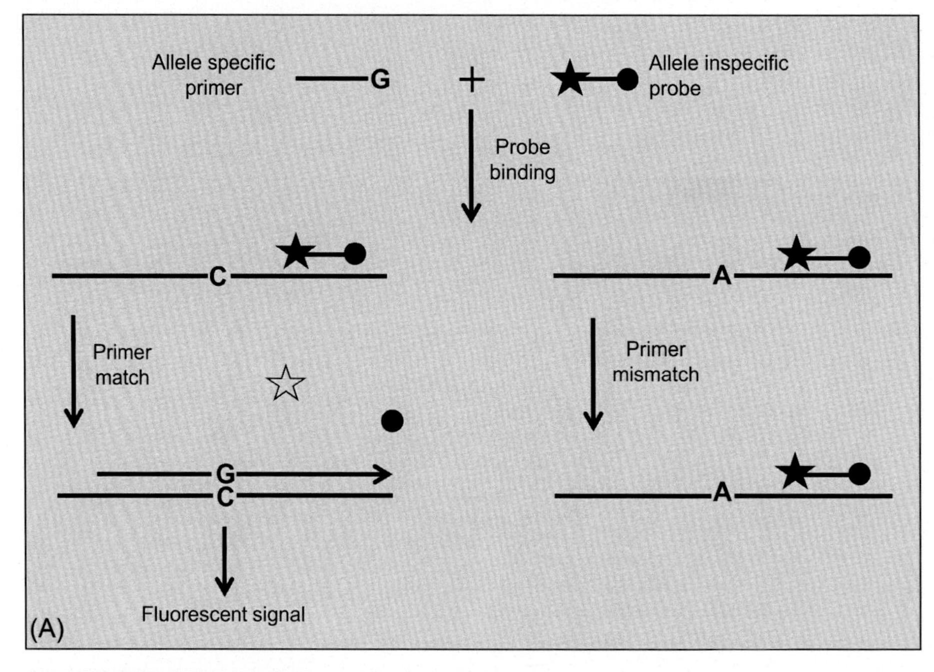

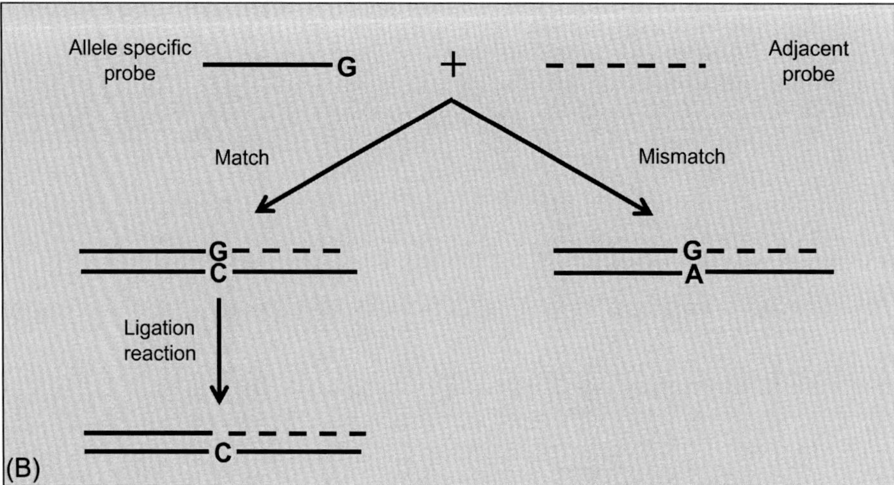

FIG. 6.4 (A) Allele-specific PCR. Black star: inactivated fluorophore. White star: active fluorophore. Black circle: quencher. (B) Ligation-based genotyping.

Hybridization-Based Techniques

For hybridization approaches, allele-specific probes must be designed so that they only hybridize with the target sequence when they match perfectly. There are several techniques available depending on the way that the hybridization event is reported [40]. Moreover, you can use allele-specific probes with different fluorophore labeling for even a quicker interpretation of the results.

- Single hybridization probes: a SNP can be detected using a single DNA probe specifically designed to hybridize only with one variant of the SNP. This is the methodology used in the most simple microarray approaches: a chip in which probes for both variants of the SNP are secured into a surface in separated positions. The target DNA is labeled, for example, with a fluorescent dye, and applied to the chip. After washing the nonhybridized DNA, you can determine the SNP variant analyzing the fluorescent emission (see Fig. 6.5A).
- Binary probes: in this approach, instead of using a single long probe, two small ones are used. Both probes are sequence-specific and bind close to each other to the DNA, one of them coinciding to the SNP position. Both probes are labeled with a part of a fluorophore so that only if both probes bind to the DNA, the fluorophore will be completed and the signal emitted (see Fig. 6.5B).

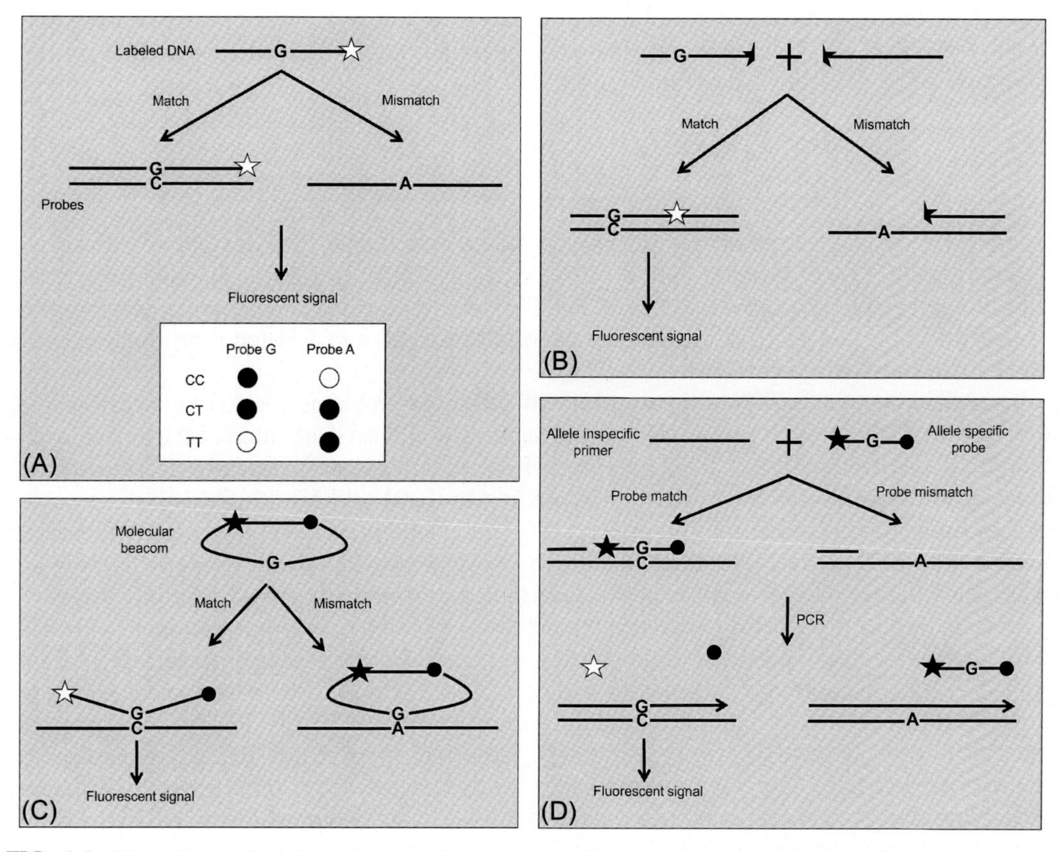

FIG. 6.5 Hybridization-based genotyping techniques. (A) Microarray chip with single hybridization probes. (B) Binary probes. (C) Molecular beacon genotyping. (D) Allele-specific probe-based RT-PCR. Black star: inactivated fluorophore. Half black star: divided inactivated fluorophore. White star: active fluorophore. Black circle: quencher.

- Molecular beacon genotyping: in this case, the probe used is a variant of the previous one that is closed into a loop, thus maintaining the fluorescent reporter and the quencher close together (see Fig. 6.5C). When the DNA sequence is complementary to the probe, it hybridizes to the DNA opening the loop, and the fluorescent signal appears [41].
- Allele-specific probe-based RT-PCR: the sample DNA is hybridized with an allele-specific probe that contains a reporter fluorescent dye in one edge and a quencher in the other. This probe will only bind to the DNA if the allele of the SNP is the one coincident with the probe sequence. In that case, when the RT-PCR is performed, the DNA polymerase will cleave the probe, freeing the dye from the quencher and thus obtaining a fluorescent signal (see Fig. 6.5D).

Ligation-Based Techniques

In ligation-based techniques, two probes are used that bind the adjacent one to the other. One of the probes has a 3′ or 5′ allele-specific nucleotide so that if the probe matches the SNP variant and binds perfectly to the target sequence, it is easily ligated to the other probe by a ligase. In contrast, the ligation reaction is inhibited if the probe does not match the SNP variant (see Fig. 6.4B). The ligation reaction can be monitored by different methods including colorimetric assays or the use of fluorophores [42,43].

CONCLUSIONS

Infertility is an extremely important problem nowadays in developed countries. Approximately, 15% of the population finds problems when trying to achieve a pregnancy. This situation added to the delay of the age for parenthood has caused an exponential increase during the last years in the number of couples that turn to assisted reproduction treatments in order to get offspring.

It is well known that in 50% of the couples with infertility problems, there is a male factor and, in 30% of them, the cause is exclusively a male factor. The clinical evaluation of the infertility situation in males includes the study of the clinical history, an hormonal test, a physical evaluation, and a regular sperm analysis performed according to the World Health Organization guidelines (volume, concentration, motility, morphology, and viability) [44]. However, in an increasing number of cases, that evaluation is insufficient to give an explanation to the cause of the situation of infertility in the male leaving it to be named as idiopathic or unexplained infertility [45,46].

During the recent years, the experts on the field have pointed to two new parameters in order to try to explain some of the cases of idiopathic infertility: the oxidative stress and the DNA damage [47,48]. One of the multiple situations that can cause sperm DNA damage is a defective protamination. As we have explained along the chapter, abnormal PRM1/PRM2 ratios have been repeatedly correlated with sperm DNA damage, abnormal sperm head shapes, and as a consequence male infertility. Also, some studies have already began to try to find a genetic explanation underlying abnormal protamination and have identified some polymorphisms within the genes of the protamines with interesting consequences over protamine expression or function.

According to the information currently available among the literature, it could be said that PRM1 is more important than PRM2. First, there are species that produce perfectly functional sperm without the use of PRM2. Second, even in humans, there are known cases of patients that not only produce sperm without PRM2 in its chromatin but also sperm that is able to achieve a fecundation with the help of an ICSI treatment, while there are none known cases of patients with sperm lacking PRM1. Also, among the SNPs that have already been described in infertile patients, the ones present in the PRM1 gene appear to be much more frequent than the ones present in PRM2. That could be explained because as the effect of a decrease in the quantity of PRM2 in the sperm DNA is less detrimental than the effect of a decrease in PRM1, there could be cases of males carrying polymorphisms in the gene of the PRM2 who do not need assisted reproduction procedures and are thus not included in the studies.

In concordance with the assumption that PRM1 is more important than PRM2, it has been observed that reduced PRM1/PRM2 ratios, which are normally due to a reduced expression of PRM1, are more detrimental for male infertility than increased ones, which are due to a reduced expression of PRM2. Moreover, it has also been demonstrated that patients showing abnormal PRM1/PRM2 ratios get better outcomes when they are derived to ICSI procedures than when they are derived to IVF procedures.

In conclusion, we believe that there are enough reasons to say that there are some polymorphism in the genes of the protamines that correlate with an abnormal PRM1/PRM2 ratio and thus with male infertility. More research should be done around this topic in order to get a better understanding of the situation that could lead to the design of a system to asses PRM1/PRM2 ratio and the putative genetic cause of the variations in that ratio so that same cases of male idiopathic infertility could be explained and thus treated more efficiently.

References

[1] F. Miescher, Das Protamin, eine neue organische Base aus den Samenfäden des Rheinlachses, Ber. Dtsch. Chem. Ges. 7 (1874) 376–379.

[2] A. Kossel, Ueber die basischen Stoffe des Zellkerns, Hoppe-Seyler's Z. Physiol. Chem. 22 (1897) 176–187.

[3] M. Dunn, The nitrogen distribution and the percentages of some amino acids in the protamine of the sardine, J. Biol. Chem. 70 (1926) 697–703.

[4] J.D. Lewis, Y. Song, M.E. de Jong, S.M. Bagha, J. Ausió, A walk though vertebrate and invertebrate protamines, Chromosoma 111 (2003) 473–482.

[5] R. Oliva, Protamines and male infertility, Hum. Reprod. Update 12 (2006) 417–435.

[6] J. Ausió, Histone H1 and evolution of sperm nuclear basic proteins, J. Biol. Chem. 274 (1999) 31115–31118.

[7] M. Jodar, R. Oliva, Protamine alterations in human spermatozoa, Adv. Exp. Med. Biol. 791 (2014) 83–102.

[8] R. Balhorn, The protamine family of sperm nuclear proteins, Genome Biol. 8 (2007) 227.

[9] J. Gosálvez, C. López-Fernández, J.L. Fernández, A. Gouraud, W.V. Holt, Relationships between the dynamics of iatrogenic DNA damage and genomic design in mammalian spermatozoa from eleven species, Mol. Reprod. Dev. 78 (2011) 951–961.

[10] E.I. Cortés-Gutiérrez, C. López-Fernández, J.L. Fernández, M.I. Dávila-Rodríguez, S.D. Johnston, J. Gosálvez, Interpreting sperm DNA damage in a diverse range of mammalian sperm by means of the two-tailed comet assay, Front. Genet. 5 (2014) 404.

[11] M. Corzett, J. Mazrimas, R. Balhorn, Protamine 1: protamine 2 stoichiometry in the sperm of eutherian mammals, Mol. Reprod. Dev. 61 (2002) 519–527.

[12] N. Hetch, Mammalian protamines and their expression, in: H. Hinilica (Ed.), Histones and Other Basic Nuclear Proteins, CRC Press, 1989, pp. 347–373.

[13] R. Oliva, G.H. Dixon, Vertebrate protamine genes and the histone-to-protamine replacement reaction, Prog. Nucleic Acid Res. Mol. Biol. 40 (1991) 25–94.

[14] V.W. Aoki, D.T. Carrell, Human protamines and the developing spermatid: Their structure, function, expression and relationship with male infertility, Asian J. Androl. 5 (2003) 315–324.

[15] R.P. Martins, G.C. Ostermeier, S.A. Krawetz, Nuclear matrix interactions at the human protamine domain, J. Biol. Chem. 279 (2004) 51862.

[16] K. Steger, L. Fink, T. Klonisch, R.M. Bohle, M. Bergmann, Protamine-1 and -2 mRNA in round spermatids is associated with RNA-binding proteins, Histochem. Cell Biol. 117 (2002) 227–234.
[17] V.W. Aoki, L. Liu, K.P. Jones, H.H. Hatasaka, M. Gibson, C.M. Peterson, et al., Sperm protamine 1/protamine 2 ratios are related to in vitro fertilization pregnancy rates and predictive of fertilization ability, Fertil. Steril. 86 (2006) 1408–1415.
[18] V.W. Aoki, S.I. Moskovtsev, J. Willis, L. Liu, J.B.M. Mullen, D.T. Carrell, DNA integrity is compromised in protamine-deficient human sperm, J. Androl. 26 (2005) 741–748.
[19] V.W. Aoki, B.R. Emery, L. Liu, D.T. Carrell, Protamine levels vary between individual sperm cells of infertile human males and correlate with viability and DNA integrity, J. Androl. 27 (2006) 890–898.
[20] L. Simon, J. Castillo, R. Oliva, S.E.M. Lewis, Relationships between human sperm protamines, DNA damage and assisted reproduction outcomes, Rep. BioMed. 23 (2011) 724–734.
[21] F.G. Iranpour, The effects of protamine deficiency on ultrastructure of human sperm nucleus, Adv. Biomed. Res. 3 (2014) 24.
[22] D.T. Carrell, B.R. Emery, S. Hammoud, Altered protamine expression and diminished spermatogenesis: what is the link? Hum. Reprod. Update 13 (2007) 313–327.
[23] D.T. Carrell, L.H. Liu, Altered protamine 2 expression is uncommon in donors of known fertility, but common among men with poor fertilizing capacity, and may reflect other abnormalities of spermiogenesis, J. Androl. 22 (2001) 604–610.
[24] S. de Mateo, C. Gazquez, M. Guimera, J. Balasch, M.L. Meistrich, J. Luis Ballesca, et al., Protamine 2 precursors (pre-P2), protamine 1 to protamine 2 ratio (P1/P2), and assisted reproduction outcome, Fertil. Steril. 91 (2009) 715–722.
[25] D.T. Carrell, B.R. Emery, S. Hammoud, The aetiology of sperm protamine abnormalities and their potential impact on the sperm epigenome, Int. J. Androl. 31 (2008) 537–545.
[26] K. Lee, H.S. Haugen, C.H. Clegg, R.E. Braun, Premature translation of protamine 1 mRNA causes precocious nuclear condensation and arrests spermatid differentiation in mice, Proc. Natl. Acad. Sci. U. S. A. 92 (1995) 12451–12455.
[27] J.M. Jodar, G. Oriola, J. Mestre, A. Castillo, J.M. Giwercman, Vidal-Taboada, et al., Polymorphisms, haplotypes and mutations in the protamine 1 and 2 genes, Int. J. Androl. 34 (2011) 470–485.
[28] L. Imken, H. Rouba, B. El Houate, N. Louanjli, A. Barakat, A. Chafik, et al., Mutations in the protamine locus: association with spermatogenic failure? Mol. Hum. Reprod. 15 (2009) 733–738.
[29] C. Ravel, S. Chantot-Bastaraud, B. El Houate, I. Berthaut, L. Verstraete, V. De Larouziere, et al., Mutations in the protamine 1 gene associated with male infertility, Mol. Hum. Reprod. 13 (2007) 461–464.
[30] C. Gazquez, J. Oriola, S. de Mateo, J.M. Vidal-Taboada, J. Luis Ballesca, R. Oliva, A common protamine 1 promoter polymorphism (−190 c → A) correlates with abnormal sperm morphology and increased protamine P1/P2 ratio in infertile patients, J. Androl. 29 (2008) 540–548.
[31] S. Jamali, M. Karimian, H. Nikzad, Y. Aftabi, The c.−190 C>A transversion in promoter region of protamine1 gene as a genetic risk factor for idiopathic oligozoospermia, Mol. Biol. Rep. 43 (2016) 795–802.
[32] S.E. Kichine, A.D. Msaidie, A. Bokilo, A. Ducourneau, N. Navarro, Levy, et al., Low-frequency protamine 1 gene transversions c.102G→T and c. −107G→C do not correlate with male infertility, J. Med. Genet. 45 (2008) 255–256.
[33] N. Iguchi, S. Yang, D.J. Lamb, N.B. Hecht, An SNP in protamine 1: a possible genetic cause of male infertility? J. Med. Genet. 43 (2006) 382–384.
[34] X. He, J. Ruan, W. Du, G. Chen, Y. Zhou, S. Xu, et al., PRM1 variant rs35576928 (Arg > Ser) is associated with defective spermatogenesis in the Chinese Han population, Rep. BioMed. 25 (2012) 627–634.
[35] H. Tanaka, Y. Miyagawa, A. Tsujimura, K. Matsumiya, A. Okuyama, Y. Nishimune, Single nucleotide polymorphisms in the protamine-1 and-2 genes of fertile and infertile human male populations, Mol. Hum. Reprod. 9 (2003) 69–73.
[36] S. Venkatesh, R. Kumar, D. Deka, M. Deecaraman, R. Dada, Analysis of sperm nuclear protein gene polymorphisms and DNA integrity in infertile men, Syst. Biol. Reprod. Med. 57 (2011) 124–132.
[37] P. Kwok, Methods for genotyping single nucleotide polymorphisms, Annu. Rev. Genom. Hum. Gen. 2 (2001) 235–258.
[38] A. Syvänen, Accessing genetic variation: genotyping single nucleotide polymorphisms, Nat. Rev. Gen. 2 (2001) 930–942.

[39] M. Olivier, The Invader® assay for SNP genotyping, Mutat. Res.—Fundam. Mol. Mech. Mutagen. 573 (2005) 103–110.
[40] K. Knez, D. Spasic, K.P.F. Janssen, J. Lammertyn, Emerging technologies for hybridization based single nucleotide polymorphism detection, Analyst 139 (2014) 353–370.
[41] F.R. Kramer, S. Tyagi, D.P. Bratu, Multicolor molecular beacons for allele discrimination, Nat. Biotechnol. 16 (1998) 49–53.
[42] M. Samiotaki, M. Samiotaki, M. Kwiatkowski, M. Kwiatkowski, J. Parik, J. Parik, et al., Dual-color detection of DNA sequence variants by ligase-mediated analysis, Genomics 20 (1994) 238–242.
[43] H. Wang, W. Liu, Z. Wu, L. Tang, X. Xu, R. Yu, et al., Homogeneous label-free genotyping of single nucleotide polymorphism using ligation-mediated strand displacement amplification with DNAzyme-based chemiluminescence detection, Anal. Chem. 83 (2011) 1883–1889.
[44] World Health Organization, WHO Laboratory Manual for the Examination and Processing of Human Semen, fifth ed., 2010.
[45] C. De Jonge, Semen analysis: looking for an upgrade in class, Fertil. Steril. 97 (2012) 260–266.
[46] S.E.M. Lewis, Is sperm evaluation useful in predicting human fertility? Reproduction 134 (2007) 31–40.
[47] A. Agarwal, K. Makker, R. Sharma, Clinical relevance of oxidative stress in male factor infertility: an update, Am. J. Reprod. Immunol. 59 (2008) 2–11.
[48] L. Simon, G. Brunborg, M. Stevenson, D. Lutton, J. McManus, S.E.M. Lewis, Clinical significance of sperm DNA damage in assisted reproduction outcome, Hum. Reprod. 25 (2010) 1594–1608.

CHAPTER

7

Small RNAs Present in Semen and Their Role in Reproduction

Meritxell Jodar, Ester Anton†*

*Molecular Biology of Reproduction and Development Research Group, August Pi i Sunyer Biomedical Research Institute (IDIBAPS), Clínic Foundation for Biomedical Research, University of Barcelona, Barcelona, Spain †Genetics of Male Fertility Group, Cell Biology Unit, Autonomous University of Barcelona, Cerdanyola del Vallès, Spain

INTRODUCTION

Small noncoding RNAs (sncRNAs) comprise a family of noncoding RNAs with sizes ranging ~21–34 nucleotides (nt) that play a critical role in the regulation of gene expression [1]. Several types of sncRNAs have been described, being microRNAs (miRNAs), endogenous small interfering RNAs (endo-siRNAs), and Piwi-interacting RNAs (piRNAs) the most extensively studied. Each subtype of sncRNAs differs in biogenesis, length, and mechanisms to accomplish their biological function (Fig. 7.1) [2].

The canonical biogenesis of miRNAs starts with the transcription of longer primary transcripts (3–4 kb in length) encoding multiple miRNAs (named pri-miRNAs) [3]. These pri-miRNAs are processed in the nucleus in shorter miRNA precursors (pre-miRNA; 41–180 nt) by the RNase III DROSHA and a double-stranded RNA-binding protein, DiGeorge syndrome critical region 8 (DGCR8) [4]. Once these pre-miRNAs are exported to the cytosol, they are processed again by another type of RNase III enzyme named DICER, resulting in a double-stranded processed RNA. Subsequently, the two semicomplementary sequences can be unwounded producing two mature single-stranded miRNAs of 16–27 nt [4]. Mature miRNAs could then associate with several members of the Argonaute (AGO) family proteins to form an effector complex named RNA-induced silencing complex (RISC). Mature miRNAs act as a guide by base pairing with its target mRNAs, and AGO proteins are the effectors inducing translational repression, mRNA deadenylation, and mRNA decay [5]. Recent evidences also show that the complexes miRNA-RISC are found not only in the cytoplasm but also in the nucleus, although miRNA-RISC assembly still occurs outside the nucleus [6]. Similarly, miRNA-RISC complex imported into the nucleus has the ability to regulate

https://doi.org/10.1016/B978-0-12-812571-7.00008-3

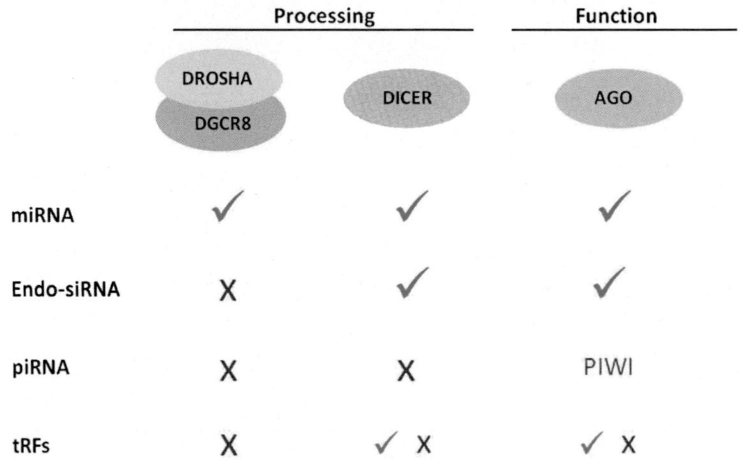

	Processing		Function
	DROSHA DGCR8	DICER	AGO
miRNA	✓	✓	✓
Endo-siRNA	X	✓	✓
piRNA	X	X	PIWI
tRFs	X	✓ X	✓ X

FIG. 7.1 Proteins involved in the biogenesis and function of sncRNAs. The canonical biogenesis of miRNAs is dependent of DROSHA, DGCR8, and DICER. In contrast, the biogenesis of endo-siRNAs is independent of DROSHA and DGCR8 but DICER-dependent. Although DROSHA, DGCR8, and DICER usually do not participate in the biogenesis of piRNAs and tRFs, some tRFs seem to be processed in a DICER-dependent manner. miRNA, endo-siRNA, and some tRFs associate with AGO proteins to form an effector complex able to regulate gene expression at transcriptional and posttranscriptional level. Specifically, piRNA associate with germ line-specific subfamily members of the AGO protein family named Piwi to accomplish their functions.

posttranscriptionally small and long noncoding RNAs [7–9]. But most interestingly, it seems that nuclear miRNA-RISC complex could also modulate gene expression activating [10,11] or repressing [12,13] the gene transcription through a mechanism that is not yet well elucidated.

Likewise to miRNAs, endo-siRNAs are also derived from double-stranded RNA precursors, and their biogenesis is dependent of DICER but DROSHA/DGCR8 independent [14]. Mature endo-siRNAs (~21 nt) form not only complexes with AGO proteins able to target mRNAs but also specific genomic regions, therefore regulating gene expression at both posttranscriptional and transcriptional level, respectively [14,15].

Another major class of sncRNAs is the piRNAs. These molecules have been well studied in the germ line being indispensable for successful spermatogenesis [16]. piRNAs have some singular features including a larger size (24–34 nt) and a strong bias for the 5′ uridine residue [17,18]. In contrast to miRNA and endo-siRNAs, piRNAs are mostly transcribed from genomic piRNA clusters ranging from 1 to 127 kb and with a density of 40–4000 piRNAs per cluster. Moreover, these piRNA clusters are processed in an RNase III-type enzyme-independent manner from single-stranded precursors, although the mechanism is not fully understood [19]. Mature piRNAs in the cytoplasm are associated with germ line-specific subfamily members of the AGO protein family named Piwi [20,21]. Not only the most well-characterized function of Piwi-piRNA complex is to suppress the activities of transposable elements [19,22–24], but also other functions have been ascribed such as regulating chromatin architecture to control genomic stability [25] and modulating the stability and translation of mRNAs [26,27].

Another class of sncRNAs called tRNA-derived fragments (tRFs) has also gained attention recently owing to play important biological functions such as in cell proliferation [28].

tRFs (16–40 nt) do not seem to be just products from random cleavage of mature tRNA; instead, they are derived from a precise processing at the 5′ or 3′ tRNA ends and are highly conserved among both prokaryotes and eukaryotes. Although the knowledge about their biogenesis, roles, and function remains limited, some tRFs are processed in a DICER-dependent manner and could associate with different types of AGO proteins suggesting that some tRFs could have similar functions to endo-siRNAs and miRNAs [28].

In relationship to the importance of sncRNAs in the sperm formation and in male fertility, detailed below, there is a revision of the published data regarding their crucial roles in spermatogenesis and sperm maturation, their content in sperm and seminal plasma from fertile and infertile males, and their potential role as fertility biomarkers. Moreover, the most significant findings related to the roles of sperm sncRNA in early embryogenesis and carriers of epigenetic information related to intergenerational epigenetic inheritance of acquired traits are described.

ROLE OF sncRNA DURING SPERMATOGENESIS AND SPERM MATURATION

Spermatogenesis and sncRNAs

Spermatogenesis is a complex developmental process that results in the highly differentiated sperm cell. Spermatogenesis could be divided in three major stages: (i) proliferation and maintenance of the spermatogonial population, (ii) reduction of genetic information from diploid to haploid state by meiosis process, and (iii) differentiation and maturation of the haploid sperm cell. All these stages are highly regulated at both transcriptional and posttranscriptional level resulting in an accurate gene expression pattern depending on both space and time [29]. The posttranscriptional gene regulation is essential during the late stages of the spermatogenesis when the transcription is blocked due to the replacement of the sperm histones with the more basic protamines [30–32]. This accurate regulation allows that proteins transcribed during meiotic stages but required in the last steps could be stored and translationally regulated until needed [30]. The ability of sncRNAs in modulating gene expression [33] and their high and precise expression during spermatogenesis suggest a pivotal role of these molecules in the specific spermatogenesis gene expression pattern. This idea is reinforced by the severe spermatogenic disruptions and fertility problems observed in testicular conditional knockouts of genes involved in the biogenesis and function of sncRNAs such as DROSHA, DICER, DGCR8, MILI, and MIWI [34–41] (Table 7.1). The overall consistency of phenotypes among all the conditional knockouts mentioned above indicates that the absence of sncRNAs rather than the loss of those specific proteins is the real cause of the spermatogenic defects observed. It is interesting to note that the conditional inactivation of DICER results in more severe spermatogenic disruptions than those ones observed in DGCR8 conditional knockouts, suggesting that endo-siRNAs are also important during spermatogenesis [36,37]. However, discrepant results were found comparing the testicular phenotype of conditional ablation of DICER and DROSHA in postnatal germ cells, since more severe testicular defects were observed in the absence of DROSHA [34]. These discrepancies in the phenotypic results may be due to technical issues or may indicate some unknown biogenesis pathways

TABLE 7.1 Summary of Studies Assessing Testicular Phenotypes From Conditional Knockouts of Genes Involved in the Biogenesis or Function of sncRNAs

Knockout Gene	Conditional Cell	Phenotype	References
Related to miRNAs			
DROSHA	Spermatogonia at postnatal day 3	Disrupted spermatogenesis. Depletion of spermatocytes and early spermatids. Azoospermia or oligoasthenozoospermia	[34,42]
DGCR8	Spermatogonia at embryonic day 18	Cumulative defects in meiotic and haploid phases of spermatogenesis, resulting in oligo-, terato-, and azoospermia	[36]
	Spermatogonia at embryonic day 8	Spermatogenic arrest, very reduced number of sperm counts with abnormal morphology, infertility	[37]
DICER	Spermatogonia at embryonic day 18	Defects in meiosis and spermiogenesis, the absence of spermatozoa, and infertility	[35]
	Spermatogonia at postnatal day 5	Spermatogonial differentiation unaffected, spermatogenesis disrupted specially during late haploid differentiation, very low number of sperm with pronounced morphological abnormalities	[38]
	Spermatogonia at postnatal day 3	Disrupted spermatogenesis. Depletion of spermatocytes and early spermatids. Azoospermia or oligoasthenozoospermia	[34,42]
	Spermatogonia at postnatal day 18	Disruption of meiotic progression and male infertility	[42a]
	Spermatogonia at embryonic day 8	Spermatogenic arrest, very reduced number of sperm counts with abnormal morphology, and infertility	[37]
Related to piRNAs			
MIWI	Global	Spermatogenic arrest at the beginning of the round spermatid stage	[39]
MIWI2	Spermatogonia at embryonic day 15	Early meiotic arrest, infertility	[40]
	Pachytene spermatocytes	No effects on spermatogenesis and other testicular functions, fertility	
	Spermatids	No effects on spermatogenesis and other testicular functions, fertility	
MILI	Postnatal spermatogonia	Spermatogenic arrest at various stages from meiosis to elongating spermatids	[41]

of sncRNAs that need to be addressed. Regarding the proteins involved in piRNAs function, the inactivation of MIWI2 in postnatal germ cells has no effects on spermatogenesis or other testicular functions indicating that MIWI2 is important during male primordial germ cell reprogramming but is not necessary for developing postnatal male germ line [40].

Several specific miRNAs have been described as crucial in different steps of the spermatogenesis such as (i) miR-20, miR-21, and miR-106a in the renewal of spermatogonial stem cells

[43,44]; (ii) miR-146 modulating spermatogonial differentiation [45]; (iii) miR-29 family as global regulators of the meiosis [46]; and (iv) miR-17-92 cluster, miR-34b/c, miR-449, and miR-210 as critical players of spermatogenesis [47–49].

Sperm Maturation and sncRNAs

After successive stages of differentiation in the testes, the resultant spermatozoon is still an immature and immotile germ cell lacking the ability to fertilize the oocyte on its own. Testicular sperm are stored in the epididymis until the ejaculation takes place, when sperm will travel from the epididymis to outside the body. In the epididymis and all along the following genital male tract, the testicular sperm come into contact with secretions from the different accessory sex glands, which contain a high number (10^{11}–10^{13}) of extracellular vesicles (EV) rich in lipids, proteins, and RNAs [50]. Growing evidences in other fields suggest the participation of EV in the intercellular communication through the selective incorporation of their cargo including proteins, RNA, and lipids to the target cell. In the same way, the interaction of seminal EV with the sperm membrane results in changes of the sperm surface and most likely in the incorporation of new proteins and RNAs that contribute to the acquisition of the potential for sperm motility and the ability to fertilize the oocyte [51]. Integrative analysis of proteins and RNAs associated to ejaculated sperm and seminal plasma has allowed inferring the transcriptional origin of the sperm proteins showing that around 9% of them seem to have an extratesticular origin [52–54]. The relevance of the sncRNAs during sperm maturation has been elucidated by the altered phenotype observed in the specific inactivation of DICER in the epididymal cells [55]. Besides the changes observed in the lipid homeostasis of epididymal cells that directly affect the lipid composition of sperm membrane and male fertility, the analysis of the sperm miRNA profile at different sections of the epididymis has revealed the loss and acquisition of 113 and 115 miRNAs during epididymal maturation, respectively [56,57]. Recent findings also show that during epididymal maturation, sperm gain other types of sncRNAs such as tRFs through epididymosomes (EV from epididymis), which seem to be involved in the epigenetic inheritance of the paternal low-protein-diet phenotype [58].

SEMEN sncRNAs AND MALE FERTILITY

The study of the sncRNA content variation in sperm/seminal plasma from individuals with different fertility impairments could provide insights into the basic molecular mechanisms that regulate spermatogenesis. In addition, this information could help to understand which pathways are affected in some physiopathologic mechanisms involved in male infertility. The accessibility in the obtainment of semen samples and the improvement of high-throughput technologies to assess the abundance of small amounts of sncRNAs sets the sncRNA semen profile up as a good source of fertility biomarkers that could eventually be used in clinics.

Sperm sncRNAs and Male Fertility

Ejaculated sperm contain a broad variety of sncRNA molecules (approximately 0.3 fg) [59,60], which, in order of abundance, they include repetitive elements (including various members of the LTR, SINE, and LINE families), piRNAs, transcription start sites

(TSS)/promoter-associated RNAs, and miRNAs [61]. Other less frequent categories include small nuclear RNAs (snRNAs), small nucleolar RNAs (snoRNAs), tRFs, Y RNAs (components of the Ro60 ribonucleoprotein particle), transcripts derived from unannotated regions of the genome, and portions of coding and noncoding transcripts [60,61].

Among them, miRNAs are the best characterized and mostly studied sncRNAs in spermatozoa. Some studies have allowed the identification of the sperm miRNA profiles in fertile individuals using either microarrays [62], qRT-PCR [63], or RNA-seq [61,64]. These studies have provided a description of the most frequent sperm miRNAs in samples of fertile individuals and a list of the miRNAs that must be constantly present in all of them. Interestingly, those miRNAs show a preference to target genes involved in spermatogenesis and embryogenesis.

On the search for new fertility biomarkers, several authors have focused their studies in analyzing the sperm miRNA content of infertile individuals. This branch of studies has been promoted by the need of finding other fertility indicators beyond the classical seminogram parameters [65–67]. In this sense, some sperm miRNAs have shown a differential and specific expression in infertile individuals. Abu-Halima et al. described a pool of miRNAs down- and upregulated in asthenozoospermic and oligoasthenozoospermic individuals [68] and proposed a panel of five miRNAs as indicators of these spermatogenic impairments (miR-34b*, miR-34b, miR-34c-5p, miR-429, and miR-122) [69]. Other single-sperm miRNAs have been associated to fertility problems by some studies: an altered expression of miR-27b has been associated to asthenozoospermia [70], while altered levels of miR-15a have been detected in varicocele patients [71]. Moreover, particular miRNA expression profiles have been described in infertile individuals with a single seminal parameter altered (i.e., teratozoospermia, oligozoospermia, or asthenozoospermia) [72]. Interestingly, the pool of differentially expressed miRNAs in each one of these groups was associated to an enrichment of genes ontologically involved with the particular seminal alterations present in the populations analyzed [73]. A similar study performed in infertile individuals with a normal seminogram revealed sperm miRNA profiles clearly differentiated from the fertile normozoospermic individuals [74]. In this case, the differentially expressed miRNAs displayed an association with targets involved in processes related to embryogenesis. The authors hypothesize that the disturbance of these targets could be related with the incapacity of these individuals to father a newborn.

On the other hand, piRNAs have been observed to be more abundant than miRNAs in sperm from fertile individuals [61,64]. Nevertheless, no studies describing their expression variations in infertile populations have been published so far.

Seminal Plasma sncRNAs and Male Fertility

Whereas several studies have attempted to identify sperm molecular clinical biomarkers, only few studies have targeted the seminal plasma. In fact, spermatozoa only account for the 5% of the ejaculate, while the remaining 95% corresponds to secretions from the different male accessory sex glands including the epididymis, prostate, and seminal vesicles. Remarkably, seminal plasma seems to play a much greater role than simply being a medium to carry the spermatozoa through the female reproductive tract, but it seems to be crucial for sperm function, fertilization, early embryogenesis, and even modulation of the offspring phenotype [75].

The sncRNA profile of seminal plasma of healthy fertile donors has been assessed by RNA-seq revealing a heterogeneous sncRNA population, including, in order of abundance, transcripts derived from unannotated regions of the genome (31%), ribosomal RNAs (rRNAs) (16.8%), tRNAs (15.52%), repeated elements (13.9%), miRNAs (13.8%), intron and exon expressed regions (6.5% and 1.5%, respectively), and piRNAs (0.9%), among other less known types such as snRNA, snoRNAs, signal recognition particle RNA (srpRNAs), and small conditional RNAs (scRNAs) [76]. The comparisons of seminal plasma sncRNA signatures between healthy fertile donors and successfully vasectomized men that do not include any secretion from the testis or epididymis have allowed discerning those miRNAs and piRNAs originated from testis and epididymis [77]. The sncRNAs present in the seminal plasma secreted from the testis and epididymis could be suitable biomarkers to assess alterations in spermatogenesis and epididymal sperm maturation. One of these testes and epididymal-specific miRNAs is the miR-34c-5p, known to have a crucial role in late stages of spermatogenesis [78]. miR-34c-5p has been found significantly decreased in the seminal plasma of patients with nonobstructive azoospermia [79] and varicocele-associated oligoasthenoteratozoospermia [80] but significantly increased in asthenozoospermic patients [79]. These data reflect the crucial role of miR-34c-5p in late steps of spermatogenesis and sperm maturation. Other sncRNAs from seminal plasma whose testicular or epididymal origin could not be assessed have been found in altered abundance mainly in patients with nonobstructive azoospermia. There is a particular interest to explore whether some specific sncRNAs from seminal plasma could be predictive for the presence of sperm in the testis in patients with azoospermia. For instance, increased levels of miR-19b, let-7a, miR-141, miR-429, and miR-7-1-3p and markedly decreased levels of miR-122, miR-146b-5p, miR-181a, miR-374b, miR-509-5p, and miR-513a-5p in those types of patients have been observed [42,81]. Similarly, five piRNAs were found significantly decreased in seminal plasma from nonobstructive azoospermic patients. For now, it remains to be established if some of these sncRNAs in seminal plasma could predict the presence of sperm in the testis from azoospermic patients.

Unfortunately, the analysis and interpretation of RNA studies from human seminal plasma are complex since it contains a mixture of secretions and EV originating from different tissues. In addition, seminal plasma also contains remnants of sperm cells that underwent apoptosis and cytoplasmic droplets released during semen sample processing [82], which could mask the functional analysis. In order to avoid the contamination from sperm leftovers, several groups have started focusing into the study of the seminal EV content. The majority of seminal EV are secreted by epithelial cells from the epididymis and the accessory sex glands, such as the seminal vesicles and prostate (the prostate being the major contributor) [51]. The sncRNA profile of seminal EV has been evaluated by RNA-seq revealing their composition: 21.7% of miRNA, 20% Y RNAs, 18.5% ribosomal RNA, 16% of tRNAs, 9% of protein-coding-derived RNAs, and 2.7% of piRNAs, among other less abundant types [50]. To the best of our knowledge, just one study has compared the miRNA content of seminal EV between normozoospermic fertile men and oligoasthenozoospermic subfertile patients [83]. The aforementioned study found an alteration in the expression levels of 36 miRNAs in the infertile patients by using a microarray strategy.

Moreover, it is interesting to mention that seminal plasma and seminal EV are emerging as a clinically relevant "liquid biopsy" with a high potential as a source of specific biomarkers for prostate cancer [84].

ROLE OF PATERNAL sncRNAs IN EMBRYOGENESIS AND EPIGENETIC INHERITANCE OF ACQUIRED TRAITS

Viable and fertile mouse parthenotes created in vitro by introducing one female pronucleus modified with specific paternal imprint marks into mature mouse eggs [85] suggested that the only paternal element required for embryogenesis is the differential methylation pattern of sperm genome. However, the very low birth rates from parthenotes and the recent findings presented below suggest that sperm-borne RNAs could have crucial roles in the confrontation and consolidation of paternal and maternal genomes, early embryogenesis, and modulation of the offspring phenotype [60,86,87].

Paternal sncRNAs and Embryogenesis

Recently, the importance of sperm-borne sncRNAs in embryogenesis have been demonstrated by the injection into wild-type oocytes of sperm with altered miRNA and endo-siRNA profiles rescued from DICER and DROSHA testicular conditional knockouts [88]. Sperm with partially altered sncRNA profiles (about 45% and 52% of miRNAs were deregulated in DICER and DROSHA conditional knockouts, respectively) could fertilize the oocyte by intracytoplasmic sperm injection (ICSI) technologies, but those embryos displayed significantly reduced developmental potential at early stages. Interestingly, better ratios of fertilization and development of zygote up to two-pronucleus stage but not in later stages were observed in embryos derived from sperm rescued from DROSHA conditional knockout. However, these deleterious effects in fertilization and early embryo development observed could be restored to the same levels of success than those observed in wild-type mice by the injection of wild-type sperm-borne RNAs. These data reflect that the tiny amounts of sperm sncRNAs delivered to the oocyte are essential for very early stages of embryogenesis before implantation. Li et al. reported that one of the most abundant miRNAs in human spermatozoa, the mir-34c, is essential for the first cleavage of mouse zygote based on the injection of miR-34c inhibitor into zygotes [89]. Moreover, this claim was supported by the correlation found between miR-34c levels in spermatozoa and ICSI outcomes [90]. But now, these findings are controversial since mir-34c knockout male mice are fertile [49,91].

Not only the sperm have a role in early embryogenesis, but also the contact of seminal plasma with female tract during intercourse may help a woman's immune system to prepare for conception and pregnancy. For instance, the excision of the seminal vesicle glands in mice, which are the major fluid contributors to the seminal plasma (~70% of volume), results in low pregnancy rates, placental hypertrophy, and metabolic alterations in the progeny [92]. New hypothesis suggests that sncRNAs present in semen (whether in sperm, encapsulated in seminal EV, or cell-free) are delivered to the female reproductive tract and modulate the immune response in woman promoting maternal tolerance to the embryo [93].

Paternal sncRNAs and Epigenetic Inheritance of Acquired Traits

Life-history experiences from male partner have a greater influence on the future health of the offspring than it was previously thought. Different studies have shown that paternal lifestyle including paternal age [94], diet [95], and toxic habits such as smoking and alcohol consumption [96,97] may end up conditioning their future offspring. Several examples of harmful phenotypes related to paternal lifestyle including diet, stress, and toxicant exposures have been observed in rodents. Examples include altered metabolism of rodent offspring due to paternal diet (calorie-restricted, low-protein, or high-fat diet) [98–102], observation of a decreased fear response and the presence of depressive symptoms in offspring of traumatized male mice [103–105], and cognitive impairment and higher anxiety and aggression levels in offspring from males with postnatal exposures to nicotine and heroin, respectively [106,107]. In some cases, these harmful phenotypes could be observed not only in the offspring but also in further generations, known as transgenerational inheritance. While these examples described above were associated to rather undesirable effects, some authors also suggested that transgenerational inheritance could reconcile the adaptation of species to new environments that parents were exposed. This is exemplified in the transmission of a traumatic olfactory experience to the offspring, which could act as a defense strategy in front of potential depredators suggesting a potential mechanism for innate behaviors in animals [108]. Few human data, mainly based on epidemiological studies, also suggest the existence of transgenerational effects due to paternal life-history experiences. These studies suggest that male overfeeding during slow growth period before pubertal peak (9–11 years) has a deleterious effect on the survival of sons and grandsons (only by paternal line) increasing their risk to suffer from diabetes [109,110]. Growing evidence points to sperm sncRNAs as causing phenotypic variations in the progeny reflective of the father's life experience (Table 7.2). The role of noncoding RNAs in the transgenerational epigenetic inheritance was demonstrated for the first time in mammals with phenotypic transgenerational effects due to ancestral genetic variants. This was exemplified by the presence of the same altered phenotype in both animals with heterozygote disruption of mouse Kit (mast/stem cell growth factor receptor Kit) gene and their offspring with normal genotype (paramutated animal) [113]. It was also observed that sperm from heterozygote and paramutated progeny were enriched with truncated Kit RNAs. Interestingly, microinjection into fertilized eggs of sperm RNAs from heterozygote mice or miRNAs that target Kit (miR-221 and miR-222) induced the heterozygote phenotype [113]. Similarly, the injection in normal zygotes of the total sperm RNAs or a subset of sperm RNAs (i.e., miRNA or tRFs from males exposed to mental stress or high-fat diet) could recapitulate the paternal phenotype [111,114,115]. Interestingly, it was observed that obese males restore the sperm sncRNA profile and achieve offspring with normal metabolic phenotypes after the intervention through diet and exercise program during two complete spermatogenic cycles [112,116].

Additionally, there are plenty of evidences showing sperm gain of specific noncoding RNAs in the epididymis through extracellular vesicles that are involved in the epigenetic transgenerational inheritance of paternal low-protein-diet phenotype [114]. Since sperm are not able to transcribe new RNAs, the importation of sncRNAs from secretions of accessory sex glands could be a good strategy to maintain the sperm sncRNA profile in optimal conditions for the next events in early embryogenesis. Moreover, these imported sncRNAs from accessory sex glands might provide environmental epigenetic information without the need to overcome the hematotesticular barrier.

TABLE 7.2 Summary of Studies Assessing the Epigenetic Inheritance of Acquired Traits by the Male Germ Line Including Paternal Exposure, Offspring Phenotype, and Changes in Sperm sncRNA Population

Species	Paternal Exposure	Offspring Phenotype	Altered Sperm sncRNAs in Founder Males	Technology	References
Paternal diet studies					
Mice	High-fat diet	Obesity and insulin resistance	miR-133b-3p, miR-196a-5p, miR-205-5p, and miR-340-5p	qRT-PCR	[102]
Mice	Western-like diet	Increased body weight and impaired glucose tolerance	miR-182-5p, miR-29a-5p, miR-340-5p, miR-19b-3p, miR-19a-3p	qRT-PCR	[111]
Rat	High-fat diet	Reduced body weight and pancreatic β-cell mass. Glucose-intolerant adult females and resistant to HFD-induced weight gain	15 miRNAs, 41 tRFs, and 1092 piRNAs	RNA-seq	[101]
Mice	Low-protein diet	Altered hepatic cholesterol biosynthesis	tRF-Gly, tRF-Lys, tRF-His, let-7 family	RNA-seq	[58]
Mice	High-fat diet	Impaired glucose tolerance and insulin resistance	78 of miRNAs and 503 tRF	RNA-seq	[115]
Paternal stress studies					
Mice	Chronic paternal stress before breeding	Reduced hypothalamic-pituitary-adrenal axis stress response	miR-29c, miR-30a, miR-30c, miR-32, miR-193-5p, miR-204, miR-375, miR-532-3p, and miR-698	Microarray	[104]
Mice	Traumatic stress in early life	Depressive-like behaviors and altered metabolism	73 miRNA and 110 piRNAs	RNA-seq	[103]
Paternal lifestyle studies					
Mice	Paternal voluntary exercise	Suppressed reinstatement of juvenile fear memory and reduced anxiety in adulthood male offspring	miR-190b, miR-19b-2, miR-133a-1, miR-133a-2, miR-455, and tRF-Gly-GCC	RNA-seq	[112]

CONCLUSIONS

Fertility evaluation in men is currently limited to standard tests based on the analysis of seminal parameters that are able to reveal gross deficiencies in count, motility, or morphology, but they have a limited capacity to discern male factor infertility and thereby predict the success of fertility treatments. The poor current knowledge in male infertility diagnosis, prognosis, and therapy is mainly due to the incomplete understanding on the real contribution of the father to the new individual. The sncRNA profile of semen including spermatozoa and seminal plasma represents the final picture of events occurring during spermatogenesis and sperm maturation opening a window to discover pathogenic mechanisms of infertility. Furthermore, semen sncRNAs also appear to play a fundamental role in early embryogenesis and modulating the phenotype of offspring, opening up new clinical opportunities to predict fertility treatment outcome and also providing information about the future health of the progeny.

Acknowledgment

M.J. is granted by the Government of Catalonia (Generalitat de Catalunya, Pla Estratègic de Recerca i Innovació en Salut, PERIS, 2016-20).

References

[1] P.D. Zamore, B. Haley, Ribo-gnome: the big world of small RNAs, Science 309 (2005) 1519–1524.
[2] V.N. Kim, J. Han, M.C. Siomi, Biogenesis of small RNAs in animals, Nat. Rev. Mol. Cell Biol. 10 (2009) 126–139.
[3] H.K. Saini, S. Griffiths-Jones, A.J. Enright, Genomic analysis of human microRNA transcripts, Proc. Natl. Acad. Sci. U. S. A. 104 (2007) 17719–17724.
[4] Z. Fang, R. Du, A. Edwards, et al., The sequence structures of human microRNA molecules and their implications, PLoS One 8 (2013) e54215.
[5] E. Huntzinger, E. Izaurralde, Gene silencing by microRNAs: contributions of translational repression and mRNA decay, Nat. Rev. Genet. 12 (2011) 99–110.
[6] C. Catalanotto, C. Cogoni, G. Zardo, MicroRNA in control of gene expression: an overview of nuclear functions, Int. J. Mol. Sci. 17 (2016) E1712.
[7] E. Leucci, F. Patella, J. Waage, et al., microRNA-9 targets the long non-coding RNA MALAT1 for degradation in the nucleus, Sci. Rep. 3 (2013) 2535.
[8] T. Chiyomaru, S. Fukuhara, S. Saini, et al., Long non-coding RNA hotair is targeted and regulated by MIR-141 in human cancer cells, J. Biol. Chem. 289 (2014) 12550–12565.
[9] R. Tang, L. Li, D. Zhu, et al., Mouse miRNA-709 directly regulates miRNA-15a/16-1 biogenesis at the posttranscriptional level in the nucleus: evidence for a microRNA hierarchy system, Cell Res. 22 (2012) 504–515.
[10] R.F. Place, L.-C. Li, D. Pookot, et al., MicroRNA-373 induces expression of genes with complementary promoter sequences, Proc. Natl. Acad. Sci. U. S. A. 105 (2008) 1608–1613.
[11] S. Majid, A.A. Dar, S. Saini, et al., MicroRNA-205-directed transcriptional activation of tumor suppressor genes in prostate cancer, Cancer 116 (2010) 5637–5649.
[12] D.H. Kim, P. Saetrom, O. Snøve, et al., MicroRNA-directed transcriptional gene silencing in mammalian cells, Proc. Natl. Acad. Sci. U. S. A. 105 (2008) 16230–16235.
[13] L. Miao, H. Yao, C. Li, et al., A dual inhibition: microRNA-552 suppresses both transcription and translation of cytochrome P450 2E1, Biochim. Biophys. Acta: Gene Regul. Mech. 1859 (2016) 650–662.
[14] R. Song, G.W. Hennig, Q. Wu, et al., Male germ cells express abundant endogenous siRNAs, Proc. Natl. Acad. Sci. U. S. A. 108 (2011) 13159–13164.

[15] S. Hilz, A.J. Modzelewski, P.E. Cohen, et al., The roles of microRNAs and siRNAs in mammalian spermatogenesis, Development 143 (2016) 3061–3073.

[16] W.S.S. Goh, I. Falciatori, O.H. Tam, et al., PiRNA-directed cleavage of meiotic transcripts regulates spermatogenesis, Genes Dev. 29 (2015) 1032–1044.

[17] A. Aravin, D. Gaidatzis, S. Pfeffer, et al., A novel class of small RNAs bind to MILI protein in mouse testes, Nature 442 (2006) 203–207.

[18] A. Girard, R. Sachidanandam, G.J. Hannon, et al., A germline-specific class of small RNAs binds mammalian Piwi proteins, Nature 442 (2006) 199–202.

[19] J. Brennecke, A.A. Aravin, A. Stark, et al., Discrete small RNA-generating loci as master regulators of transposon activity in Drosophila, Cell 128 (2007) 1089–1103.

[20] H. Lin, piRNAs in the germ line, Science 316 (2007) 397.

[21] M.C. Siomi, K. Sato, D. Pezic, et al., PIWI-interacting small RNAs: the vanguard of genome defence, Nat. Rev. Mol. Cell Biol. 12 (2011) 246–258.

[22] A.a. Aravin, G.J. Hannon, J. Brennecke, The Piwi-piRNA pathway provides an adaptive defense in the transposon arms race, Science 318 (2007) 761–764.

[23] J.C. Peng, H. Lin, Beyond transposons: the epigenetic and somatic functions of the Piwi-piRNA mechanism, Curr. Opin. Cell Biol. 25 (2013) 190–194.

[24] C.D. Malone, G.J. Hannon, Small RNAs as guardians of the genome, Cell 136 (2009) 656–668.

[25] S.R. Mani, C.E. Juliano, Untangling the web: the diverse functions of the PIWI/piRNA pathway, Mol. Reprod. Dev. 80 (2013) 632–664.

[26] S.T. Grivna, E. Beyret, Z. Wang, et al., A novel class of small RNAs in mouse spermatogenic cells, Genes Dev. 20 (2006) 1709–1714.

[27] C. Rouget, C. Papin, A. Boureux, et al., Maternal mRNA deadenylation and decay by the piRNA pathway in the early Drosophila embryo, Nature 467 (2010) 1128–1132.

[28] S. Keam, G. Hutvagner, tRNA-derived fragments (tRFs): emerging new roles for an ancient RNA in the regulation of gene expression, Life 5 (2015) 1638–1651.

[29] S. Kimmins, P. Sassone-Corsi, Chromatin remodelling and epigenetic features of germ cells, Nature 434 (2005) 583–589.

[30] R.P. Yadav, N. Kotaja, Small RNAs in spermatogenesis, Mol. Cell. Endocrinol. 382 (2014) 498–508.

[31] M. Jodar, R. Oliva, Protamine alterations in human spermatozoa. in: E. Baldi, M. Muratori (Eds.), Genetic Damage in Human Spermatozoa, Springer, New York, 2014, pp. 83–102, https://doi.org/10.1007/978-1-4614-7783-9_6.

[32] R. Oliva, Protamines and male infertility, Hum. Reprod. Update 12 (2006) 417–435.

[33] M. Ghildiyal, P.D. Zamore, Small silencing RNAs: an expanding universe, Nat. Rev. Genet. 10 (2009) 94–108.

[34] Q. Wu, R. Song, N. Ortogero, et al., The RNase III enzyme DROSHA is essential for MicroRNA production and spermatogenesis, J. Biol. Chem. 287 (2012) 25173–25190.

[35] Y. Romero, O. Meikar, M.D. Papaioannou, et al., Dicer1 depletion in male germ cells leads to infertility due to cumulative meiotic and spermiogenic defects, PLoS One 6 (2011) e25241.

[36] C. Zimmermann, Y. Romero, M. Warnefors, et al., Germ cell-specific targeting of DICER or DGCR8 reveals a novel role for endo-siRNAs in the progression of mammalian spermatogenesis and male fertility, PLoS One 9 (2014) e107023.

[37] A.J. Modzelewski, S. Hilz, E.A. Crate, et al., Dgcr8 and Dicer are essential for sex chromosome integrity in male meiosis, J. Cell Sci. 128 (2015) 2314–2327.

[38] H.M. Korhonen, O. Meikar, R.P. Yadav, et al., Dicer is required for haploid male germ cell differentiation in mice, PLoS One 6 (2011) e24821.

[39] W. Deng, H. Lin, miwi, a murine homolog of piwi, encodes a cytoplasmic protein essential for spermatogenesis, Dev. Cell 2 (2002) 819–830.

[40] J. Bao, Y. Zhang, A. Schuster, et al., Conditional inactivation of Miwi2 reveals that MIWI2 is only essential for prospermatogonial development in mice, Cell Death Differ. 21 (2014) 783–796.

[41] M. DiGiacomo, S. Comazzetto, H. Saini, et al., Multiple epigenetic mechanisms and the piRNA pathway enforce LINE1 silencing during adult spermatogenesis, Mol. Cell 50 (2013) 601–608.

[42] W. Wu, Z. Hu, Y. Qin, et al., Seminal plasma microRNAs: potential biomarkers for spermatogenesis status, Mol. Hum. Reprod. 18 (2012) 489–497.

[42a] A.R. Greenlee, M.S. Shiao, E. Snyder, et al., Deregulated sex chromosome gene expression with male germ cell-specific loss of *Dicer1*, PLoS One 7 (2012) e46359.

[43] Z. Niu, S.M. Goodyear, S. Rao, et al., MicroRNA-21 regulates the self-renewal of mouse spermatogonial stem cells, Proc. Natl. Acad. Sci. U. S. A. 108 (2011) 12740–12745.
[44] Z. He, J. Jiang, M. Kokkinaki, et al., MiRNA-20 and MiRNA-106a regulate spermatogonial stem cell renewal at the post-transcriptional level via targeting STAT3 and Ccnd1, Stem Cells 31 (2013) 2205–2217.
[45] J.M. Huszar, C.J. Payne, MicroRNA 146 (Mir146) modulates spermatogonial differentiation by retinoic acid in mice, Biol. Reprod. 88 (2013) 15.
[46] S. Hilz, E.A. Fogarty, A.J. Modzelewski, et al., Transcriptome profiling of the developing male germ line identifies the miR-29 family as a global regulator during meiosis, RNA Biol. 14 (2017) 219–235.
[47] D. Tang, Y. Huang, W. Liu, et al., Up-regulation of microRNA-210 is associated with spermatogenesis by targeting IGF2 in male infertility, Med. Sci. Monit. 22 (2016) 2905–2910.
[48] R. Xie, X. Lin, T. Du, et al., Targeted disruption of miR-17-92 impairs mouse spermatogenesis by activating mTOR signaling pathway, Medicine (Baltimore) 95 (2016) e2713.
[49] J. Wu, J. Bao, M. Kim, et al., Two miRNA clusters, miR-34b/c and miR-449, are essential for normal brain development, motile ciliogenesis, and spermatogenesis, Proc. Natl. Acad. Sci. U. S. A. 111 (2014) E2851–E2857.
[50] L. Vojtech, S. Woo, S. Hughes, et al., Exosomes in human semen carry a distinctive repertoire of small noncoding RNAs with potential regulatory functions, Nucleic Acids Res. 42 (2014) 7290–7304.
[51] R. Sullivan, F. Saez, Epididymosomes, prostasomes and liposomes; their role in mammalian male reproductive physiology, Reproduction 146 (2013) R21–R35.
[52] G.D. Johnson, P. Mackie, M. Jodar, et al., Chromatin and extracellular vesicle associated sperm RNAs, Nucleic Acids Res. 43 (2015) 6847–6859.
[53] M. Jodar, A. Soler-Ventura, R. Oliva, Semen proteomics and male infertility, J. Proteomics (2016), https://doi.org/10.1016/j.jprot.2016.08.018.
[54] M. Jodar, E. Sendler, S.A. Krawetz, The protein and transcript profiles of human semen, Cell Tissue Res. 363 (2016) 85–96.
[55] I. Björkgren, H. Gylling, H. Turunen, et al., Imbalanced lipid homeostasis in the conditional Dicer1 knockout mouse epididymis causes instability of the sperm membrane, FASEB J. (2014), https://doi.org/10.1096/fj.14-259382.
[56] B. Nixon, S.J. Stanger, B.P. Mihalas, et al., The microRNA signature of mouse spermatozoa is substantially modified during epididymal maturation, Biol. Reprod. 93 (2015) 91.
[57] B. Nixon, S.J. Stanger, B.P. Mihalas, et al., Next generation sequencing analysis reveals segmental patterns of microRNA expression in mouse epididymal epithelial cells, PLoS One 10 (2015) e0135605.
[58] U. Sharma, C.C. Conine, J.M. Shea, et al., Biogenesis and function of tRNA fragments during sperm maturation and fertilization in mammals, Science 351 (2016) 391–396.
[59] R.J. Goodrich, E. Anton, S.A. Krawetz, Isolating mRNA and small non-coding RNAs from human sperm, Methods Mol. Biol. 927 (2013) 385–396.
[60] M. Jodar, S. Selvaraju, E. Sendler, et al., The presence, role and clinical use of spermatozoal RNAs, Hum. Reprod. Update 19 (2013) 604–624.
[61] S.A. Krawetz, A. Kruger, C. Lalancette, et al., A survey of small RNAs in human sperm, Hum. Reprod. 26 (12) (2011) 3401.
[62] G.C. Ostermeier, R.J. Goodrich, J.S. Moldenhauer, et al., A suite of novel human spermatozoal RNAs, J. Androl. 26 (2005) 70–74.
[63] A. Salas-Huetos, J. Blanco, F. Vidal, et al., New insights into the expression profile and function of microribonucleic acid in human spermatozoa, Fertil. Steril. 102 (2014) 213–222.
[64] L. Pantano, M. Jodar, M. Bak, et al., The small RNA content of human sperm reveals pseudogene-derived piRNAs complementary to protein-coding genes, RNA 21 (2015) 1085–1095.
[65] L. Lefièvre, K. Bedu-Addo, S.J. Conner, et al., Counting sperm does not add up any more: time for a new equation? Reproduction 133 (2007) 675–684.
[66] S.E.M. Lewis, Is sperm evaluation useful in predicting human fertility? Reproduction 134 (2007) 31–40.
[67] E. Anton, S. Krawetz, Spermatozoa as biomarkers for the assessment of human male infertility and genotoxicity, Syst. Biol. Reprod. Med. 58 (2012) 41–50.
[68] M. Abu-Halima, M. Hammadeh, J. Schmitt, et al., Altered microRNA expression profiles of human spermatozoa in patients with different spermatogenic impairments, Fertil. Steril. 99 (2013) 1249–1255.
[69] M. Abu-Halima, M. Hammadeh, C. Backes, et al., Panel of five microRNAs as potential biomarkers for the diagnosis and assessment of male infertility, Fertil. Steril. 102 (2014) 989–997. e1.

[70] J.H. Zhou, Q.Z. Zhou, X.M. Lyu, et al., The expression of cysteine-rich secretory protein 2 (CRISP2) and its specific regulator miR-27b in the spermatozoa of patients with asthenozoospermia, Biol. Reprod. 92 (2015) 28.
[71] Z. Ji, R. Lu, L. Mou, et al., Expressions of miR-15a and its target gene HSPA1B in the spermatozoa of patients with varicocele, Reproduction 147 (2014) 693–701.
[72] A. Salas-Huetos, J. Blanco, F. Vidal, et al., Spermatozoa from patients with seminal alterations exhibit a differential micro-ribonucleic acid profile, Fertil. Steril. 104 (2015) 591–601.
[73] M. Jodar, E. Anton, S.A. Krawetz, et al., Sperm RNA and its use as a clinical marker. in: C.J. De Jonge, C.L.R. Barratt (Eds.), The Sperm Cell, Cambridge University Press, 2017, pp. 59–72, https://doi.org/10.1017/9781316411124.006.
[74] A. Salas-Huetos, J. Blanco, F. Vidal, et al., Spermatozoa from normozoospermic fertile and infertile individuals convey a distinct miRNA cargo, Andrology (2016), https://doi.org/10.1111/andr.12276.
[75] J.J. Bromfield, Seminal fluid and reproduction: much more than previously thought, J. Assist. Reprod. Genet. 31 (2014) 627–636.
[76] Y. Hong, C. Wang, Z. Fu, et al., Systematic characterization of seminal plasma piRNAs as molecular biomarkers for male infertility, Sci. Rep. 6 (2016) 24229.
[77] C. Xiong, Identification of microRNAs predominately derived from testis and epididymis in human seminal plasma, Clin. Biochem. 47 (2014) 967–972.
[78] F. Bouhallier, N. Allioli, F. Lavial, et al., Role of miR-34c microRNA in the late steps of spermatogenesis, RNA 16 (2010) 720–731.
[79] C. Wang, C. Yang, X. Chen, et al., Altered profile of seminal plasma microRNAs in the molecular diagnosis of male infertility, Clin. Chem. 57 (2011) 1722–1731.
[80] T. Mostafa, L.A. Rashed, N.I. Nabil, et al., Seminal miRNA relationship with apoptotic markers and oxidative stress in infertile men with varicocele, Biomed. Res. Int. 2016 (2016) Article ID 4302754.
[81] W. Wu, Y. Qin, Z. Li, et al., Genome-wide microRNA expression profiling in idiopathic non-obstructive azoospermia: significant up-regulation of miR-141, miR-429 and miR-7-1-3p, Hum. Reprod. 28 (2013) 1827–1836.
[82] T.G. Cooper, Cytoplasmic droplets: the good, the bad or just confusing? Hum. Reprod. 20 (2005) 9–11.
[83] M. Abu-Halima, N. Ludwig, M. Hart, et al., Altered micro-ribonucleic acid expression profiles of extracellular microvesicles in the seminal plasma of patients with oligoasthenozoospermia, Fertil. Steril. 106 (2016) 1061–1069.e3.
[84] C. Zijlstra, W. Stoorvogel, Prostasomes as a source of diagnostic biomarkers for prostate cancer, J. Clin. Investig. 126 (2016) 1144–1151.
[85] T. Kono, Y. Obata, Q. Wu, et al., Birth of parthenogenetic mice that can develop to adulthood, Nature 428 (2004) 860–864.
[86] D. Miller, Confrontation, consolidation, and recognition: the oocyte's perspective on the incoming sperm, Cold Spring Harb. Perspect. Med. 5 (2015) a023408.
[87] O.J. Rando, Daddy issues: paternal effects on phenotype, Cell 151 (2012) 702–708.
[88] S. Yuan, A. Schuster, C. Tang, et al., Sperm-borne miRNAs and endo-siRNAs are important for fertilization and preimplantation embryonic development, Development 143 (2016) 635–647.
[89] W.-M. Liu, R.T.K. Pang, P.C.N. Chiu, et al., Sperm-borne microRNA-34c is required for the first cleavage division in mouse, Proc. Natl. Acad. Sci. U. S. A. 109 (2012) 490–494.
[90] L. Cui, L. Fang, B. Shi, et al., Spermatozoa micro ribonucleic acid-34c level is correlated with intracytoplasmic sperm injection outcomes, Fertil. Steril. 104 (2015) 312–317. e1.
[91] S. Yuan, C. Tang, Y. Zhang, et al., mir-34b/c and mir-449a/b/c are required for spermatogenesis, but not for the first cleavage division in mice, Biol. Open 4 (2015) 212–223.
[92] J.J. Bromfield, J.E. Schjenken, P.Y. Chin, et al., Maternal tract factors contribute to paternal seminal fluid impact on metabolic phenotype in offspring, Proc. Natl. Acad. Sci. U. S. A. 111 (2014) 2200–2205.
[93] J.E. Schjenken, S.A. Robertson, Seminal fluid and immune adaptation for pregnancy—comparative biology in mammalian species, Reprod. Domest. Anim. 49 (2014) 27–36.
[94] R. Ramasamy, K. Chiba, P. Butler, et al., Male biological clock: a critical analysis of advanced paternal age, Fertil. Steril. 103 (2015) 1402–1406.
[95] U. Schagdarsurengin, K. Steger, Epigenetics in male reproduction: effect of paternal diet on sperm quality and offspring health, Nat. Rev. Urol. 13 (2016) 584–595.
[96] P. Esakky, K.H. Moley, Paternal smoking and germ cell death: a mechanistic link to the effects of cigarette smoke on spermatogenesis and possible long-term sequelae in offspring, Mol. Cell. Endocrinol. 435 (2016) 85–93.

[97] A. Finegersh, G.R. Rompala, D.I.K. Martin, et al., Drinking beyond a lifetime: new and emerging insights into paternal alcohol exposure on subsequent generations, Alcohol 49 (2015) 461–470.
[98] B.R. Carone, L. Fauquier, N. Habib, et al., Paternally induced transgenerational environmental reprogramming of metabolic gene expression in mammals, Cell 143 (2010) 1084–1096.
[99] S.-F. Ng, R.C.Y. Lin, D.R. Laybutt, et al., Chronic high-fat diet in fathers programs β-cell dysfunction in female rat offspring, Nature 467 (2010) 963–966.
[100] L.M. Anderson, L. Riffle, R. Wilson, et al., Preconceptional fasting of fathers alters serum glucose in offspring of mice, Nutrition 22 (2006) 327–331.
[101] T. de Castro Barbosa, L.R. Ingerslev, P.S. Alm, et al., High-fat diet reprograms the epigenome of rat spermatozoa and transgenerationally affects metabolism of the offspring, Mol. Metab. 5 (2016) 184–197.
[102] T. Fullston, E.M.C.O. Teague, N.O. Palmer, et al., Paternal obesity initiates metabolic disturbances in two generations of mice with incomplete penetrance to the F2 generation and alters the transcriptional profile of testis and sperm microRNA content, FASEB J. 27 (2013) 4226–4243.
[103] K. Gapp, A. Jawaid, P. Sarkies, et al., Implication of sperm RNAs in transgenerational inheritance of the effects of early trauma in mice, Nat. Neurosci. 17 (2014) 667–669.
[104] A.B. Rodgers, C.P. Morgan, S.L. Bronson, et al., Paternal stress exposure alters sperm microRNA content and reprograms offspring HPA stress axis regulation, J. Neurosci. 33 (2013) 9003–9012.
[105] D.M. Dietz, Q. Laplant, E.L. Watts, et al., Paternal transmission of stress-induced pathologies, Biol. Psychiatry 70 (2011) 408–414.
[106] S.M. Renaud, S.B. Fountain, Transgenerational effects of adolescent nicotine exposure in rats: Evidence for cognitive deficits in adult female offspring, Neurotoxicol. Teratol. 56 (2016) 47–54.
[107] M.Z. Farah Naquiah, R.J. James, S. Suratman, et al., Transgenerational effects of paternal heroin addiction on anxiety and aggression behavior in male offspring, Behav. Brain Funct. 12 (2016) 23.
[108] B.G. Dias, K.J. Ressler, Parental olfactory experience influences behavior and neural structure in subsequent generations, Nat. Neurosci. 17 (2014) 89–96.
[109] L.O. Bygren, G. Kaati, S. Edvinsson, Longevity determined by paternal ancestors' nutrition during their slow growth period, Acta Biotheor. 49 (2001) 53–59.
[110] G. Kaati, L. Bygren, S. Edvinsson, Cardiovascular and diabetes mortality determined by nutrition during parents' and grandparents' slow growth period, Eur. J. Hum. Genet. 10 (2002) 682–688.
[111] V. Grandjean, S. Fourré, D.A.F. De Abreu, et al., RNA-mediated paternal heredity of diet-induced obesity and metabolic disorders, Sci. Rep. 5 (2015) 18193.
[112] A.K. Short, S. Yeshurun, R. Powell, et al., Exercise alters mouse sperm small noncoding RNAs and induces a transgenerational modification of male offspring conditioned fear and anxiety, Transl. Psychiatry 7 (2017) e1114.
[113] M. Rassoulzadegan, V. Grandjean, P. Gounon, et al., RNA-mediated non-mendelian inheritance of an epigenetic change in the mouse, Nature 441 (2006) 469–474.
[114] U. Sharma, C.C. Conine, J.M. Shea, et al., Biogenesis and function of tRNA fragments during sperm maturation and fertilization in mammals, Science (2015), https://doi.org/10.1126/science.aad6780.
[115] Q.Q. Chen, M. Yan, Z. Cao, et al., Sperm tsRNAs contribute to intergenerational inheritance of an acquired metabolic disorder, Science 7977 (2015) 1–8.
[116] N.O. McPherson, J.A. Owens, T. Fullston, et al., Preconception diet or exercise interventions in obese fathers normalizes sperm microRNA profile and metabolic syndrome in female offspring, Am. J. Physiol. Endocrinol. Metab. (2015), https://doi.org/10.1152/ajpendo.00013.2015.

CHAPTER

8

Altered Transcriptomic Profiles Associated With Male Infertility

*Sara Larriba**, *Lluís Bassas*†

*Molecular Genetics of Male Fertility, Human Molecular Genetics Group, Bellvitge Biomedical Research Institute (IDIBELL), Barcelona, Spain †Laboratory of Seminology and Embryology, Andrology Service, Fundació Puigvert, Barcelona, Spain

INTRODUCTION

The World Health Organization (WHO) defines human infertility as "the failure to achieve a pregnancy after 12 months or more of regular unprotected sexual intercourse." It is a prevalent disorder that affects 50 million couples (13%–15% couples) worldwide [1,2]. It is generally accepted that male factors, as either pure or contributory factors, are responsible for half of all cases of infertility; this implies that approximately 5%–7% of all men in western countries have fertility problems. Traditionally, the diagnosis of male infertility has relied upon microscopic assessment and biochemical assays to determine human semen quality [3]. The majority of cases of male infertility are due to quantitative defects leading to the absence of sperm (azoospermia) or a reduction in sperm number (oligozoospermia), while a substantial number of cases are caused by morphological or qualitative defects in sperm, which lead to defective sperm function. These failures mainly result from an impaired production of male gamete or an impaired spermatogenic process, although we should also take into account ejaculatory disorders or obstructions impacting on the transport of sperm from the testis. In terms of etiology, there is a general agreement that male infertility is a multifactorial disease and is caused by both congenital and acquired conditions. For example, decreased sperm production is closely related to urogenital abnormalities including testicular dysfunction (a large proportion of cases are caused by genetic abnormalities such as karyotype anomalies and Y chromosome microdeletions), varicocele, infections of the genital tract, immunologic problems, and/or exposure to exogenous chemical or physical agents. Severe cases of male infertility are likely to have a predominantly genetic etiology, and it is widely accepted that infertile men are at higher risk of being carriers of genetic anomalies in both their genomic DNA and gametes. This has potential consequences for the offspring born through the use

Reproductomics
https://doi.org/10.1016/B978-0-12-812571-7.00009-5

of assisted reproduction techniques. Despite many efforts directed at identifying the genetic factors involved in decreased sperm production and/or impaired sperm function, the molecular etiology of male infertility remains undefined in a large proportion of infertile patients [4], which is evidence of the molecular complexity of fertility.

The aim of this chapter is to highlight the relevance of an adequate, precise, and tightly regulated gene expression in the germ line, focusing on the alterations that have been assessed at the tissue and cellular transcriptome level to shed more light on the molecular defects involved in male infertility.

THE COMPLEXITY OF THE TRANSCRIPTOME IN SPERMATOGENESIS AND SPERM

Gene Expression Profile: Coding RNAs

Spermatogenesis is a highly orchestrated developmental process that takes place in the seminiferous tubules of the testis, through which the undifferentiated male germ cells develop into mature haploid gametes for sexual reproduction. During the course of spermatogenesis, the three major forms of cell cycle are represented: (i) mitosis of primitive spermatogonia; (ii) two rounds of meiosis, from primary spermatocytes to haploid round spermatids; and (iii) differentiation, which includes structural and nuclear changes to generate mature spermatids and spermatozoa. All these processes are tightly regulated and depend on precise gene expression specific to the developmental stage and the germ-cell type. Sperm production involves the coordinated and regulated expression of approximately 3000 genes (reviewed in Ref. [5]) that influence testicular and sperm function. Microarray technology has made it possible to identify many of these genes and characterize the transcriptional profile in germ and somatic cells at different steps of mouse testicular development [6–13]. Stage-regulated gene expression is a widespread and fundamental characteristic of the germ line, and defects in essential genes can result in impaired sperm or no sperm at all.

The male gamete, the spermatozoon, is a terminal cell; it is highly specialized and perfectly adapted for its function: fertilizing the oocyte. In addition to the haploid genetic material, sperm contain proteins [14] and an abundant number of functionally viable transcripts [15–17] that will be released into the oocyte. Increasing data support the idea that these RNAs may have a role for the spermatozoa and/or the embryo after oocyte fertilization. This conclusion comes after considering the following pieces of evidence:

- The mature sperm contains highly condensed chromatin architecture, which is enriched in protamines, but very little cytoplasm. It lacks endoplasmic reticulum, Golgi apparatus, lysosomes, and cytosolic ribosomes. Although the absence of essential components of the protein synthesis machinery would seem to question the possibility that mRNA may be newly translated de novo in sperm, recent studies have shown that polysomial complexes are present in mitochondrial ribosomes, which suggests that the translation of these mRNAs can take place in the spermatozoa [18,19]. These proteins seem to contribute to the correct maturation of the germ cell, allowing the male gamete to function properly [18], and additionally, they may be implicated in structural sperm complexes [20].

- Inhibition of transcription prior to the completion of spermatogenesis and retention of RNAs in the male germ cell is similar to a comparable process in oogenesis. However, although there is still a lot of debate surrounding the function of RNAs in the spermatozoa, the function of stored transcripts in the oocyte is clear: they generate proteins for the maturation of the oocyte itself and for the early stages of prenatal development. Human spermatozoa can deliver mRNA to the oocyte during fertilization [21]. Some of these mRNAs have been shown to be translated de novo in the oocyte after fertilization, supporting the hypothesis that at least some transcripts of paternal origin might have a function during or beyond the process of fertilization [18,22,23] and that they also contribute to zygote development prior to activation of the embryonic genome [15,24].

Noncoding RNAs and the Regulation of Gene Expression

Understanding this tightly regulated process is even more complicated than merely knowing the function of the genes. Recent publications have reported that most of the human genome is transcribed, but only a small fraction of the transcribed RNA encodes proteins [25]. Evidence is emerging that noncoding RNAs (ncRNAs), which were previously considered to be "junk" DNA, actually play a critical role in the control of gene expression. They function both at the transcriptional level, as components of chromatin remodeling complexes, or posttranscriptionally.

These regulatory ncRNAs can be classified according to their length into small ncRNAs (sncRNAs), which are shorter than 200 nucleotides (nt), and long ncRNAs (lncRNAs) with a length of more than 200 nt. Small regulatory ncRNAs, which function by guiding a crucial cofactor—an AGO or PIWI protein of the Argonaute family—to target RNAs, can be further subdivided into microRNAs (miRNAs), endogenous small interfering RNAs (endo-siRNAs), and PIWI-associated RNAs (piRNAs) based on their size, function, mode of action, and the member of the Argonaute protein family they bind to [26,27].

The regulatory network that confers specific germ-line gene expression during spermatogenesis in mammals is now being understood, because of the discovery of thousands of these noncoding RNAs (lncRNAs, miRNAs, and piRNAs, among others) that have roles in mRNA stability, protein translation, protein modification, and protection of the germ line. In a similar way to mRNAs, noncoding RNAs are differentially regulated during spermatogenesis, where they play a relevant role (Fig. 8.1). Understanding the gene expression profile and this regulatory step is essential for determining the molecular requirements for the progression of spermatogenesis, and thus, by extension, for understanding male infertility.

Similarly to what occurs in the testis, different studies provide evidence that the spermatozoa contain a more complex composition of RNAs than was initially thought. In addition to rRNA (ribosomal ribonucleic acid, which is mostly fragmented) and mRNA [15,17], the sperm RNA population is composed of sncRNAs and lncRNAs [28,29]. Recent studies have shown the presence of multiple classes of these sperm sncRNAs (18–30 nt) including miRNA (7%), piRNAs (17%), repeat-associated RNAs (65%) [29], and other gamete-specific sncRNAs designated sperm RNAs (spR)-12 and -13 [28]. They have been associated with a variety of processes including posttranscriptional regulation of gene expression, chromatin formation, and protection of the genome against transposons.

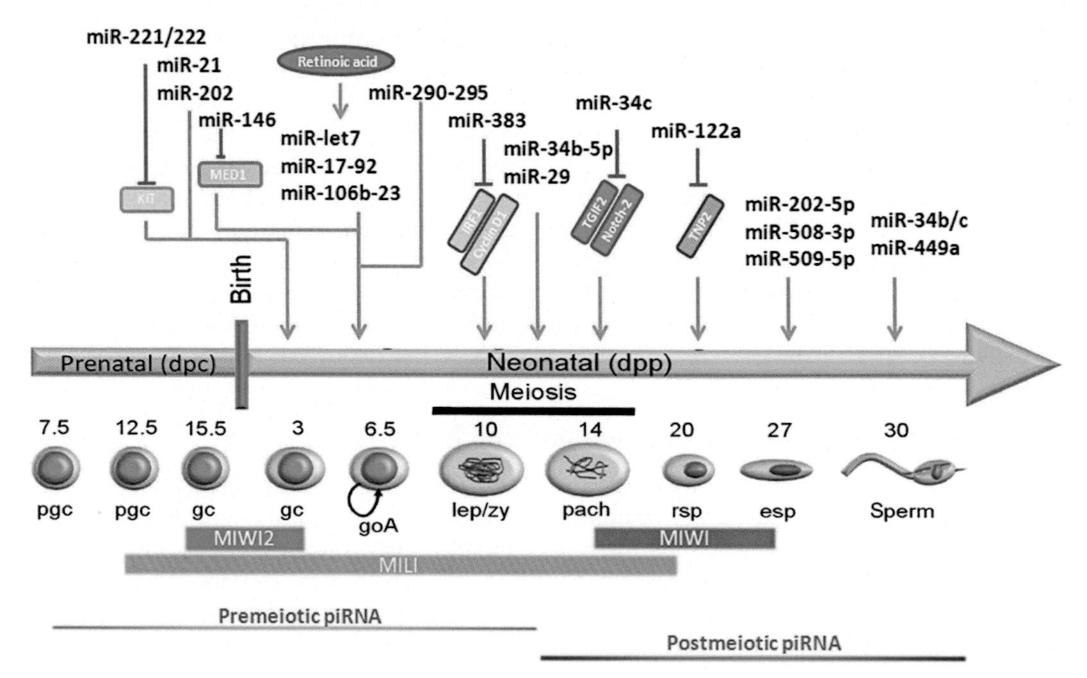

FIG. 8.1 Time course of functional miRNAs throughout mouse spermatogenesis. Noncoding RNAs are differentially regulated during spermatogenesis, and it has shown some examples. Gene targets are depicted in boxes. PIWI proteins and piRNAs expression are also annotated. *Pgc*, primordial germ cell; *gc*, gonocyte; *goA*, spermatogonia A; *rsp*, round spermatid; *esp.*, elongating spermatid; *lep*, leptotene; *zy*, zygotene; *pach*, pachytene; *dpc*, days postcoitum; and *dpp*, days postpartum.

lncRNAs

Approximately 10,000–20,000 different lncRNAs have been annotated in humans; they constitute the largest group of ncRNAs [30]. LncRNAs include long intergenic ncRNAs (lincRNA), pseudogene transcripts, antisense RNAs, and circular RNAs (circRNAs). In the same way that occurs in mRNA biogenesis, RNA polymerase is involved in lncRNA transcription; however, lncRNAs lack an open reading frame. Transcriptomic studies and genome-wide association studies have recognized that lncRNAs play multiple regulatory roles [31], modulating the transcriptional status of individual genes or even the entire chromosome [32,33]. Based on recent studies, lncRNAs can be classified according to their mode of regulation into competitors, recruiter/activators, and precursors. First as a competitor, lncRNA can bind to DNA-binding proteins, such as transcription factors [34], or miRNAs [35]. Secondly, lncRNA can activate epigenetic modifiers or recruit them to specific targets sites, regulating DNA methylation [36] and histone modifications [37,38]. Finally, lncRNAs can be fragmented into shorter RNAs involved in posttranscriptional regulation such as H19 [39]. Many lncRNAs have been identified in gametes and in developing germ cells in the testis [40]. However, very few of them have been functionally characterized (reviewed in Ref. [41]), these include Meiotic recombination hot-spot locus (*Mrhl*) [42], Testis-specific X-linked (*Tsx*) [43], and Dmrt1-related gene (*Dmr*) [35]. Additional studies show that lncRNA expression levels exhibit coordinated changes throughout spermatogenesis [40]. Only a few studies have assessed lncRNA expression in human spermatogenic disorders, such as the study by Lü and coworkers [44], which shows altered lncRNA expression profiles in patients with meiotic arrest. Much more research is needed to identify the precise role of lncRNAs in the seminal process.

miRNAs

miRNA production depends on two endonucleases: the complex Drosha/DGCR8 and Dicer. Mature miRNAs are derived from larger hairpin-forming transcripts; these are recognized by the RNA-binding protein DGCR8 and processed by the nuclease DROSHA, releasing the hairpin from the primary transcript [45–48]. Then, the nuclease DICER (DICER1) processes the hairpin precursor miRNA to produce the mature miRNA, which is loaded onto an AGO protein to generate the effector complex [27,49]. miRNAs recently emerged as important regulators of gene expression; they suppress gene expression posttranscriptionally via sequence-specific interactions with the 3′-untranslated regions (UTRs) of specific genes. The suppression occurs though the destabilization or cleavage of the mRNA and/or translational inhibition (reviewed in Ref. [50]). Interestingly, miRNAs have enormous potential to regulate the genome; they have been estimated to target over one-third of human genes [51]. The number of miRNAs being identified is constantly rising. A list of miRNAs is collected in the miRBase v21 (www.mirbase.org), a searchable database of published miRNA sequences and annotations; at the time of writing, it showed 1881 precursors and 2588 mature human miRNAs.

Several strands of evidence suggest that miRNAs could play an active and important role during spermatogenesis. Studies of miRNA expression profiles in the testis have shown elevated levels at the different germ-cell stages throughout spermatogenesis [52–54]. Furthermore, KO mouse models of Dicer1, a protein necessary for miRNA processing, and Dnd1, a protein implicated in the protection of mRNAs from miRNAs, have evidenced that miRNAs are essential for the completion of spermatogenesis [55,56], including spermatid differentiation [57,58] and Sertoli function [59]. Some examples are depicted in Fig. 8.1. Specifically, mouse studies have shown that miR-122a is predominantly expressed in postmeiotic male germ cells and that it can complementarily bind and induce Transition protein 2 (*Tp2*) mRNA cleavage in round spermatids [60]. Additionally, there is strong evidence that the miR-34b/c and miR-449 clusters are involved in spermatogenesis. They have been reported to be highly expressed in the testis in the late stages of spermatogenesis and in sperm but absent in oocytes [61–63], and the expression of both miRNA clusters is essential for a physiological spermiogenesis [64]. Both miRNA clusters have been described as functioning redundantly by targeting the E2F-pRb pathway [63]. *TGIF2* and *NOTCH2* are targets of miR-34c [62], and the function of NOTCH1, the production of which is regulated by both miR-34b and miR-34c, is critical for germ-cell differentiation and survival during spermatogenesis. Other miRNAs that have been described as having a meiotic and premeiotic germ-cell expression are (i) the miR-29 family [65] and miR-383 [66], which are global regulators during meiosis; (ii) the miRNA-17–92 cluster, miR-290–295 cluster, and miR-146 [67], which are abundantly expressed in mouse spermatogonia; [55] and (iii) miRNAs such as miR-202 [68] and miR-221/222 [69], which are responsible for maintaining spermatogonial stem cells.

The most abundant miRNAs in sperm from fertile individuals are miR-34c [29], miR-34b-3p [70], or miR-1246 [71]; the result being dependent on the technique used for the evaluation of miRNA levels in sperm (RNA-seq vs qPCR array, respectively).

There are many publications that describe miRNA profiles in male infertility. In this chapter, we review and summarize the role of miRNAs in spermatogenic disorders and discuss their potential applications as diagnostic and prognostic fertility biomarkers in more detail below.

Endo-siRNAs

Endo-siRNAs (~19–23nt in length) originate from long, perfectly complementary double-stranded RNAs that are formed through sense-antisense transcript pairs, long stem-loop structures, and transposon transcripts [72]. Unlike miRNA biogenesis, siRNAs are only processed by the enzyme DICER and neither DGCR8 nor DROSHA is involved [27,73,74]. Just like miRNAs, endo-siRNAs also associate with AGO proteins. siRNAs have been detected in mammalian embryonic stem cells (ESCs), oocytes, and spermatocytes [73–76], but their regulatory role during mammalian spermatogenesis is poorly understood. In contrast to the situation for miRNAs, mammalian siRNA-target relationships are poorly characterized. While miRNAs are absolutely essential for germ-cell development and spermatogenesis, a lesser role for endo-siRNAs in the meiotic and postmeiotic stages of spermatogenesis has been suggested [77].

piRNAs

piRNAs is the third main class of sncRNAs. They are ~24–31nt long, derived from single-stranded piRNA precursor transcripts and processed in a Dicer-independent way [78,79]. Interestingly, they are predominantly expressed in the germ line (mainly in pachytene spermatocytes and round spermatids). piRNAs interact with the AGO subfamily proteins known as PIWI proteins [80] . This PIWI-piRNA complex, known as PIWI machinery, is involved in transposon silencing through heterochromatin formation and RNA destabilization [81,82]. There are three PIWI proteins in humans; they are mainly expressed in the testis [82] in a stage-specific manner during spermatogenesis. They are PIWIL1 (known as MIWI in mice), PIWIL2 (MILI), and PIWIL4 (MIWI2). PIWIL4 expression is observed in early embryonic development and postnatally in very immature germ cells [83,84]. PIWIL2 has been described as being expressed in the germ line during early spermatogenesis [85], whereas PIWIL1 is expressed after birth in pachytene spermatocytes and spermatids, and there is a suggestion that it has a role in translational control in the latest stages of spermatogenesis [86,87] (Fig. 8.1).

Additionally, piRNAs are the most abundant class of sncRNAs in sperm, and surprisingly, many of them are processed from pseudogenes in testis [71] and are still found in sperm cells. They play a major role in male germ-cell development and fertility by repressing the transposable elements and protecting the genetic information by the RNA interference mechanism. The piRNA process acts upstream of known mediators of DNA methylation ensuring that transcriptional transposable elements are silenced throughout spermatogenesis [88]. The piRNA-directed methylation pathway is a major determinant of mammalian fertility. In line with this, we determined a decreased cellular expression level of *PIWIL2*, its associated molecule *TDRD1*, and premeiotic-piRNAs in severe spermatogenic disorders [89].

SPERMATOGENIC IMPAIRMENT

Semen Evaluation

In the clinics, the diagnosis of male infertility is first assessed by the microscopic evaluation of semen, as male factor infertility often consists of decreased semen quality in terms of number, progressive motility, and morphology of sperm. The World Health Organization (updated in 2010 [3]) publishes internationally agreed recommendations for the interpretation of sperm analyses and the reference normal value ranges for these parameters leading to the classification of male infertility:

- Taking into account the sperm concentration (due to spermatogenic failure, SpF): azoospermia (no sperm in semen), cryptozoospermia (less than 1 million spermatozoa/mL), and oligozoospermia (sperm concentration below 15 million sperm/mL; severe oligozoospermia is a concentration below 5 million sperm/mL).
- Taking into account the sperm motility: asthenozoospermia with less than 32% of progressively motile sperm.
- Taking into account sperm morphology: teratozoospermia with less than 4% of morphologically normal forms.

Histopathological Patterns of Testicular Biopsy in SpF

Testicular biopsies from infertile men with an altered sperm production usually show different pathological patterns.

Severe and homogeneous patterns are the following (also shown in Fig. 8.2):

- *Sertoli-cell-only syndrome (SCO) or germ-cell aplasia*. Only Sertoli cells are present in the seminiferous tubule. This is the most severe phenotype. In this case, the individual shows azoospermia as the consequence of a complete absence of germ cells in the tubules.
- *Germ-cell maturation failure*. The process of spermatogenesis was totally or nearly arrested at a specific cell stage:
    - Maturation failure at the spermatogonia stage *(sgMF)* displaying only immature germ cells in the tubules
    - Maturation failure at the spermatocyte stage *(scMF)* with spermatogonia and spermatocytes but few or no spermatids
    - Maturation failure at the round spermatid stage *(rsMF)* showing no elongated spermatids

Additionally, other less severe patterns are as follows:

- *Hypospermatogenesis*: The cellularity of germinal epithelium was reduced in general; all stages of germ cells (spermatogonia, spermatocytes, and spermatids) were present but reduced in number.
- *Mixed pattern*: More than one pathological pattern was seen in the same testicular biopsy.

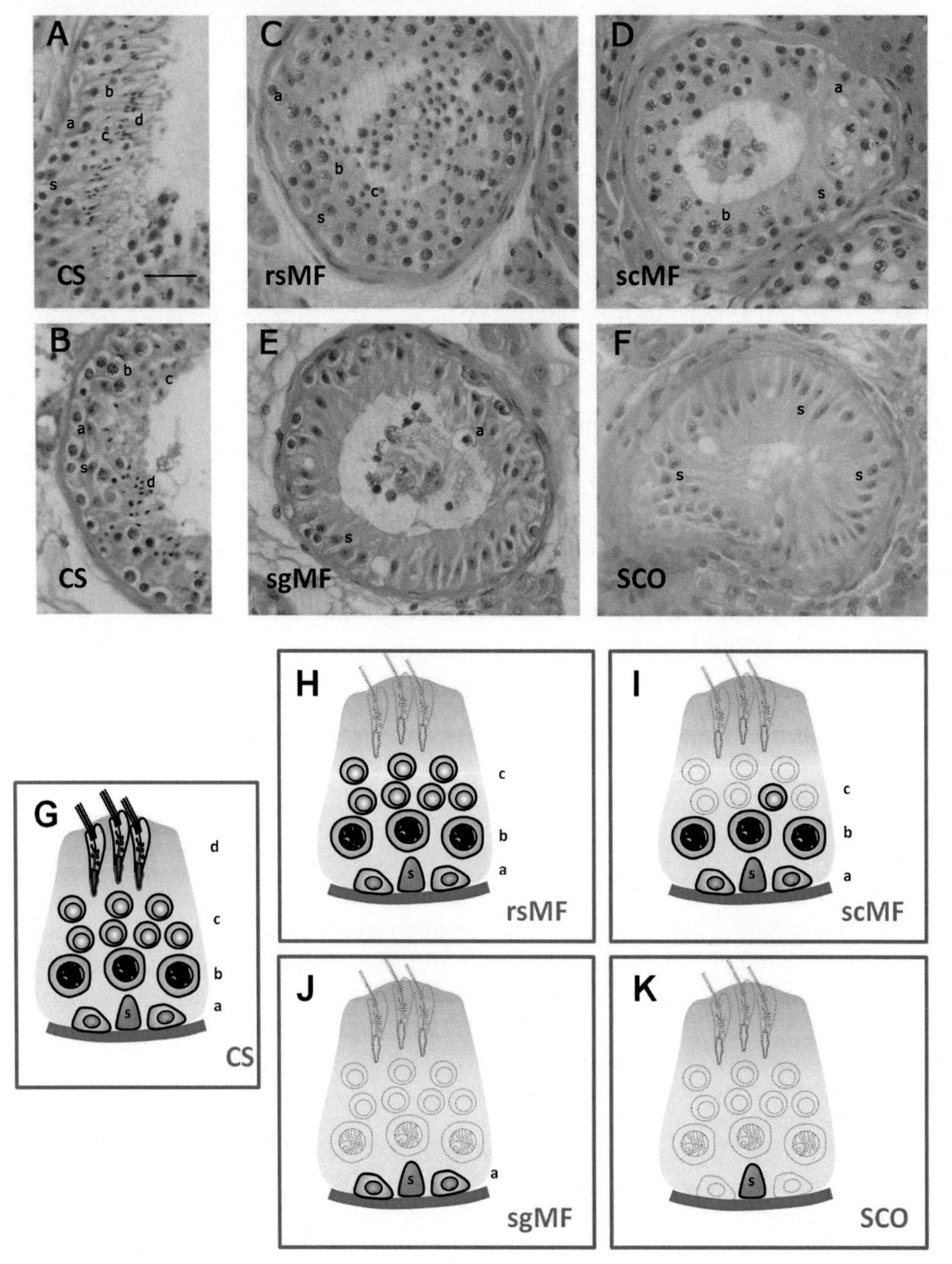

FIG. 8.2 Testicular histology of representative sections of seminiferous tubules from infertile men showing different spermatogenic patterns: conserved spermatogenesis (CS) containing all germ-cell stages (A and B), maturation failure at the round spermatid stage (rsMF) showing no elongated spermatids (C), maturation failure at the spermatocyte stage (scMF) with spermatogonia and spermatocytes but few or no spermatids (D), maturation failure at the spermatogonia stage (sgMF) displaying only these primordial cells (E), and Sertoli-cell-only phenotype or germ-cell aplasia (SCO) where seminiferous tubules contain exclusively Sertoli cells (F). Schematic diagrams, showing the epithelium of a seminiferous tubule that consists of a single Sertoli cell and the germ-cell components of the different phenotypes, are placed below the histological images: CS (G), rsMF (H), scMF (I), sgMF (J), and SCO (K). Absent cell populations are drawn as dashed line shadings. Examples of different cell types, including Sertoli cells and specific germ-cell stages, are identified with lowercase letters: (a), spermatogonia; (b), primary spermatocytes; (c), round spermatids; (d), elongated spermatids; and (s), Sertoli cell nucleus. Hematoxylin and eosin staining. Bar = 50 μm. *Reproduced from X. Muñoz, A. Mata, L. Bassas, S. Larriba, Altered miRNA signature of developing germ-cells in infertile patients relates to the severity of spermatogenic failure and persists in spermatozoa, Sci. Rep. 5 (2015) 17991.*

Gene expression studies are usually performed on samples from patients selected for having a homogeneous and severe histological pattern of tubules from the same testicular section.

TESTICULAR AND SPERM TRANSCRIPTOMIC PROFILES ARE ALTERED IN HUMAN SPERMATOGENIC DISORDERS

Testicular mRNA Expression Levels in Impaired Sperm Production

The gene expression profile in SpF can be used to identify candidate genes for spermatogenic disorders. Many mRNA profiles have been assessed for testicular samples with different histological phenotypes from infertile patients with abnormalities in sperm production, such as SpF-azoospermia and SpF-oligozoospermia, evidencing altered profiles that affect the expression of hundreds of genes (Table 8.1A) [90–95] that are thought to have a potential role in a successful spermatogenesis.

However, it is much more challenging to carry out gene expression analysis of SpF testicular samples, with a decreased number of germ cells, due to the cellular complexity of this organ, which could lead to results being misinterpreted. In particular, as the pathological seminiferous tubules lack germ cells to varying degrees, changes in gene expression at the tissue level can reflect changes in the content of mRNA in a specific cell type as well as changes in the cell type number or composition in pathological testis. It is important to bear in mind the fact that the levels of different mRNAs could change as the proportion of germ cells from different stages will be different when observing tubules in conserved spermatogenesis, in which all stages of germ cell are present, and from observing tubules with maturation arrest, where some stages are totally or partially absent (Fig. 8.2). Therefore, the reduction of gene expression in SpF patients described above could be partially explained by the decreased number of germ cells that specifically express those genes in these individuals and could represent a consequence rather than a cause. However, our group presented additional studies on cellular expression levels of germ-cell-specific genes that showed that in SpF patients, a large proportion of genes preferentially expressed in spermatocyte (MMR genes) and in spermatogonia (*CCNE1*, *DAZL*, *RBM15*, and *STRA8* genes) exhibited reduced testicular expression levels when compared with samples from individuals showing conserved spermatogenesis [89,98,105]. This underlines the determinant role of a successful meiotic and premeiotic gene expression in the progression of the spermatogenic process. Thus, the tissue gene expression differences are attributable not only to different proportions of the germ-cell stages in the tubule but also to real differences in the expression capability of the cell. More recent data from our group demonstrate that low spermatogenic efficiency in infertile men is accompanied by meiotic and premeiotic patterns of gene expression deregulation in spermatogenesis, and these are contributing to spermatogenic blockade. Furthermore, the differences in expression during the initial stages of spermatogenesis in individuals with meiotic arrest suggested that this phenotype is already determined or arises in the premeiotic stages of the germ line, something that further supports the role of proper gene expression in early germ-line stages for successful sperm production [98].

TABLE 8.1 Summary of Studies Determining the Transcriptomic Profile to Characterize the Differentially Expressed Genes (A) and miRNAs (B) in Human Altered Spermatogenesis

Assessment	Tissue/Cell Type	Testicular Phenotype	Technology	Results	References
(A)					
Global gene expression	Testis biopsies	CS ($n=3$); SCO ($n=3$)	Microarray	682 Genes downregulated in SCO	[90]
Global gene expression	Testis biopsies	ND	Microarray	19 Differentially expressed in SpF versus fertile samples	[91]
Global gene expression	Testis biopsies	CS ($n=12$); HS ($n=6$); scMF ($n=5$); SCO ($n=5$)	Microarray	1181 Genes altered in expression	[92]
188 Upregulated genes in altered spermatogenesis	Testis biopsies	OA ($n=3$); SpF ($n=10$)	Microarray	Induction of an inflammatory-like response	[93]
Global gene expression	Testis biopsies	CS, HS, scMF, SCO (total $n=69$)	Microarray	Induction of an inflammatory-like response	[94]
Global gene expression	Sperm	CS ($n=13$); Ter ($n=8$)	Microarray	Transcripts encoding components of various cellular remodeling pathways disrupted	[20]
84 Chromatin remodeling factors	Testis biopsies	CS, rsMF	qPCR arrays	1 gene upregulated; 21 genes downregulated in rsMF	[95]
Global gene expression	Sperm	OAT ($n=8$); Ft ($n=3$)	Microarray	Up to 33-fold reduction expression of genes involved in spermatogenesis and sperm motility	[96]
Global gene expression	Sperm	Ft ($n=4$); Asth ($n=4$)	Microarray	2 Genes upregulated and 17 genes downregulated in Asth	[97]
Expression of 47 genes	Testis biopsies	SCO ($n=14$); scMF ($n=7$); HS ($n=12$); CS ($n=17$)	qPCR arrays (TaqMan arrays)	Spermatogonia have reduced levels of transcripts in SpF	[98]
Global gene expression	Sperm	Ft ($n=20$); Asth ($n=20$)	Microarray	281 Genes altered in expression	[99]

TABLE 8.1 Summary of Studies Determining the Transcriptomic Profile to Characterize the Differentially Expressed Genes (A) and miRNAs (B) in Human Altered Spermatogenesis—cont'd

Assessment	Tissue/Cell Type	Testicular Phenotype	Technology	Results	References
(B)					
Global miRNA expression	Testis biopsies	CS (*n*=2); scMA (*n*=3)	Microarray	173 miRNAs were differentially expressed	[100]
Global miRNA expression	Sperm	Ft-N (*n*=9); Asth (*n*=9); OA (*n*=9)	Microarray	77 (Asth) and 86 (OA) miRNAs were altered in expression	[101]
Global miRNA expression	Testis biopsies (FFPE)	SCO (*n*=12); scMF (*n*=16); MA (*n*=12); CS (*n*=16)	Microarray	197 (SCO), 68 (MA) and 46 (scMF) miRNAs were altered in expression	[102]
Expression of 623 miRNAs	Testis biopsies / Sperm	CS (*n*=3); scMF (*n*=3); SCO (*n*=3) / SpF (*n*=23)	qPCR miRNA array (Exiqon)	186 miRNAs in SCO and/or scMF groups are altered in expression; decreased cellular miRNA content in developing and mature germ cells of the SpF patients	[103]
Expression of 736 miRNA	Sperm	Asth (*n*=10); Ter (*n*=10); Oligo (*n*=10)	qPCR miRNA array (TaqMan arrays)	32 miRNA in Asth, 19 miRNA in Ter, 18 miRNA in Oligo are altered in expression	[104]
Global mRNA and miRNA expression	Testis biopsies	OA (*n*=3); NOA (*n*=4)	Microarray	1944 Upregulated and 2768 downregulated genes and 51 upregulated and 42 downregulated miRNAs	[104a]

Only studies using extensive mRNA or miRNA analysis are included.

CS, conserved spermatogenesis; *SCO*, Sertoli-cell-only syndrome; *scMF*, maturation failure at spermatocyte level; *rsMF*, maturation failure at round spermatid level; *HS*, hypospermatogenesis; *OA*, obstructive azoospermia; *MA*, mixed atrophy; *Ft*, fertile individuals; *Asth*, asthenozoospermia; *Ter*, teratozoospermia; *Oligo*, oligozoospermia; *OA*, oligoasthenozoospermia.

Altered mRNA Expression Profiles in Sperm Related to Semen Quality

The first analyses of the sperm expression profile suggested that there were differences in composition/concentration in sperm mRNA population between samples from different individuals, possibly related to semen quality [15,17]; this hypothesis has since been corroborated. A number of extensive mRNA studies, carried out to identify spermatozoa mRNA expression profiles related to semen quality, have shown differences in expression related to the morphometric characteristics of semen, such as concentration, motility, and morphology of sperm. For concentration, this includes up to a 33-fold reduction in the expression of genes involved in spermatogenesis and sperm motility, and important decreases in the expression of genes involved in DNA repair [96]. Expression of the genes related to mobility, for example, *ANXA2*, *BRD2*, and *OAZ3*, has been shown to be altered [97], whereas *RPL24*, *HNRNPM*, *RPL4*, *PRPF8*, *and HTN3* were upregulated in another study [97,99]. Transcripts encoding components of various cellular remodeling pathways, such as the ubiquitin-proteosome pathway, are severely disrupted when the sperm present an abnormal morphology [20] (Table 8.1A), and there are differences in the concentration of certain transcripts in spermatozoa among fertile and infertile men with low seminal quality [106–109].

Testicular miRNA Expression Profile in Human Spermatogenic Disorders

Not surprisingly, in view of the enormous regulatory potential of miRNAs, deregulated miRNA expression has been revealed in various disease states, such as cancer and other diseases. Regarding fertility, miRNAs are the best characterized ncRNAs in testis and sperm, and because of this, their role in male infertility of testicular origin, they are being explored further. MiRNAs actively participate in testis differentiation in the embryo, male germ-line development, and sperm production [52,55,56,62,110]. The miRNA profile has been determined in normal human testes by next-generation sequencing [111]. Song and his collaborators demonstrated that most X-linked miRNAs escape meiotic sex chromosome inactivation (MscI) during spermatogenesis, which suggest that they have a relevant role in meiotic and postmeiotic stages of sperm production [112]. Seeming to corroborate this, altered testicular miRNA expression was found to accompany nonobstructive or secretory azoospermia [100]. However, this study did not address the histological characterization of the samples, so the authors did not consider interpreting the results in relation to the arrested maturation stage of the germ line, and it is therefore possible that the alteration of the miRNA profile was a consequence of the loss of specific germ-cell stages in the pathological tissue.

Subsequent studies (Table 8.1B) have characterized the expression profile of miRNAs in different testicular histopathologic patterns in FFPE testicular samples [102] and in frozen samples [103], and these do corroborate an altered testicular miRNA signature in severe spermatogenic disorders. These studies include descriptive information about the samples. Those miRNAs that were not expressed in SCO or SpF-scMF phenotypes are interpreted as germ-cell-specific or postmeiotic miRNAs, respectively. Several of the SCO downregulated miRNAs were located on chromosome 19 [103], mapping to the nonconserved cluster composed of 54 miRNAs [113] and cluster miR-371-3 (both described as oncogenes in testicular germ-cell tumors [114]).

Interestingly, two miRNA clusters miR-34b/c and miR-449 (consisting of three members: miR-449a, miR-449b, and miR-449c) were downregulated in two different studies [102,103]. These miRNAs presented the greatest fold-change reduction in expression in SCO testis, and three of the miRNAs contained in these clusters, miR-34b, miR-34b*, and miR-449a, were also differentially reduced in the SpF-scMF phenotype [103] and in the azoospermic patients [102]. Another notable finding was that a decreased expression level of miR-34c-5p was also found in seminal plasma from histologically uncharacterized men with azoospermia [115]; this is probably related to the absence of the germ line in the testis, corroborating the testicular results. All together, these findings suggest that a widely altered miRNA molecular signature in germ cells is associated with severe spermatogenic disorders. There were, however, specific individual miRNAs that were poorly correlated between both studies; these differences may arise from the method of sample conservation (FFPE samples vs frozen samples) and/or the use of different strategies of miRNA quantification in both studies. The two strategies (microarray-based analysis vs RT-qPCR study) have different specificities and sensitivities [116,117].

Given this evidence, the testicular miRNA expression profile is a potential tool for diagnosing male infertility that could supplement histopathologic diagnosis [102] and also be used as a means of predicting the availability of sperm in the biopsy for assisted reproduction [103].

Further interesting conclusions can be drawn from the study by Muñoz et al. [103]. In spermatogenic disorders, the somatic and germ cells of the testis exhibit associated patterns of miRNA expression deregulation, and the earlier the stage of germ-cell differentiation that is affected by maturation failure, the more severe this deregulation is, which in turn contributes to spermatogenic blockade. This means that the miRNA signature in germ cells not only plays a role in the spermatogenic process but also is an indicator of its likelihood of success.

Aberrant Content of miRNA in Spermatozoa From Patients With Seminal Alterations

The results from our group additionally supported the idea that spermatozoa from patients with severe spermatogenic disorders that have nevertheless fulfilled the differentiation process still retain the altered miRNA pattern observed in the developing SpF germ cells [103]. The aberrant miRNA content of these spermatozoa was confirmed in extensive miRNA studies showing that spermatozoa from patients with seminal alterations exhibit a differential miRNA profile (Table 8.1B) [101,104].

Additional results in mice corroborate these results. Dicer and Drosha cKO male mice were infertile as a result of low sperm counts, low sperm motility, and abnormal sperm morphology, resembling oligoasthenoteratozoospermia (OAT) in humans [55,118]. Additionally, Dicer and Drosha cKO sperm were deficient in sncRNAs (nearly 50% of miRNAs were dysregulated) confirming that miRNA expression alteration initiates in early stages of spermatogenesis and remains in the mature germ cell.

TRANSCRIPTOMIC STUDIES RELATED TO THE FERTILIZING POTENTIAL OF SPERM

It is well known that 15% of couples are infertile despite presenting normal clinical values and are consequently diagnosed as having unexplained infertility. A man is traditionally considered fertile when the number of motile and morphological normal sperm is adequate, although the function of sperm is not further assessed and fertilization potential is still in question. Deciphering the underlying causes of unexplained infertility and finding new biomarkers to improve the diagnosis are two of the main challenges in reproductive medicine today. Data on differences in the levels of individual or groups of transcripts between unexplained

TABLE 8.2 Summary of Studies Determining the Sperm Transcriptomic Profile to Characterize the Differentially Expressed Genes and miRNAs in Human Altered Sperm Function

Assessment	Tissue/Cell Type	Phenotype	Technology	Results	References
Global gene expression	Sperm	UIFt-P ($n=10$ pooled samples); UIFt-NP ($n=10$ pooled samples)	Microarray	Pregnant versus nonpregnant IUI. Near 2500 genes presented differences in expression	[119]
Global gene expression	Sperm	UIFt-P ($n=5+5$, pooled); UIFt-NP ($n=5+5$, pooled)	Microarray	Pregnant versus nonpregnant ICSI. Fresh versus frozen samples. Near 200 genes presented differences in expression	[120]
Expression of 95 genes	Sperm	Normozoospermic donors ($n=68$)	qPCR mRNA array (TaqMan arrays)	Significant differences in the expression of individual genes were observed between groups of donors with the lowest and highest pregnancy rates after IUI. Developed a model based on gene expression to classify the fertility status of semen samples for IUI	[121]
Global RNA expression	Sperm	UIFt ($n=72$)	NGS-RNA-seq	648 sperm RNA elements (SRE) are required for a successful IUI. Forty percent of the SREs are within exonic regions of genes that are known to be involved in spermatogenesis, sperm motility, fertilization, and the first steps of embryogenesis prior to implantation	[122]
Global gene expression	Sperm	Ft ($n=20$); UIFt ($n=20$)	Microarray	281 Genes altered in expression	[99]
Expression of 736 miRNA	Sperm	UIFt ($n=8$); Ft ($n=10$)	qPCR miRNA array (TaqMan arrays)	57 miRNAs are differentially expressed between populations	[122a]

Only studies using extensive mRNA or miRNA analysis are included.
UIFt, unexplained infertile patients; *FT*, fertile individuals; *UIFt-P*, normozoospermic infertile individuals that achieved pregnancy after ART; *UIFt-NP*, normozoospermic infertile individuals that did not achieve pregnancy after ART.

infertile patients and fertile individuals (Table 8.2) are providing tools to decipher the molecular pathways involved in successful oocyte fertilization. Assessment of gene expression and its regulatory network in the sperm is considered crucial for this task. Furthermore, the assessment of sperm transcriptome may identify potential fertility biomarkers, complementing the basic sperm analysis to predict the fertility capability of sperm.

Gene Expression Profile Related to Sperm Function

Sperm transcript profiles may reflect, to some extent, the functional characteristics of the cells and be indicators of future reproductive success. One approach focuses on determining sperm gene expression profiles with the purpose of predicting the success of assisted reproductive techniques (ART). Studies in this area have compared sperm samples with or without ART reproductive success from couples with unexplained infertility, showing different transcript profiles associated with successful pregnancy depending on the fertility ART treatment used [119,120]. Some examples of transcripts overexpressed in samples that achieved pregnancy by the intrauterine insemination technique (IUI) were *AYTL1*, *ADAM9*, *FCGR3A*, and *OSM* among 1500 sequences (many of them are involved in testicular function, spermatogenesis, or sperm physiology) [119], whereas in pregnancies achieved by intracytoplasmic sperm injection (ICSI), *APOE*, *APOC1*, and *CFD* were among a total of 1400 upregulated genes, and only four genes were involved in sperm motility [120]. This is not surprising, because when the reproductive technique is less invasive (IUI < IVF < ICSI), the functional requirements for sperm to fertilize are higher. However, this type of analysis provides relevant information that can help to decipher the physiological pathways that are negatively affecting fertility outcomes and can help the clinician to select the ART technique that has more chances of success with a given semen.

Additional studies have provided interesting results. The approach used by our group differed from those proposed by other groups in that we used normozoospermic donors with different pregnancy rates when used in the less invasive reproductive technique: the IUI [121]. Recruitment of semen donors was carried out among young university students with unknown fertility status at the time of donation, so they were representatives of the normozoospermic general population. These results appear to confirm that sperm RNA expression differences are not only related to the morphometric characteristics of spermatozoa contained in semen but also related to sperm function and fertilization capacity. From the initial study of 85 genes in semen from normozoospermic donors with different IUI pregnancy rates, we showed a fingerprint gene expression (based on the combination of data from four genes—*EIF5A, RPL13, RPL23A, and RPS27A*) that is related to very low rates of pregnancy, and therefore, it constitutes a marker of low quality and poor fecundation capacity of the spermatozoa. This genetic fingerprint can be used as an additional screening test for semen donors (discriminating those with the worst ability to fertilize oocyte) to improve the outcome of the fertility treatments for women with suboptimal fertility, thus contributing to an increase in the effectiveness of the technique and the chances of pregnancy success.

Recent technological advances have stimulated the search for genes involved in unexplained male infertility. Jodar and collaborators identified 648 sperm RNA elements required for a successful IUI by using RNA-seq [122], and this further helps to understand the pathophysiological processes of impaired sperm function.

miRNAs Altered in Sperm From Men With Unexplained Infertility

The argument for a potential role of paternal sncRNAs in fertilization and postfertilization events [123] has been reinforced in recent years. An example is miR-34c that has been described as being essential in spermatozoa for the first cell division of the mouse embryo [124] as it is involved in the regulation of *Bcl-2* expression. Subsequently, Yuan and collaborators used ICSI mouse spermatozoa that were partially deficient in miRNAs and endo-siRNAs to study whether ablation of paternal sncRNAs could lead to defects in fertilization and/or preimplantation embryonic development [118]. These authors showed that eggs fertilized by spermatozoa with aberrant miRNA and endo-siRNA contents through ICSI display reduced preimplantation development potential and that this can be reversed when these fertilized eggs are supplemented with WT sperm-borne total RNA. These results suggested that paternal sncRNAs are important for initiating the normal development program during early preimplantation development, especially during fertilization and zygote to two-cell transition.

Future research on sperm miRNA content of infertile couples in an ART context must be encouraged, so we can determine its potential as a diagnostic and/or prognostic biomarker.

References

[1] M.N. Mascarenhas, S.R. Flaxman, T. Boerma, S. Vanderpoel, G.A. Stevens, National, regional, and global trends in infertility prevalence since 1990: a systematic analysis of 277 health surveys, PLoS Med. 9 (2012) e1001356.

[2] J. Datta, M.J. Palmer, C. Tanton, L.J. Gibson, K.G. Jones, W. Macdowall, et al., Prevalence of infertility and help seeking among 15 000 women and men, Hum. Reprod. 31 (2016) 2108–2118.

[3] T.G. Cooper, E. Noonan, S. von Eckardstein, J. Auger, H.W. Baker, H.M. Behre, et al., World Health Organization reference values for human semen characteristics, Hum. Reprod. Update 16 (2010) 231–245.

[4] A. Ferlin, F. Raicu, V. Gatta, D. Zuccarello, G. Palka, C. Foresta, Male infertility: role of genetic background, Reprod. BioMed. Online 14 (2007) 734–745.

[5] M.M. Matzuk, D.J. Lamb, The biology of infertility: research advances and clinical challenges, Nat. Med. 14 (2008) 1197–1213.

[6] J. Sha, Z. Zhou, J. Li, L. Yin, H. Yang, G. Hu, et al., Identification of testis development and spermatogenesis-related genes in human and mouse testes using cDNA arrays, Mol. Hum. Reprod. 8 (2002) 511–517.

[7] A.L. Pang, H.C. Taylor, W. Johnson, S. Alexander, Y. Chen, Y.A. Su, et al., Identification of differentially expressed genes in mouse spermatogenesis, J. Androl. 24 (2003) 899–911.

[8] N. Schultz, F.K. Hamra, D.L. Garbers, A multitude of genes expressed solely in meiotic or postmeiotic spermatogenic cells offers a myriad of contraceptive targets, Proc. Natl. Acad. Sci. U. S. A. 100 (2003) 12201–12206.

[9] U. Schlecht, P. Demougin, R. Koch, L. Hermida, C. Wiederkehr, P. Descombes, et al., Expression profiling of mammalian male meiosis and gametogenesis identifies novel candidate genes for roles in the regulation of fertility, Mol. Biol. Cell 15 (2004) 1031–1043.

[10] J.E. Shima, D.J. McLean, J.R. McCarrey, M.D. Griswold, The murine testicular transcriptome: characterizing gene expression in the testis during the progression of spermatogenesis, Biol. Reprod. 71 (2004) 319–330.

[11] S. Diederichs, N. Baumer, N. Schultz, F.K. Hamra, M.G. Schrader, M.L. Sandstede, et al., Expression patterns of mitotic and meiotic cell cycle regulators in testicular cancer and development, Int. J. Cancer 116 (2005) 207–217.

[12] S.H. Namekawa, P.J. Park, L.F. Zhang, J.E. Shima, J.R. McCarrey, M.D. Griswold, et al., Postmeiotic sex chromatin in the male germline of mice, Curr. Biol. 16 (7) (2006) 660.

[13] F. Chalmel, A.D. Rolland, C. Niederhauser-Wiederkehr, S.S. Chung, P. Demougin, A. Gattiker, et al., The conserved transcriptome in human and rodent male gametogenesis, Proc. Natl. Acad. Sci. U. S. A. 104 (2007) 8346–8351.

[14] J. Parrington, M.L. Jones, R. Tunwell, C. Devader, M. Katan, K. Swann, Phospholipase C isoforms in mammalian spermatozoa: potential components of the sperm factor that causes Ca2+ release in eggs, Reproduction 123 (2002) 31–39.
[15] G.C. Ostermeier, D.J. Dix, D. Miller, P. Khatri, S.A. Krawetz, Spermatozoal RNA profiles of normal fertile men, Lancet 360 (2002) 772–777.
[16] S.A. Krawetz, Paternal contribution: new insights and future challenges, Nat. Rev. Genet. 6 (2005) 633–642.
[17] Y. Zhao, Q. Li, C. Yao, Z. Wang, Y. Zhou, Y. Wang, et al., Characterization and quantification of mRNA transcripts in ejaculated spermatozoa of fertile men by serial analysis of gene expression, Hum. Reprod. 21 (2006) 1583–1590.
[18] Y. Gur, H. Breitbart, Mammalian sperm translate nuclear-encoded proteins by mitochondrial-type ribosomes, Genes Dev. 20 (2006) 411–416.
[19] Y. Gur, H. Breitbart, Protein synthesis in sperm: dialog between mitochondria and cytoplasm, Mol. Cell. Endocrinol. 282 (2008) 45–55.
[20] A.E. Platts, D.J. Dix, H.E. Chemes, K.E. Thompson, R. Goodrich, J.C. Rockett, et al., Success and failure in human spermatogenesis as revealed by teratozoospermic RNAs, Hum. Mol. Genet. 16 (2007) 763–773.
[21] G.C. Ostermeier, D. Miller, J.D. Huntriss, M.P. Diamond, S.A. Krawetz, Reproductive biology: delivering spermatozoan RNA to the oocyte, Nature 429 (2004) 154.
[22] P. Braude, V. Bolton, S. Moore, Human gene expression first occurs between the four- and eight-cell stages of preimplantation development, Nature 332 (1988) 459–461.
[23] J.P. Siffroi, J.P. Dadoune, Accumulation of transcripts in the mature human sperm nucleus: implication of the haploid genome in a functional role, Ital. J. Anat. Embryol. 106 (2001) 189–197.
[24] A. Boerke, S.J. Dieleman, B.M. Gadella, A possible role for sperm RNA in early embryo development, Theriogenology 68 (Suppl. 1) (2007) S147–S155.
[25] P.P. Amaral, M.E. Dinger, T.R. Mercer, J.S. Mattick, The eukaryotic genome as an RNA machine, Science 319 (2008) 1787–1789.
[26] T.A. Farazi, S.A. Juranek, T. Tuschl, The growing catalog of small RNAs and their association with distinct Argonaute/Piwi family members, Development 135 (2008) 1201–1214.
[27] V.N. Kim, J. Han, M.C. Siomi, Biogenesis of small RNAs in animals, Nat. Rev. Mol. Cell Biol. 10 (2009) 126–139.
[28] M. Kawano, H. Kawaji, V. Grandjean, J. Kiani, M. Rassoulzadegan, Novel small noncoding RNAs in mouse spermatozoa, zygotes and early embryos, PLoS One 7 (2012) e44542.
[29] S.A. Krawetz, A. Kruger, C. Lalancette, R. Tagett, E. Anton, S. Draghici, et al., A survey of small RNAs in human sperm, Hum. Reprod. 26 (12) (2011) 3401.
[30] T. Derrien, R. Johnson, G. Bussotti, A. Tanzer, S. Djebali, H. Tilgner, et al., The GENCODE v7 catalog of human long noncoding RNAs: analysis of their gene structure, evolution, and expression, Genome Res. 22 (2012) 1775–1789.
[31] S. Djebali, C.A. Davis, A. Merkel, A. Dobin, T. Lassmann, A. Mortazavi, et al., Landscape of transcription in human cells, Nature 489 (2012) 101–108.
[32] C.P. Ponting, P.L. Oliver, W. Reik, Evolution and functions of long noncoding RNAs, Cell 136 (2009) 629–641.
[33] A. Gabory, H. Jammes, L. Dandolo, The H19 locus: role of an imprinted non-coding RNA in growth and development, BioEssays 32 (2010) 473–480.
[34] T. Hung, Y. Wang, M.F. Lin, A.K. Koegel, Y. Kotake, G.D. Grant, et al., Extensive and coordinated transcription of noncoding RNAs within cell-cycle promoters, Nat. Genet. 43 (2011) 621–629.
[35] L. Zhang, H. Lu, D. Xin, H. Cheng, R. Zhou, A novel ncRNA gene from mouse chromosome 5 trans-splices with Dmrt1 on chromosome 19, Biochem. Biophys. Res. Commun. 400 (2010) 696–700.
[36] E.G. Berghoff, M.F. Clark, S. Chen, I. Cajigas, D.E. Leib, J.D. Kohtz, Evf2 (Dlx6as) lncRNA regulates ultraconserved enhancer methylation and the differential transcriptional control of adjacent genes, Development 140 (2013) 4407–4416.
[37] K.L. Yap, S. Li, A.M. Munoz-Cabello, S. Raguz, L. Zeng, S. Mujtaba, et al., Molecular interplay of the noncoding RNA ANRIL and methylated histone H3 lysine 27 by polycomb CBX7 in transcriptional silencing of INK4a, Mol. Cell 38 (2010) 662–674.
[38] C.A. Klattenhoff, J.C. Scheuermann, L.E. Surface, R.K. Bradley, P.A. Fields, M.L. Steinhauser, et al., Braveheart, a long noncoding RNA required for cardiovascular lineage commitment, Cell 152 (2013) 570–583.
[39] A. Keniry, D. Oxley, P. Monnier, M. Kyba, L. Dandolo, G. Smits, et al., The H19 lincRNA is a developmental reservoir of miR-675 that suppresses growth and Igf1r, Nat. Cell Biol. 14 (2012) 659–665.

[40] M. Liang, W. Li, H. Tian, T. Hu, L. Wang, Y. Lin, et al., Sequential expression of long noncoding RNA as mRNA gene expression in specific stages of mouse spermatogenesis, Sci. Rep. 4 (2014) 5966.
[41] A.C. Luk, W.Y. Chan, O.M. Rennert, T.L. Lee, Long noncoding RNAs in spermatogenesis: insights from recent high-throughput transcriptome studies, Reproduction 147 (2014) R131–R141.
[42] K.T. Nishant, H. Ravishankar, M.R. Rao, Characterization of a mouse recombination hot spot locus encoding a novel non-protein-coding RNA, Mol. Cell. Biol. 24 (2004) 5620–5634.
[43] M.C. Anguera, W. Ma, D. Clift, S. Namekawa, R.J. Kelleher 3rd, J.T. Lee, Tsx produces a long noncoding RNA and has general functions in the germline, stem cells, and brain, PLoS Genet. 7 (2011) e1002248.
[44] M. Lü, H. Tian, Y.X. Cao, X. He, L. Chen, X. Song, et al., Downregulation of miR-320a/383-sponge-like long non-coding RNA NLC1-C (narcolepsy candidate-region 1 genes) is associated with male infertility and promotes testicular embryonal carcinoma cell proliferation, Cell Death Dis. 6 (2015) e1960.
[45] A.M. Denli, B.B. Tops, R.H. Plasterk, R.F. Ketting, G.J. Hannon, Processing of primary microRNAs by the microprocessor complex, Nature 432 (2004) 231–235.
[46] M. Faller, D. Toso, M. Matsunaga, I. Atanasov, R. Senturia, Y. Chen, et al., DGCR8 recognizes primary transcripts of microRNAs through highly cooperative binding and formation of higher-order structures, RNA 16 (2010) 1570–1583.
[47] R.I. Gregory, K.P. Yan, G. Amuthan, T. Chendrimada, B. Doratotaj, N. Cooch, et al., The microprocessor complex mediates the genesis of microRNAs, Nature 432 (2004) 235–240.
[48] M. Landthaler, A. Yalcin, T. Tuschl, The human DiGeorge syndrome critical region gene 8 and its *D. melanogaster* homolog are required for miRNA biogenesis, Curr. Biol. 14 (2004) 2162–2167.
[49] E. Lund, J.E. Dahlberg, Substrate selectivity of exportin 5 and Dicer in the biogenesis of microRNAs, Cold Spring Harb. Symp. Quant. Biol. 71 (2006) 59–66.
[50] V.K. Gangaraju, H. Lin, MicroRNAs: key regulators of stem cells, Nat. Rev. Mol. Cell Biol. 10 (2009) 116–125.
[51] B.P. Lewis, C.B. Burge, D.P. Bartel, Conserved seed pairing, often flanked by adenosines, indicates that thousands of human genes are microRNA targets, Cell 120 (2005) 15–20.
[52] S. Ro, C. Park, K.M. Sanders, J.R. McCarrey, W. Yan, Cloning and expression profiling of testis-expressed microRNAs, Dev. Biol. 311 (2007) 592–602.
[53] E. Marcon, T. Babak, G. Chua, T. Hughes, P.B. Moens, miRNA and piRNA localization in the male mammalian meiotic nucleus, Chromosom. Res. 16 (2008) 243–260.
[54] L. Smorag, Y. Zheng, J. Nolte, U. Zechner, W. Engel, D.V. Pantakani, MicroRNA signature in various cell types of mouse spermatogenesis: evidence for stage-specifically expressed miRNA-221, -203 and -34b-5p mediated spermatogenesis regulation, Biol. Cell 104 (2012) 677–692.
[55] K. Hayashi, S.M. Chuva de Sousa Lopes, M. Kaneda, F. Tang, P. Hajkova, K. Lao, et al., MicroRNA biogenesis is required for mouse primordial germ cell development and spermatogenesis, PLoS One 3 (2008) e1738.
[56] D.M. Maatouk, K.L. Loveland, M.T. McManus, K. Moore, B.D. Harfe, Dicer1 is required for differentiation of the mouse male germline, Biol. Reprod. 79 (2008) 696–703.
[57] H.M. Korhonen, O. Meikar, R.P. Yadav, M.D. Papaioannou, Y. Romero, M. Da Ros, et al., Dicer is required for haploid male germ cell differentiation in mice, PLoS One 6 (2011) e24821.
[58] Y. Romero, O. Meikar, M.D. Papaioannou, B. Conne, C. Grey, M. Weier, et al., Dicer1 depletion in male germ cells leads to infertility due to cumulative meiotic and spermiogenic defects, PLoS One 6 (2011) e25241.
[59] M.D. Papaioannou, J.L. Pitetti, S. Ro, C. Park, F. Aubry, O. Schaad, et al., Sertoli cell Dicer is essential for spermatogenesis in mice, Dev. Biol. 326 (2009) 250–259.
[60] Z. Yu, T. Raabe, N.B. Hecht, MicroRNA Mirn122a reduces expression of the posttranscriptionally regulated germ cell transition protein 2 (Tnp2) messenger RNA (mRNA) by mRNA cleavage, Biol. Reprod. 73 (2005) 427–433.
[61] O. Barad, E. Meiri, A. Avniel, R. Aharonov, A. Barzilai, I. Bentwich, et al., MicroRNA expression detected by oligonucleotide microarrays: system establishment and expression profiling in human tissues, Genome Res. 14 (2004) 2486–2494.
[62] F. Bouhallier, N. Allioli, F. Lavial, F. Chalmel, M.H. Perrard, P. Durand, et al., Role of miR-34c microRNA in the late steps of spermatogenesis, RNA 16 (2010) 720–731.
[63] J. Bao, D. Li, L. Wang, J. Wu, Y. Hu, Z. Wang, et al., MicroRNA-449 and microRNA-34b/c function redundantly in murine testes by targeting E2F transcription factor-retinoblastoma protein (E2F-pRb) pathway, J. Biol. Chem. 287 (2012) 21686–21698.
[64] S. Yuan, C. Tang, Y. Zhang, J. Wu, J. Bao, H. Zheng, et al., mir-34b/c and mir-449a/b/c are required for spermatogenesis, but not for the first cleavage division in mice, Biol. Open 4 (2015) 212–223.

[65] S. Hilz, E.A. Fogarty, A.J. Modzelewski, P.E. Cohen, A. Grimson, Transcriptome profiling of the developing male germ line identifies the miR-29 family as a global regulator during meiosis, RNA Biol. 14 (2017) 219–235.
[66] H. Tian, Y.X. Cao, X.S. Zhang, W.P. Liao, Y.H. Yi, J. Lian, et al., The targeting and functions of miRNA-383 are mediated by FMRP during spermatogenesis, Cell Death Dis. 4 (2013) e617.
[67] J.M. Huszar, C.J. Payne, MicroRNA 146 (Mir146) modulates spermatogonial differentiation by retinoic acid in mice, Biol. Reprod. 88 (2013) 15.
[68] J. Chen, T. Cai, C. Zheng, X. Lin, G. Wang, S. Liao, et al., MicroRNA-202 maintains spermatogonial stem cells by inhibiting cell cycle regulators and RNA binding proteins, Nucleic Acids Res. (2016).
[69] Q.E. Yang, K.E. Racicot, A.V. Kaucher, M.J. Oatley, J.M. Oatley, MicroRNAs 221 and 222 regulate the undifferentiated state in mammalian male germ cells, Development 140 (2013) 280–290.
[70] A. Salas-Huetos, J. Blanco, F. Vidal, J.M. Mercader, N. Garrido, E. Anton, New insights into the expression profile and function of micro-ribonucleic acid in human spermatozoa, Fertil. Steril. 102 (2014) 213–222 (e4).
[71] L. Pantano, M. Jodar, M. Bak, J.L. Ballesca, N. Tommerup, R. Oliva, et al., The small RNA content of human sperm reveals pseudogene-derived piRNAs complementary to protein-coding genes, RNA 21 (2015) 1085–1095.
[72] D.E. Golden, V.R. Gerbasi, E.J. Sontheimer, An inside job for siRNAs, Mol. Cell 31 (2008) 309–312.
[73] O.H. Tam, A.A. Aravin, P. Stein, A. Girard, E.P. Murchison, S. Cheloufi, et al., Pseudogene-derived small interfering RNAs regulate gene expression in mouse oocytes, Nature 453 (2008) 534–538.
[74] T. Watanabe, Y. Totoki, A. Toyoda, M. Kaneda, S. Kuramochi-Miyagawa, Y. Obata, et al., Endogenous siRNAs from naturally formed dsRNAs regulate transcripts in mouse oocytes, Nature 453 (2008) 539–543.
[75] J.E. Babiarz, J.G. Ruby, Y. Wang, D.P. Bartel, R. Blelloch, Mouse ES cells express endogenous shRNAs, siRNAs, and other microprocessor-independent, Dicer-dependent small RNAs, Genes Dev. 22 (2008) 2773–2785.
[76] R. Song, G.W. Hennig, Q. Wu, C. Jose, H. Zheng, W. Yan, Male germ cells express abundant endogenous siRNAs, Proc. Natl. Acad. Sci. U. S. A. 108 (2011) 13159–13164.
[77] C. Zimmermann, Y. Romero, M. Warnefors, A. Bilican, C. Borel, L.B. Smith, et al., Germ cell-specific targeting of DICER or DGCR8 reveals a novel role for endo-siRNAs in the progression of mammalian spermatogenesis and male fertility, PLoS One 9 (2014) e107023.
[78] V.V. Vagin, A. Sigova, C. Li, H. Seitz, V. Gvozdev, P.D. Zamore, A distinct small RNA pathway silences selfish genetic elements in the germline, Science 313 (2006) 320–324.
[79] X.Z. Li, C.K. Roy, M.J. Moore, P.D. Zamore, Defining piRNA primary transcripts, Cell Cycle 12 (2013) 1657–1658.
[80] A.A. Aravin, R. Sachidanandam, D. Bourc'his, C. Schaefer, D. Pezic, K.F. Toth, et al., A piRNA pathway primed by individual transposons is linked to de novo DNA methylation in mice, Mol. Cell 31 (2008) 785–799.
[81] C. Klattenhoff, W. Theurkauf, Biogenesis and germline functions of piRNAs, Development 135 (2008) 3–9.
[82] E.M. Weick, E.A. Miska, piRNAs: from biogenesis to function, Development 141 (2014) 3458–3471.
[83] M.A. Carmell, A. Girard, H.J. van de Kant, D. Bourc'his, T.H. Bestor, D.G. de Rooij, et al., MIWI2 is essential for spermatogenesis and repression of transposons in the mouse male germline, Dev. Cell 12 (2007) 503–514.
[84] J. Bao, Y. Zhang, A.S. Schuster, N. Ortogero, E.E. Nilsson, M.K. Skinner, et al., Conditional inactivation of Miwi2 reveals that MIWI2 is only essential for prospermatogonial development in mice, Cell Death Differ. 21 (2014) 783–796.
[85] Y. Unhavaithaya, Y. Hao, E. Beyret, H. Yin, S. Kuramochi-Miyagawa, T. Nakano, et al., MILI, a PIWI-interacting RNA-binding protein, is required for germ line stem cell self-renewal and appears to positively regulate translation, J. Biol. Chem. 284 (2009) 6507–6519.
[86] W. Deng, H. Lin, miwi, a murine homolog of piwi, encodes a cytoplasmic protein essential for spermatogenesis, Dev. Cell 2 (2002) 819–830.
[87] T. Watanabe, E.C. Cheng, M. Zhong, H. Lin, Retrotransposons and pseudogenes regulate mRNAs and lncRNAs via the piRNA pathway in the germline, Genome Res. 25 (2015) 368–380.
[88] S. Kuramochi-Miyagawa, T. Watanabe, K. Gotoh, Y. Totoki, A. Toyoda, M. Ikawa, et al., DNA methylation of retrotransposon genes is regulated by Piwi family members MILI and MIWI2 in murine fetal testes, Genes Dev. 22 (2008) 908–917.
[89] H. Heyn, H.J. Ferreira, L. Bassas, S. Bonache, S. Sayols, J. Sandoval, et al., Epigenetic disruption of the PIWI pathway in human spermatogenic disorders, PLoS One 7 (2012) e47892.
[90] M.S. Fox, V.X. Ares, P.J. Turek, C. Haqq, R.A. Reijo Pera, Feasibility of global gene expression analysis in testicular biopsies from infertile men, Mol. Reprod. Dev. 66 (2003) 403–421.

[91] J.C. Rockett, P. Patrizio, J.E. Schmid, N.B. Hecht, D.J. Dix, Gene expression patterns associated with infertility in humans and rodent models, Mutat. Res. 549 (2004) 225–240.
[92] C. Feig, C. Kirchhoff, R. Ivell, O. Naether, W. Schulze, A.N. Spiess, A new paradigm for profiling testicular gene expression during normal and disturbed human spermatogenesis, Mol. Hum. Reprod. 13 (2007) 33–43.
[93] P.J. Ellis, R.A. Furlong, S.J. Conner, J. Kirkman-Brown, M. Afnan, C. Barratt, et al., Coordinated transcriptional regulation patterns associated with infertility phenotypes in men, J. Med. Genet. 44 (2007) 498–508.
[94] A.N. Spiess, C. Feig, W. Schulze, F. Chalmel, H. Cappallo-Obermann, M. Primig, et al., Cross-platform gene expression signature of human spermatogenic failure reveals inflammatory-like response, Hum. Reprod. 22 (2007) 2936–2946.
[95] C. Steilmann, M.C. Cavalcanti, M. Bergmann, S. Kliesch, W. Weidner, K. Steger, Aberrant mRNA expression of chromatin remodelling factors in round spermatid maturation arrest compared with normal human spermatogenesis, Mol. Hum. Reprod. 16 (2010) 726–733.
[96] D. Montjean, P. De La Grange, D. Gentien, A. Rapinat, S. Belloc, P. Cohen-Bacrie, et al., Sperm transcriptome profiling in oligozoospermia, J. Assist. Reprod. Genet. 29 (2012) 3–10.
[97] M. Jodar, S. Kalko, J. Castillo, J.L. Ballesca, R. Oliva, Differential RNAs in the sperm cells of asthenozoospermic patients, Hum. Reprod. 27 (2012) 1431–1438.
[98] S. Bonache, F. Algaba, E. Franco, L. Bassas, S. Larriba, Altered gene expression signature of early stages of the germ line supports the pre-meiotic origin of human spermatogenic failure, Andrology 2 (2014) 596–606.
[99] S.K. Bansal, N. Gupta, S.N. Sankhwar, S. Rajender, Differential genes expression between fertile and infertile spermatozoa revealed by transcriptome analysis, PLoS One 10 (2015) e0127007.
[100] J. Lian, X. Zhang, H. Tian, N. Liang, Y. Wang, C. Liang, et al., Altered microRNA expression in patients with non-obstructive azoospermia, Reprod. Biol. Endocrinol. 7 (2009) 13.
[101] M. Abu-Halima, M. Hammadeh, J. Schmitt, P. Leidinger, A. Keller, E. Meese, et al., Altered microRNA expression profiles of human spermatozoa in patients with different spermatogenic impairments, Fertil. Steril. 99 (2013) 1249–1255. e16.
[102] M. Abu-Halima, C. Backes, P. Leidinger, A. Keller, A.M. Lubbad, M. Hammadeh, et al., MicroRNA expression profiles in human testicular tissues of infertile men with different histopathologic patterns, Fertil. Steril. 101 (2014) 78–86. e2.
[103] X. Muñoz, A. Mata, L. Bassas, S. Larriba, Altered miRNA signature of developing germ-cells in infertile patients relates to the severity of spermatogenic failure and persists in spermatozoa, Sci. Rep. 5 (2015) 17991.
[104] A. Salas-Huetos, J. Blanco, F. Vidal, A. Godo, M. Grossmann, M.C. Pons, et al., Spermatozoa from patients with seminal alterations exhibit a differential micro-ribonucleic acid profile, Fertil. Steril. 104 (2015) 591–601.
[104a] X. Zhuang, Z. Li, H. Lin, L. Gu, Q. Lin, Z. Lu, C.M. Tzeng, Integrated miRNA and mRNA expression profiling to identify mRNA targets of dysregulated miRNAs in non-obstructive azoospermia, Sci. Rep. 5 (2015) 7922.
[105] E. Terribas, S. Bonache, M. Garcia-Arevalo, J. Sanchez, E. Franco, L. Bassas, et al., Changes in the expression profile of the meiosis-involved mismatch repair genes in impaired human spermatogenesis, J. Androl. 31 (2010) 346–357.
[106] X. Guo, Y.T. Gui, A.F. Tang, L.H. Lu, X. Gao, Z.M. Cai, Differential expression of VASA gene in ejaculated spermatozoa from normozoospermic men and patients with oligozoospermia, Asian J. Androl. 9 (2007) 339–344.
[107] S. Carreau, S. Lambard, L. Said, A. Saad, I. Galeraud-Denis, RNA dynamics of fertile and infertile spermatozoa, Biochem. Soc. Trans. 35 (2007) 634–636.
[108] K. Steger, J. Wilhelm, L. Konrad, T. Stalf, R. Greb, T. Diemer, et al., Both protamine-1 to protamine-2 mRNA ratio and Bcl2 mRNA content in testicular spermatids and ejaculated spermatozoa discriminate between fertile and infertile men, Hum. Reprod. 23 (2008) 11–16.
[109] C. Avendano, A. Franchi, E. Jones, S. Oehninger, Pregnancy-specific {beta}-1-glycoprotein 1 and human leukocyte antigen-E mRNA in human sperm: differential expression in fertile and infertile men and evidence of a possible functional role during early development, Hum. Reprod. 24 (2009) 270–277.
[110] J. Rakoczy, S.L. Fernandez-Valverde, E.A. Glazov, E.N. Wainwright, T. Sato, S. Takada, et al., MicroRNAs-140-5p/140-3p modulate Leydig cell numbers in the developing mouse testis, Biol. Reprod. 88 (2013) 143.
[111] Q. Yang, J. Hua, L. Wang, B. Xu, H. Zhang, N. Ye, et al., MicroRNA and piRNA profiles in normal human testis detected by next generation sequencing, PLoS One 8 (2013) e66809.
[112] R. Song, S. Ro, J.D. Michaels, C. Park, J.R. McCarrey, W. Yan, Many X-linked microRNAs escape meiotic sex chromosome inactivation, Nat. Genet. 41 (2009) 488–493.

[113] I. Bentwich, A. Avniel, Y. Karov, R. Aharonov, S. Gilad, O. Barad, et al., Identification of hundreds of conserved and nonconserved human microRNAs, Nat. Genet. 37 (2005) 766–770.
[114] P.M. Voorhoeve, C. le Sage, M. Schrier, A.J. Gillis, H. Stoop, R. Nagel, et al., A genetic screen implicates miRNA-372 and miRNA-373 as oncogenes in testicular germ cell tumors, Cell 124 (2006) 1169–1181.
[115] C. Wang, C. Yang, X. Chen, B. Yao, C. Yang, C. Zhu, et al., Altered profile of seminal plasma microRNAs in the molecular diagnosis of male infertility, Clin. Chem. 57 (2011) 1722–1731.
[116] Y. Chen, J.A. Gelfond, L.M. McManus, P.K. Shireman, Reproducibility of quantitative RT-PCR array in miRNA expression profiling and comparison with microarray analysis, BMC Genomics 10 (2009) 407.
[117] J. Koshiol, E. Wang, Y. Zhao, F. Marincola, M.T. Landi, Strengths and limitations of laboratory procedures for microRNA detection, Cancer Epidemiol. Biomark. Prev. 19 (2010) 907–911.
[118] S. Yuan, A. Schuster, C. Tang, T. Yu, N. Ortogero, J. Bao, et al., Sperm-borne miRNAs and endo-siRNAs are important for fertilization and preimplantation embryonic development, Development 143 (2016) 635–647.
[119] S. Garcia-Herrero, M. Meseguer, J.A. Martinez-Conejero, J. Remohi, A. Pellicer, N. Garrido, The transcriptome of spermatozoa used in homologous intrauterine insemination varies considerably between samples that achieve pregnancy and those that do not, Fertil. Steril. 94 (2010) 1360–1373.
[120] S. Garcia-Herrero, N. Garrido, J.A. Martinez-Conejero, J. Remohi, A. Pellicer, M. Meseguer, Differential transcriptomic profile in spermatozoa achieving pregnancy or not via ICSI, Reprod. BioMed. Online 22 (2011) 25–36.
[121] S. Bonache, A. Mata, M.D. Ramos, L. Bassas, S. Larriba, Sperm gene expression profile is related to pregnancy rate after insemination and is predictive of low fecundity in normozoospermic men, Hum. Reprod. 27 (2012) 1556–1567.
[122] M. Jodar, E. Sendler, S.I. Moskovtsev, C.L. Librach, R. Goodrich, S. Swanson, et al., Absence of sperm RNA elements correlates with idiopathic male infertility, Sci. Transl. Med. 7 (2015) 295re6.
[122a] A. Salas-Huetos, J. Blanco, F. Vidal, M. Grossmann, M.C. Pons, N. Garrido, E. Anton, Spermatozoa from normozoospermic fertile and infertile individuals convey a distinct miRNA cargo, Andrology 4 (2016) 1028–1036.
[123] M. Amanai, M. Brahmajosyula, A.C. Perry, A restricted role for sperm-borne microRNAs in mammalian fertilization, Biol. Reprod. 75 (2006) 877–884.
[124] W.M. Liu, R.T. Pang, P.C. Chiu, B.P. Wong, K. Lao, K.F. Lee, et al., Sperm-borne microRNA-34c is required for the first cleavage division in mouse, Proc. Natl. Acad. Sci. U. S. A. 109 (2012) 490–494.

CHAPTER

9

Proteomics and Metabolomics Studies and Clinical Outcomes

Giulia Mariani[*], *José Bellver*[†,‡,§]

[*]Department of Biomedical, Experimental and Clinical Sciences, Division of Obstetrics and Gynecology, University of Florence, Florence, Italy [†]Instituto Valenciano de Infertilidad, University of Valencia, Valencia, Spain [‡]Department of Pediatrics, Obstetrics and Gynecology, Faculty of Medicine, University of Valencia, Valencia, Spain [§]IVI Foundation, Catedrático Agustín Escardino 9, Parc Científic, University of Valencia, Valencia, Spain

INTRODUCTION

The increasing use of OMICS technologies in the field of reproductive biology comes from an urgent need to find new diagnostic and therapeutic tools that improve the efficacy of assisted reproductive techniques. These new fields of molecular biology provide a large amount of information regarding biological processes at a relatively low cost and with little effort. This allows us to better understand, the molecular physiology of germ cells, embryos, and the endometrium, which are the three main players responsible for the success of reproductive treatments [1]. With the widespread use of OMICS technologies, the current assessment of germ cells, embryos, and the endometrium quality, which are mainly based on subjective morphological criteria, could be replaced by biomolecular analysis. By identifying the best gametes, which, in turn, produced the best embryos, it is possible to increase fertilization, implantation, and live birth rates in assisted reproductive treatments. The new concept of "reproductomics" has been coined to refer to the application of these technologies in the field of human reproduction [2].

Following the sequencing of the human genome, it has been clear that genetic information alone is not sufficient to provide a complete description of the cellular and biochemical mechanisms of complex biological systems. Moreover, it has been demonstrated that mRNA is not always translated entirely into proteins [3] and that DNA-RNA may not be strictly related [4]. For these reasons, the field of molecular biology has seen the development of new

Reproductomics
https://doi.org/10.1016/B978-0-12-812571-7.00010-1

research methods such as proteomics and metabolomics that, through the systematic analysis of proteins and metabolites present in biological samples, cells, tissues, or organisms, can significantly improve our understanding of a biological system with respect to genomics. Proteomics studies all the identifying features of a protein: the amino acid sequence, three-dimensional structure, function, posttranslational modifications, quantity, and interaction with other proteins or with DNA or RNA. Metabolomics studies the entire set of metabolites present within an organism, cell, or tissue and the interactions between them. Differences in protein and metabolite expression levels may be related to pathological conditions or alterations of the metabolism. Identifying specific biomarkers could be helpful in the diagnosis and treatment of human reproductive pathologies [5] such as infertility and has the potential to be useful in the selection of the best gametes and embryos, as well as the most receptive endometrium.

Proteomics in Reproductive Medicine

Proteomics, a term coined due to the analogy with genomics, studies the changes in all proteins expressed and translated from a single genome [6]. The term proteome, a fusion between protein and genome, refers to all of the proteins produced by the cells and released into the surrounding biological fluid. It represents a constantly changing entity depending on the stimuli the cell receives. It is interesting to note that the biological fluid consists of proteins that have spilled out from the cells in response to certain stimuli and that the same cells are simultaneously influenced by the concentration and type of proteins present in the surrounding medium. The set of proteins expressed by a cell, an organism, or a biological fluid are enormous, varying from cell to cell and from moment to moment, and depend on the interaction of multiple factors. Therefore, proteomics is far more complex than genomics, since an organism's genome remains almost unchanged. Proteome complexity is also related to posttranslational modifications of proteins expressed by thousands of genes. These modifications consist of structural and functional changes after protein synthesis, with phosphorylation, glycosylation, acetylation, and methylation constituting the main players [7].

Several proteomic techniques are available for investigating cellular functions in human reproduction [8] (Fig. 9.1). Traditional technologies involve two-dimensional electrophoresis (2-D), gel-based methods such as one-dimensional sodium dodecyl sulfate-polyacrylamide gel electrophoresis (1-D-SDS-PAGE) [9,10], 2-D-SDS-PAGE [11,12], and 2-D differential in-gel electrophoresis (2-D-DIGE) [13,14]. New proteomic techniques include mass spectrometry (MS) [15], high-performance liquid chromatography (HPLC) and ultra performance liquid chromatography (UPLC) [16,17], reverse-phase liquid chromatography-mass spectrometry (RP-LC-MS/MS) [18], isotope-coded affinity tags (ICAT), proteome arrays and proteome bioinformatics [19], and soft ionization methods such as surface-enhanced laser desorption/ionization time of flight (SELDI-TOF) and matrix-assisted laser desorption ionization time of flight (MALDI-TOF) [20].

Several studies regarding male and female reproductive function have been carried out to define the physiological role of every protein at the tissue level and to identify the proteins involved in various reproductive health disorders and IVF failures.

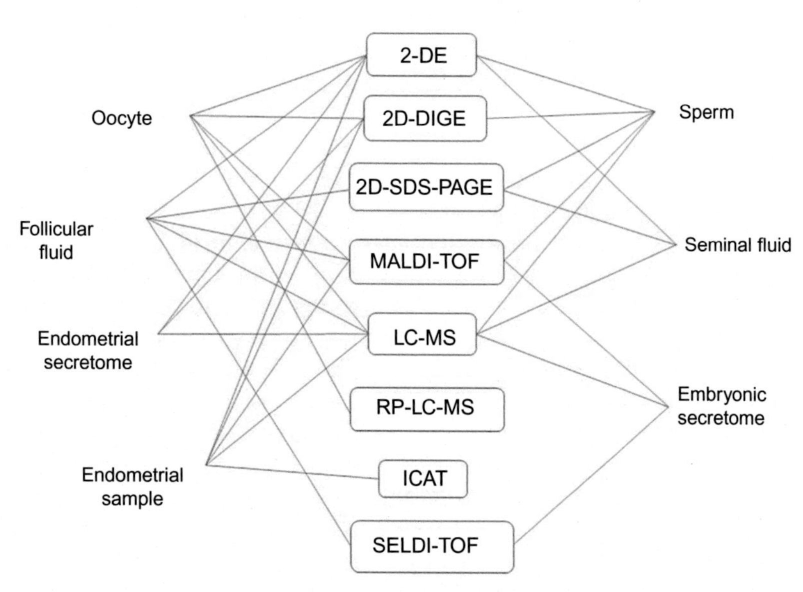

FIG. 9.1 Schematic representation of proteomic techniques in human reproduction. 2-D, Two-dimensional electrophoresis; 2-D-DIGE, two-dimensional differential in-gel electrophoresis; 2-D-SDS-PAGE, two-dimensional sodium dodecyl sulfate-polyacrylamide gel electrophoresis; MALDI-TOF, matrix-assisted laser desorption ionization coupled to time of flight; LC-MS, liquid chromatography-mass spectrometry; RP-LC-MS, reverse-phase liquid chromatography-mass spectrometry; ICAT, isotope-coded affinity tags; SELDI-TOF, surface-enhanced laser desorption/ionization coupled to time of flight. *Adapted from R.D. Upadhyay, N.H. Balasinor, A.V. Kumar, G. Sachdeva, P. Parte, K. Dumasia, Proteomics in reproductive biology: beacon for unraveling the molecular complexities, Biochim. Biophys. Acta. 1834 (2013) 8–15.*

Oocyte

Recently, many efforts have been made in assisted reproduction to identify noninvasive oocyte selection methods. Currently, the best oocyte for insemination is selected using morphological assessment. Since the oocyte proteome can represent a direct marker of oocyte maturation and competency, proteomics has been proposed as a more objective technique for gamete selection. The identification of cumulus-oocyte complex (COC) biomarkers and the definition of their biological role are mandatory steps to better understand molecular mechanisms of reproductive processes such as oocyte maturation and follicle development. Despite growing interest in the application of proteomics to reproductive medicine, little is still known about human oocyte proteins and their functions. The main limitation is related to the large amount of material required to perform a molecular analysis. The little information currently available has been obtained from mammalian species, mainly mice. It is interesting to note that a mature oocyte can express the whole complex of maternal proteins necessary to complete the fertilization process, the zygotic gene activation, and the initial stages of embryogenesis [21].

Analyzing the protein profiles of a mature mouse COC using 2-D associated with MS, Meng et al. [22] identified 156 proteins with different biological functions, such as gene/protein expression, cell metabolism, cell/organism defense, and cell signaling/communication. Combining 2-D, MS, and phosphoprotein staining, Ma et al. [21] detected 380 proteins in metaphase II (MII) mature mouse oocytes without zona pellucida. Thanks to the use of a more effective proteomic approach, 1-D-SDS-PAGE and RP-LC-MS/MS, Zhang et al. [23] identified 625 different proteins extracted from 2700 mature mouse oocytes lacking their zona pellucida. Bioinformatic analysis has demonstrated a correlation between different protein compositions and oocyte characteristics at different developmental stages. More recently, the number of proteins identified in MII mouse oocytes has grown to 3699, of which 28 have also been detected in the proteome of undifferentiated mouse embryo stem cells. The majority of proteins are associated with nuclear reprogramming of proteins, RNAs, lipids, and small molecules [24]. The results of these studies have allowed the characterization of important maternal proteins involved in oogenesis, fertilization, and early embryonic development. Analysis of the mouse oocyte proteome at different developmental stages using semiquantitative MS analysis has revealed 2781 proteins in germinal vesicle oocytes, 2973 in MII oocytes, and 2082 proteins in zygotes [25]. Bioinformatic analysis has revealed a correlation between different protein compositions and oocyte maturation characteristics. For example, MII oocytes contain proteins that participate in the regulation of cell-cycle events and epigenetic modifications.

Sheehan et al. [26] demonstrated the impact of ovarian stimulation with exogenous gonadotropins on the proteome of maturing mouse oocytes. Minimal differences were observed between the proteome of MII oocytes primed only with human chorionic gonadotropin (hCG) and naturally ovulated oocytes. However, the proteome of MII oocytes collected after ovarian stimulation with exogenous gonadotropins was significantly different from that of MII oocytes from females primed only with hCG and naturally ovulated oocytes. These results revealed that, in the mouse model, exogenous gonadotropins influence the circulating follicular endocrine environment and hence can affect oocyte quality. In humans, the comparison of the protein expression of the cumulus cells (CCs) of MII oocytes obtained after two different ovarian stimulation protocols (recombinant follicle-stimulating hormone (rFSH) vs. human menopausal gonadotropin (hMG)) in the same patient showed that the type of hormonal treatment influences CC protein expression [27]. Proteomic analysis has also revealed an impact of maternal aging on the protein profile of human CCs. A total of 1423 proteins were identified in CCs of eight infertile 40–45-year-old women, and in CCs of eight young oocyte donors, 110 were significantly differentially expressed [28].

Proteomic analysis has been used to examine the effects of the type of cryopreservation procedure on oocyte physiology. In the mouse model, the two most common oocyte cryopreservation techniques, vitrification and slow freezing, were compared in in vivo-derived MII oocytes. Results showed that vitrified MII oocytes had a similar proteome to noncryopreserved controls, whereas MII oocytes that underwent slow freezing exhibited a different protein profile, with up- and downregulation of 19 positively charged and 21 negatively charged proteins [29]. The identification of these differentially expressed proteins is the goal of current research. Only by clarifying the effects of oocyte cryopreservation at the cellular level will it be possible to improve cryopreservation techniques and to obtain firm evidence of how the slow freezing process affects oocyte cryopreservation.

Follicular fluid

Follicular fluid (FF) is a complex dynamic biological fluid that surrounds the developing oocyte. It is produced in growing antral follicles and contains a variety of molecules, such as proteins, steroid hormones, polysaccharides, metabolites, reactive oxygen species, and antioxidants. The primary role of these molecules is to modulate oocyte and follicular maturation [30] and protect follicular cells from physical or oxidative damage [31]. They are also responsible for the communication between somatic and germ cells, which is necessary for the oocyte to acquire its competence. Proteomic analysis of FF is a noninvasive method for obtaining information about follicular development and oocyte quality. The fluid is easily aspirated during oocyte retrieval without compromising the health of the oocyte. Although the FF proteome has a simpler pattern than that of somatic cells, proteomic analysis is difficult to apply because of the high abundance of albumin, immunoglobulin, and other abundant serum proteins that have a masking effect on less abundant proteins [32]. In order to better define the normal intrafollicular environment, Zamah et al. [33] used the MS approach to perform a detailed analysis of FF from three healthy ovum donors. They identified a total of 742 different proteins, of which 413 had not been described before. In a subsequent quantitative analysis of FF samples extracted before and after hCG administration from a second set of patients, they discovered 17 FF proteins that had significantly changed their expression levels during follicular maturation. In clinical practice, the identification of specific biomarkers of premature or normal luteinization can improve ovarian stimulation outcomes. This comprehensive proteomic analysis of human FF has allowed the creation of a database of FF proteins that should serve as a useful basis for the study of FF proteomes in patients with ovarian disorders of various etiologies. Indeed, a more detailed knowledge of the constituents of a normal FF proteome could provide a better understanding of ovarian physiology and of all ovarian disorders related to infertility.

Several studies have been carried out with different proteomic approaches to identify the FF proteome during folliculogenesis in fertile and infertile women undergoing ovarian stimulation during in vitro fertilization (IVF) [34–37], and these studies have shown differentially expressed proteins between the groups. Currently, the aim of proteomic research is not only to identify FF protein components but also to study their role in ovarian physiology and their pathways related to IVF outcomes. After a rigorous PubMed search, Bianchi et al. [38] reported 617 nonambiguous FF proteins, of which 337 were clustered into five different functional groups by the DAVID analysis tool. The most representative groups were inflammation and wound response, which represented 28% and 22% of the FF proteins, respectively. This is not surprising, since ovulation is an acute inflammatory process followed by a tissue repair processes to support corpus luteum formation.

Recent studies have shown that the FF proteome is characterized by alterations of its molecular constituents in advanced maternal aging [39] and in certain medical conditions such as ovarian hyperstimulation syndrome (OHSS) [40], in reproductive disorders including endometriosis [41], in polycystic ovarian syndrome (PCOS) [42], and in recurrent spontaneous miscarriage [43]. Furthermore, quantitative proteomic analysis of FF has identified potential biomarkers that can predict poor ovarian response. In a recent study, Oh et al. [44] described a total of 1079 proteins after comparing FF samples of women undergoing assisted reproductive treatment who are poor responders with those of normal responders.

Despite this progress, only a small part of the entire human FF proteome is currently known [45]. Therefore, there is a real need to discover new proteins that might provide a better understanding of the biological processes underlying fertilization and oocyte development.

Embryo

Advances in proteomic technologies have led to the development of new noninvasive methods of embryo selection. Limited by legal and ethical reasons, embryo quality in assisted reproductive technology (ART) cycles has until now been determined mainly through morphological evaluation [46]. However, this approach, which is based on subjective criteria, is unreliable for determining embryo implantation potential. Protein characterization is required to really understand the physiological processes that are critical for implantation. The identification of proteins related to viable embryos could lead to an increase in pregnancy rate in IVF and to a reduction in the number of embryos transferred and early pregnancy losses. Several studies have described the embryonic secretome, which is the set of the proteins produced by the embryo and secreted into the surrounding medium [47,48]. Increasing our knowledge about the human embryonic secretome may lead to a better understanding of the cellular processes of embryonic development and to the identification of those embryos with the highest developmental and implantation potential.

In the past, proteomic research has mainly focused on identifying a single protein factor, that is, it did not consider that biological processes typically involve more than one biochemical pathway. One of the first factors to be recognized as part of the human embryonic secretome is 1-o-alkyl-2-acetyl-sn-glycero-3-phosphocholine (paf), which produced and released preimplantation in mammalian embryos [49]. Paf acts in an autocrine and paracrine manner as a survival factor during embryonic development. Its release in human embryos is commonly compromised by assisted reproductive technology, leading to a reduced survival of the embryos generated with these techniques. Leptin is another molecule secreted into the surrounding medium by human blastocysts. The leptin system is thought to represent an important mechanism of the molecular dialogue between the receptive endometrium and the embryo during the window of implantation (WOI) [50]. Noci et al. [51] have proposed the soluble human leukocyte antigen-G (sHLA-G) as a potential marker of embryo development, with higher pregnancy rates being observed when the medium of day 3 embryos contained sHLA-G [52]. However, these results were not conclusive, since pregnancies were also achieved in sHLA-G-negative embryos.

The protein profile of in vivo embryos has been compared with that of in vitro cultured embryos. Katz-Jaffe et al. [53] investigated the effects of oxygen concentration on the mouse blastocyst proteome and detected a downregulation of 10 proteins in in vitro embryos exposed to high oxygen concentrations (20%) compared with embryos cultured at lower concentrations (5%). Furthermore, the embryos generated under reduced oxygen concentration secreted a proteome with greater similarity to that of embryos developed in vivo. Later, Katz-Jaffe et al. [54] used TOF-MS to analyze the proteome of individual embryos and observed differences in protein profiles depending on the developmental stage of the blastocysts. A different protein composition was found between early and expanded blastocysts and between developing blastocysts and degenerating embryos. Analyzing the secretome of both human and mouse embryos, Katz-Jaffe et al. [55] confirmed that each embryonic developmental stage is characterized by a specific protein composition, independent of morphology. In fact, different protein profiles are secreted by embryos at each 24h developmental stage, from the moment of fertilization to the blastocyst stage. Furthermore, a correlation has been reported between a protein biomarker (ubiquitin) that was significantly upregulated in the day 5 secretome and an ongoing blastocyst development. In contrast, this protein has showed limited expression in the degenerating embryo secretome. The blastocyst secretome has been investigated by Domínguez et al. [56], who compared successfully implanted human blastocysts with those that did not. Two proteins were found to be decreased in the secretome of implanted blastocysts. These were granulocyte-macrophage colony-stimulating factor (GM-CSF) and CXCL13, and therefore, these proteins may be considered biomarkers of blastocyst viability. In a retrospective cohort study combining proteomic and time-lapse morphokinetic analysis, Domínguez et al. [57] compared the spent media of 16 embryos that had implanted with the spent media of 12 embryos that had not implanted. The results showed that the presence of interleukin (IL) 6 and cell-cycle duration was the most important feature for selecting embryos for transfer. The association of both these technologies constitutes a strong diagnostic tool to predict embryo implantation potential.

Secretome analysis has been applied to the study of embryonic chromosomal abnormalities in order to replace the embryo biopsy with a noninvasive technique [58]. Extremely different secretome profiles have been described between hatching blastocyst euploid for 10 chromosomes and hatching blastocyst aneuploid for 10 chromosomes. By integrating proteomic profiling of spent culture media from individual blastocysts and comprehensive chromosomal analysis of all 23 pairs of chromosomes, Katz-Jaffe et al. [59] were able to discriminate between euploid and aneuploid blastocysts. Using an LC-MS/MS platform, McReynolds et al. [60] identified the first potential biomarker for noninvasive aneuploidy screening, lipocalin-1, a protein overproduced under conditions of stress, infection, and inflammation. Large-scale prospective studies that can confirm reproducibility are required to validate the application of this noninvasive tool to distinguish between euploid and aneuploid blastocysts.

Poli et al. [48] attempted to identify a contaminant-free medium with the potential to mask embryonic molecules and developed a new procedure—called blastocentesis—to analyze the protein composition of the blastosol, which represents a highly purified sample of embryo secretions. Through this new technique, which employs an integrated proteomic, genomic, and embryological approach, they have quantified proteins secreted by the early human embryo just before implantation from a total of 145 surplus human embryos donated by patients undergoing IVF treatment cycles. The identification and quantification of embryo-specific proteins have revealed a correlation between target proteins present in the blastosol with the chromosomal status of the embryo. Further validation is needed to improve the development of this technique as a noninvasive method of selecting euploid blastocysts.

The lack of sensitivity combined with variability in data interpretation, methodology, sample collection, and storage and with deficiencies in follow-up studies constitutes the main limitations of current proteomic platforms [61]. Nevertheless, in spite of these limitations, the embryonic secretome, together with morphology assessment, represents a promising platform aimed to improve embryo selection techniques and ART success.

Semen

Routine assessment of semen quality, based on sperm concentration, motility, and morphology, is not sufficient to investigate male infertility. Although these clinical parameters can be useful for identifying some cases of severe male infertility, they have little predictive power regarding the probability of achieving pregnancy in reproductive treatments. Proteomics allows a more objective identification of potential biomarkers of male infertility. Spermatozoon is very accessible for proteomic analysis, as it is easily purified from semen and is transcriptionally and translationally silent [62]. In contrast, the extremely heterogeneous composition of seminal fluid, with a mixture of proteins released from different sex glands, results in a more complex analysis. Many published proteomic studies have aimed to identify and quantify the proteins present in sperm [63–67], the testis [68], and seminal fluid [69], while others have compared abnormal sperm samples with those of normozoospermic controls [70,71]. A recent comprehensive compilation of all human sperm proteomic studies identified a list of 6198 different proteins present in the normal mature ejaculated human sperm cell [72]. Assuming that the complete human sperm proteome is composed of at least 7500 proteins, the ones identified so far represent about 78%. An altered proteomic profile with overexpression of proteins that regulate apoptosis in the seminal plasma has recently been described in adolescent males presenting varicocele and abnormal semen quality [73].

Only a few studies have focused on the identification of potential sperm protein biomarkers predictive of pregnancy outcome in ART treatments [74–77]. One of the first studies on the utility of proteomic analysis in the study of sperm function and male reproductive disorders is the case report of Pixton et al. [12]. By combining 2-D with in-gel digestion and MS peptide identification, they compared the sperm proteome from one normozoospermic patient who experienced failed fertilization at IVF with that of three fertile donors and detected 20 differences in protein expression. A similar result, with 14 differentially expressed proteins, was obtained by Frapsauce et al. [78] using a different 2-D-DIGE proteomic approach to compare the sperm protein profiles of three patients with a complete failure of fertilization after IVF with that of three fertile donors. More recently, using MALDI-TOF/TOF analysis, Xu et al. [74] compared 10 sperm samples from normozoospermic infertile patients with failed fertilization after artificial insemination and 10 sperm samples from fertile sperm donor controls and identified 24 differentially abundant proteins related to signaling and metabolic pathways. Molecular sperm alterations with changes in sperm protein profile may be responsible for poor ART outcomes, especially in normozoospermic patients whose female partners do not become pregnant. Azpiazu et al. [79] compared the proteome of sperm samples from 31 normozoospermic patients with different ART outcomes using isobaric tandem mass tags (TMT) and LC-MS/MS and detected 66 differentially expressed proteins that may be responsible for epigenetic errors during spermatogenesis through preventing a correct embryonic development. Studying the sperm proteome could lead to the development of potential biomarkers for use in the diagnosis and treatment of male reproductive disorders. Furthermore, proteomic studies on sperm function might help physicians in their therapeutic decisions. For example, this could be by predicting the success of sperm retrieval procedures after testicular biopsy for azoospermic patients [80] or for selecting patients who are more likely to benefit from varicocelectomy due to the consequent improvement in sperm quality and increased likelihood of pregnancy [81–83].

Wang et al. [84] performed a proteomic analysis using the 2-D-PAGE technique to investigate the effect of sperm cryopreservation on spermatozoon function, with the ultimate aim of identifying biomarkers for cryodamage. They detected a different protein characterization between freeze-thawed and fresh sperm samples obtained from nine normozoospermic donors.

The 27 proteins that were differentially expressed between the two groups are involved in the regulation of motility, capacitation, and fertilization capacity. Changes in human sperm proteome throughout the cryopreservation process have also been demonstrated by Bogle et al. [85]. They compared protein levels in fresh versus cryopreserved semen using TMT technology coupled with LC-MS/MS. The authors recommended prudent manipulation of semen, as the impact of some of these proteins on the offspring's health is still unknown. In this context, the identification of factors that may impair the capacity of cryopreserved spermatozoa to properly fertilize an oocyte could be of great benefit in improving ART protocols.

Further validation through big data studies involving multi-institutional cohorts of sperm and seminal plasma proteomes is needed to convert deregulated proteins into clinically relevant biomarkers that can be useful in the diagnosis and treatment of male reproductive disorders. Thus, this will hopefully improve the results for couples undergoing assisted reproductive techniques.

Endometrium

With the advent of transcriptomics, genetic analysis involving RNA sequencing has been employed to study the endometrium. The endometrial transcriptome throughout the menstrual cycle has been widely investigated, and a new molecular method based on the transcriptomic signature of the WOI, known as endometrial receptivity array (ERA), has been developed for the diagnosis of endometrial receptivity [86]. However, little is known about proteins translated from the transcripts expressed during the WOI. Proteomic analysis of endometrial biopsies obtained from different phases of the menstrual cycle [87–90] and from the prereceptive and receptive endometrium [91,92] has led to the identification of several differentially expressed proteins that would seem to be involved in human endometrial receptivity and embryo implantation. Endometrial receptivity has also been investigated through the exploration of the endometrial secretome, which includes all uterine secretions present in the uterine cavity [93,94]. The collection of uterine fluid by lavage or aspiration allows real-time analysis of a small biological sample (generally, the uterine fluid volume retrieved by aspiration from a single woman is not more than 10 mL) [95]. Uterine fluid contains every soluble factor released by the cells into the extracellular space, which includes high levels of locally secreted proteins. Analysis of the uterine fluid provides a less invasive and traumatic analysis than tissue biopsy [96,97]. In a prospective cohort study, Boomsma et al. [98] used immunoassay that investigates the endometrial secretions aspirated immediately before embryo transfer in 210 women undergoing IVF. A cytokine profile predictive of implantation and clinical pregnancy was identified, with the ratio between tumor necrosis factor-α (TNF-α) and interleukin-1β (IL-1β) proving to be a potential indicator of endometrial receptivity. The comparison between proteomic data obtained from endometrial secretions and tissue molecular analysis has shown that only one protein, apolipoprotein H (APOH), also known as beta-2 glycoprotein 1, is upregulated in both samples [91,97]. This protein has been implicated in a variety of pathways including lipoprotein metabolism, coagulation, and the production of antiphospholipid autoantibodies, which may be associated with recurrent fetal loss [99].

Endometrial biopsy analysis has the advantage of providing information about the cells that compose the tissue, but this method has limitations. In fact, endometrial tissue is subject to changes in its structure, cellular composition, and morphology due to menstrual cycle phases. Domínguez et al. [91] identified 32 proteins differentially expressed during the human endometrium's transition from the prereceptive (2 days after the luteinizing hormone (LH) surge) to the receptive phase (7 days after the LH surge) in endometrial biopsies obtained during the same cycle of eight healthy fertile women. Only stathmin 1 and annexin A2, which are two cytoskeleton-related proteins, were regulated in a strongly opposing way. In fact, while stathmin 1 was downregulated in the receptive endometrium, annexin A2 was upregulated. These two differentially expressed proteins, which are both implicated in the regulation of the decidualization process (in which the cell cytoskeleton is remodeled), may be considered potential biomarkers for endometrial receptivity. Similarly, Li et al. [14], when comparing the proteomic patterns between prereceptive (2 days after the LH surge) and receptive (7 days after the LH surge) endometria in the same spontaneous menstrual cycle of four fertile women, identified 31 differentially expressed proteins using 2-D-DIGE and MALDI-TOF/MS. They have discovered that annexin A4 was significantly upregulated in receptive samples compared with prereceptive samples. This protein plays a crucial role in the receptive process and seems to provoke cell apoptosis during the endometrial transformation, which occurs between the proliferative and the midsecretory phase of the menstrual cycle. More recently, Garrido-Gomez et al. [100] used DIGE and MS to analyze the proteomic profiles of 12 human endometrial biopsies from patients undergoing IVF treatment. They reported different proteomic signatures, with 24 differentially expressed proteins between receptive and nonreceptive ERA-diagnosed endometria obtained on the fifth day of a hormone replacement therapy treatment cycle. Surprisingly, there was no correlation between the 238 genes expressed in the receptive endometrium and the 24 deregulated proteins, which was probably due to the intrinsic limitations of the DIGE technique. The study results were validated by western blots and immunohistochemistry using two proteins: progesterone receptor membrane component 1 (PGRMC1) and annexin A6 (ANXA6). The expression levels of both proteins were increased in ERA-nonreceptive versus ERA-receptive endometrial samples, suggesting a role for these molecules in endometrial receptive status. Furthermore, two statistically significant pathways, "carbohydrate biosynthetic process" and "nuclear mRNA splicing via spliceosome," were identified in the proteomic network of deregulated proteins, with both being implicated in endometrial receptivity.

The proteome and secretome of in vitro decidualized endometrial stromal cells were investigated by Garrido-Gomez et al. [95] in a prospective case-control study. The decidualization of the human endometrium constitutes the basis for the embryo implantation process. After collecting endometrial biopsies of 16 ovum donors on the day of oocyte retrieval, they analyzed, by combining gel-based methods and protein arrays, the mechanisms involved in the decidualization process. Furthermore, they have created an interactome model by the integration of proteome and secretome profiles together with the data present in public databases.

A proteomic approach has also been used to investigate possible alterations during the putative window of endometrial receptivity in patients with unexplained recurrent implantation failure (RIF). This is defined as a failure to achieve a positive pregnancy test after three or more consecutive IVF treatment cycles with the transfer of good-quality embryos. Interestingly, midsecretory endometrial samples obtained from women with RIF showed a different protein profile compared with fertile controls, characterized by a potent anti-inflammatory molecule with anti-implantation properties known as apolipoprotein A-I (apoA-I) [101]. Furthermore, apoA-I expression was highly increased in midsecretory eutopic endometrial tissue from patients with endometriosis. Together, these results suggest that the dysregulation of apoA-1 is responsible for various pathologies associated with infertility, such as endometriosis and RIF.

In conclusion, the correct determination of endometrial receptivity status is currently based mainly on molecular signatures rather than histological examination. Molecular tools should be considered as an objective approach to better understand implantation failure related to a poor-quality endometrium.

METABOLOMICS IN REPRODUCTIVE MEDICINE

Metabolomics is the science that explores the dynamic interactions between multiple low-molecular-weight molecules (amino acids, lipids, nucleotides, signaling molecules, etc.) found in biological samples, with the aim of understanding complex biological processes [102]. Metabolites potentially hold more information about a biological system than the study of gene expression, mRNAs, or proteins, because they represent the end products of cell regulatory processes. While the genome is representative of what it might be, the proteome defines what has been expressed, and it is the metabolome that reveals the functional state of a system and its cells. Current research is moving toward the study of groups of target metabolites and their interactions, rather than the selective study of single metabolites [103]. Since biofluids are in dynamic equilibrium with cells, the identification and quantification of all the metabolites present in a fluid medium allow the metabolic status of cells to be determined, in both physiological and pathological conditions [104]. The metabolic study of FF, seminal plasma, embryo culture medium, or endometrial fluid might be useful for determining oocyte, sperm, embryo viability, and uterine functional competence, respectively, and thus the probability of success during an IVF procedure.

The major limitations of metabolomics are due to the labile nature, chemical complexity, and heterogeneity of the metabolites. In fact, unlike genomics and proteomics, which only deal with one class of compounds, metabolomic analysis focuses on various types of molecules that are produced at different stages of metabolic pathways [105].

Currently, assessment of the metabolite profile of human reproduction is carried out by spectroscopic/spectrometric and chromatographic techniques, such as proton nuclear magnetic resonance (^{1}H NMR) spectroscopy [106] and MS techniques [107], alone or in combination with LC, gas chromatography, capillary electrophoresis, HPLC, near-infrared (NIR) optical spectroscopy [108], or Raman spectroscopy [109] (Fig. 9.2).

Ooctye and FF

Fertilization, early embryonic development, and pregnancy maintenance are all strictly dependent on oocyte quality. Oocyte selection is currently based on morphological criteria that are highly subjective and therefore often inaccurate [110]. The identification of noninvasive biomarkers of oocyte competence is one of the major targets of current research in the field of human reproduction. Selecting the best oocyte to inseminate or to freeze has two main advantages: the limitation of embryo overproduction and the improvement of oocyte cryostorage results. Metabolomics has contributed to clarifying the mechanisms underlying oocyte

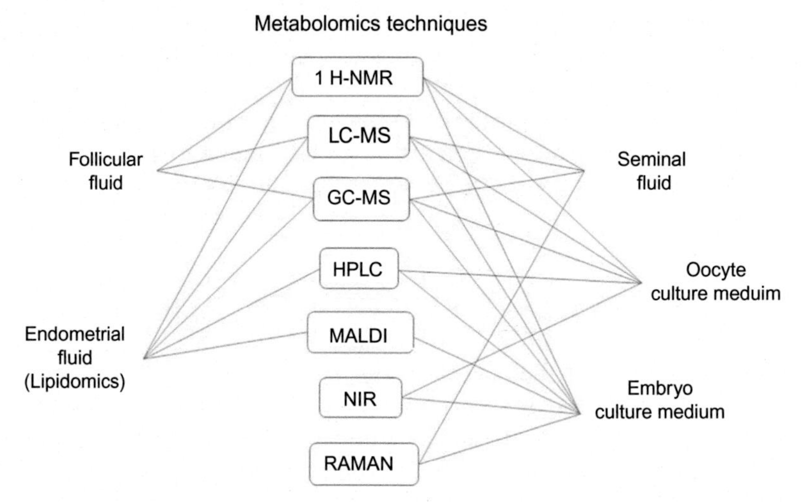

FIG. 9.2 Schematic representation of metabolomic techniques in human reproduction. *^{1}H NMR*, Proton nuclear magnetic resonance; *GS-MS*, gas chromatography–mass spectrometry; HPLC, high-performance liquid chromatography; NIR, near-infrared spectroscopy.

growth and maturation and the link between oocyte metabolism and reproductive outcomes. Metabolomic analysis is mainly focused on the biochemical constituents of the follicular environment, composed not only of secretions from the oocyte but also by surrounding cumulus, granulosa, and theca cells, which represent an ideal source of noninvasive predictive factors of oocyte quality [103]. A useful approach to metabolomic assessment of oocyte viability is the study of oocyte culture medium. A recent study of 412 oocyte culture samples collected from 43 patients undergoing IVF has demonstrated a correlation between metabolomic profiling of oocyte spent culture medium and nuclear maturity status, developmental viability, and implantation rate of derived embryos [111]. In fact, metabolomic profiles of metaphase I, metaphase II, and prophase I oocytes differed significantly, as did the metabolites identified in culture samples of oocytes that developed into grade A embryos and those of grade C/D. Finally, the oocytes that resulted in pregnancy had distinct profiles compared with those that did not.

Most previous studies have focused on FF metabolomic analysis [103,112,113]. For instance, Wallace et al. [114] have recently investigated the relationship between the metabolic profile of human FF samples from patients undergoing stimulated IVF and oocyte quality. Using ^{1}H NMR spectroscopy, they discovered that lower levels of lactate and choline/phosphocholine and higher levels of glucose and high-density lipoproteins in the FF were associated with oocytes that failed to cleave as an embryo in comparison with oocytes that developed into early cleavage-stage embryos. In regard to pregnancy outcome, women who did not achieve pregnancy were associated with a lower FF level of glucose and higher levels of proline, lactate, leucine, and isoleucine. Furthermore, subsequent work in the same laboratory showed that the fatty acid composition of FF from follicles in fertilized oocytes that developed into early cleavage-state embryos significantly differed in nine fatty acids from that of oocytes that fertilized normally but failed to cleave [115]. The metabolic profile of FF, reflecting oocyte developmental potential, could thus be considered a noninvasive method of oocyte/embryo selection.

An alteration of FF metabolic profiles has been demonstrated in a variety of subfertility conditions. In a prospective study, Zhang et al. [116] measured follicular amino acid (AA) levels in 63 PCOS patients and 48 controls using LC-MS/MS. Follicular AA metabolic profiles were altered by PCOS and obesity, and this AA disruption is probably responsible for the poor oocyte quality and reduced pregnancy outcome in obese patients and increased miscarriage rate in PCOS patients. Using the GC/MS technique, Xia et al. [117] observed a difference in the metabolite profile of 20 metabolites in FF samples from 13 patients with repeated IVF failure and from 15 patients with successful pregnancy after the first IVF cycle. In a prospective study, Pacella et al. [118] examined the FF environment in women with reduced ovarian reserve or advanced maternal age and found alterations of follicular cell metabolism, FF metabolites, and progesterone production, all of which are likely to affect oocyte developmental competence.

Large-scale, prospective studies are needed to apply this noninvasive technology to IVF clinical practice.

Embryo

Embryo assessment in clinical settings is currently based on morphology and cleavage rate. This subjective evaluation has a limited predictive power due to inter- and intraobserver variability. Preimplantation genetic diagnosis cannot be used to predict the competence of the embryo. In addition, it is an invasive, expensive, time-consuming, and controversial procedure. If more objective and accurate criteria for preembryo selection were available, it would be possible to employ elective single-embryo transfer without exposing patients to recurrent failures in embryo implantation, miscarriages, and multiple IVF or to the risk of multiple pregnancies. The metabolic profile of embryo culture media has been used to predict oocyte and embryo developmental potential. The embryo collects the nutrients needed for its development from the medium in which it grows and secretes the products of its metabolism into the same medium. In the early stages of preimplantation development, pyruvate and lactate are the two main sources of energy for the embryo [119]. As embryo development progresses, glucose consumption increases, and glucose metabolism becomes predominant at the blastocyst stage. Amino acids and fatty acids are also essential for an optimal embryonic development. The degradation of these nutrients produces metabolites, and both nutrients and metabolites can be considered potential biomarkers of embryo viability [120]. Embryos remain in culture for longer periods than oocytes, so the accumulation of metabolites in the medium makes their detection easier. Most studies have focused on the measurement of uptake and secretion of single or multiple substrates in the culture medium ("targeted metabolomics") or on the untargeted metabolic profiling of embryo spent culture medium with the aim of obtaining information about all metabolites, known and unknown [121–125]. Embryo metabolic activity can also be measured directly within the embryo itself. However, this is an invasive procedure without a well-defined impact on the embryo, which limits the use of this approach.

An accurate diagnosis of embryo viability is made difficult by the presence of several factors that can affect the metabolic profile of the human embryo, such as the embryonic stage of development, sex of the embryo, medium composition, frequency of medium renewal, rate of ammonium accumulation, and oxygen concentration [126]. In addition, it has been shown that embryo metabolism varies depending on female body mass index (BMI). In fact, obese women present a specific metabolomic profile in the spent culture media of day 3 embryos that are characterized by a lower concentration of saturated fatty acids than that associated with normal weight women [127]. This metabolic alteration may be the cause of the poorer results obtained in obese patients after assisted reproduction.

Metabolomics has been applied to identify chromosomally abnormal embryos through a noninvasive approach. A specific metabolomic signature has been found in spent media from preimplantation day 3 embryos with trisomy 21/monosomy 21, suggesting a correlation between aneuploidies and early embryo metabolism [128].

Expanding our knowledge of metabolomics should lead to the formulation of improved culture media for human embryos, including specific media for particular factors, such as diabetes, elevated BMI, or advanced maternal age [124], and for the identification of the most metabolically viable embryos. Although metabolomic technology has been successfully applied to the study of embryo metabolism, high-quality evidence from randomized controlled trials is needed in order to confirm that metabolomic assessment of embryos prior to implantation has a beneficial effect on live birth, ongoing pregnancy, or miscarriage rates [129].

Please refer to the previous chapter for a wider discussion about noninvasive embryo selection methods through spent culture medium analysis.

Semen

Oxidative stress has been proposed as a potential cause of male infertility. Increased levels of reactive oxygen species (ROS) or decreased levels of antioxidants in seminal plasma can result in spermatogenic abnormalities, while an imbalance of oxidative stress has been demonstrated in men with unexplained infertility [130]. In fact, Raman spectroscopy revealed increased markers of oxidative stress in the seminal plasma metabolome of men with unexplained infertility in comparison with fertile men. Elevated levels of oxysterols such as 5α-cholesterol and 7-ketocholesterol in seminal plasma of patients with asthenozoospermia confirm once again that oxidative stress plays an important role in the induction of mechanisms of sperm quality reduction [131]. Changes in the concentration of oxidative stress biomarkers have been found in patients with idiopathic infertility, varicocele, and vasectomy reversal [132]. Patients with varicocele are characterized by higher levels of oxidative stress compared with fertile normozoospermic patients [133]. Indeed, superoxide generation in human spermatozoa has been investigated as a possible factor of male infertility. De Iuliis et al. [134] demonstrated a negative correlation between the spontaneous production of superoxide by human spermatozoa and sperm motility. This inverse association was further confirmed by the induction of severe sperm motility loss through the pharmacological stimulation of the generation of this radical in normal spermatozoa. To investigate the relationship between smoking and oxidative stress alteration, two important antioxidants, retinol and α-tocopherol, were measured in the seminal plasma of smokers and nonsmokers; however, no significant differences were reported between the two groups [135]. Using Raman spectroscopy combined with chemometrics, Gilany et al. [136] identified important changes in the metabolome of the human seminal plasma of asthenozoospermic patients versus normozoospermic controls. Significant differences have also been observed in the seminal plasma metabolome of nonobstructive azoospermic men analyzed by Fourier-transform infrared spectroscopy

(FTIR) when compared with that of normozoospermic men [137]. The authors concluded that metabolic analysis of seminal plasma could be an alternative noninvasive method to testicular biopsy when seeking to evaluate spermatogenesis status in men with azoospermia. In addition, differences in seminal plasma biomarker profiles between patients with idiopathic/male factor infertility (including normozoospermic parameters, oligozoospermia, asthenozoospermia, azoospermia, and teratozoospermia) and healthy controls with proved fertility have been demonstrated by means of NMR spectroscopy [138]. In fact, the idiopathic infertile group exhibited lower levels of valine, 2-hydroxyisovalerate, lysine, hippurate, and fructose than the healthy controls, suggesting a metabolic alteration underlies the origin of idiopathic infertility. Metabolomic analysis may, therefore, be useful in determining the cause of unexplained infertility.

In addition to seminal fluid, urine and plasma metabolic profiles have also been used to differentiate fertile from infertile patients. The urinary metabolome of oligozoospermic [139] or normozoospermic infertile men has revealed significant changes compared with fertile controls [140]. Several metabolites associated with energy production, oxidative stress, and cell apoptosis in spermatogenesis were found to be differentially expressed in the plasma of nonobstructive azoospermic men [141]. Two metabolites, 1,5-anhydro-sorbitol and α-hydroxyisovaleric acid, were identified as potential biomarkers for distinguishing between plasma samples from healthy controls and from male infertile patients [142]. In addition, lactate, glutamate, and cholesterol were screened as specific biomarker candidates of semen abnormalities and erectile dysfunction of patients.

More studies are necessary for the development of clinically useful biomarkers to aid in male infertility diagnosis, treatment, and counseling [143].

Endometrium

One of the main goals of research in the field of assisted reproduction is the identification and characterization of endometrial receptivity status with the aim of increasing implantation and pregnancy rates in IVF cycles. Lipidomics, which is a part of metabolomics, is the study of lipid species together with their networks and the metabolic pathways that are present in a cell or in any other biological system [144]. Unlike genomics and proteomics, in which a biopsy is required to confirm a diagnosis, the lipidomic approach using endometrial fluid allows a noninvasive diagnosis of endometrial receptivity. Lipid molecules such as endocannabinoids, lysophosphatidic acid, and prostaglandins (PGs) have been identified as important mediators of embryo implantation in animal models [145]. The important role of lipids in predicting the state of endometrial receptivity has been shown in two recent studies [146,147]. A significant increase in the concentration of prostaglandins E2 and F2a has been demonstrated in the endometrial fluid during the WOI, specifically from days 19 to 21, compared with any other period throughout the menstrual cycle [146]. Elevated concentrations of these PGs during the WOI have been found not only in natural cycles but also in stimulated and ovum recipient cycles [147]. Furthermore, PGE2 and PGF2 levels from endometrial fluid aspirated 24h prior to embryo transfer have been considered as predictors of implantation outcome.

Metabolomics can be a useful tool to define the molecular mechanisms that regulate altered endometrial receptivity status. Poor PG synthesis in the human endometrium has been associated with RIF in patients undergoing IVF treatments [148]. RoyChoudhury et al. [149] have attempted to explain why a large cohort of women failed to conceive in three or more consecutive IVF attempts, despite the transfer of good-quality embryos. Blood samples collected during the WOI of 28 women with RIF were compared with those of 24 women with recurrent implantation success using ^{1}H NMR and advanced statistical analysis. Eight metabolites, including valine, adipic acid, L-lysine, creatine, ornithine, glycerol, D-glucose, and urea, were found to be significantly upregulated in the RIF group. The identification of these markers, along with understanding of their metabolic functions, in relation to energy metabolism, lipid metabolism, and arginine metabolic pathway, may lead to early therapeutic strategies to improve implantation rates in IVF cycles. The authors stated that, despite the validity of the results obtained, as long as large-scale, multicenter, and international studies are not performed, the usefulness of these results would be confined to the research field. In fact, because of the small number of patient samples and the inter- and intraindividual metabolite variance between the two groups, the use of these results is not yet recommended in clinical practice.

A metabolic disruption has been proved to be responsible for poor endometrial receptivity and pregnancy loss. A combination of ^{1}H NMR spectroscopy and multivariate analysis was used to compare the serum metabolic profiles of 36 women with idiopathic recurrent miscarriage during the WOI with those of 28 fertile controls [150]. Differently expressed metabolites were reported between the two groups, with a total of seven metabolites upregulated in women with recurrent miscarriage. From a functional aspect, the altered metabolites are probably involved in exaggerated inflammatory response and vascular dysfunction.

CONCLUSION

The new "omics" technologies, in particular proteomics and metabolomics, have allowed us to expand our knowledge of human reproduction and, hence, of human infertility. A greater comprehension of the molecular mechanisms underlying reproductive physiology has led to the discovery of relevant biomarkers that are useful in the diagnosis, treatment, and counseling of several infertility-related conditions and for the development of molecular strategies to select the best gametes and embryos and the most receptive endometrium (Table 9.1). The study of all the molecular factors involved in reproductive function is advantageous as it provides a large amount of useful information in exchange for relatively little effort. Despite the promising use of these technologies, their application to clinical practice remains limited. These techniques are usually expensive and require trained staff and specially equipped laboratories, and the results are often not obtained as quickly as is needed to transfer the best available embryo into a receptive endometrium. In addition, their clinical applicability is also limited by the lack of reproducibility. Further validation with prospective clinical trials using unified protocols is clearly needed.

In the near future, thanks to well-powered studies employing new-generation, highly sensitive technologies and validated biomarkers, as well as improved reproducibility and a greater uniformity in data collection. Omics technologies will undoubtedly be widely employed for the investigation, diagnostics, and therapeutics in ART. Table 9.2 shows the benefits achieved through the application of proteomics and metabolomics in reproductive medicine.

TABLE 9.1 Main Findings in Proteomics and Metabolomics in the Field of Human Reproduction

Proteomics	Metabolomics
The oocyte proteome represents a direct marker of oocyte maturation and competence	The oocyte metabolome reflects oocyte growth and developmental competence
The FF proteome provides information about follicular development and oocyte quality. It has contributed to a better understanding of the biological processes underlying ovarian disorders related to infertility	The FF metabolome provides information about oocyte developmental potential and viability. It has contributed to a better understanding of the mechanisms involved in determining poor oocyte quality
The sperm proteome has helped to clarify sperm function and to identify molecular alterations present in male reproductive disorders responsible for poor ART outcomes	The sperm metabolome can explain normal spermatogenesis and, through the analysis of protein profile changes, the mechanisms underlying the reduction in sperm quality
The embryonic secretome has allowed a better understanding of cellular processes of embryonic development and enables identification of the embryos with the highest developmental and implantation potential	The metabolic profile of embryo culture media is useful in predicting embryo developmental potential
The endometrium proteome can reflect endometrial receptivity status and explains the alterations that occur during the WOI that can cause implantation failures and infertility	The endometrium metabolome can predict the outcome of implantation and can define the molecular mechanisms responsible for poor endometrial receptivity and pregnancy loss

TABLE 9.2 Key Issues

- Proteomics and metabolomics represent noninvasive oocyte selection methods
- Proteomics and metabolomics provide potential biomarkers for male infertility
- Proteomics and metabolomics are promising alternatives for the development of noninvasive methods of embryo selection
- Proteomics and metabolomics constitute noninvasive predictors of endometrial receptivity (secretomics-lipidomics)

References

[1] R. Rivera Egea, N. Garrido Puchalt, M. Meseguer Escrivá, A.C. Varghese, OMICS: current and future perspectives in reproductive medicine and technology, J. Hum. Reprod. Sci. 7 (2014) 73–92.

[2] J. Bellver, M. Mundi, F. Esteban, S. Mosquera, J. Horcajadas, '-omics' technology and human reproduction: reproductomics, Exp. Rev. Obstet. Gynecol. 7 (2012) 493–506.

[3] E. Lundberg, L. Fagerberg, D. Klevebring, I. Matic, T. Geiger, J. Cox, C. Algenäs, J. Lundeberg, M. Mann, M. Uhlen, Defining the transcriptome and proteome in three functionally different human cell lines, Mol. Syst. Biol. 6 (2010) 450.

[4] M. Li, I.X. Wang, Y. Li, A. Bruzel, A.L. Richards, J.M. Toung, V.G. Cheung, Widespread RNA and DNA sequence differences in the human transcriptome, Science 333 (2011) 53–58.

[5] N. Verrills, Clinical proteomics: present and future prospects, Clin. Biochem. Rev. 27 (2006) 99–116.

[6] P. James, Protein identification in the post-genome era: the rapid rise of proteomics, Q. Rev. Biophys 30 (1997) 279–331.

[7] E. Witze, W. Old, K. Resing, N. Ahn, Mapping protein post-translational modifications with mass spectrometry, Nat. Methods 4 (2007) 798–806.

[8] P. Wright, J. Noirel, S.Y. Ow, A. Fazeli, A review of current proteomics technologies with a survey on their widespread use in reproductive biology investigations, Theriogenology 77 (2012) 738–765.
[9] R. Oliva, S. de Mateo, J. Estanyol, Sperm cell proteomics, Proteomics 9 (2009) 1004–1017.
[10] A.D. Rolland, R. Lavigne, C. Dauly, P. Calvel, C. Kervarrev, T. Freour, B. Evrard, N. Rioux-Leclercq, J. Auger, C. Pineau, Identification of genital tract markers in the human seminal plasma using an integrative genomics approach, Hum. Reprod. 28 (2013) 199–209.
[11] S. de Mateo, J. Martínez-Heredia, J. Estanyol, D. Domínguez-Fandos, J. Vidal-Taboada, J. Ballescà, R. Oliva, Marked correlations in protein expression identified by proteomic analysis of human spermatozoa, Proteomic. 7 (2007) 4264–4277.
[12] K. Pixton, E. Deeks, F. Flesch, F. Moseley, L. Björndahl, P.R. Ashton, C.L. Barratt, I.A. Brewis, Sperm proteome mapping of a patient who experienced failed fertilization at IVF reveals altered expression of at least 20 proteins compared with fertile donors: case report, Hum. Reprod. 19 (2004) 1438–1447.
[13] S. du Plessis, A. Kashou, D. Benjamin, S. Yadav, A. Agarwal, Proteomics: a subcellular look at spermatozoa, Reprod. Biol. Endocrinol. 9 (2011) 36.
[14] J. Li, Z. Tan, M. Li, T. Xia, P. Liu, W. Yu, Proteomic analysis of endometrium in fertile women during the prereceptive and receptive phases after luteinizing hormone surge, Fertil. Steril. 95 (2011) 1161–1163.
[15] N. Treff, J. Su, N. Kasabwala, X. Tao, K. Miller, R.J. Scott, Robust embryo identification using first polar body single nucleotide polymorphism microarray-based DNA fingerprinting, Fertil. Steril. 93 (2010) 2453–2455.
[16] S. Cortezzi, J. Garcia, C. Ferreira, D. Braga, R. Figueira, A. Iaconelli, G. Souza, E. Borges Jr., M. Eberlin, Secretome of the preimplantation human embryo by bottom-up label-free proteomics, Anal. Bioanal. Chem. 401 (2011) 1331–1339.
[17] J. Twigt, R. Steegers-Theunissen, K. Bezstarosti, J. Demmers, Proteomic analysis of the microenvironment of developing oocytes, Proteomics 12 (2012) 1463–1471.
[18] P. Zhang, X. Ni, Y. Guo, X. Guo, Y. Wang, Z. Zhou, R. Huo, J. Sha, Proteomic-based identification of maternal proteins in mature mouse oocytes, B.M.C. Genomics 10 (2009) 348.
[19] T. Garrido-Gomez, F. Dominguez, J.A. Lopez, E. Camafeita, A. Quiñonero, J. Martinez-Conejero, A. Pellicer, A. Conesa, C. Simón, Modeling human endometrial decidualization from the interaction between proteome and secretome, J. Clin. Endocrinol. Metab. 96 (2011) 706–716.
[20] S. Thacker, S. Yadav, R. Sharma, A. Kashou, B. Willard, D. Zhang, A. Agarwal, Evaluation of sperm proteins in infertile men: a proteomic approach, Fertil. Steril. 95 (2011) 2745–2748.
[21] M. Ma, X. Guo, F. Wang, C. Zhao, Z. Liu, Z. Shi, Y. Wang, P. Zhang, K. Zhang, N. Wang, M. Lin, Z. Zhou, J. Liu, Q. Li, L. Wang, R. Huo, J. Sha, Q. Zhou, Protein expression profile of the Mouse Metaphase II oocyte, J. Proteome Res. 7 (2008) 4821–4830.
[22] Y. Meng, X. Liu, X. Ma, Y. Shen, L. Fan, J. Leng, J.Y. Liu, J.H. Sha, The protein profile of mouse mature cumulus-oocyte complex, Biochim. Biophys. Acta. 1774 (2007) 1477–1490.
[23] P. Zhang, X. Ni, Y. Guo, X. Guo, Y. Wang, Z. Zhou, R. Huo, J. Sha, Proteomic based identification of maternal proteins in mature mouse oocytes, B.M.C. Genomics 10 (2009) 348.
[24] M. Pfeiffer, M. Siatkowski, Y. Paudel, S. Balbach, N. Baeumer, N. Crosetto, H.C. Drexler, G. Fuellen, M. Boiani, Proteomic analysis of mouse oocytes reveals 28 candidate factors for the "Reprogammome", J. Proteomic Res. 10 (2011) 2140–2153.
[25] S. Wang, Z. Kou, Z. Jing, Y. Zhang, X. Guo, M. Dong, I. Wilmut, S. Gao, Proteome of mouse oocytes at different developmental stages, Proc. Natl. Acad. Sci. 107 (2010) 17639–17644.
[26] C. Sheehan, M. Katz-Jaffe, W. Schoolcraft, D. Gardner, Exogenous gonadotropins alter the proteome of metaphase II mouse oocytes, Fertil. Steril. 86 (2006) 387.
[27] S. Hamamah, V. Matha, C. Berthenet, T. Anahory, V. Loup, H. Dechaud, B. Hedon, A. Fernandez, N. Lamb, Comparative protein expression profiling in human cumulus cells in relation to oocyte fertilization and ovarian stimulation protocol, Reprod. Biomed. Online 13 (2006) 807–814.
[28] S. McReynolds, M. Dzieciatkowska, B. McCallie, S. Mitchell, J. Stevens, K. Hansen, W.B. Schoolcraft, M.G. Katz-Jaffe, Impact of maternal aging on the molecular signature of human cumulus cells, Fertil. Steril. 98 (2012) 1574–1805.
[29] M. Larman, M.G. Katz-Jaffe, C. Sheehan, D. Gardner, 1,2-propanediol and the type of cryopreservation procedure adversely affect mouse oocyte physiology, Hum. Reprod. 22 (2007) 250–259.
[30] L. Bianchi, A. Gagliardi, C. Landi, R. Focarelli, V. De Leo, A. Luddi, L. Bini, P. Piomboni, Protein pathways working in human follicular fluid: the future for tailored IVF? Expert. Rev. Mol. Med. 18 (2016) e9.

[31] A. Ambekar, R. Nirujogi, S. Srikanth, S. Chavan, D. Kelkar, I. Hinduja, K. Zaveri, T.S. Prasad, H.C. Harsha, A. Pandey, S. Mukherjee, Proteomic analysis of human follicular fluid: a new perspective toward understanding folliculogenesis, J. Proteom. 87 (2013) 68–77.
[32] E. Seli, C. Robert, M. Sirard, OMICS in assisted reproduction: possibilities and pitfalls, Mol. Hum. Reprod. 16 (2010) 513–530.
[33] A. Zamah, M. Hassis, M. Albertolle, K. Williams, Proteomic analysis of human follicular fluid from fertile women, Clinical Proteomics 12 (2015) 5.
[34] S.J. Estes, B. Ye, W. Qiu, D. Cramer, M. Hornstein, S. Missmer, A proteomic analysis of IVF follicular fluid in women <or = 32 years old, Fertil. Steril. 92 (2009) 1569–1578.
[35] M. Kushnir, T. Naessén, K. Wanggren, A. Rockwood, D. Crockett, J. Bergquist, Protein and steroid profiles in follicular fluid after ovarian hyperstimulation as potential biomarkers of IVF outcome, J. Proteome Res. 11 (2012) 5090–5100.
[36] Y. Wu, Y. Wu, J. Zhang, N. Hou, A. Liu, J. Pan, J. Lu, J. Sheng, H. Huang, Preliminary proteomic analysis on the alterations in follicular fluid proteins from women undergoing natural cycles or controlled ovarian hyperstimulation, J. Assist. Reprod. Genet. 32 (2015) 417–427.
[37] M. Hashemitabar, M. Bahmanzadeh, A. Mostafaie, M. Orazizadeh, M. Farimani, R. Nikbakht, A proteomic analysis of human follicular fluid: comparison between younger and older women with normal FSH levels, Int. J. Mol. Sci. 15 (2014) 17518–17540.
[38] L. Bianchi, A. Gagliardi, C. Landi, R. Focarelli, V. De Leo, A. Luddi, L. Bini, P. Piomboni, Protein pathways working in human follicular fluid: the future for tailored IVF? Expert. Rev. Mol. Med. 18 (2016) e9.
[39] D. Dumesic, D. Meldrum, M. Katz-Jaffe, R. Krisher, W. Schoolcraft, Oocyte environment: follicular fluid and cumulus cells are critical for oocyte health, Fertil. Steril. 103 (2015) 303–316.
[40] K. Jarkovska, H. Kupcova Skalnikova, P. Halada, R. Hrabakova, J. Moos, K. Rezabek, S.J. Gadher, H. Kovarova, Development of ovarian hyperstimulation syndrome: interrogation of key proteins and biological processes in human follicular fluid of women undergoing in vitro fertilization, Mol. Hum. Reprod. 17 (2011) 679–692.
[41] T. Regiani, F. Cordeiro, L.V. da Costa, J. Salgueiro, K. Cardozo, V. Carvalho, K.J. Perkel, D.S. Zylbersztejn, A.P. Cedenho, E.G. Lo Turco, Follicular fluid alterations in endometriosis: label-free proteomics by MS(E) as a functional tool for endometriosis, Syst. Biol. Reprod. Med. 61 (2015) 263–276.
[42] A. Ambekar, D. Kelkar, S. Pinto, R. Sharma, I. Hinduja, K. Zaveri, A. Pandey, T.S. Prasad, H. Gowda, S. Mukherjee, Proteomics of follicular fluid from women with polycystic ovary syndrome suggests molecular defects in follicular development, J. Clin. Endocrinol. Metab. 100 (2015) 744–753.
[43] Y. Kim, M. Kim, S. Lee, B. Choi, J. Lim, K. Cha, K.H. Baek, Proteomic analysis of recurrent spontaneous abortion: identification of an inadequately expressed set of proteins in human follicular fluid, Proteomics 6 (2006) 3445–3454.
[44] J. Oh, S. Kim, K. Cho, M. Kim, C. Suh, J. Lee, K.P. Kim, Proteomic analysis of human follicular fluid in poor ovarian responders during in vitro fertilization, Proteomics 17 (2017).
[45] M. Benkhalifa, A. Madkour, N. Louanjli, N. Bouamoud, B. Saadani, I. Kaarouch, H. Chahine, O. Sefrioui, P. Merviel, H. Copin, From global proteome profiling to single targeted molecules of follicular fluid and oocyte: contribution to embryo development and IVF outcome, Exp. Rev. Proteomics 12 (2015) 407–423.
[46] T. Ebner, M. Moser, M. Sommergruber, G. Tews, Selection based on morphological assessment of oocytes and embryos at different stages of preimplantation development: a review, Hum. Reprod. Update 9 (2003) 251–262.
[47] M. Katz-Jaffe, S. McReynolds, D. Gardner, W. Schoolcraft, The role of proteomics in defining the human embryonic secretome, Mol. Hum. Reprod. 15 (2009) 271–277.
[48] M. Poli, A. Ori, T. Child, S. Jaroudi, K. Spath, M. Beck, D. Wells, Characterization and quantification of proteins secreted by single human embryos prior to implantation, EMBO Molecular Medicine 7 (2015) 1465–1479.
[49] C. O'Neill, The role of paf in embryo physiology, Hum. Reprod. Update. 11 (2005) 215–228.
[50] A. Cervero, J. Horcajadas, F. Dominguez, A. Pellicer, C. Simón, Leptin system in embryo development and implantation: a protein in search of a function, Reprod. Biomed. Online 10 (2005) 217–223.
[51] I. Noci, B. Fuzzi, R. Rizzo, L. Melchiorri, L. Criscuoli, S. Dabizzi, R. Biagiotti, S. Pellegrini, A. Menicucci, O.R. Baricordi, Embryonic soluble HLA-G as a marker of developmental potential in embryos, Hum. Reprod. 20 (2005) 138–146.
[52] G. Sher, L. Keskintepe, J. Fisch, B. Acacio, P. Ahlering, J. Batzofin, M. Ginsburg, Soluble human leukocyte antigen G expression in phase I culture media at 46 h after fertilization predicts pregnancy and implantation from day 3 embryo transfer, Fertil. Steril. 83 (2005) 1410–1413.

[53] M. Katz-Jaffe, D. Linck, W. Schoolcraft, D. Gardner, A proteomic analysis of mammalian preimplantation embryonic development, Reproduction 130 (2005) 899–905.
[54] M. Katz-Jaffe, D. Gardner, W. Schoolcraft, Proteomic analysis of individual human embryos to identify novel biomarkers of development and viability, Fertil. Steril. 85 (2006) 101–107.
[55] M. Katz-Jaffe, W. Schoolcraft, D. Gardner, Analysis of protein expression (secretome) by human and mouse preimplantation embryos, Fertil. Steril. 86 (2006) 678–685.
[56] F. Domínguez, B. Gadea, F. Esteban, J. Horcajadas, A. Pellicer, C. Simón, Comparative protein-profile analysis of implanted versus non-implanted human blastocysts, Hum. Reprod. 23 (2008) 1993–2000.
[57] F. Dominguez, M. Meseguer, B. Aparicio-Ruiz, P. Piqueras, A. Quiñonero, C. Simón, New strategy for diagnosing embryo implantation potential by combining proteomics and time-lapse technologies, Fertil. Steril. 104 (2015) 908–914.
[58] M. Katz-Jaffe, J. Stevens, W. Kearns, D. Gardner, W. Schoolcraft, Relationship between embryonic secretome and chromosomal abnormalities in human IVF, Fertil. Steril. 86 (2006) S57.
[59] M. Katz-Jaffe, E. Fagouli, J. Fillipovits, D. Wells, W. Schoolcraft, Relationship between the human blastocyst secretome and chromosomal constitution, Fertil. Steril. 90 (2008) S80.
[60] S. McReynolds, L. Vanderlinden, J. Stevens, K. Hansen, W. Schoolcraft, M. Katz-Jaffe, Lipocalin-1: a potential marker for noninvasive aneuploidy screening, Fertil. Steril. 95 (2011) 2631–2633.
[61] M. Katz-Jaffe, S. McReynolds, Embryology in the era of proteomics, Fertil. Steril. 99 (2013) 1073–1077.
[62] M. Jodar, S. Selvaraju, E. Sendler, M. Diamond, S. Krawetz, Reproductive Medicine Network. The presence, role and clinical use of spermatozoal RNAs, Hum. Reprod. Update 19 (2013) 604–624.
[63] A. Amaral, J. Castillo, J. Estanyol, J. Ballesca, J. Ramalho-Santos, R. Oliva, Human sperm tail proteome suggests new endogenous metabolic pathways, Mol. Cell. Proteomics 12 (2013) 330–342.
[64] M. Baker, N. Naumovski, L. Hetherington, A. Weinberg, T. Velkov, R. Aitken, Head and flagella subcompartmental proteomic analysis of human spermatozoa, Proteomics 13 (2013) 61–74.
[65] G. Wang, Y. Guo, T. Zhou, X. Shi, J. Yu, Y. Yang, Y. Wu, J. Wang, M. Liu, X. Chen, W. Tu, Y. Zeng, M. Jiang, S. Li, P. Zhang, Q. Zhou, B. Zheng, C. Yu, Z. Zhou, X. Guo, J. Sha, In-depth proteomic analysis of the human sperm reveals complex protein compositions, J. Proteomics 79 (2013) 114–122.
[66] M. Codina, J. Estanyol, M. Fidalgo, J. Ballescà, R. Oliva, Advances in sperm proteomics: best-practise methodology and clinical potential, Exp. Rev. Proteomics 12 (2015) 255–277.
[67] A. Agarwal, R. Bertolla, L. Samanta, Sperm proteomics: potential impact on male infertility treatment, Exp. Rev. Proteomics 13 (2016) 285–296.
[68] M. Liu, Z. Hu, L. Qi, J. Wang, T. Zhou, Y. Guo, Y. Zeng, B. Zheng, Y. Wu, P. Zhang, X. Chen, W. Tu, T. Zhang, Q. Zhou, M. Jiang, X. Guo, Z. Zhou, J. Sha, Scanning of novel cancer/testis proteins by human testis proteomic analysis, Proteomics 13 (2013) 1200–1210.
[69] K. Gilany, A. Minai-Tehrani, E. Savadi-Shiraz, H. Rezadoost, N. Lakpour, Exploring the human seminal plasma proteome: an unexplored gold mine of biomarker for male infertility and male reproduction disorder, J. Reprod. Infertil. 16 (2015) 61–71.
[70] M. Jodar, A. Soler-Ventura, R. Oliva, Group MBoRaDR, Semen proteomics and male infertility, J. Prot. (2016), https://doi.org/10.1016/j.jprot.2016.08.018.
[71] M. Alikhani, M. Mirzaei, M. Sabbaghian, P. Parsamatin, R. Karamzadeh, S. Adib, N. Sodeifi, M.A. Gilani, M. Zabet-Moghaddam, L. Parker, Y. Wu, V. Gupta, P.A. Haynes, H. Gourabi, H. Baharvand, G.H. Salekdeh, Quantitative proteomic analysis of human testis reveals system-wide molecular and cellular pathways associated with non-obstructive azoospermia, J. Prot. (2017), https://doi.org/10.1016/j.jprot.2017.02.007.
[72] A. Amaral, J. Castillo, J. Ramalho-Santos, R. Oliva, The combined human sperm proteome: cellular pathways and implications for basic and clinical science, Hum. Reprod. Update 20 (2014) 40–62.
[73] D. Zylbersztejn, C. Andreoni, P. Del Giudice, D. Spaine, L. Borsari, G. Souza, R. Bertolla, R. Fraietta, Proteomic analysis of seminal plasma in adolescents with and without varicocele, Fertil. Steril. 99 (2013) 92–98.
[74] W. Xu, H. Hu, Z. Wang, X. Chen, F. Yang, Z. Zhu, P. Fang, J. Dai, L. Wang, H. Shi, Z. Li, Z. Qiao, Proteomic characteristics of spermatozoa in normozoospermic patients with infertility, J. Proteome 75 (2012) 5426–5436.
[75] S. McReynolds, M. Dzieciatkowska, J. Stevens, K. Hansen, W. Schoolcraft, M. Katz-Jaffe, Toward the identification of a subset of unexplained infertility: a sperm proteomic approach, Fertil. Steril. 102 (2014) 692–699.
[76] Y. Zhu, Y. Wu, K. Jin, H. Lu, F. Liu, Y. Guo, F. Yan, W. Shi, Y. Liu, X. Cao, H. Hu, H. Zhu, X. Guo, J. Sha, Z. Li, Z. Zhou, Differential proteomic profiling in human spermatozoa that did or did not result in pregnancy via IVF and AID, Proteomics Clin. Appl. 7 (2013) 850–858.

[77] C. Frapsauce, C. Pionneau, J. Bouley, V. Delarouziere, I. Berthaut, C. Ravel, J.M. Antoine, F. Soubrier, J. Mandelbaum, Proteomic identification of target proteins in normal but nonfertilizing sperm, Fertil. Steril. 102 (2014) 372–380.

[78] C. Frapsauce, C. Pionneau, J. Bouley, V. de Larouzière, I. Berthaut, C. Ravel, J.M. Antoine, F. Soubrier, J. Mandelbaum, Unexpected in vitro fertilization failure in patients with normal sperm: a proteomic analysis, Gynecol. Obstet. Fertil. (2009) 37.

[79] R. Azpiazu, A. Amaral, J. Castillo, J. Estanyol, M. Guimer, J. Ballescà, J. Balasch, R. Oliva, High-throughput sperm differential proteomics suggests that epigenetic alterations contribute to failed assisted reproduction, Hum. Reprod. 29 (2014) 1225–1237.

[80] A. Drabovich, A. Dimitromanolakis, P. Saraon, A. Soosaipillai, I. Batruch, B. Mullen, K. Jarvi, E.P. Diamandis, Differential diagnosis of azoospermia with proteomic biomarkers ECM1 and TEX101 quantified in seminal plasma, Sci. Transl. Med. 5 (2013) 212ra160.

[81] A. Agarwal, R. Sharma, D. Durairajanayagam, Z. Cui, A. Ayaz, S. Gupta, B. Willard, B. Gopalan, E. Sabanegh, Differential proteomic profiling of spermatozoal proteins of infertile men with unilateral or bilateral varicocele, Urology 85 (2015) 580–588.

[82] A. Agarwal, R. Sharma, L. Samanta, D. Durairajanayagam, E. Sabanegh, Proteomic signatures of infertile men with clinical varicocele and their validation studies reveal mitochondrial dysfunction leading to infertility, Asian J. Androl. 18 (2016) 282–291.

[83] P. Del Giudice, L. Belardin, M. Camargo, D. Zylbersztejn, V. Carvalho, K.H. Cardozo, R.P. Bertolla, A.P. Cedenho, Determination of testicular function in adolescents with varicocoele—a proteomics approach, Andrology 4 (2016) 447–455.

[84] S. Wang, W. Wang, Y. Xu, M. Tang, J. Fang, H. Sun, Y. Sun, M. Gu, Z. Liu, Z. Zhang, F. Lin, T. Wu, N. Song, Z. Wang, W. Zhang, C. Yin, Proteomic characteristics of human sperm cryopreservation, Proteomics 14 (2014) 298–310.

[85] O. Bogle, K. Kumar, C. Attardo-Parrinello, S. Lewis, J. Estanyol, J. Ballescà, R. Oliva, Identification of protein changes in human spermatozoa throughout the cryopreservation process, Andrology 5 (2017) 10–22.

[86] P. Díaz-Gimeno, J. Horcajadas, J. Martínez-Conejero, F. Esteban, P. Alamá, A. Pellicer, C. Simón, A genomic diagnostic tool for human endometrial receptivity based on the transcriptomic signature, Fertil. Steril. 95 (2011) 50–60.

[87] J. Chen, N. Hannan, Y. Mak, P. Nicholls, J. Zhang, A. Rainczuk, P.G. Stanton, D.M. Robertson, L.A. Salamonsen, A.N. Stephens, Proteomic Characterization of mid-proliferative and mid-secretory human endometrium, J. Proteome Res. 8 (2009) 2032–2044.

[88] L. DeSouza, G. Diehl, E. Yang, J. Guo, M. Rodrigues, A. Romaschin, T. Colgan, K. Siu, Proteomic analysis of the proliferative and secretory phases of the human endometrium: protein identification and differential protein expression, Proteomics 5 (2005) 270–281.

[89] T. Parmar, S. Gadkar-Sable, L. Savardekar, R. Katkam, S. Dharma, P. Meherji, C. Puri, G. Sachdeva, Protein profiling of human endometrial tissues in the midsecretory and proliferative phases of the menstrual cycle, Fertil. Steril. 92 (2009) 1091–1103.

[90] P. Rai, V. Kota, C. Sundaram, M. Deendayal, S. Shivaji, Proteome of human endometrium: identification of differentially expressed proteins in proliferative and secretory phase endometrium, Proteomics Clin. Appl. 4 (2010) 48–59.

[91] F. Domínguez, T. Garrido-Gómez, J. López, E. Camafeita, A. Quiñonero, A. Pellicer, C. Simón, Proteomic analysis of the human receptive versus non-receptive endometrium using differential in-gel electrophoresis and MALDI-MS unveils stathmin 1 and annexin A2 as differentially regulated, Hum. Reprod. 24 (2009) 2607–2617.

[92] J. Li, Z. Tan, M. Li, T. Xia, P. Liu, W. Yu, Proteomic analysis of endometrium in fertile women during the prereceptive and receptive phases after luteinizing hormone surge, Fertil. Steril. 95 (2011) 1161–1163.

[93] J. Casado-Vela, E. Rodriguez-Suarez, I. Iloro, A. Ametzazurra, N. Alkorta, J. Garcia-Velasco, R. Matorras, B. Prieto, S. González, D. Nagore, L. Simón, F. Elortza, Comprehensive proteomic analysis of human endometrial fluid aspirate, J. Proteome Res. 8 (2009) 4622–4632.

[94] N. Hannan, A. Stephens, A. Rainczuk, C. Hincks, L. Rombauts, L. Salamonsen, 2D-DiGE analysis of the human endometrial secretome reveals differences between receptive and nonreceptive states in fertile and infertile women, J. Proteome Res. 9 (2010) 6256–6264.

[95] T. Garrido-Gomez, F. Dominguez, J. Lopez, E. Camafeita, A. Quiñonero, J. Martinez-Conejero, A. Pellicer, A. Cones, C. Simón, Modeling human endometrial decidualization from the interaction between proteome and secretome, J. Clin. Endocrinol. Metab. 96 (2011) 706–716.

[96] C. Boomsma, A. Kavelaars, M. Eijkemans, K. Amarouchi, G. Teklenburg, D. Gutknecht, B. Fauser, C. Heijnen, N. Macklon, Cytokine profiling in endometrial secretions: a non-invasive window on endometrial receptivity, Reprod. Biomed. Online 18 (2009) 85–94.
[97] J. Scotchie, M. Fritz, M. Mocanu, B. Lessey, S. Young, Proteomic analysis of the luteal endometrial secretome, Reprod. Sci. 16 (2009) 883–893.
[98] C. Boomsma, A. Kavelaars, M. Eijkemans, E. Lentjes, B. Fauser, C. Heijnen, N.S. Macklon, Endometrial secretion analysis identifies a cytokine profile predictive of pregnancy in IVF, Hum. Reprod. 24 (2009) 1427–1435.
[99] N. Di Simone, E. Raschi, C. Testoni, R. Castellani, M. D'Asta, T. Shi, S.A. Krilis, A. Caruso, P.L. Meroni, Pathogenic role of anti-beta 2-glycoprotein I antibodies in antiphospholipid associated fetal loss: characterization of beta 2-glycoprotein I binding to trophoblast cells and functional effects of anti-beta 2-glycoprotein I antibodies in vitro, Ann. Rheum. Dis. 64 (2005) 462–467.
[100] T. Garrido-Gómez, A. Quiñonero, O. Antúnez, P. Díaz-Gimeno, J. Bellver, C. Simón, F. Domínguez, Deciphering the proteomic signature of human endometrial receptivity, Hum. Reprod. 29 (2014) 1957–1967.
[101] J. Brosens, A. Hodgetts, F. Feroze-Zaidi, J. Sherwin, L. Fusi, M. Salker, J. Higham, G.L. Rose, T. Kajihara, S.L. Young, B.A. Lessey, P. Henriet, P.R. Langford, A.T. Fazleabas, Proteomic analysis of endometrium from fertile and infertile patients suggests a role for apolipoprotein A-I in embryo implantation failure and endometriosis, Mol. Hum. Reprod. 16 (2010) 273–285.
[102] D. Wishart, Proteomics and the human metabolome project, Exp. Rev. Proteomics 4 (2007) 333–335.
[103] A. Revelli, L. Delle Piane, S. Casano, E. Molinari, M. Massobrio, P. Rinaudo, Follicular fluid content and oocyte quality: from single biochemical markers to metabolomics, Reprod. Biol. Endocrinol. 7 (2009) 40.
[104] J. Nicholson, J. Lindon, E. Holmes, 'Metabonomics': understanding the metabolic responses of living systems to pathophysiological stimuli via multivariate statistical analysis of biological NMR spectroscopic data, Xenobiotica 29 (1999) 1181–1189.
[105] P. Whitfield, A. German, P. Noble, Metabolomics: an emerging post-genomic tool for nutrition, Br. J. Nutr. 92 (2004) 549–555.
[106] E. Seli, L. Botros, D. Sakkas, D. Burns, Noninvasive metabolomic profiling of embryo culture media using proton nuclear magnetic resonance correlates with reproductive potential of embryos in women undergoing in vitro fertilization, Fertil. Steril. 90 (2008) 2183–2189.
[107] S. Villas-Boas, S. Mas, M. Akesson, J. Smedsgaard, J. Nielsen, Mass spectrometry in metabolome analysis, Mass Spectrom. Rev. 24 (2005) 613–646.
[108] Z. Nagy, D. Sakkas, B. Behr, Symposium: innovative techniques in human embryo viability assessment. Non-invasive assessment of embryo viability by metabolomic profiling of culture media ('metabolomics'), Reprod. Biomed. Online 17 (2008) 502–507.
[109] R. Scott, E. Seli, K. Miller, D. Sakkas, K. Scott, D. Burns, Noninvasive metabolomic profiling of human embryo culture media using Raman spectroscopy predicts embryonic reproductive potential: a prospective blinded pilot study, Fertil. Steril. 90 (2008) 77–83.
[110] Q. Wang, Q. Sun, Evaluation of oocyte quality: morphological, cellular and molecular predictors, Reprod. Fertil. Dev. 19 (1) (2007) 12.
[111] Z. Nagy, S. Jones-Colon, P. Roos, L. Botros, E. Greco, J. Dasig, B. Behr, Metabolomic assessment of oocyte viability, Reprod. Biomed. Online 18 (2009) 219–225.
[112] Y. Wu, L. Tang, J. Cai, X. Lu, J. Xu, X. Zhu, Q. Luo, H.F. Huang, High bone morphogenic protein-15 level in follicular fluid is associated with high quality oocyte and subsequent embryonic development, Hum. Reprod. 22 (2007) 1526–1531.
[113] E. Piñero-Sagredo, S. Nunes, M. de los Santos, B. Celda, V. Esteve, NMR metabolic profile of human follicular fluid, NMR Biomed. 23 (2010) 485–495.
[114] M. Wallace, E. Cottell, M. Gibney, F. McAuliffe, M. Wingfield, L. Brennan, An investigation into the relationship between the metabolic profile of follicular fluid, oocyte developmental potential, and implantation outcome, Fertil. Steril. 97 (2012) 1078–1084.
[115] A. O'Gorman, M. Wallace, E. Cottell, M. Gibney, F. McAuliffe, M. Wingfield, L. Brennan, Metabolic profiling of human follicular fluid identifies potential biomarkers of oocyte developmental competence, Reproduction 146 (2013) 389–395.
[116] C. Zhang, Y. Zhao, R. Li, Y. Yu, L. Yan, L. Li, N. Liu, P. Liu, J. Qiao, Metabolic heterogeneity of follicular amino acids in polycystic ovary syndrome is affected by obesity and related to pregnancy outcome, BMC Pregnancy Childbirth 14 (2014) 11.

[117] L. Xia, X. Zhao, Y. Sun, Y. Hong, Y. Gao, S. Hu, Metabolomic profiling of human follicular fluid from patients with repeated failure of in vitro fertilization using gas chromatography/mass spectrometry, Int. J. Clin. Exp. Pathol. 7 (2014) 7220–7229.

[118] L. Pacella, D. Zander-Fox, D. Armstrong, M. Lane, Women with reduced ovarian reserve or advanced maternal age have an altered follicular environment, Fertil. Steril. 98 (2012). 986-94.e1-2.

[119] D. Gardner, H. Leese, The role of glucose and pyruvate transport in regulating nutrient utilization by preimplantation mouse embryos, Development 104 (1988) 423–429.

[120] E. Seli, C. Robert, M. Sirard, OMICS in assisted reproduction: possibilities and pitfalls, Mol. Hum. Reprod. 16 (2010) 513–530.

[121] L. Nel-Themaat, Z. Nagy, A review of the promises and pitfalls of oocyte and embryo metabolomics, Placenta 32 (2011) S257–S263.

[122] H. Leese, Metabolism of the preimplantation embryo: 40 years on, Reproduction 143 (2012) 417–427.

[123] D. Gardner, A. Harvey, Blastocyst metabolism, Reprod. Fertil. Dev. 27 (2015) 638–654.

[124] R. Krisher, W. Schoolcraft, M. Katz-Jaffe, Omics as a window to view embryo viability, Fertil. Steril. 103 (2015) 333–341.

[125] J. Thompson, H. Brown, M. Sutton-McDowall, Measuring embryo metabolism to predict embryo quality, Reprod. Fertil. Dev. 28 (2016) 41–50.

[126] D. Gardner, M. Meseguer, C. Rubio, R. Treff, Diagnosis of human preimplantation embryo viability, Hum. Reprod. Update 21 (2015) 727–747.

[127] J. Bellver, M. De los Santos, P. Alamá, D. Castelló, L. Privitera, D. Galliano, E. Labarta, C. Vidal, A. Pellicer, F. Dominguez, Day-3 embryo metabolomics in the spent culture media is altered in obese women undergoing in vitro fertilization, Fertil. Steril. 103 (2015) 1407–1415.

[128] I. Sánchez-Ribas, M. Riqueros, P. Vime, L. Puchades-Carrasco, T. Jönsson, A. Pineda-Lucena, A. Ballesteros, F. Dominguez, C. Simón, Differential metabolic profiling of non-pure trisomy 21 human preimplantation embryos, Fertil. Steril. 98 (2012) 1157–1164.

[129] C. Siristatidis, E. Sertedaki, D. Vaidakis, Metabolomics for improving pregnancy outcomes in women undergoing assisted reproductive technologies, Cochrane Database of Syst. Rev. 5 (2017). CD011872.

[130] N. Jafarzadeh, A. Mani-Varnosfaderani, A. Minai-Tehrani, E. Savadi-Shiraz, M. Sadeghi, K. Gilany, Metabolomics fingerprinting of seminal plasma from unexplained infertile men: a need for novel diagnostic biomarkers, Mol. Reprod. Dev. 82 (2015).

[131] X. Zhang, R. Diao, X. Zhu, Z. Li, Z. Cai, Metabolic characterization of asthenozoospermia using nontargeted seminal plasma metabolomics, Clinica Chimica Acta 450 (2015) 254–261.

[132] F. Deepinder, H. Chowdary, A. Agarwal, Role of metabolomic analysis of biomarkers in the management of male infertility, Exp. Rev. Mol. Diagn. 7 (2007) 351–358.

[133] T. Mostafa, T. Anis, H. Imam, A. El-Nashar, I. Osman, Seminal reactive oxygen species-antioxidant relationship in fertile males with and without varicocele, Andrologia 41 (2009) 125–129.

[134] G. De Iuliis, J. Wingate, A. Koppers, E. McLaughlin, R. Aitken, Definitive evidence for the nonmitochondrial production of superoxide anion by human spermatozoa, J. Clin. Endocrinol. Metab. 91 (2006) 1968–1975.

[135] R. Kandar, P. Drabkova, K. Myslikova, R. Hampl, Determination of retinol and alpha-tocopherol in human seminal plasma using an HPLC with UV detection, Andrologia 46 (2014) 472–478.

[136] K. Gilany, R. Moazeni-Pourasil, N. Jafarzadeh, E. Savadi-Shiraz, Metabolomics fingerprinting of the human seminal plasma of asthenozoospermic patients, Mol. Reprod. Dev. 81 (2014) 84–86.

[137] K. Gilany, R. Pouracil, M. Sadeghi, Fourier transform infrared spectroscopy: a potential technique for noninvasive detection of spermatogenesis, Avicenna J. Med. Biotechnol. 6 (2014) 47–52.

[138] V. Jayaraman, S. Ghosh, A. Sengupta, S. Srivastava, H. Sonawat, P. Narayan, Identification of biochemical differences between different forms of male infertility by nuclear magnetic resonance (NMR) spectroscopy, J. Assist. Reprod. Genet. 31 (2014) 1195–1204.

[139] J. Zhang, Z. Huang, M. Chen, Y. Xia, F. Martin, W. Hang, H. Shen, Urinary metabolome identifies signatures of oligozoospermic infertile men, Fertil. Steril. 102 (2014) 44–53.

[140] J. Zhang, X. Mu, Y. Xia, F. Martin, W. Hang, L. Liu, M. Tian, Q. Huang, H. Shen, Metabolomic analysis reveals a unique urinary pattern in normozoospermic infertile men, J. Proteome Res. 13 (2014) 3088–3099.

[141] Z. Zhang, Y. Zhang, C. Liu, M. Zhao, Y. Yang, H. Wu, H. Zhang, H. Lin, L. Zheng, H. Jiang, Serum metabolomic profiling identifies characterization of non-obstructive azoospermic men, Int. J. Mol. Sci. 18 (2017).

[142] X. Zhou, Y. Wang, Y. Yun, Z. Xia, H. Lu, J. Luo, Y. Liang, A potential tool for diagnosis of male infertility: plasma metabolomics based on GC–MS, Talanta 147 (2016) 82–89.

[143] J. Bieniek, A. Drabovich, K. Lo, Seminal biomarkers for the evaluation of male infertility, Asian Journal of Andrology. 18 (2016) 426–433.

[144] M. Lagarde, A. Géloën, M. Record, D. Vance, F. Spener, Lipidomics is emerging, Biochim. Biophys. Acta 1634 (2003) 61.

[145] M. Sordelli, J. Beltrame, M. Cella, M. Gervasi, S. Perez Martinez, J. Burdet, E. Zotta, A.M. Franchi, M.L. Ribeiro, Interaction between lysophosphatidic acid, prostaglandins and the endocannabinoid system during the window of implantation in the rat uterus, PLoS One 7 (2012) e46059.

[146] F. Vilella, L. Ramirez, C. Simón, Lipidomics as an emerging tool to predict endometrial receptivity, Fertil. Steril. 99 (2013) 1100–1106.

[147] F. Vilella, L. Ramirez, O. Berlanga, S. Martínez, P. Alamá, M. Meseguer, A. Pellicer, C. Simón, PGE2 and PGF2 Concentrations in human endometrial fluid as biomarkers for embryonic implantation, J. Clin. Endocrinol. Metab. 98 (2013) 4123–4132.

[148] H. Achache, A. Tsafrir, D. Prus, R. Reich, A. Revel, Defective endometrial prostaglandin synthesis identified in patients with repeated implantation failure undergoing in vitro fertilization, Fertil. Steril. 94 (2010) 1271–1278.

[149] S. RoyChoudhury, A. Singh, N. Gupta, S. Srivastava, M. Joshi, B. Chakravarty, K. Chaudhury, Repeated implantation failure versus repeated implantation success: discrimination at a metabolomic level, Hum. Reprod. 31 (2016) 1265–1274.

[150] P. Banerjee, M. Dutta, S. Srivastava, M. Joshi, B. Chakravarty, K. Chaudhury, ^{1}H NMR serum metabonomics for understanding metabolic dysregulation in women with idiopathic recurrent spontaneous miscarriage during implantation window, J. Proteome Res. 13 (2014) 3100–3106.

Further Reading

[151] R.D. Upadhyay, N.H. Balasinor, A.V. Kumar, G. Sachdeva, P. Parte, K. Dumasia, Proteomics in reproductive biology: beacon for unraveling the molecular complexities, Biochim. Biophys. Acta. 1834 (2013) 8–15.

CHAPTER

10

Epigenetics, Spermatogenesis, and Male Infertility

Sezgin Gunes, Ahmet Kablan*,†, Ashok Agarwal‡, Ralf Henkel*,§*

***Department of Medical Biology, Ondokuz Mayis University, Samsun, Turkey †Nuclear Medicine, Ondokuz Mayis University, Samsun, Turkey ‡American Center for Reproductive Medicine, Cleveland Clinic, Cleveland, OH, United States §Department of Medical Bioscience, University of the Western Cape, Bellville, South Africa**

INTRODUCTION

Worldwide, almost one in six couples is infertile, and male factor infertility comprises about half of all infertility cases with a prevalence for infertility of approximately 7% [1]. Infertility has a wide range of causes including genetics and environmental effects or both [1,2]. The most common and well-known genetic conditions of infertility can be explained with karyotypic abnormalities, Y chromosome microdeletions, and cystic fibrosis transmembrane conductance regulator (*CFTR*) gene mutations, which is especially the case in men with azoospermia or severe oligozoospermia [3,4].

Recently, copy number variations have been linked to some of the infertility cases of severe oligozoospermia or Sertoli-cell-only (SCO) syndrome or both [5]. Additionally, some autosomal deletions, DNA repair mechanism defects, Y-linked syndromes, rare X-linked copy number variation (CNV), and some single-nucleotide polymorphisms (SNPs) have been found to be related with male factor infertility [6–12]. Apparently, the known genetic causes of male infertility account for up to 30% of the cases, while the causes for an estimate of 30% of the infertility cases are still unknown [13,14] highlighting the urgent need of new and reliable methods to identify male infertility causes.

Epigenetics is defined as heritable and reversible DNA modification in gene activity or expression without any alteration in DNA sequences [15]. These modifications are inherited by both mitotic and meiotic ways. Male infertility is of heterogeneous nature, which can be caused by multiple gene aberrations and gene-gene interactions [16]. Epigenetic causes identified in infertile man are observed in around 15%–30% of all cases [17,18]. Major epigenetic

Reproductomics
https://doi.org/10.1016/B978-0-12-812571-7.00011-3

modifications are DNA methylation in imprinted and developmental genes; histone tail modifications and short noncoding RNAs in spermatozoa may have a role in idiopathic male infertility [19]. Recent studies have shown that DNA methylation aberrations of reproduction- and development-related and imprinted genes or even the whole genome may have a role in unknown infertility [17,18,20–22]. DNA methylation is important in development, aging, and many other health conditions besides infertility [23–28]. Apart from aberrant DNA methylation, several recent studies have reported the role of short noncoding RNAs and various histone tail modifications to play a role in infertility [19,29–32].

Epigenetics explains the reversible manner of gene expression and regulation without any alterations of DNA sequences. These epigenetic changes can be inherited through both mitotic and meiotic cell divisions. Therefore, investigating the role of epigenetic alterations in idiopathic male infertility has increasingly been a promising area of research. Recent studies have shown that aberrant DNA methylation of reproduction-related genes and imprinted genes can be useful to explain idiopathic male infertility [23–28]. Sperm cells have a characteristic way of DNA methylation during spermatogenesis, which is necessary for proper sperm production [33]. Hence, DNA methylation seems most relevant to epigenetic anomalies in spermatozoa. Therefore, in this chapter, we will explain the impact of epigenetics including DNA methylation, histone alteration, and noncoding RNAs on male infertility.

SPERM EPIGENOME

Spermatozoa have a unique morphology with specific function and are very different from any other cell type in order to fulfill their function, fertilization [34]. Fertilization involves numerous physiological processes including movement of the male germ cells all along the female reproductive system; attachment to and penetration of the zona pellucida; and binding to, fusion with, and penetration into the oocyte. During differentiation and the development of mature spermatozoa, extensive epigenetic modifications take place [19]. These epigenetic modifications are substantial regulators of spermatogenesis and consist of sperm DNA methylation, histone tail modifications, and chromatin remodeling [34]. Histone proteins package the DNA of both oocyte and somatic cells. However, in normal mature spermatozoa, approximately 85% of the DNA is not packed by histones, but packed by nuclear proteins called protamines (P) that are rich in arginine and smaller than histones [34]. The replacement of histones with protamines is a multistep process during spermatogenesis whereby most somatic histones are first substituted by testis-specific histones H1t, TH2A, TH2B, and TH3. Subsequently, during spermiogenesis, these testis-specific histones and remaining somatic histones are replaced in a two-step process with transition proteins (TP). During this process, the sperm nucleus is losing its nucleosomal structure. Finally, the TPs are replaced by P1 and P2 to build a tight toroidal structure that compact the sperm nucleus about 6–20 times tighter than the nucleus of somatic cell [35,36].

SPERM DNA METHYLATION

DNA methylation is a well-characterized epigenetic mechanism that transfers a methyl group (——CH_3) from *S*-adenosyl methionine to the fifth carbon atom of the cytosine ring (5meC) in the cytosine-phosphate-guanine (CpG) islands (CGIs). CGIs are localized in promoters or regulatory regions of most developmental and housekeeping genes and some tissue-specific genes and are C+G-rich DNA sequences [29,37]. Cytosine methylation in CGIs causes silencing or inactivation of relevant genes, whereas hypomethylation is correlated with increased gene expression by either preventing the binding of transcription factors (TF) to the promoter or repression-mediated methyl-CpG-binding proteins [29,38].

DNA methylation is catalyzed by a group of DNA methyltransferases (DNMTs) that are maintenance DNA methyltransferases (DNMT1) and *de novo* DNMTs (DNMT3A, DNMT3B, and DNMT3L) [39]. DNMT1 leads to maintenance of DNA methylation during the DNA replication, and the lack of this enzyme results in damages to spermatogenesis and loss of methylation especially on paternally imprinted genes [40]. On the other hand, DNMT3A, DNMT3B, and DNMT3L are responsible for methylation of DNA during embryonic development of germ cells. While DNMT3A and DNMT3B provide catalytic activity, DNMT3L is a cofactor for DNMT3A. All of these enzymes are necessary for an optimal methylation process. The function of these enzymes is crucial for proper spermatogenesis.

Impaired spermatogenesis and aberrant paternal imprinting in spermatogonia were observed in conditional Dnmt3a knockout mice studies. However, Dnmt3l-null male mice germ cells demonstrated hypomethylation at several GC-poor regions and imprinted the maternally expressed transcript (*H19*) differentially in methylated regions (DMR) with delayed entry into meiosis [41].

After fertilization, the DNA of embryonal cells first undergoes whole genome-wide demethylation with the exception of transposable repetitive elements and imprinted genes [42]. Methylation marks of primordial germ cells disappear during the embryogenesis (8–13.5 days) [42–44]. After this demethylation process, a specified remethylation program starts in spermatogonia and type I spermatocytes; therefore, paternal imprints are transmitted to the spermatozoa [45,46].

Passive and active demethylations are the two ways of DNA demethylation. Passive demethylation occurs in mammalian DNA replication processes during the interphase of the cell cycle as a result of losing maintenance by defective DNMTS and can cause a loss of 5mC [47]. On the other hand, active demethylation demands a family of DNA hydroxylases called ten-eleven-translocation (TET) proteins to catalyze the conversion process of 5mC to 5-formylcytosine (5fC), 5-hydroxymethylcytosine (5hmC), and 5-carboxylcytosine (5caC) in three consecutive oxidation reactions [48]. Thereafter, a series of enzymatic oxidation and DNA repair processes take a role to accomplish DNA demethylation [49]. After this demethylation, spermatogonia and type I spermatocytes proceed a specific remethylation process at about 15.5 days post coitus that allows spermatozoa to transmit the paternal imprint [45,46].

Some genes are imprinted differentially by DNA methylation depending on which parent they are inherited from, which lead to gene expression alterations depending on the allele transmitted from the mother or the father. Many imprinted genes are essential for proper growth and development [50]. Imprinted genes are biallelic genes expressed in a monoallelic pattern depending on the allele transmitted from the mother or the father. These genes have a different path of methylation that causes modifications of gene expressions [51]. Some regions of the genome are DMR and act in a different way of methylation depending on the parental origin [19]. For example, in the human, mature sperm DNA should be methylated in paternal DMRs, but not in maternal DMRs [52]. Therefore, the *Dlk1-Gtl2*, *Igf2/H19*, *Rasgrf1*, and *Zdbf2* loci of the spermatozoan genome are methylated only in male germ cells and not expressed in male cells but expressed only in females [53]. There are also several genes that are methylated only in female, X-bearing germ cells, and expressed only in males [54]. Mesodermal-specific transcript (MEST) (also known as a paternally expressed gene 1) encodes various hydrolytic enzymes and takes an important role of the development of embryonic mesoderm with paternal monoallelic expression [55]. In addition, paternally expressed genes have different members like ZAC1 protein that takes a role in apoptosis in the G1 cell cycle and PEG 3 that has a role in the p53-dependent apoptosis pathway [56,57].

The relationship between aberrant sperm DNA methylation of imprinted genes and abnormal sperm parameters including sperm count, morphology, and motility has been studied repeatedly [25,26,52,54,58–62]. Aberrant methylation of these imprinted genes might disrupt proper sperm production and result in oligoasthenoteratozoospermia (OAT). The strong association between hypomethylation of paternally imprinted gene *H19/IGF2* and hypermethylation of maternally imprinted gene *MEST* significantly correlate to abnormal decreased sperm motility, sperm count, and abnormal sperm morphology. Hypomethylation of paternal genes could be explained by low function of DNMT, and the reason for hypermethylation of maternally imprinted genes could be de novo methylation or failure in demethylation of maternally imprinted genes in male germ cells [63].

GENE-SPECIFIC AND GENOME-WIDE/GLOBAL DNA METHYLATION

Besides the imprinting genes, methylation of nonimprinting genes specific to spermatogenesis and the global sperm DNA methylation status might also be an important factor for oligozoospermia, abnormal sperm morphology, and sperm motility. A recent global DNA methylation study has indicated that locus-dependent DNA methylation shows significant differences in infertile men compared with the fertile control group. In addition, this study has found that sperm cells show a lower methylation level in specific gene regions than somatic cells [28].

Methylenetetrahydrofolate reductase (MTHFR) has an important role in the methylation processes and DNA synthesis. Animal studies have demonstrated a correlation between decreased MTHFR activity and male infertility [64]. In addition, the methylation status of this gene has been found significantly related with infertility in nonobstructive azoospermic males with poor sperm quality and recurrent spontaneous miscarriages [65–68].

Discoidin domain receptor 1 (DDR1) is a tyrosine kinase that plays a role in postmeiotic germ cells for proliferation, apoptosis, differentiation, and cell morphogenesis. One study showed that *DDR1* promoter methylation and the expression level are associated with nonobstructive azoospermia (NOA) patients compared with control group. The authors suggested that *DDR1* expression levels are important in proper migration and the development of primordial germ cells [69].

HISTONE MODIFICATIONS AND MALE INFERTILITY

Histones are small nuclear basic proteins containing lysine and arginine and play a role in packing of the genome. Histones are subject to covalent and generally enzymatic posttranslational modifications on their amino- and carboxy-terminal tails [70]. These chemical modifications change the binding capacity of regulatory factors to DNA, which, in turn, results in alterations in gene expression and activity. Histones act to modify the gene expression. Certain changes in histone tail structures such as acetylation, methylation, ubiquitination, and phosphorylation lead to alterations in gene activity and generally interact with lysine [70]. Acetylation of H3 and H4, methylation of H3K4, and ubiquitination of H2B—which is specific for testicular tissue—result in gene activation and increased expression. On the other hand, ubiquitination of H2A and methylation of H3K9 and H3K27 generally cause gene silencing. The function of methylation of H3K4 and H3K27 is dual. Thus, all these modifications may cause activation or inhibition depending on place and time [71].

Histone acetyl transferases (HATs) are responsible for acetylation of histone proteins, a process that causes loosening of the DNA and gene activation [72]. On the contrary, histone deacetylases (HDACs) remove the acetyl groups and cause inactivation of gene expression [72]. Besides acetylation, methylation of histones also alters the gene activity [73]. Methylation and demethylation of H3 and H4 are controlled by histone methyl transferases (HMTs) and demethylases (HDMs) with variable results. Histone methylation may have a role in both activation and inactivation of transcription [74]. While H3K4 methylation causes gene activation and expression, H3K9 and H3K27 methylation inhibits or silences the gene activation [75]. These methylation processes are regulated by specific enzymes for proper spermatogenesis [76]. The decrease of the HMT MII2 activity leads to a significantly decreased number of spermatocytes by apoptosis in mice [77]. On the other hand, if LSD1/KDM1 (HDM), an enzyme responsible for demethylation of H3K4, is inhibited, sterility and germ cell apoptosis are inevitable during the meiosis. JHDM2A (HDM) demethylases the H3K9 and causes transcriptional activation [65] (Table 10.1).

In general, acetylation of lysine (K) residues of histone 3 (H3) and histone 4 (H4) is associated with transcriptional activation through triggering chromatin in an open configuration and therefore facilitating binding of transcription factors in spermatogonial DNA [78,79]. However, deacetylation leads to transcriptional inactivation and mostly links with histone methylation [80]. Although trichostatin A (TSA) is an inhibitor of histone deacetylase (HDAC) and is able to trigger cell cycle arrest in immortal somatic cells [81], TSA-treated mice demonstrated no significant effects in mitotically active spermatogonia on either proliferation or apoptosis compared with controls [82]. Moreover, retraction of TSA results in complete seminiferous epithelium regeneration in fertility assays [82]. On the contrary, increasing doses of TSA cause a significant increase in apoptosis of both spermatocytes and spermatids, suggesting a repressive function of TSA primarily on meiosis but not mitosis [82,83].

During spermatogenesis, histone methyltransferases (HTM) are responsible for methylation of H3K and H4K histone tails, a process that is regulated by histone demethylases (HDM) [84,85]. Acetylation of H2A, H2B, H3, and H4 was demonstrated to be elevated in mouse spermatogonia. These histone tails are deacetylated throughout meiosis in round spermatids and reacetylated in elongating spermatids. H4K hyperacetylation has been demonstrated to be responsible for the exchange of histones by protamines in elongating spermatids [79]. The functions of histone tail modifications relevant to spermatogenesis are summarized in Table 10.2.

Recently, the role of histone tail modification was investigated in spermatogenesis. The methylation of H3K4Me, H3K4Me3, H3K9Me2, H3K79Me2, and H3K36Me3 and acetylation of H3K4Ac and H4K5Ac in normal and abnormal human sperm were evaluated. Results have shown heterogeneous modifications of histones and the existence of H3K4Me1,

TABLE 10.1 DNA Modifying Enzymes, Their Function and Timing During Spermatogenesis

Name of the Enzyme	Function	Steps in Spermatogenesis
DNMT1	Maintenance of DNA methylation	Mitosis
DNMT3A	De novo DNA methylation	
DNMT3B	De novo DNA methylation	
DNMT3L	DNMT3a cofactor	
HAT	Acetylation of H3 and H4	
MII2 (HMT)	Methylation of H3K4	
TET	Conversion of 5mC into 5hmC, 5fC, and 5caC	
HDAC	Deacetylation of H3 and H4	Meiosis
LSD1/KDM1	Demethylation of H3K4	
HMT	Methylation of H3K9 and H3K27	
HAT	Hyperacetylation of H4	Spermiogenesis
JHDM2A	Demethylation of H3K9	

Abbreviations: *DNMT*, DNA methyl transferases; *TET*, ten-eleven translocation; *5hmC*, 5-hydroxymethylcytosine; *5fC*, 5-formylcytosine; *5caC*, 5-carboxylcytosine; *5mC*, 5-methylcytosine; *HAT*, histone acetyl transferases; *HDAC*, histone deacetylases.

TABLE 10.2 Functions of Histone Tail Modifications Relevant to Sperm

Histone Tail Modifications	Silencing Modifications	Activating Modifications	Bivalency
Acetylation of H3		+	
Acetylation of H4		+	
H3K4me2		+	
H3K4me3		+	
H3K9me	+		
H3K27me3	+		
H3K4me3 and H3K27me3			+
Ubiquitination of H2B		+	
Ubiquitination of H2A	+		

H3K9Me2, H3K4Me3, H3K79Me2, and H3K36Me3 marks in poor human sperm samples [86]. Furthermore, studies conducted in a knockout mouse model for the histone variant H3.3, *H3f3b*, have indicated that loss of the *H3f3b* gene induces aberrations in sperm and testes morphology leading to infertility [87,88]. Additionally, mice with *H3f3b*-null testes showed chromatin organization abnormalities in germ cells with increased apoptosis and decreased protamine incorporation [32]. Further studies are needed to assess the effect of sperm histone tail modifications in human infertility.

PACKAGING OF SPERM DNA WITH PROTAMINES

Packaging of sperm DNA with protamines is an essential transition that functions to arrange tremendous amounts of DNA into a small sperm nucleus. This process is accomplished by a regulatory mechanism that controls the exchange of 85%–95% of the histones by protamines. These proteins are synthesized during the later stages of spermatogenesis and are located in the sperm nucleus. Protamination protects the sperm genome from oxidation, allows nuclear compaction, is essential for sperm motility, and also protects sperm DNA from detrimental molecules within the female reproductive tract [89]. Histone-bound sperm DNA in fertile man consists of developmental genes, genes encoding microRNAs, and imprinted loci. After fertilization, the sperm chromatin sustains decondensation, and the protamines are translocated by oocyte-derived histone pronuclear development [90,91].

The replacement process of histones by protamines involves two steps. In the first step, histones are replaced by selected histone variants that are expressed during spermatogenesis. This step is accomplished by hyperacetylation of histone tails that cause loosening of sperm chromatin and induce DNA strand breaks by DNA topoisomerase that results in segregation of histones and translocation by transition proteins (TPs) [92,93]. The second step is characterized by the replacement of TP1 and TP2 by protamines. Transition proteins have an important function in histone separation and sperm chromatin condensation by protamines during this second step of histone replacement [34,92] (Fig. 10.1).

Two protamines (P1 and P2) are equally expressed in human sperm, and the P1/P2 ratio is equal to one in fertile men [94]. Errors in the protamine processing cause an increase in the immature P2 precursor production associated with subfertility [95,96]. Despite the controversial reports, skewing in the protamine ratio might be correlated with diverse phenotypes including a reduction in sperm function and counts, elevated DNA damage, decreased fertilization and implantation rates, and poor embryonic quality [95,97,98].

THE ROLE OF MICRO RNAs (miRNAs) IN GAMETOGENESIS

miRNAs are 21–25 nucleotide long, endogenous noncoding RNAs that suppress gene activation by binding to their mRNA targets, leading to either translational repression or mRNA degradation [99]. Recent data have indicated that short noncoding RNAs called miRNAs are essential in posttranscriptional regulation at each stage of differentiation of male germ cells [100]. miRNA biogenesis is a multistep process starting with pri-miRNAs that are large precursor RNA molecules [101]. These pri-miRNAs are then processed to become pre-miRNAs by a type III RNase called Drosha and its cofactor Pasha/DGCR8 in the nucleus [102]. Pre-miRNAs are transported from the nucleus into the cytoplasm by exportin 5 [103]. Pre-miRNAs are cleaved by type III RNase called dicer in order to obtain mature double-stranded miRNA molecules in the cytoplasm [104]. Either strand of this double-stranded miRNA may be incorporated into the RNA-induced silencing complex (RISC). Therefore, the target mRNA is downregulated via complementary base pairing at its 3′-untranslated region (3′UTR). The extent of the complementarity of mRNA-miRNA is regarded to have a role in

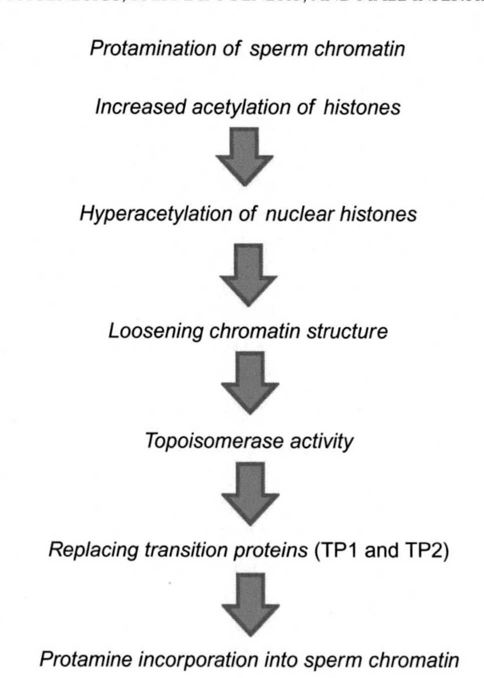

FIG. 10.1 Protamine incorporation into sperm chromatin.

the posttranscriptional mechanism employed by miRNAs [102]. Partial pairing leads to translational repression/silencing, whereas complete base pairing results in cleavage of mRNA and its subsequent degradation.

Recent studies have demonstrated that miRNAs are expressed in all phases of spermatogenesis and are necessary for male germ cell differentiation in mammals [100]. In two different mouse models, the requirement for miRNAs for spermatogenesis has been shown by two initial studies where the *Dicer1* gene was knocked out [105,106]. Germ-cell-specific DICER1 deletion in these mouse models causes alterations in meiotic progression, failure of haploid germ cell differentiation, increased apoptosis, and infertility [106]. Drosha has a function in miRNA biogenesis for dissection of miRNAs. Wu et al. [31] have used *Dicer* and *Drosha* conditional knockout mouse models to investigate the function of miRNA in spermatogenesis. The authors found that for both knockout systems, male mice showed impaired spermatogenesis characterized by azoospermia or oligoteratozoospermia due to exhaustion of spermatocytes and spermatids that results in infertility. Spermatogenesis was more severely disrupted in *Drosha* knockouts compared with *Dicer* knockouts demonstrating the importance of miRNAs in spermatogenesis and male fertility [31].

Several studies have reported that investigating the function of miRNAs in human spermatogenesis and the results of these studies encourage the findings obtained from animal models. Human spermatogenic cell populations from different stages of spermatogenesis have been isolated, and their miRNA expression profiles have been identified by microarray assay [107]. A total of 559 miRNAs were identified to be distinctively expressed in spermatogonia, pachytene spermatocytes, and round spermatids. Comparative analyses showed 78 miRNAs to be downregulated while 32 miRNAs to be significantly upregulated in human spermatogonia and pachytene spermatocytes. The results indicated the potential role of miRNAs in mitotic and meiotic stages of spermatogenesis. Furthermore, 29 miRNAs were downregulated, and 144 miRNAs were found to be upregulated between pachytene spermatocytes and round spermatids, suggesting the function of these miRNAs in the regulation of spermiogenesis [107]. The results of next generation sequencing analysis of short RNA transcriptome from three normal human testes have demonstrated a total of 770 known and 5 novel human miRNAs, suggesting high miRNA complexity in the human testis [108]. This study also demonstrated that abundantly detected miRNAs are let-7 family members, miR-34c-5p, miR-103a-3p, miR-202-5p, miR-508-3p, and miR-509-3-5p. These miRNAs target gene transcripts are involved in regulation of spermatogenesis, germ cell testicular development, meiosis, apoptosis, homologous recombination, and p53-related pathways [108].

Recent data indicated the essential roles of miRNAs in the regulation of gene expression, regulating certain transitions between the main phases of spermatogenesis in mammals [100,104,109]. Finally, environmental factors appear to have critical contributions to the control of the epigenetic regulation of spermatogenesis. The male sperm epigenome is subject to change due to certain environmental factors, and environmentally induced epigenetic alterations could be inherited to the next generations even though they are not exposed to the stressors themselves [110]. A miRNA microarray study carried out with spermatozoa from adult men living in an environmentally polluted region has demonstrated 109 downregulated miRNAs and 73 significantly upregulated compared with the control group [111]. Since environmental pollution is one of the well-known factors accounting for poor sperm quality in humans [112], these data suggest that differentially expressed miRNAs under such conditions might help explain the harmful effect of environmental pollution on sperm count and quality.

ENVIRONMENTAL EFFECT AND MALE INFERTILITY

Environmental factors comprise a diverse range of exposures, such as ultraviolet light, pesticides, infectious agents, or traffic pollution. In addition, occupational exposures (workplace exposures), medical treatments (medicines and radiation including chemotherapy), exposures resulting from lifestyle choices (smoking, cosmetics, nutrition, and physical activity), anabolic hormone, or drugs suppressing the immune system [112] have significant negative effects on male fertility. These factors might cause alterations in the regulation of genetic and epigenetic mechanisms. Air pollution is a major environmental factor associated with human diseases, and World Health Organization (WHO) data demonstrate that about 25% of all human diseases are caused by air pollution. In addition to air pollution, increased cell phone use; consumption of opioids, marijuana, and cigarettes; the lack of exercise; and obesity could be accepted as environmental dangers. These changes increased enormously over the past four decades. Both epidemiological and experimental studies showed that environmental contaminants are significant risk factors for male infertility [113], and the reproductive system, especially the male reproductive system, seems to be quite vulnerable to such factors [114–116]. Besides air pollution, other environmental factors could show detrimental effects on spermatogenesis and cause male infertility, particularly in industrialized countries. From this perspective, it is crucial to understand the genetic and epigenetic mechanisms of environmental alterations on reproductive system [112]. These studies will help us identifying male infertility and even prevent or treat the condition.

In the last three decades, studies have shown that infertility rates have been rising particularly in industrialized countries worldwide [117,118]. Environmental changes effect spermatogenesis at different levels and cause variable consequences including abnormal morphology, decreased sperm count, and volume [119–121].

Environmental pollutants cause loss of DNA integrity, DNA fragmentation, and many other effects including epigenetic aberrations in male germ cells [122]. A few studies reported that air pollution has been linked to increased DNA damage without any changes in sperm quality or motility etc. [123]. On the contrary, during the past decades, some studies on subjects exposed to environmental pollutants have shown high infertility rates, abnormal sperm morphology, and DNA fragmentation rates compared with the control groups [124–126]. Higher traffic emissions or heavy metals caused a decline in sperm motility and increase in sperm DNA damage even in fertile men living in a polluted area ([127–129]). However, the exact mechanism of how air pollutants cause these damages to the genome is not fully lightened, and it is merely suggested that it might be related to reactive oxidation species (ROS) with oxidative stress, abnormal DNA chromatin packaging, or apoptosis [122]. Furthermore, heavy metals in polluted air could enter a reaction with protamines and cause fragmented sperm DNA [130–134].

Although the studies investigating the effects of toxic environmental pollutants on epigenetic mechanisms in the human are limited, *in vivo* studies suggested that environmental factors could alter the gene expression via epigenetic mechanisms [135]. Recently, an *in vivo* study showed that exposure to toxicants might change the miRNA expression that plays an important epigenetic role for proper sperm production [136]. It is quite possible that environmental changes affect epigenetic mechanisms that may cause infertility or subclinical abnormal sperm production along with other diseases.

Lifestyle changes including the lack of physical activity, smoking, alcohol consumption, or stress affect both male and female fertility [137]. These changes affect both genetic and epigenetic mechanisms. Another report indicated that alterations like changes in miRNA status or the histones might be inherited to the offspring [138]. DNA methylation is a well-studied epigenetic mechanism, and studies demonstrate that maternal diet, exposure to chemicals, alcohol consumption, or smoking affects fetal cells [139,140].

The effects of paternal exposure to certain chemicals on the offspring have been studied over the years [141,142]. Paternal exposure to pesticides was found to be related with an elevated risk of nervous system tumors in the offspring [141]. Although animal models in general showed that changes in the DNA methylation pattern that was caused by exposure to pesticides and chemicals could be inherited to the next generation [143,144], the effects of particular lifestyle changes still remain uncertain and are a huge promising area to study.

CONCLUSION AND FUTURE PERSPECTIVE

Advancement of assisted reproductive techniques (ARTs) leads to epigenetic changes that may alter the normal gene imprinting mechanisms. Therefore, understanding of the epigenetic mechanisms that regulate spermatogenesis and sperm abnormalities is important. Studies reported aberrant methylation in both females who undergo hormone therapy and males with infertility [145]. The combination of these studies may lead to deeper understanding of the origin of pregnancy loss and abnormalities in embryonic development seen both in normal conceptions and *in vitro* fertilization clinics.

Appropriate diagnostic and treatment algorithms of male infertility became important issues of contemporary andrology. Despite strict criteria for morphological analysis are used, routine semen analysis still shows a poor predictive value. Accordingly, considerable efforts have been made in the search to define a link between certain diseases and ARTs. Among them, epigenetic mechanisms have been suggested in a large number of studies of animal models. Yet, consequences of epigenetic mechanisms on human germ cells still present a puzzle since the relevant diseases are rare. Thus, epigenetics is a promising area that can be helpful to understand the reasons of idiopathic male infertility.

References

[1] J. Boivin, L. Bunting, J.A. Collins, K.G. Nygren, International estimates of infertility prevalence and treatment-seeking: potential need and demand for infertility medical care, Hum. Reprod. 22 (2007) 1506–1512.

[2] N. Rives, Y chromosome microdeletions and alterations of spermatogenesis, patient approach and genetic counseling, Ann. Endocrinol. (Paris) 75 (2014) 112–114.

[3] S. Gunes, R. Asci, G. Okten, F. Atac, O.E. Onat, G. Ogur, O. Aydin, T. Ozcelik, H. Bagci, Two males with SRY-positive 46,XX testicular disorder of sex development, Syst. Biol. Reprod. Med. 59 (2013) 42–47.

[4] J.M. Hotaling, Genetics of male infertility, Urol. Clin. North Am. 41 (2014) 1–17.

[5] F. Tuttelmann, M. Simoni, S. Kliesch, S. Ledig, B. Dworniczak, P. Wieacker, A. Ropke, Copy number variants in patients with severe oligozoospermia and Sertoli-cell-only syndrome, PLoS One 6 (2011) e19426.

[6] A. Abhari, N. Zarghami, L. Farzadi, M. Nouri, V. Shahnazi, Altered of microRNA expression level in oligospermic patients, Iran J. Reprod. Med. 12 (2014) 681–686.

[7] S. Gunes, M. Al-Sadaan, A. Agarwal, Spermatogenesis, DNA damage and DNA repair mechanisms in male infertility, Reprod. Biomed. Online 31 (2015) 309–319.

[8] N. Gupta, S. Gupta, M. Dama, A. David, G. Khanna, A. Khanna, S. Rajender, Strong association of 677 C>T substitution in the MTHFR gene with male infertility—a study on an indian population and a meta-analysis, PLoS One 6 (2011) e22277.

[9] X.J. He, J. Ruan, W.D. Du, G. Chen, Y. Zhou, S. Xu, X.B. Zuo, Y.X. Cao, X.J. Zhang, PRM1 variant rs35576928 (Arg>Ser) is associated with defective spermatogenesis in the Chinese Han population, Reprod. Biomed. Online 25 (2012) 627–634.

[10] A.M. Lopes, K.I. Aston, E. Thompson, F. Carvalho, J. Goncalves, N. Huang, R. Matthiesen, M.J. Noordam, I. Quintela, A. Ramu, C. Seabra, A.B. Wilfert, J. Dai, J.M. Downie, S. Fernandes, X. Guo, J. Sha, A. Amorim, A. Barros, A. Carracedo, Z. Hu, M.E. Hurles, S. Moskovtsev, C. Ober, D.A. Paduch, J.D. Schiffman, P.N. Schlegel, M. Sousa, D.T. Carrell, D.F. Conrad, Human spermatogenic failure purges deleterious mutation load from the autosomes and both sex chromosomes, including the gene DMRT1, PLoS Genet. 9 (2013) e1003349.

[11] Y.N. Teng, Y.P. Chang, J.T. Tseng, P.H. Kuo, I.W. Lee, M.S. Lee, P.L. Kuo, A single-nucleotide polymorphism of the DAZL gene promoter confers susceptibility to spermatogenic failure in the Taiwanese Han, Hum. Reprod. 27 (2012) 2857–2865.

[12] W. Wu, J. Lu, Q. Tang, S. Zhang, B. Yuan, J. Li, W. Di, H. Sun, C. Lu, Y. Xia, D. Chen, J. SHA, X. Wang, GSTM1 and GSTT1 null polymorphisms and male infertility risk: an updated meta-analysis encompassing 6934 subjects, Sci. Rep. 3 (2013) 2258.

[13] G.L. Harton, H.G. Tempest, Chromosomal disorders and male infertility, Asian J. Androl. 14 (2012) 32–39.
[14] A. Jungwirth, A. Giwercman, H. Tournaye, T. Diemer, Z. Kopa, G. Dohle, C. Krausz, European Association of Urology Working Group on Male Infertility, European Association of Urology guidelines on Male Infertility: the 2012 update, Eur. Urol. 62 (2012) 324–332.
[15] T. Tollefsbol, Transgenerational Epigenetics, Academic Press, Elsevier, 2014.
[16] S. Laurentino, J. Borgmann, J. Gromoll, On the origin of sperm epigenetic heterogeneity, Reproduction 151 (2016) R71–R78.
[17] D.J. Amor, J. Halliday, A review of known imprinting syndromes and their association with assisted reproduction technologies, Hum. Reprod. 23 (2008) 2826–2834.
[18] G. Pliushch, E. Schneider, D. Weise, N. El Hajj, A. Tresch, L. Seidmann, W. Coerdt, A.M. Muller, U. Zechner, T. Haaf, Extreme methylation values of imprinted genes in human abortions and stillbirths, Am. J. Pathol. 176 (2010) 1084–1090.
[19] S. Gunes, M.A. Arslan, G.N.T. Hekim, R. Asci, The role of epigenetics in idiopathic male infertility, J. Assist. Reprod. Genet. 33 (2016) 553–569.
[20] S.C. Esteves, A clinical appraisal of the genetic basis in unexplained male infertility, J. Hum. Reprod. Sci. 6 (2013) 176–182.
[21] T.G. Jenkins, K.I. Aston, J.M. Hotaling, M.B. Shamsi, L. Simon, D.T. Carrell, Teratozoospermia and asthenozoospermia are associated with specific epigenetic signatures, Andrology 4 (2016) 843–849.
[22] A. Soubry, Epigenetic inheritance and evolution: a paternal perspective on dietary influences, Prog. Biophys. Mol. Biol. 118 (2015) 79–85.
[23] F. Ferfouri, F. Boitrelle, I. Ghout, M. Albert, D. Molina Gomes, R. Wainer, M. Bailly, J. Selva, F. Vialard, A genome-wide DNA methylation study in azoospermia, Andrology 1 (2013) 815–821.
[24] S.S. Hammoud, D.A. Nix, A.O. Hammoud, M. Gibson, B.R. Cairns, D.T. Carrell, Genome-wide analysis identifies changes in histone retention and epigenetic modifications at developmental and imprinted gene loci in the sperm of infertile men, Hum. Reprod. 26 (2011) 2558–2569.
[25] S. Houshdaran, V.K. Cortessis, K. Siegmund, A. Yang, P.W. Laird, R.Z. Sokol, Widespread epigenetic abnormalities suggest a broad DNA methylation erasure defect in abnormal human sperm, PLoS One 2 (2007) e1289.
[26] J.C. Rotondo, R. Selvatici, M. Di Domenico, R. Marci, F. Vesce, M. Tognon, F. Martini, Methylation loss at H19 imprinted gene correlates with methylenetetrahydrofolate reductase gene promoter hypermethylation in semen samples from infertile males, Epigenetics 8 (2013) 990–997.
[27] B. Schutte, N. El Hajj, J. Kuhtz, I. Nanda, J. Gromoll, T. Hahn, M. Dittrich, M. Schorsch, T. Muller, T. Haaf, Broad DNA methylation changes of spermatogenesis, inflammation and immune response-related genes in a subgroup of sperm samples for assisted reproduction, Andrology 1 (2013) 822–829.
[28] R.G. Urdinguio, G.F. Bayon, M. Dmitrijeva, E.G. Torano, C. Bravo, M.F. Fraga, L. Bassas, S. Larriba, A.F. Fernandez, Aberrant DNA methylation patterns of spermatozoa in men with unexplained infertility, Hum. Reprod. 30 (2015) 1014–1028.
[29] A.M. Deaton, A. Bird, CpG islands and the regulation of transcription, Genes Dev. 25 (2011) 1010–1022.
[30] R. Song, G.W. Hennig, Q. Wu, C. Jose, H. Zheng, W. Yan, Male germ cells express abundant endogenous siRNAs, Proc. Natl. Acad. Sci. U. S. A. 108 (2011) 13159–13164.
[31] Q. Wu, R. Song, N. Ortogero, H. Zheng, R. Evanoff, C.L. Small, M.D. Griswold, S.H. Namekawa, H. Royo, J.M. Turner, W. Yan, The RNase III enzyme DROsha is essential for microRNA production and spermatogenesis, J. Biol. Chem. 287 (2012) 25173–25190.
[32] B.T. Yuen, K.M. Bush, B.L. Barrilleaux, R. Cotterman, P.S. Knoepfler, Histone H3.3 regulates dynamic chromatin states during spermatogenesis, Development 141 (2014) 3483–3494.
[33] R. Klaver, F. Tuttelmann, A. Bleiziffer, T. Haaf, S. Kliesch, J. Gromoll, DNA methylation in spermatozoa as a prospective marker in andrology, Andrology 1 (2013) 731–740.
[34] S. Gunes, T. Kulac, The role of epigenetics in spermatogenesis, Turk. J. Urol. 39 (2013) 181–187.
[35] R.J. Oko, V. Jando, C.L. Wagner, W.S. Kistler, L.S. Hermo, Chromatin reorganization in rat spermatids during the disappearance of testis-specific histone, H1t, and the appearance of transition proteins TP1 and TP2, Biol. Reprod. 54 (1996) 1141–1157.
[36] W.S. Ward, D.S. Coffey, DNA packaging and organization in mammalian spermatozoa: comparison with somatic cells, Biol. Reprod. 44 (1991) 569–574.
[37] J. Zhu, F. He, S. Hu, J. Yu, On the nature of human housekeeping genes, Trends Genet. 24 (2008) 481–484.

[38] R.J. Klose, A.P. Bird, Genomic DNA methylation: the mark and its mediators, Trends Biochem. Sci. 31 (2006) 89–97.
[39] S. Eden, H. Cedar, Role of DNA methylation in the regulation of transcription, Curr. Opin. Genet. Dev. 4 (1994) 255–259.
[40] P. Cheng, H. Chen, R.P. Zhang, S.R. Liu, A. Zhou-Cun, Polymorphism in DNMT1 may modify the susceptibility to oligospermia, Reprod. Biomed. Online 28 (2014) 644–649.
[41] S. LA Salle, C.C. Oakes, O.R. Neaga, D. Bourc'his, T.H. Bestor, J.M. Trasler, Loss of spermatogonia and widespread DNA methylation defects in newborn male mice deficient in DNMT3L, BMC Dev. Biol. 7 (2007) 104.
[42] P. Hajkova, S. Erhardt, N. Lane, T. Haaf, O. El-Maarri, W. Reik, J. Walter, M.A. Surani, Epigenetic reprogramming in mouse primordial germ cells, Mech. Dev. 117 (2002) 15–23.
[43] E. Li, Chromatin modification and epigenetic reprogramming in mammalian development, Nat. Rev. Genet. 3 (2002) 662–673.
[44] F. Santos, W. Dean, Epigenetic reprogramming during early development in mammals, Reproduction 127 (2004) 643–651.
[45] C.C. Boissonnas, P. Jouannet, H. Jammes, Epigenetic disorders and male subfertility, Fertil. Steril. 99 (2013) 624–631.
[46] T.L. Davis, G.J. Yang, J.R. Mccarrey, M.S. Bartolomei, The H19 methylation imprint is erased and re-established differentially on the parental alleles during male germ cell development, Hum. Mol. Genet. 9 (2000) 2885–2894.
[47] M. Tahiliani, K.P. Koh, Y. Shen, W.A. Pastor, H. Bandukwala, Y. Brudno, S. Agarwal, L.M. Iyer, D.R. Liu, L. Aravind, A. Rao, Conversion of 5-methylcytosine to 5-hydroxymethylcytosine in mammalian DNA by MLL partner TET1, Science 324 (2009) 930–935.
[48] G. Ficz, New insights into mechanisms that regulate DNA methylation patterning, J. Exp. Biol. 218 (2015) 14–20.
[49] H. Wu, Y. Zhang, Mechanisms and functions of Tet protein-mediated 5-methylcytosine oxidation, Genes Dev. 25 (2011) 2436–2452.
[50] B. Horsthemke, In brief: genomic imprinting and imprinting diseases, J. Pathol. 232 (2014) 485–487.
[51] H.A. Lawson, J.M. Cheverud, J.B. Wolf, Genomic imprinting and parent-of-origin effects on complex traits, Nat. Rev. Genet. 14 (2013) 609–617.
[52] C. Camprubi, M. Pladevall, M. Grossmann, N. Garrido, M.C. Pons, J. Blanco, Semen samples showing an increased rate of spermatozoa with imprinting errors have a negligible effect in the outcome of assisted reproduction techniques, Epigenetics 7 (2012) 1115–1124.
[53] P. Arnaud, Genomic imprinting in germ cells: imprints are under control, Reproduction 140 (2010) 411–423.
[54] C.J. Marques, F. Carvalho, M. Sousa, A. Barros, Genomic imprinting in disruptive spermatogenesis, Lancet 363 (2004) 1700–1702.
[55] H.Y. Zheng, X.Y. Shi, F.R. Wu, Y.Q. Wu, L.L. Wang, S.L. Chen, Assisted reproductive technologies do not increase risk of abnormal methylation of PEG1/MEST in human early pregnancy loss, Fertil. Steril. 96 (2011) 84–89. e2.
[56] K.D. Broad, J.P. Curley, E.B. Keverne, Increased apoptosis during neonatal brain development underlies the adult behavioral deficits seen in mice lacking a functional paternally expressed gene 3 (Peg3), Dev. Neurobiol. 69 (2009) 314–325.
[57] M. Kamiya, H. Judson, Y. Okazaki, M. Kusakabe, M. Muramatsu, S. Takada, N. Takagi, T. Arima, N. Wake, K. Kamimura, K. Satomura, R. Hermann, D.T. Bonthron, Y. Hayashizaki, The cell cycle control gene ZAC/PLAGL1 is imprinted—a strong candidate gene for transient neonatal diabetes, Hum. Mol. Genet. 9 (2000) 453–460.
[58] N. El Hajj, U. Zechner, E. Schneider, A. Tresch, J. Gromoll, T. Hahn, M. Schorsch, T. Haaf, Methylation status of imprinted genes and repetitive elements in sperm DNA from infertile males, Sex. Dev. 5 (2011) 60–69.
[59] H. Kobayashi, A. Sato, E. Otsu, H. Hiura, C. Tomatsu, T. Utsunomiya, H. Sasaki, N. Yaegashi, T. Arima, Aberrant DNA methylation of imprinted loci in sperm from oligospermic patients, Hum. Mol. Genet. 16 (2007) 2542–2551.
[60] C.J. Marques, P. Costa, B. Vaz, F. Carvalho, S. Fernandes, A. Barros, M. Sousa, Abnormal methylation of imprinted genes in human sperm is associated with oligozoospermia, Mol. Hum. Reprod. 14 (2008) 67–74.
[61] C.J. Marques, T. Francisco, S. Sousa, F. Carvalho, A. Barros, M. Sousa, Methylation defects of imprinted genes in human testicular spermatozoa, Fertil. Steril. 94 (2010) 585–594.

[62] A. Minor, V. Chow, S. Ma, Aberrant DNA methylation at imprinted genes in testicular sperm retrieved from men with obstructive azoospermia and undergoing vasectomy reversal, Reproduction 141 (2011) 749–757.
[63] A. Poplinski, F. Tuttelmann, D. Kanber, B. Horsthemke, J. Gromoll, Idiopathic male infertility is strongly associated with aberrant methylation of MEST and IGF2/H19 ICR1, Int. J. Androl. 33 (2010) 642–649.
[64] T.L. Kelly, O.R. Neaga, B.C. Schwahn, R. Rozen, J.M. Trasler, Infertility in 5,10-methylenetetrahydrofolate reductase (MTHFR)-deficient male mice is partially alleviated by lifetime dietary betaine supplementation, Biol. Reprod. 72 (2005) 667–677.
[65] A. Botezatu, R. Socolov, D. Socolov, I.V. Iancu, G. Anton, Methylation pattern of methylene tetrahydrofolate reductase and small nuclear ribonucleoprotein polypeptide N promoters in oligoasthenospermia: a case-control study, Reprod. Biomed. Online 28 (2014) 225–231.
[66] N. Khazamipour, M. Noruzinia, P. Fatehmanesh, M. Keyhanee, P. Pujol, MTHFR promoter hypermethylation in testicular biopsies of patients with non-obstructive azoospermia: the role of epigenetics in male infertility, Hum. Reprod. 24 (2009) 2361–2364.
[67] J.C. Rotondo, S. Bosi, E. Bazzan, M. Di Domenico, M. de Mattei, R. Selvatici, A. Patella, R. Marci, M. Tognon, F. Martini, Methylenetetrahydrofolate reductase gene promoter hypermethylation in semen samples of infertile couples correlates with recurrent spontaneous abortion, Hum. Reprod. 27 (2012) 3632–3638.
[68] W. Wu, O. Shen, Y. Qin, X. Niu, C. Lu, Y. Xia, L. Song, S. Wang, X. Wang, Idiopathic male infertility is strongly associated with aberrant promoter methylation of methylenetetrahydrofolate reductase (MTHFR), PLoS One 5 (2010) e13884.
[69] R. Ramasamy, A. Ridgeway, L.I. Lipshultz, D.J. Lamb, Integrative DNA methylation and gene expression analysis identifies discoidin domain receptor 1 association with idiopathic nonobstructive azoospermia, Fertil. Steril. 102 (2014) 968–973. e3.
[70] E.I. Campos, D. Reinberg, Histones: annotating chromatin, Annu. Rev. Genet. 43 (2009) 559–599.
[71] M. Werner, A.J. Ruthenburg, The United States of histone ubiquitylation and methylation, Mol. Cell 43 (2011) 5–7.
[72] E. Seto, M. Yoshida, Erasers of histone acetylation: the histone deacetylase enzymes, Cold Spring Harb. Perspect. Biol. 6 (2014) a018713.
[73] H. Hashimoto, P.M. Vertino, X. Cheng, Molecular coupling of DNA methylation and histone methylation, Epigenomics 2 (2010) 657–669.
[74] D.T. Carrell, Epigenetics of the male gamete, Fertil. Steril. 97 (2012) 267–274.
[75] S. Rajender, K. Avery, A. Agarwal, Epigenetics, spermatogenesis and male infertility, Mutat. Res. 727 (2011) 62–71.
[76] C.C. Boissonnas, H.E. Abdalaoui, V. Haelewyn, P. Fauque, J.M. DuPont, I. Gut, D. Vaiman, P. Jouannet, J. Tost, H. Jammes, Specific epigenetic alterations of IGF2-H19 locus in spermatozoa from infertile men, Eur. J. Hum. Genet. 18 (2010) 73–80.
[77] S. Glaser, S. Lubitz, K.L. Loveland, K. Ohbo, L. Robb, F. Schwenk, J. Seibler, D. Roellig, A. Kranz, K. Anastassiadis, A.F. Stewart, The histone 3 lysine 4 methyltransferase, Mll2, is only required briefly in development and spermatogenesis, Epigenetics Chromatin 2 (2009) 5.
[78] B.R. Cairns, The logic of chromatin architecture and remodelling at promoters, Nature 461 (2009) 193–198.
[79] M. Hazzouri, C. Pivot-Pajot, A.K. Faure, Y. Usson, R. Pelletier, B. Sele, S. Khochbin, S. Rousseaux, Regulated hyperacetylation of core histones during mouse spermatogenesis: involvement of histone deacetylases, Eur. J. Cell Biol. 79 (2000) 950–960.
[80] L. Peng, E. Seto, Deacetylation of nonhistone proteins by HDACs and the implications in cancer, Handb. Exp. Pharmacol. 206 (2011) 39–56.
[81] D.M. Vigushin, S. Ali, P.E. Pace, N. Mirsaidi, K. Ito, I. Adcock, R.C. Coombes, Trichostatin A is a histone deacetylase inhibitor with potent antitumor activity against breast cancer in vivo, Clin. Cancer Res. 7 (2001) 971–976.
[82] I. Fenic, V. Sonnack, K. Failing, M. Bergmann, K. Steger, In vivo effects of histone-deacetylase inhibitor trichostatin-A on murine spermatogenesis, J. Androl. 25 (2004) 811–818.
[83] I. Fenic, H.M. Hossain, V. Sonnack, S. Tchatalbachev, F. Thierer, J. Trapp, K. Failing, K.S. Edler, M. Bergmann, M. Jung, T. Chakraborty, K. Steger, In vivo application of histone deacetylase inhibitor trichostatin-a impairs murine male meiosis, J. Androl. 29 (2008) 172–185.
[84] D.T. Carrell, B.R. Emery, S. Hammoud, The aetiology of sperm protamine abnormalities and their potential impact on the sperm epigenome, Int. J. Androl. 31 (2008) 537–545.

[85] M. Lachner, T. Jenuwein, The many faces of histone lysine methylation, Curr. Opin. Cell Biol. 14 (2002) 286–298.
[86] F.A. La Spina, M. Romanato, S. Brugo-Olmedo, S. de Vincentiis, V. Julianelli, R.M. Rivera, M.G. Buffone, Heterogeneous distribution of histone methylation in mature human sperm, J. Assist. Reprod. Genet. 31 (2014) 45–49.
[87] K.M. Bush, B.T. Yuen, B.L. Barrilleaux, J.W. Riggs, H. O'geen, R.F. Cotterman, P.S. Knoepfler, Endogenous mammalian histone H3.3 exhibits chromatin-related functions during development, Epigenetics Chromatin 6 (2013) 7.
[88] C. Couldrey, M.B. Carlton, P.M. Nolan, W.H. Colledge, M.J. Evans, A retroviral gene trap insertion into the histone 3.3A gene causes partial neonatal lethality, stunted growth, neuromuscular deficits and male sub-fertility in transgenic mice, Hum. Mol. Genet. 8 (1999) 2489–2495.
[89] N. Torregrosa, D. Dominguez-Fandos, M.I. Camejo, C.R. Shirley, M.L. Meistrich, J.L. Ballesca, R. Oliva, Protamine 2 precursors, protamine 1/protamine 2 ratio, DNA integrity and other sperm parameters in infertile patients, Hum. Reprod. 21 (2006) 2084–2089.
[90] R. Balhorn, The protamine family of sperm nuclear proteins, Genome Biol. 8 (2007) 227.
[91] C. Rathke, W.M. Baarends, S. Awe, R. Renkawitz-Pohl, Chromatin dynamics during spermiogenesis, Biochim. Biophys. Acta 1839 (2014) 155–168.
[92] M.L. Meistrich, B. Mohapatra, C.R. Shirley, M. Zhao, Roles of transition nuclear proteins in spermiogenesis, Chromosoma 111 (2003) 483–488.
[93] S. Rousseaux, F. Boussouar, J. Gaucher, N. Reynoird, E. Montellier, S. Curtet, A.L. Vitte, S. Khochbin, Molecular models for post-meiotic male genome reprogramming, Syst. Biol. Reprod. Med. 57 (2011) 50–53.
[94] L. Nanassy, D.T. Carrell, Abnormal methylation of the promoter of CREM is broadly associated with male factor infertility and poor sperm quality but is improved in sperm selected by density gradient centrifugation, Fertil. Steril. 95 (2011) 2310–2314.
[95] V.W. Aoki, L. Liu, D.T. Carrell, Identification and evaluation of a novel sperm protamine abnormality in a population of infertile males, Hum. Reprod. 20 (2005) 1298–1306.
[96] I.A. Belokopytova, E.I. Kostyleva, A.N. Tomilin, V.I. Vorob'ev, Human male infertility may be due to a decrease of the protamine P2 content in sperm chromatin, Mol. Reprod. Dev. 34 (1993) 53–57.
[97] S. de Mateo, C. Gazquez, M. Guimera, J. Balasch, M.L. Meistrich, J.L. Ballesca, R. Oliva, Protamine 2 precursors (pre-P2), protamine 1 to protamine 2 ratio (P1/P2), and assisted reproduction outcome, Fertil. Steril. 91 (2009) 715–722.
[98] M.H. Nasr-Esfahani, M. Salehi, S. Razavi, M. Mardani, H. Bahramian, K. Steger, F. Oreizi, Effect of protamine-2 deficiency on ICSI outcome, Reprod. Biomed. Online 9 (2004) 652–658.
[99] D. Bourc'his, O. Voinnet, A small-RNA perspective on gametogenesis, fertilization, and early zygotic development, Science 330 (2010) 617–622.
[100] C. Yao, Y. Liu, M. Sun, M. Niu, Q. Yuan, Y. Hai, Y. Guo, Z. Chen, J. Hou, Y. Liu, Z. He, MicroRNAs and DNA methylation as epigenetic regulators of mitosis, meiosis and spermiogenesis, Reproduction 150 (2015) R25–R34.
[101] Z. He, M. Kokkinaki, D. Pant, G.I. Gallicano, M. Dym, Small RNA molecules in the regulation of spermatogenesis, Reproduction 137 (2009) 901–911.
[102] J. Krol, I. Loedige, W. Filipowicz, The widespread regulation of microRNA biogenesis, function and decay, Nat. Rev. Genet. 11 (2010) 597–610.
[103] A. Kohler, E. Hurt, Exporting Rna from the nucleus to the cytoplasm, Nat. Rev. Mol. Cell Biol. 8 (2007) 761–773.
[104] B.J. Hale, A.F. Keating, C.X. Yang, J.W. Ross, Small RNAs: their possible roles in reproductive failure, Adv. Exp. Med. Biol. 868 (2015) 49–79.
[105] H.M. Korhonen, O. Meikar, R.P. Yadav, M.D. Papaioannou, Y. Romero, M. Da Ros, P.L. Herrera, J. Toppari, S. Nef, N. Kotaja, Dicer is required for haploid male germ cell differentiation in mice, PLoS One 6 (2011) e24821.
[106] Y. Romero, O. Meikar, M.D. Papaioannou, B. Conne, C. Grey, M. Weier, F. Pralong, B. de Massy, H. Kaessmann, J.D. Vassalli, N. Kotaja, S. Nef, Dicer1 depletion in male germ cells leads to infertility due to cumulative meiotic and spermiogenic defects, PLoS One 6 (2011) e25241.
[107] Y. Liu, M. Niu, C. Yao, Y. Hai, Q. Yuan, Y. Guo, Z. Li, Z. He, Fractionation of human spermatogenic cells using STA-PUT gravity sedimentation and their miRNA profiling, Sci. Rep. 5 (2015) 8084.
[108] Q. Yang, J. Hua, L. Wang, B. Xu, H. Zhang, N. Ye, Z. Zhang, D. Yu, H.J. Cooke, Y. Zhang, Q. Shi, MicroRNA and piRNA profiles in normal human testis detected by next generation sequencing, PLoS One 8 (2013) e66809.

[109] L. Wang, C. Xu, Role of microRNAs in mammalian spermatogenesis and testicular germ cell tumors, Reproduction 149 (2015) R127–R137.

[110] A.M. O'doherty, P.A. Mcgettigan, Epigenetic processes in the male germline, Reprod. Fertil. Dev. 27 (2014) 725–738.

[111] Y. Li, M. Li, Y. Liu, G. Song, N. Liu, A microarray for microRNA profiling in spermatozoa from adult men living in an environmentally polluted site, Bull. Environ. Contam. Toxicol. 89 (2012) 1111–1114.

[112] R.M. Sharpe, Environmental/lifestyle effects on spermatogenesis, Philos. Trans. R. Soc. Lond. B Biol. Sci. 365 (2010) 1697–1712.

[113] Y. Barazani, B.F. Katz, H.M. Nagler, D.S. Stember, Lifestyle, environment, and male reproductive health, Urol. Clin. North Am. 41 (2014) 55–66.

[114] R. Hauser, R. Sokol, Science linking environmental contaminant exposures with fertility and reproductive health impacts in the adult male, Fertil. Steril. 89 (2008) e59–e65.

[115] F.P. Manfo, E.A. Nantia, P.P. Mathur, Effect of environmental contaminants on mammalian testis, Curr. Mol. Pharmacol. 7 (2014) 119–135.

[116] E.W. Wong, C.Y. Cheng, Impacts of environmental toxicants on male reproductive dysfunction, Trends Pharmacol. Sci. 32 (2011) 290–299.

[117] A.M. Andersson, N. Jorgensen, K.M. Main, J. Toppari, E. Rajpert-De Meyts, H. Leffers, A. Juul, T.K. Jensen, N.E. Skakkebaek, Adverse trends in male reproductive health: we may have reached a crucial 'tipping point', Int. J. Androl. 31 (2008) 74–80.

[118] S.H. Swan, E.P. Elkin, L. Fenster, The question of declining sperm density revisited: an analysis of 101 studies published 1934-1996, Environ. Health Perspect. 108 (2000) 961–966.

[119] E. Carlsen, A. Giwercman, N. Keiding, N.E. Skakkebaek, Evidence for decreasing quality of semen during past 50 years, BMJ 305 (1992) 609–613.

[120] S. Irvine, E. Cawood, D. Richardson, E. Macdonald, J. Aitken, Evidence of deteriorating semen quality in the United Kingdom: birth cohort study in 577 men in Scotland over 11 years, BMJ 312 (1996) 467–471.

[121] S.H. Swan, E.P. Elkin, L. Fenster, Have sperm densities declined? A reanalysis of global trend data, Environ. Health Perspect. 105 (1997) 1228–1232.

[122] C. Vecoli, L. Montano, M.G. Andreassi, Environmental pollutants: genetic damage and epigenetic changes in male germ cells, Environ. Sci. Pollut. Res. Int. 23 (2016) 23339–23348.

[123] J. Rubes, S.G. Selevan, D.P. Evenson, D. Zudova, M. Vozdova, Z. Zudova, W.A. Robbins, S.D. Perreault, Episodic air pollution is associated with increased DNA fragmentation in human sperm without other changes in semen quality, Hum. Reprod. 20 (2005) 2776–2783.

[124] S.G. Selevan, L. Borkovec, V.L. Slott, Z. Zudova, J. Rubes, D.P. Evenson, S.D. Perreault, Semen quality and reproductive health of young Czech men exposed to seasonal air pollution, Environ. Health Perspect. 108 (2000) 887–894.

[125] R.J. Sram, I. Benes, B. Binkova, J. Dejmek, D. Horstman, F. Kotesovec, D. Otto, S.D. Perreault, J. Rubes, S.G. Selevan, I. Skalik, R.K. Stevens, J. Lewtas, Teplice program—the impact of air pollution on human health, Environ. Health Perspect. 104 (Suppl. 4) (1996) 699–714.

[126] L. Tamburrino, S. Marchiani, M. Montoya, F. Elia Marino, I. Natali, M. Cambi, G. Forti, E. Baldi, M. Muratori, Mechanisms and clinical correlates of sperm DNA damage, Asian J. Androl. 14 (2012) 24–31.

[127] P. Bergamo, M.G. Volpe, S. Lorenzetti, A. Mantovani, T. Notari, E. Cocca, S. Cerullo, M. Di Stasio, P. Cerino, L. Montano, Human semen as an early, sensitive biomarker of highly polluted living environment in healthy men: a pilot biomonitoring study on trace elements in blood and semen and their relationship with sperm quality and RedOx status, Reprod. Toxicol. 66 (2016) 1–9.

[128] M. de Rosa, S. Zarrilli, L. Paesano, U. Carbone, B. Boggia, M. Petretta, A. Maisto, F. Cimmino, G. Puca, A. Colao, G. Lombardi, Traffic pollutants affect fertility in men, Hum. Reprod. 18 (2003) 1055–1061.

[129] A. Mazza, P. Piscitelli, C. Neglia, G. Della Rosa, L. Iannuzzi, Illegal dumping of toxic waste and its effect on human health in Campania, Italy, Int. J. Environ. Res. Public Health 12 (2015) 6818–6831.

[130] S. Benoff, A. Jacob, I.R. Hurley, Male infertility and environmental exposure to lead and cadmium, Hum. Reprod. Update 6 (2000) 107–121.

[131] W.G. Foster, A. McMahon, D.C. Rice, Sperm chromatin structure is altered in cynomolgus monkeys with environmentally relevant blood lead levels, Toxicol. Ind. Health 12 (1996) 723–735.

[132] I. Hernandez-Ochoa, M. Sanchez-Gutierrez, M.J. Solis-Heredia, B. Quintanilla-Vega, Spermatozoa nucleus takes up lead during the epididymal maturation altering chromatin condensation, Reprod. Toxicol. 21 (2006) 171–178.

[133] P.C. Hsu, H.Y. Chang, Y.L. Guo, Y.C. Liu, T.S. Shih, Effect of smoking on blood lead levels in workers and role of reactive oxygen species in lead-induced sperm chromatin DNA damage, Fertil. Steril. 91 (2009) 1096–1103.
[134] B. Quintanilla-Vega, D. Hoover, W. Bal, E.K. Silbergeld, M.P. Waalkes, L.D. Anderson, Lead effects on protamine-DNA binding, Am. J. Ind. Med. 38 (2000) 324–329.
[135] L. Hou, X. Zhang, D. Wang, A. Baccarelli, Environmental chemical exposures and human epigenetics, Int. J. Epidemiol. 41 (2012) 79–105.
[136] B. Zhang, X. Pan, RDX induces aberrant expression of microRNAs in mouse brain and liver, Environ. Health Perspect. 117 (2009) 231–240.
[137] R. Sharma, K.R. Biedenharn, J.M. Fedor, A. Agarwal, Lifestyle factors and reproductive health: taking control of your fertility, Reprod. Biol. Endocrinol. 11 (2013) 66.
[138] J.A. Alegria-Torres, A. Baccarelli, V. Bollati, Epigenetics and lifestyle, Epigenomics 3 (2011) 267–277.
[139] A. Baccarelli, V. Bollati, Epigenetics and environmental chemicals, Curr. Opin. Pediatr. 21 (2009) 243–251.
[140] C.S. Wilhelm-Benartzi, E.A. Houseman, M.A. Maccani, G.M. Poage, D.C. Koestler, S.M. Langevin, L.A. Gagne, C.E. Banister, J.F. Padbury, C.J. Marsit, In utero exposures, infant growth, and DNA methylation of repetitive elements and developmentally related genes in human placenta, Environ. Health Perspect. 120 (2012) 296–302.
[141] M. Feychting, N. Plato, G. Nise, A. Ahlbom, Paternal occupational exposures and childhood cancer, Environ. Health Perspect. 109 (2001) 193–196.
[142] A. Soubry, C. Hoyo, R.L. Jirtle, S.K. Murphy, A paternal environmental legacy: evidence for epigenetic inheritance through the male germ line, Bioessays 36 (2014) 359–371.
[143] M.D. Anway, C. Leathers, M.K. Skinner, Endocrine disruptor vinclozolin induced epigenetic transgenerational adult-onset disease, Endocrinology 147 (2006) 5515–5523.
[144] C. Guerrero-Bosagna, T.R. Covert, M.M. Haque, M. Settles, E.E. Nilsson, M.D. Anway, M.K. Skinner, Epigenetic transgenerational inheritance of vinclozolin induced mouse adult onset disease and associated sperm epigenome biomarkers, Reprod. Toxicol. 34 (2012) 694–707.
[145] G. Lazaraviciute, M. Kauser, S. Bhattacharya, P. Haggarty, S. Bhattacharya, A systematic review and meta-analysis of DNA methylation levels and imprinting disorders in children conceived by IVF/ICSI compared with children conceived spontaneously, Hum. Reprod. Update 20 (2014) 840–852.

CHAPTER

11

Changes in DNA Methylation Related to Male Infertility

R.G. Urdinguio[*,†], M.F. Fraga*[*], A.F. Fernández*[†]

[*]Nanomaterials and Nanotechnology Research Center (CINN-CSIC), University of Oviedo, Principado de Asturias, Asturias, Spain [†]Instituto de Investigación Sanitaria del Principado de Asturias (ISPA), Hospital Universitario Central de Asturias (HUCA), University Institute of Oncology of Asturias (IUOPA), Oviedo, Spain

INTRODUCTION

We humans are made up of more than 50 billion cells. Each cell in the human body is perfectly coordinated with its surrounding cells and carries all the information that the next generation will require. This information is organized within chromosomes, enabling it to be packaged up in a 2m-long molecule in the form of DNA. Our genes are codified in our DNA and transmitted from one generation to the next. However, we are more than genes as demonstrated by examples such as the observed differences between monozygotic twins [1–3] or cloned mammals such as Dolly the sheep, whose disease phenotype varied considerably with respect to that of her genetic mother [4,5].

In order to refer to the specific phenotypes defined by the interaction between genes and the environment, the embryologist and geneticist Sir Conrad Hal Waddington established the term *epigenetics* in 1942 [6–8]. It was the first attempt to understand the development of cells from common progenitors with the same genetic material that had completely different phenotypes, such as neurons in the brain, muscle cells in the heart, and blood cells in the bone marrow of an individual.

Currently, there are several definitions of epigenetics, but a useful general working definition is the "study of heritable changes in genome function that occur without a change in DNA sequence" [9]. There are three main epigenetic mechanisms that participate in gene regulation: DNA methylation, posttranslational histone modifications, and noncoding RNAs. These mechanisms have an important role in the regulation of gene expression (turning genes ON and OFF, for instance) and in the maintenance of chromatin structure and stability, that is, they are responsible for the maintenance of active euchromatin and inactive heterochromatin [10] (Fig. 11.1).

Reproductomics
https://doi.org/10.1016/B978-0-12-812571-7.00012-5

Up to now, the most widely studied epigenetic mechanism is DNA methylation that, in mammals, consists of the addition of a methyl group to the cytosines immediately upstream of guanines, which transforms them into 5-methylcytosines (5mC), in the so-called CpG dinucleotides [11–13]. Although in mammals, DNA methylation usually occurs in CpG dinucleotides, such a pattern is not exclusive. Plants, for instance, have been found to present cytosine methylation not only in CG sequences but also in CHG and CHH (H=A, C, or T) [14,15]. DNA methylation is carried out by DNA methyltransferases (DNMTs) that, in mammals, comprise DNMT3a and DNMT3b, as de novo methyltransferases, and DNMT1, which is known as the maintenance DNA methyltransferase. These proteins transfer a methyl group to DNA using *S*-adenosylmethionine as donor [16].

As previously mentioned, DNA methylation in mammals mainly occurs in cytosines at CpG dinucleotides, but these CpG sites are not always randomly distributed throughout the genome, and they are sometimes concentrated in CpG-rich regions called CpG islands [13]. Actually, 60% of genes have CpG islands at promoter regions that, moreover, are frequently unmethylated, allowing gene expression [17]. DNA methylation changes in these regions are usually accompanied by gene expression changes. However, those CpGs that are spread out along the gene body and particularly in genomic repetitive regions are generally methylated. While the function of DNA methylation in gene bodies is more complex than initially expected and can also be associated with gene regulation, it seems that the loss of DNA methylation in repetitive regions is related to the loss of chromosome stability [17–19].

DNA methylation is involved in many biological and physiological processes. Many methylation changes have been found during development, and their role in the control of gene expression, X-chromosome inactivation in females, and establishment of imprinting has been extensively demonstrated [20–22]. The importance of DNA methylation was confirmed in a 1992 work where Li and collaborators demonstrated that DNMT knockout mice showed such severe alterations during embryo development that they were not able to survive [23]. Apart from the role that DNA methylation plays during development, its involvement in differentiation processes has also been proved. Recently, a work analyzing the DNA methylation levels

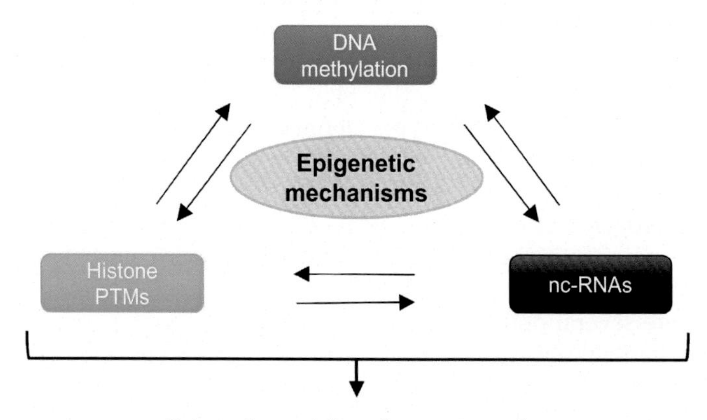

FIG. 11.1 Epigenetic mechanisms contributing to gene regulation. DNA methylation, histone posttranslational modifications (PTMs), and noncoding RNAs.

of more than 14,000 genes in various different hematologic cell types allowed the identification of gene-specific methylation changes associated with specific differentiation processes from each of the hematopoietic cell linages studied, and more importantly, those methylation changes were associated with changes in levels of expression of cell-type specific genes [24]. This work shows that DNA methylation loss at different specific gene promoters of the embryonic cell and hematopoietic progenitors plays an important role in the establishment of differentiated cell types such as neutrophils and several types of lymphocytes [24].

Many disorders have been linked to alterations in DNA methylation patterns, such as cardiovascular diseases, autoimmune diseases, metabolic diseases, fertility problems, and a variety of neurologic disorders [25–29]. In particular, a great number of alterations in DNA methylation levels have been described in cancer from the 1980s onward, following the identification of a global loss of DNA methylation associated with this disease [11,30,31].

From a clinical point of view, alterations of DNA methylation have been associated not only with disease diagnosis, comparing normal with tumoral tissues, but also with prognosis in terms of patient survival [32,33]. Also, specific alterations of DNA methylation have been associated with response to treatment, meaning that they could be used as predictive biomarkers [32,33]. Finally, altered methylation patterns have been used to identify tumors of unknown primary origin. In order to achieve this, methylation patterns from the tumor of unknown origin (metastasis) are analyzed and compared with the patterns of previously identified tumors [34]. Recently, a kit to perform an epigenetic test has been developed and commercialized with that purpose (EPICUP) [34].

Taking into account the relevance of DNA methylation with respect to different diseases, the discovery of 5-hydroxymethylcytosine (5hmC) as a new and different epigenetic mark [35,36] has changed our perspective on the cancer epigenome. 5hmC is a product of 5mC oxidation carried out by ten-eleven translocation (TET) enzymes [36,37]. A priori, 5hmC seems to be a simple intermediate of DNA methylation loss, the process whereby 5hmC undergoes sequential oxidations, deriving to 5-formyl methylcytosine (5fC) and 5-carboxy cytosine (5caC) [36,37]. After a glycosylation process and base excision repair (BER), it is finally transformed into cytosine. However, there are tissue-specific levels of 5hmC, and there is a debate as to whether 5hmC may be not only a demethylation intermediate but also a stable mark with a specific functional role [37] (Fig. 11.2).

ENVIRONMENTAL AND EXTERNAL FACTORS AFFECTING EMBRYONIC DEVELOPMENT AND POSTNATAL STAGES

In general, these natural changes in DNA methylation patterns during the development mentioned above stabilize at a given time and then remain stable over the life of an organism [38]. However, certain loci may undergo DNA methylation alterations over time. This so-called epigenetic drift depends on both intrinsic and environmental factors and may have consequences on gene expression [38].

Environmental factors play an important role in the appearance of epigenetic alterations, and depending on the stage of development, these may affect the individual to a lesser or greater extent. If alterations appear during embryonic development, over time, these alterations could be transmitted to a large number of tissues. If, on the contrary, a change occurs

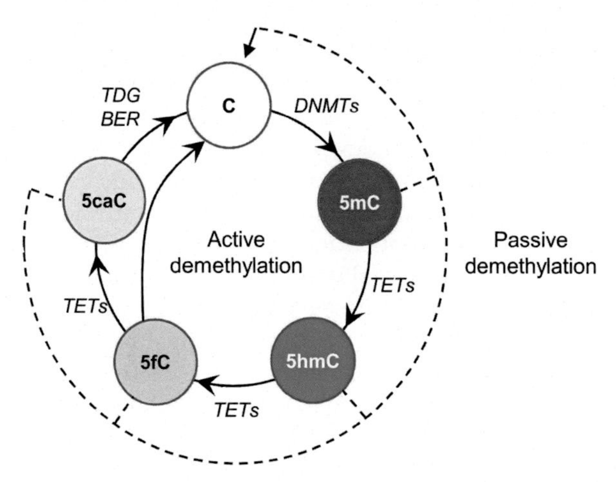

FIG. 11.2 DNA methylation and plausible demethylation mechanisms. DNA methylation is catalyzed by DNA methyltransferases (DNMTs), and the loss of DNA methylation could be explained by two different mechanisms. Passive DNA demethylation of 5mC, 5hmC, 5fC, and 5caC is represented by the *dotted line*. Active DNA demethylation is catalyzed by the ten-eleven translocation enzymes (TETs), and finally, 5fC and 5caC are excised by the thymine DNA glycosylase (TDG) during the base excision repair (BER) process; *C*, cytosine; *5mC*, 5-methylcytosine; *5hmC*, 5-hydroxymethylcytosine; *5fC*, 5-formylcytosine; and *5caC*, 5-carboxylcytosine.

in adult life, the alteration is likely to be restricted to a specific tissue and, a priori, be more controlled from the clinical point of view.

A classic example of how the environment can alter the epigenome during embryonic development is found in the agouti mouse model [39]. The agouti gene determines the color of the mouse's coat and is regulated by the DNA methylation status of an intracisternal A particle (IAP) inserted upstream of the canonical transcription start site (TSS). Using this model, it was observed that methyl-donor supplementation during midgestation was associated with the DNA methylation of the IAP and consequently there was no ectopic expression of the gene. In contrast, the lack of methyl group donors facilitated the loss of methylation of this particle and promoted ectopic expression of the gene, resulting in a yellow coloration of the mouse coat. In addition, these epigenetic changes induced by maternal diet supplementation not only affected the fetus (F1) but also can affect the epigenetic status of fetal germ cells (F2) and thus be transmitted to the progeny [39].

It seems that the gestational period is especially susceptible to epigenetic alterations as a consequence of external, or environmental, factors. Using animal models, different types of compounds, such as bisphenol A (BPA) and phthalates (DEHP/MEHP and DBP/BBP/MBP), have been identified that may affect development and be related to certain diseases [40]. It has also been demonstrated in animal models that exposure to the endocrine disruptor BPA during pregnancy is associated with the development of various types of cancer, such as prostate or breast, in adulthood, and that it may be mediated by epigenetic mechanisms, such as alterations of DNA methylation [41,42].

Although in humans these associations are not as well studied as in animal models, there are several studies that demonstrate the effect that adverse environmental conditions during the embryonic stage could have on the epigenome and that have been associated with diseases during adult life [43–45]. One of the most cited works makes use of data from the so-called Hunger Winter, the Dutch famine during the winter of 1944–1945 caused by the German army's siege of the Netherlands during World War II. The results showed that adverse environmental conditions (in this case lack of food) during the early stages of life caused epigenetic alterations, such as the alteration of imprinted genes like *IGF2*, which were able to persist throughout life, and may be associated with the onset of disease in adult life [44].

Another recent work has demonstrated how periconceptional dietary conditions define methylation patterns in metastable epialleles may be related to interindividual epigenetic variability and have a role in the onset of diseases in humans [45]. Subsequently, it has been verified that the arrangement of methyl donors, which is dependent on climatic conditions during the early stages of pregnancy, can define patterns of DNA methylation of genes such as proopiomelanocortin (*POMC*), whose hypermethylation variant leads to increased risk for obesity development [46].

Another interesting example highlighting the effect of environmental conditions on the epigenome during the embryonic stage comes from studies of pregnant women subjected to prenatal stress [43]. In this study called "Project Ice Storm," analyses were performed on the DNA methylation of blood cells and saliva from children of women who were pregnant during or conceived within 3 months of the 1998 Quebec ice storm, one of the worst natural disasters in Canadian history. In the winter of 1998, after the passage of freezing rain storms, much of southern Quebec was covered by ice, and more than 3 million people were isolated for 6 weeks. The results demonstrated that in utero exposure to maternal stress was associated with specific DNA methylation signatures in the offspring [43].

Along the same lines, particularly in animal models, there is currently great research interest in how the social status and the care received by the mother impacts on the epigenome of the fetus and/or newborn and how it is related to the development of the brain and the possible effects on mental health and cognition. For example, in rats, maternal care has been shown to be related to the epigenetic regulation of the expression of the gene that encodes the glucocorticoid receptor (*Nr3c1*) [47]. The absence of methylation of the gene promoter and increased expression of this receptor that is associated with increased maternal care was found to be related to decreased cortisone levels and decreased anxiety in offspring [47]. And more importantly, these associations between DNA methylation and gene expression are known to be maintained over time and be related to the stress response of the adult individual [48]. Interestingly, it appears that the behavior of mothers in relation to baby care in turn has an effect on epigenetic patterns, and, for example, specific patterns have been identified in some imprinted genes such as *Mest* and *Peg3* [27].

In humans, it also seems that epigenetic mechanisms are able to modulate the effect of socioeconomic status on disease predisposition, that is, high social capital is related to better health (the Roseto effect). The most representative example of this is the Italian-American community of Roseto in Pennsylvania (United States), where in the 1950s, it was noted that this group had much lower levels of cardiovascular problems compared with the surrounding communities. However, in the late 1960s, the young people from Roseto had to move out of the community to seek work and their social capital declined, and there was an accompanying increase in deaths from heart disease. These associations between social capital and health cannot be explained exclusively by genetic issues, but it seems that epigenetics may be playing an important role [49].

ENVIRONMENTAL AND EXTERNAL FACTORS AFFECTING THE EPIGENOME DURING ADULTHOOD

Although there are not many human studies on how environmental factors impact during the embryonic stages, there are many animal studies showing the importance of exposure to different types of compounds and external factors and the effects that these can cause in the epigenome. The subsequent alterations or changes in epigenetic marks can have repercussions in diverse biological processes and even be associated with the appearance of different diseases in the adulthood. Numerous studies have shown how chemical and physical environmental stressors, diet, life habits, and pharmacological treatments can alter the epigenome during adulthood and how these alterations may be determining healthy and diseased phenotypes [50,51]. Most of these studies have focused on DNA methylation, which is the epigenetic mark that is easiest to analyze. It has been observed that a large variety of metals, such as arsenic, mercury, nickel, lead, and cadmium, which may be widespread in the environment, impact in several ways on human health [50,51]. Chronic exposure to many of these compounds has been associated with different types of diseases, including cancer, as well as with alterations in DNA methylation and various posttranslational marks of histones, suggesting that epigenetic alterations could be mediating the effect of exposure to these compounds on human health [50,51].

In addition to metals, particle pollution, also known as atmospheric particulate matter (PM), can also affect the epigenome and human health. Exposure to atmospheric pollutants, which may be traffic-related or due to common compounds in certain industries or jobs, such as asbestos or benzene, is known to be able to alter the epigenome at different levels and is, in turn, associated with diseases related to the cardiovascular system or to the appearance of different types of cancers, particularly lung cancer [50,51]. In this section, we must emphasize the importance of endocrine disruptors, chemical pollutants in adults, where they affect the endocrine system and produce adverse health effects. Among compounds of this type, such as pesticides, fungicides, herbicide, industrial chemicals, and plant hormones, we should highlight plastics and specifically bisphenol A (BPA) and the phthalates, as mentioned with respect to pregnant women, which have been frequently associated with health risks. Specifically, in adult individuals, it has been found that BPA can affect epigenetic marks in male germ cells and is associated with susceptibility to cancer or decreased fertility in females [50,51]. Another environmental stressor, electromagnetic radiation, has also been found to affect the epigenome, being associated with DNA damage, immune system alterations, and skin lesions that may reach the level of skin cancer [50,51]. And finally, in this section, we should include nanoparticles and nanomaterials as other potential environmental stressors, since, although they have not as yet been clearly demonstrated to affect human health, they are known to affect the epigenome in other mammals [52].

Another external factor that can affect the epigenome in adulthood and in turn impact on health is diet. The effect on the epigenome of dietary methyl donors such as folates and methionine or dietary compounds such as selenium and polyphenols has been found to be associated with modifications of epigenetic status and have an effect on health and aging [50,51]. And finally, it is important to mention the effects that unhealthy lifestyle habits can have on epigenetic marks and the health. There are a large number of studies focused on the role of tobacco smoking on epigenetic alterations, specifically at the DNA methylation level, and the effect it has on health and especially on predisposition to diseases such as cancer [50,51,53]. Along with tobacco, excessive consumption of alcohol is known to be harmful to health and affects the epigenome through alterations in various epigenetic marks, for example, hypomethylation of DNA, important because this compound can interfere with methionine metabolism and thus affect the availability of methyl groups [50,51]. Apart from good eating habits and the potential effects they may have on the epigenome, we are beginning to understand the mechanisms through which physical exercise is beneficial to health, and it is becoming clear that exercise in humans is associated with epigenetic changes in different types of tissues and cells, of specific note being the effect it can have on germ cells [50,51].

TRANSGENERATIONAL EPIGENETIC INHERITANCE

One of the issues that has created most interest in recent years is the possibility that epigenetic alterations related to external factors can be transmitted from one generation to another. At this point, we must make clear how transgenerational epigenetic inheritance is defined, so as not to confuse it with transgenerational epigenetic effects. The difference is related to whether the alterations are transmitted by direct environmental exposure (the latter case) or if they are transmitted through the germinal route without direct exposure (the former) [54–56]. In order to define a change as transgenerational epigenetic inheritance exposure to an external factor would have to affect F2 or F3, depending on whether it is an adult individual or a pregnant female, respectively, that is, the generation is not directly exposed (Fig. 11.3). Although transgenerational epigenetic inheritance has been demonstrated in animal models, in humans, there are currently few papers confirming these questions, mainly due to the difficulty of studying several generations [55,57]. Most studies to demonstrate

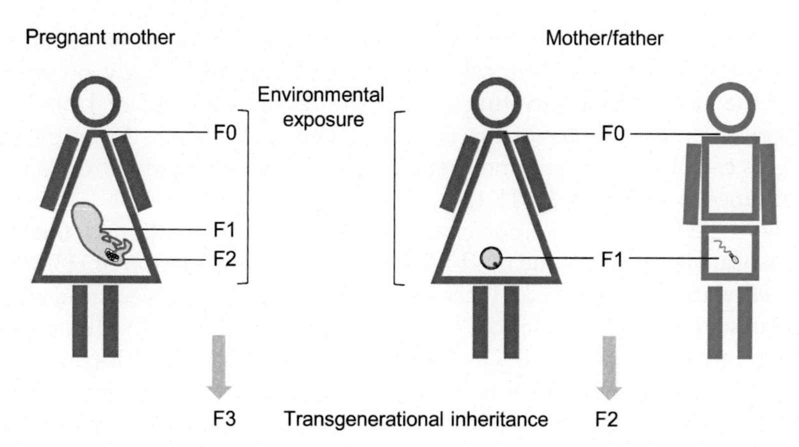

FIG. 11.3 Schemes showing the two manners of transgenerational epigenetic inheritance. (1) Exposure to environmental factors during pregnancy can affect the epigenome of the F0 (mother), F1 (embryo), and F2 (germ line), and transgenerational inheritance refers to the transfer of epigenetic changes to the F3 generation (which has no direct exposure). (2) Exposure to environmental factors in adults can affect the epigenome of the mother/father (F0) or the germ line (F1), and in this case, transgenerational inheritance refers to the transfer of epigenetic changes to the F2 generation.

transgenerational epigenetic inheritance have focused on the effects of adverse environmental conditions during the embryonic stages. Several of these examples, including exposures to plastics, caloric restriction, or heat, were summarized recently in an article by Skinner [55]. There are also several studies that demonstrate the transmission of alterations associated with environmental exposures to certain compounds (toxic) or situations of ancestral stress through the male or female germ line. Intrauterine exposures to toxins such as the fungicide vinclozolin, pesticides, BPA, or poor nutritional situations can signal alterations in responses to stressful situations or an increased risk of metabolic diseases in offspring [58–63].

The need to study the sperm epigenome arises from the fact that the effect of external factors that affect the epigenome can be transmitted through the germinal line. In addition to associations with potential fertility problems, alterations of the sperm epigenome, although still under debate, appear to be associated with propensity for certain diseases in offspring [64–68]. Future studies in this field will no doubt clarify these issues.

EPIGENETIC MECHANISMS IN GERM CELLS DURING DEVELOPMENT AND SPERMATOGENESIS

In mammals, epigenetic alterations can be transmitted from one generation to the next even though there is genome-wide epigenetic reprogramming during two of the phases of embryonic development. These processes are considered to be similar to an epigenetic resetting to avoid (as far as possible) the transmission of epigenetic alterations to the offspring. The first reprogramming wave happens/begins after fertilization and lasts until implantation, and it involves the progressive loss of DNA methylation (except in imprinted genes) after which DNA methylation patterns are reestablished. The second wave occurs during the migration of germ progenitors to the developing gonads, when imprinting is deleted, and this is followed by a remethylation phase. This second reprogramming wave is fundamental for the correct establishment of epigenetic marks in the developing germ cell. Epigenetic reprogramming of the primordial germ cells (PGCs) starts, approximately, on embryonic day (E) 7.25. After the erasure of imprinting marks, the reestablishment of DNA methylation, which will depend on the gender of the germ cell, begins on E12.5. As an example, paternal imprinting is established during the first phases of spermatogenesis from approximately E14.5 and continues until birth (Fig. 11.4).

In concert with the DNA methylation changes that is already mentioned, there is also a reordering of posttranslational histone modifications such as methylation of H3K9 and H3K27. While DNA methylation decreases, an increase in H3K27me3 is observed until approximately E13.5. Similar to with DNA methylation, histone modifications slightly differ depending on the gender of the emergent germ cell. During spermatogenesis and the formation of primary spermatocytes, these changes in histone marks are maintained. However, at the beginning of meiosis, H3K27me3 progressively decreases, and H3K9 methylation gradually increases. Additionally, in the final phases of spermatogenesis, approximately 85% of histones are exchanged for protamines during spermatozoid formation [69,70].

In summary, there are two phases where there are crucial epigenetic changes during spermatogenesis that could suffer alterations leading to functional defects. During the first phases of mitosis, the main epigenetic event is DNA methylation, and errors may occur in

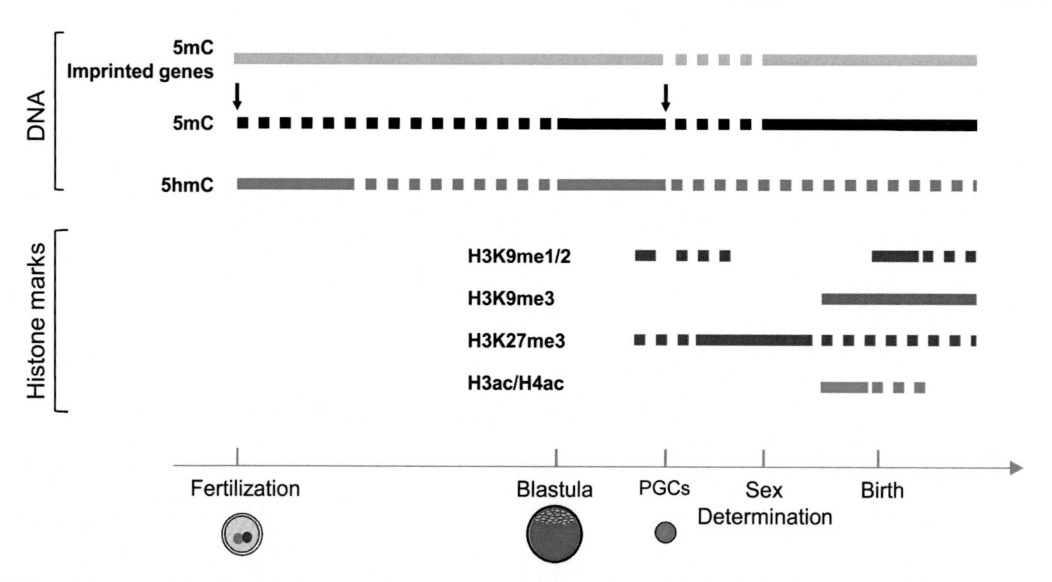

FIG. 11.4 Representative dynamic epigenetic changes during early embryo development and the germ cell epigenetic reprogramming in mammals. For clarity, only the male gamete is represented. Black arrows define the start of each of the two waves of DNA demethylation. *Dashed lines* indicate that the level of epigenetic modification is lower and solid lines that is higher during these periods. *PGCs*, primordial germ cells.

the reestablishment of methylation patterns, leading to various disorders. The other critical moment occurs during the first phases of meiosis and the differentiation process that leads to mature sperm, that is, when there are changes in histone marks and histones transition to protamines (necessary for the correct packaging of DNA) takes place [69,70].

SPERM EPIGENOME AND ALTERATIONS OF DNA METHYLATION ASSOCIATED WITH HUMAN DISEASE

As can be concluded from the previous section, the epigenome of mature spermatozoa is reached after several cycles of epigenetic reprogramming during embryonic development and gametogenesis, at the levels of both DNA and chromatin markers, making it easy to imagine that these processes may give rise to epigenetic variability of the germ cell [70–72].

Partial genome methylation patterns of human spermatozoa have been described in several works using methylation arrays or methylated DNA immunoprecipitation (MeDIP), and recently, Molaro and collaborators have described DNA methylation profiles at the complete genome level using shotgun bisulfite sequencing [71–74]. Apart from some interindividual variability in DNA methylation patterns, these studies have shown that global DNA methylation and specifically DNA methylation in certain repetitive regions of the genome are significantly lower when compared with somatic cells [71,72,74], and it has been suggested that this could be related to the increased risk of structural mutations that has been found in germ cells [75].

In addition to specific DNA epigenetic profiles, posttranslational histone marks have also been characterized in mature sperm. During spermiogenesis, histones are replaced almost entirely by protamines, which play a major role in the compaction of sperm DNA, facilitating both DNA protection and sperm motility [69]. Although the presence of some nucleosomes was initially thought to be due to inefficient replacement by protamines, it is now known that these nucleosomes are enriched with specific histone marks associated with developmental loci [69,71–73].

Alterations of the paternally methylated genomic regions have been associated with several imprinting diseases such as Beckwith-Wiedemann syndrome (BWS) and Silver-Russell syndrome (SRS) [76]. Precisely, this type of imprinting disease has been associated with problems related to assisted reproduction practices (ART) and methods of sperm selection, although there are contradictory results, and in some cases, it is not known whether the epigenetic alterations are due to *in vitro* manipulations or the underlying problems of subfertility of parents (see review [26]).

Although the data obtained so far are not conclusive, it has also been suggested that DNA methylation changes associated with sperm age may be associated with the onset of disease in offspring [77,78]. But what has been studied in greater depth are the alterations of the epigenetic marks and specifically of DNA methylation related to reproductive success and in particular of the male fertility, which we will discuss more extensively in the following section.

DNA METHYLATION ALTERATIONS ASSOCIATED WITH HUMAN MALE INFERTILITY

Even with the double epigenetic reprogramming that occurs during development as described in a previous section, normal patterns of DNA methylation of the germ cell might not become reestablished correctly. Although both male and female gametes may be susceptible to DNA methylation alterations, in practice, most studies have been conducted on male gametes, principally due to the low numbers of and the limitations of accessing female gametes.

The first studies searching for alterations of DNA methylation patterns of the male germ cell, related to the alterations in the spermatogenesis, were performed in specific genes, particularly in imprinted genes. In one of these studies, Marques and collaborators analyzed the DNA methylation patterns of two imprinted genes, *MEST* and *H19*, using sperm samples from normal individuals and individuals with different degrees of oligospermia (Fig. 11.5). Using bisulfite genomic sequencing, they found that in a high percentage of oligospermic patients, DNA methylation alterations of the imprinted gene *H19* appeared [79], which in addition to being related to fertility problems could also affect embryo development should this sperm be used in ART. In fact, the same research group a few years later verified these epigenetic alterations in *H19* and found new alterations of *MEST* associated with oligospermia in a new series of patients by using bisulfite sequencing of multiple clones. In this case, *MEST*, which is not methylated in sperm of normal individuals, presented increases in DNA methylation in certain patients [80]. Subsequently, other research groups have verified these same associations between defects in imprinting and low numbers of spermatozoa using large cohorts of patients. One example is the work of Poplinski and collaborators, who observed that as the number of spermatozoa in the patients decreased, there was a higher probability of finding lower levels of DNA methylation in *H19* and higher methylation in *MEST* [81]. Apart from these associations found between the DNA methylation patterns of these two imprinted genes and low numbers of sperm, similar DNA methylation alterations have also been found in samples of infertile individuals with sperm with low mobility and abnormal morphology, that is, in cases of asthenozoospermia and teratospermia [81].

Other alterations in DNA methylation associated with male infertility, in addition to the genes mentioned initially, have been described in other imprinted genes, genes that are not imprinted, and even in repetitive genome sequences such as LINE-1 and Alu sequences [72,82,83].

By using candidate gene approaches, nonimprinted genes such as that encoding the enzyme methylenetetrahydrofolate reductase (*MTHFR*) have been studied. This gene is associated with folate metabolism and methylation reactions, and it has been found that its mutation is linked to male infertility. It has been observed that in the testis of azoospermic individuals (nonobstructive azoospermia), this gene was hypermethylated and was not expressed, leading to the conclusion that it could play a very important role in this type of infertility [84]. Subsequently, this gene was also found to be hypermethylated in sperm in a high percentage of cases of patients with idiopathic infertility and especially in cases of oligospermia, which has revealed its potential as a biomarker for the identification of individuals at risk of infertility [85].

Interestingly, alterations of DNA methylation related to male infertility have also been identified in patients with alterations in the process of replacing protamines with histones during spermiogenesis. Specifically, alterations of methylation have been observed in the promoter of the cAMP-gene-responsive element modulator (*CREM*) in some patients with protamination abnormalities [86]. This gene is especially interesting because it codes for an important transcription factor during spermiogenesis, although, the results obtained to date with regard to linking the aberrant DNA methylation of this gene to cases of male infertility need to be validated in a greater number of samples [86]. In the same vein, the key spermatogenesis gene termed deleted in azoospermia like (*DAZL*) has been found aberrantly methylated in oligoasthenoteratozoospermic men [87], which highlights the importance of aberrant methylation in key genes in spermatogenesis for reproductive success (Fig. 11.5).

In recent years, the emergence of new DNA methylation analysis technologies has allowed the study of hundreds or thousands of genes at the same time, and in this way, it has been possible to identify the aberrant DNA methylation of many new genes in samples from infertile individuals with alterations related to sperm concentration, mobility, and/or morphology. These new technologies include methylation arrays, which have made it possible to simultaneously analyze the DNA methylation patterns of a large number of genes, indeed increasing from a few thousand initially to the point now where all the genes of the human genome can be analyzed simultaneously [28,88–92] (Fig. 11.5). One of the earliest studies using these microarray-based approaches [93,94] analyzed over 14,000 human genes, comparing the DNA methylation profiles of sperm from fertile individuals with that of infertile individuals with low mobility sperm [92]. In this work, alterations of DNA methylation patterns were found in more than 9000 sequences, of which almost 200 corresponded to imprinted genes, thus demonstrating and confirming that epigenetic alterations of imprinted genes, as identified by candidate gene approaches, were common in idiopathic infertile individuals [92].

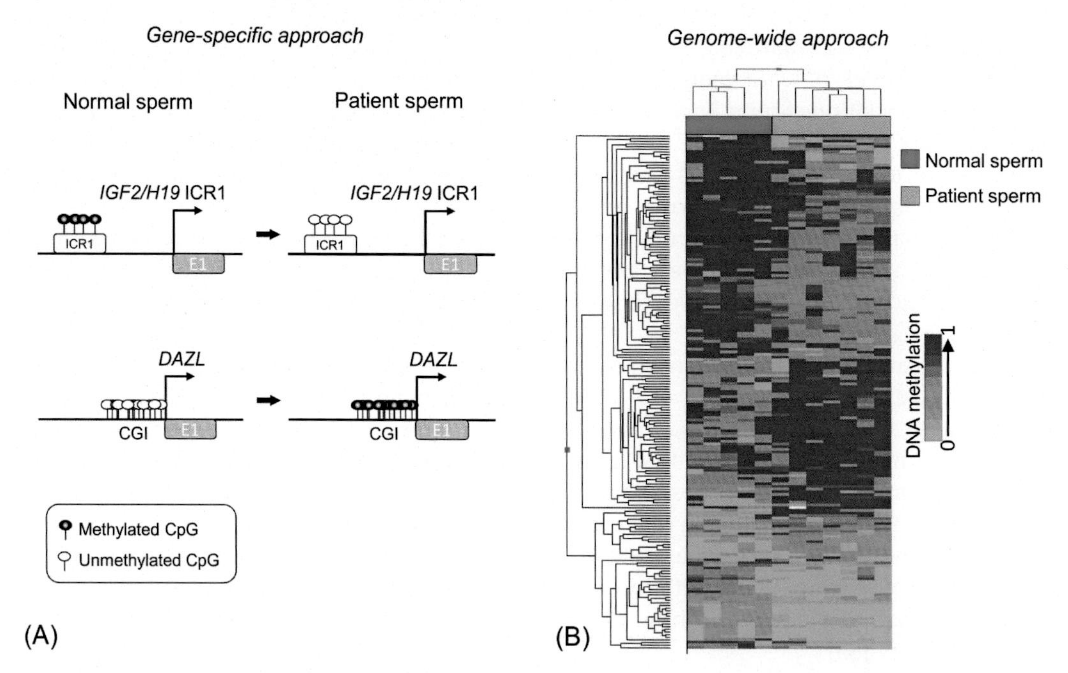

FIG. 11.5 Approaches to identify DNA methylation alterations in sperm from subfertile male patients. (A) Candidate gene approaches used to analyze imprinted (upper pictures) or nonimprinted (lower pictures) genes. *ICR1*, imprinting control region 1; *CGI*, CpG island. (B) Genome-wide approaches, where thousands of genes were analyzed at once.

Besides studies in patients with low seminal quality, the same type of array has been used to compare the DNA methylation profiles of samples of testis from individuals with the absence of germ cells and failures in spermatogenesis with individuals with conserved spermatogenesis. In this way, epigenetic alterations in hundreds of genes have been found, including those for piwi like RNA-mediated gene silencing (*PIWI*) and tudor domain containing (*TDRD*), both strongly related to the process of spermatogenesis and that have been found to be hypermethylated in samples of testis of individuals with defects in germ cell production [88].

More recently, and in order to further study the epigenetic alterations associated with male infertility in humans, genome-wide analysis tools have been used. Among the most recent technologies to analyze DNA methylation in human samples are the 450K Illumina arrays [93,95]. The results found using this type of analysis tool have found thousands of alterations in the DNA methylation of males with fertility problems [72,90,91,96]. Surprisingly, alterations of sperm DNA methylation patterns also appear to be behind the fertility problems of individuals who apparently have good seminal quality, at least in terms of sperm number, mobility, and morphology [72]. The application of next generation technologies, such as whole genome bisulfite sequencing, will allow us to address and verify all these issues in future studies.

ENVIRONMENTAL CAUSES OF EPIGENETIC ALTERATIONS ASSOCIATED WITH MALE INFERTILITY

In the previous sections of this chapter, the importance of epigenetic mechanisms in the germinal cell has been described, as well as how the alteration of these mechanisms has been associated with various cases of male infertility. The possibility that alterations of epigenetic marks are mediating the adverse effects of several environmental factors during the embryonic stage has also been commented on, as well as that the effects could even be transmitted to the next generation if the alterations affect the germinal line. In this section, we will describe several studies that show several external factors that may be behind alterations in sperm DNA methylation patterns that may compromise reproductive success and that may also be transgenerationally transmitted. Most of these studies have been done in animal models and in general are aimed at the analysis of specific regions of the genome. One of these studies showed how neonatal exposure to BPA in male rats was related to alterations in the DNA methylation of the *H19* imprinting control region (ICR) in sperm, and more importantly, when these males were crossed with normal females, these alterations were transmitted to the embryo, leading to postimplantation miscarriage [97].

As was mentioned in previous sections of this chapter, during embryonic development, one of the critical periods in epigenetic reprogramming comprises the stage of germ cell development. During this phase in the uterus, any disturbance could affect the epigenome of the male germ cell and could potentially be transmitted to offspring. Diet is one such factor that determines the correct establishment of the epigenetic patterns of the spermatozoa, the reason why several works have looked at this factor. One of these works using animal models has shown how diet can alter the sperm epigenome and that the change is associated with congenital defects in offspring [98]. In this recent work using mice, a folate-deficient diet (folate metabolism is indispensable for DNA methylation processes) was supplied to males during the embryonic stage (i.e., pregnant females). The sperm of these animals were analyzed in adulthood, and although the number was normal when compared with control males, a genome-wide level epigenetic analysis (using methylated DNA immunoprecipitation (MeDIP) array) revealed DNA methylation alterations in a large number of genes associated with both development and several diseases, such as diabetes, autism, schizophrenia, and cancer, which appeared to be behind major birth defects presented by the offspring [98]. Another recent study showed that males similarly exposed over their lifetime to either folate-deficient or folate-supplemented diets had altered DNA methylation patterns in imprinted genes with evidence of the transmission of adverse effects to the offspring [99].

Another issue that raises further interest is the transmission of these epigenetic alterations of the sperm and consequently the associated diseases, to following generations. In this sense, several studies have been carried out in animal models, and it has been observed that, for example, caloric restriction during embryonic development alters the normal methylation patterns of sperm DNA and might contribute to the intergenerational transmission of diseases [100]. As well as diet, the effect of several environmental compounds, including BPA, phthalates, and pesticides on the sperm epigenome, has been studied [101]. It would of course be of great interest to analyze whether any epigenetic alterations related to those compounds could be perpetuated in subsequent generations and how this might affect the health of the individual. This question was recently studied in a murine model where gestating mice were exposed to a mixture of plastic derived endocrine disruptor compounds during the critical development of gonadal sex determination. The sperm of the unexposed generation of male descendants was then analyzed, and alterations of DNA methylation patterns of nearly 200 genes were found, which in turn were correlated with the occurrence of several disease phenotypes [102].

As discussed at the beginning of this section, most studies aimed at studying the alterations of sperm DNA methylation patterns caused by external factors have been performed, for obvious ethical reasons, in animal models. However, there are some epidemiological studies in humans where results similar to those found in animals have been observed. In one of these studies, the effect of high-dose folic acid supplementation on normozoospermic men presenting with idiopathic infertility was analyzed [103]. Sperm DNA methylation alterations were observed in genes related to cancer and neurobehavioral disorders, and these alterations were more pronounced in patients homozygous for a specific polymorphism of *MTHFR* [103], an enzyme required for folate metabolism, which as previously mentioned, is often genetically and epigenetically altered in cases of infertility [84]. In addition to the importance of diet in adults, it has been observed that medication may also have effects on the sperm epigenome and might affect fertility or even be transmitted to offspring [104].

Another issue currently being clarified is whether epigenetic alterations of the sperm are behind the effect that unhealthy lifestyle habits of the father have on the health of his children, for example, whether epigenetic alterations could explain that high-fat diet or smoking during prepuberty (10–11 years), a critical stage in sperm development, was associated with being overweight in the offspring [105,106]. Indeed, a recent work showed that alterations of DNA methylation in imprinted genes in human sperm of obese men could affect embryonic and fetal growth [107]. Moreover, another work using genome-wide approaches showed aberrant DNA methylation in the sperm of cigarette smokers could affect semen quality and ultimately could compromise reproductive success [108]. More studies are needed to clarify which environmental factors associated with lifestyle might be behind the epigenetic alterations of sperm in patients with reduced fertility although there is already evidence to suggest that habits such as smoking, drinking carbonated beverages, and the lack of exercise are related to aberrant methylation patterns that may explain some cases of severe oligospermia [83].

References

[1] J.T. Bell, T.D. Spector, A twin approach to unraveling epigenetics, Trends Genet. 27 (2011) 116–125.

[2] M.F. Fraga, E. Ballestar, M.F. Paz, S. Ropero, F. Setien, M.L. Ballestar, D. Heine-Suner, J.C. Cigudosa, M. Urioste, J. Benitez, M. Boix-Chornet, A. Sanchez-Aguilera, C. Ling, E. Carlsson, P. Poulsen, A. Vaag, Z. Stephan, T.D. Spector, Y.Z. Wu, C. Plass, M. Esteller, Epigenetic differences arise during the lifetime of monozygotic twins, Proc. Natl. Acad. Sci. U. S. A. 102 (2005) 10604–10609.

[3] M. Ollikainen, J.M. Craig, Epigenetic discordance at imprinting control regions in twins, Epigenomics 3 (2011) 295–306.

[4] H. Kiefer, L. Jouneau, E. Campion, D. Rousseau-Ralliard, T. Larcher, M.L. Martin-Magniette, S. Balzergue, M. Ledevin, A. Prezelin, P. Chavatte-Palmer, Y. Heyman, C. Richard, D. Le Bourhis, J.P. Renard, H. Jammes, Altered DNA methylation associated with an abnormal liver phenotype in a cattle model with a high incidence of perinatal pathologies, Sci. Rep. 6 (2016) 38869.

[5] I. Wilmut, A.E. Schnieke, J. McWhir, A.J. Kind, K.H. Campbell, Viable offspring derived from fetal and adult mammalian cells, Cloning Stem Cells 9 (2007) 3–7.

[6] D. Noble, Conrad Waddington and the origin of epigenetics, J. Exp. Biol. 218 (2015) 816–818.
[7] D.M. Ruden, M.D. Garfinkel, V.E. Sollars, X. Lu, Waddington's widget: Hsp90 and the inheritance of acquired characters, Semin. Cell Dev. Biol. 14 (2003) 301–310.
[8] C.H. Waddington, The epigenotype, Endeavour 1 (1942) 2.
[9] A. Bird, Perceptions of epigenetics, Nature 447 (2007) 396–398.
[10] G. Egger, G. Liang, A. Aparicio, P.A. Jones, Epigenetics in human disease and prospects for epigenetic therapy, Nature 429 (2004) 457–463.
[11] M. Esteller, Epigenetics in cancer, N. Engl. J. Med. 358 (2008) 1148–1159.
[12] J.G. Herman, S.B. Baylin, Gene silencing in cancer in association with promoter hypermethylation, N. Engl. J. Med. 349 (2003) 2042–2054.
[13] P.A. Jones, D. Takai, The role of DNA methylation in mammalian epigenetics, Science 293 (2001) 1068–1070.
[14] E.J. Finnegan, R.K. Genger, W.J. Peacock, E.S. Dennis, DNA methylation in plants, Annu. Rev. Plant. Physiol. Plant. Mol. Biol. 49 (1998) 223–247.
[15] M. Tariq, J. Paszkowski, DNA and histone methylation in plants, Trends Genet. 20 (2004) 244–251.
[16] T.H. Bestor, The DNA methyltransferases of mammals, Hum. Mol. Genet. 9 (2000) 2395–2402.
[17] M.M. Suzuki, A. Bird, DNA methylation landscapes: provocative insights from epigenomics, Nat. Rev. Genet. 9 (2008) 465–476.
[18] M. Kulis, A.C. Queiros, R. Beekman, J.I. Martin-Subero, Intragenic DNA methylation in transcriptional regulation, normal differentiation and cancer, Biochim. Biophys. Acta 1829 (2013) 1161–1174.
[19] E.L. Putiri, K.D. Robertson, Epigenetic mechanisms and genome stability, Clin. Epigenetics 2 (2011) 299–314.
[20] W. Doerfler, Patterns of DNA methylation--evolutionary vestiges of foreign DNA inactivation as a host defense mechanism. A proposal, Biol. Chem. Hoppe Seyler 372 (1991) 557–564.
[21] A.P. Feinberg, H. Cui, R. Ohlsson, DNA methylation and genomic imprinting: insights from cancer into epigenetic mechanisms, Semin. Cancer Biol. 12 (2002) 389–398.
[22] B. Payer, J.T. Lee, X chromosome dosage compensation: how mammals keep the balance, Annu. Rev. Genet. 42 (2008) 733–772.
[23] E. Li, T.H. Bestor, R. Jaenisch, Targeted mutation of the DNA methyltransferase gene results in embryonic lethality, Cell 69 (1992) 915–926.
[24] V. Calvanese, A.F. Fernandez, R.G. Urdinguio, B. Suarez-Alvarez, C. Mangas, V. Perez-Garcia, C. Bueno, R. Montes, V. Ramos-Mejia, P. Martinez-Camblor, C. Ferrero, Y. Assenov, C. Bock, P. Menendez, A.C. Carrera, C. Lopez-Larrea, M.F. Fraga, A promoter DNA demethylation landscape of human hematopoietic differentiation, Nucleic Acids Res. 40 (2012) 116–131.
[25] E. Ballestar, Epigenetic alterations in autoimmune rheumatic diseases, Nat. Rev. Rheumatol. 7 (2011) 263–271.
[26] A.F. Fernandez, E.G. Torano, R.G. Urdinguio, A.G. Lana, I.A. Fernandez, M.F. Fraga, The epigenetic basis of adaptation and responses to environmental change: perspective on human reproduction, Adv. Exp. Med. Biol. 753 (2014) 97–117.
[27] J. Peters, The role of genomic imprinting in biology and disease: an expanding view, Nat. Rev. Genet. 15 (2014) 517–530.
[28] R.G. Urdinguio, J.V. Sanchez-Mut, M. Esteller, Epigenetic mechanisms in neurological diseases: genes, syndromes, and therapies, Lancet Neurol. 8 (2009) 1056–1072.
[29] P. Valencia-Morales Mdel, S. Zaina, H. Heyn, F.J. Carmona, N. Varol, S. Sayols, E. Condom, J. Ramirez-Ruz, A. Gomez, S. Moran, G. Lund, D. Rodriguez-Rios, G. Lopez-Gonzalez, M. Ramirez-Nava, C. de la Rocha, A. Sanchez-Flores, M. Esteller, The DNA methylation drift of the atherosclerotic aorta increases with lesion progression, BMC Med. Genomics 8 (2015) 7.
[30] A.F. Fernandez, Y. Assenov, J.I. Martin-Subero, B. Balint, R. Siebert, H. Taniguchi, H. Yamamoto, M. Hidalgo, A.C. Tan, O. Galm, I. Ferrer, M. Sanchez-Cespedes, A. Villanueva, J. Carmona, J.V. Sanchez-Mut, M. Berdasco, V. Moreno, G. Capella, D. Monk, E. Ballestar, S. Ropero, R. Martinez, M. Sanchez-Carbayo, F. Prosper, X. Agirre, M.F. Fraga, O. Grana, L. Perez-Jurado, J. Mora, S. Puig, J. Prat, L. Badimon, A.A. Puca, S.J. Meltzer, T. Lengauer, J. Bridgewater, C. Bock, M. Esteller, A DNA methylation fingerprint of 1628 human samples, Genome Res. 22 (2012) 407–419.
[31] P.A. Jones, S.B. Baylin, The fundamental role of epigenetic events in cancer, Nat. Rev. Genet. 3 (2002) 415–428.
[32] H. Heyn, M. Esteller, DNA methylation profiling in the clinic: applications and challenges, Nat. Rev. Genet. 13 (2012) 679–692.
[33] BLUEPRINT consortium, Quantitative comparison of DNA methylation assays for biomarker development and clinical applications, Nat. Biotechnol. 34 (2016) 726–737.

[34] S. Moran, A. Martinez-Cardus, S. Sayols, E. Musulen, C. Balana, A. Estival-Gonzalez, C. Moutinho, H. Heyn, A. Diaz-Lagares, M.C. de Moura, G.M. Stella, P.M. Comoglio, M. Ruiz-Miro, X. Matias-Guiu, R. Pazo-Cid, A. Anton, R. Lopez-Lopez, G. Soler, F. Longo, I. Guerra, S. Fernandez, Y. Assenov, C. Plass, R. Morales, J. Carles, D. Bowtell, L. Mileshkin, D. Sia, R. Tothill, J. Tabernero, J.M. Llovet, M. Esteller, Epigenetic profiling to classify cancer of unknown primary: a multicentre, retrospective analysis, Lancet Oncol. 17 (2016) 1386–1395.

[35] S. Kriaucionis, N. Heintz, The nuclear DNA base 5-hydroxymethylcytosine is present in Purkinje neurons and the brain, Science 324 (2009) 929–930.

[36] M. Tahiliani, K.P. Koh, Y. Shen, W.A. Pastor, H. Bandukwala, Y. Brudno, S. Agarwal, L.M. Iyer, D.R. Liu, L. Aravind, A. Rao, Conversion of 5-methylcytosine to 5-hydroxymethylcytosine in mammalian DNA by MLL partner TET1, Science 324 (2009) 930–935.

[37] V. Lopez, A.F. Fernandez, M.F. Fraga, The role of 5-hydroxymethylcytosine in development, aging and age-related diseases, Ageing Res. Rev. 37 (2017) 28–38.

[38] R. Feil, M.F. Fraga, Epigenetics and the environment: emerging patterns and implications, Nat. Rev. Genet. 13 (2012) 97–109.

[39] J.E. Cropley, C.M. Suter, K.B. Beckman, D.I. Martin, Germ-line epigenetic modification of the murine A vy allele by nutritional supplementation, Proc. Natl. Acad. Sci. U. S. A. 103 (2006) 17308–17312.

[40] S. Singh, S.S. Li, Epigenetic effects of environmental chemicals bisphenol A and phthalates, Int. J. Mol. Sci. 13 (2012) 10143–10153.

[41] E. Dhimolea, P.R. Wadia, T.J. Murray, M.L. Settles, J.D. Treitman, C. Sonnenschein, T. Shioda, A.M. Soto, Prenatal exposure to BPA alters the epigenome of the rat mammary gland and increases the propensity to neoplastic development, PLoS One 9 (2014) e99800.

[42] S.M. Ho, W.Y. Tang, J. Belmonte de Frausto, G.S. Prins, Developmental exposure to estradiol and bisphenol A increases susceptibility to prostate carcinogenesis and epigenetically regulates phosphodiesterase type 4 variant 4, Cancer Res. 66 (2006) 5624–5632.

[43] L. Cao-Lei, R. Massart, M.J. Suderman, Z. Machnes, G. Elgbeili, D.P. Laplante, M. Szyf, S. King, DNA methylation signatures triggered by prenatal maternal stress exposure to a natural disaster: Project Ice Storm, PLoS One 9 (2014) e107653.

[44] B.T. Heijmans, E.W. Tobi, A.D. Stein, H. Putter, G.J. Blauw, E.S. Susser, P.E. Slagboom, L.H. Lumey, Persistent epigenetic differences associated with prenatal exposure to famine in humans, Proc. Natl. Acad. Sci. U. S. A. 105 (2008) 17046–17049.

[45] R.A. Waterland, R. Kellermayer, E. Laritsky, P. Rayco-Solon, R.A. Harris, M. Travisano, W. Zhang, M.S. Torskaya, J. Zhang, L. Shen, M.J. Manary, A.M. Prentice, Season of conception in rural gambia affects DNA methylation at putative human metastable epialleles, PLoS Genet. 6 (2010) e1001252.

[46] P. Kuhnen, D. Handke, R.A. Waterland, B.J. Hennig, M. Silver, A.J. Fulford, P. Dominguez-Salas, S.E. Moore, A.M. Prentice, J. Spranger, A. Hinney, J. Hebebrand, F.L. Heppner, L. Walzer, C. Grotzinger, J. Gromoll, S. Wiegand, A. Gruters, H. Krude, Interindividual variation in DNA methylation at a putative POMC metastable epiallele is associated with obesity, Cell Metab. 24 (2016) 502–509.

[47] D.A. Hackman, M.J. Farah, M.J. Meaney, Socioeconomic status and the brain: mechanistic insights from human and animal research, Nat. Rev. Neurosci. 11 (9) (2010) 651.

[48] J. Graff, D. Kim, M.M. Dobbin, L.H. Tsai, Epigenetic regulation of gene expression in physiological and pathological brain processes, Physiol. Rev. 91 (2011) 603–649.

[49] H. Sehgal, W. Toscano, Social Determinants of Health and Epigenetics: A New Tool for Health Policy, (Preprints) 2017, 2017030235.

[50] O. Aguilera, A.F. Fernandez, A. Munoz, M.F. Fraga, Epigenetics and environment: a complex relationship, J. Appl. Physiol. 109 (2010) 243–251.

[51] E.G. Torano, M.G. Garcia, J.L. Fernandez-Morera, P. Nino-Garcia, A.F. Fernandez, The impact of external factors on the epigenome: in utero and over lifetime, Biomed. Res. Int. 2016 (2016) 2568635.

[52] M.I. Sierra, A. Valdes, A.F. Fernandez, R. Torrecillas, M.F. Fraga, The effect of exposure to nanoparticles and nanomaterials on the mammalian epigenome, Int. J. Nanomedicine 11 (2016) 6297–6306.

[53] S. Ambatipudi, C. Cuenin, H. Hernandez-Vargas, A. Ghantous, F. Le Calvez-Kelm, R. Kaaks, M. Barrdahl, H. Boeing, K. Aleksandrova, A. Trichopoulou, P. Lagiou, A. Naska, D. Palli, V. Krogh, S. Polidoro, R. Tumino, S. Panico, B. Bueno-de-Mesquita, P.H. Peeters, J.R. Quiros, C. Navarro, E. Ardanaz, M. Dorronsoro, T. Key, P. Vineis, N. Murphy, E. Riboli, I. Romieu, Z. Herceg, Tobacco smoking-associated genome-wide DNA methylation changes in the EPIC study, Epigenomics 8 (2016) 599–618.

[54] L. Daxinger, E. Whitelaw, Understanding transgenerational epigenetic inheritance via the gametes in mammals, Nat. Rev. Genet. 13 (2012) 153–162.
[55] M.K. Skinner, Environmental stress and epigenetic transgenerational inheritance, BMC Med. 12 (2014) 153.
[56] S.D. van Otterdijk, K.B. Michels, Transgenerational epigenetic inheritance in mammals: how good is the evidence? FASEB J. 30 (2016) 2457–2465.
[57] Y. Wei, H. Schatten, Q.Y. Sun, Environmental epigenetic inheritance through gametes and implications for human reproduction, Hum. Reprod. Update 21 (2015) 194–208.
[58] D. Crews, R. Gillette, S.V. Scarpino, M. Manikkam, M.I. Savenkova, M.K. Skinner, Epigenetic transgenerational inheritance of altered stress responses, Proc. Natl. Acad. Sci. U. S. A. 109 (2015) 9143–9148.
[59] K. Gapp, A. Jawaid, P. Sarkies, J. Bohacek, P. Pelczar, J. Prados, L. Farinelli, E. Miska, I.M. Mansuy, Implication of sperm RNAs in transgenerational inheritance of the effects of early trauma in mice, Nat. Neurosci. 17 (2014) 667–669.
[60] M. Manikkam, M.M. Haque, C. Guerrero-Bosagna, E.E. Nilsson, M.K. Skinner, Pesticide methoxychlor promotes the epigenetic transgenerational inheritance of adult-onset disease through the female germline, PLoS One 9 (2014) e102091.
[61] E.E. Nilsson, M.D. Anway, J. Stanfield, M.K. Skinner, Transgenerational epigenetic effects of the endocrine disruptor vinclozolin on pregnancies and female adult onset disease, Reproduction 135 (2008) 713–721.
[62] M.K. Skinner, M. Manikkam, R. Tracey, C. Guerrero-Bosagna, M. Haque, E.E. Nilsson, Ancestral dichlorodiphenyltrichloroethane (DDT) exposure promotes epigenetic transgenerational inheritance of obesity, BMC Med. 11 (2013) 228.
[63] M.V. Veenendaal, R.C. Painter, S.R. de Rooij, P.M. Bossuyt, J.A. van der Post, P.D. Gluckman, M.A. Hanson, T.J. Roseboom, Transgenerational effects of prenatal exposure to the 1944-45 Dutch famine, BJOG 120 (2013) 548–553.
[64] J.E. Cropley, S.A. Eaton, A. Aiken, P.E. Young, E. Giannoulatou, J.W. Ho, M.E. Buckland, S.P. Keam, G. Hutvagner, D.T. Humphreys, K.G. Langley, D.C. Henstridge, D.I. Martin, M.A. Febbraio, C.M. Suter, Male-lineage transmission of an acquired metabolic phenotype induced by grand-paternal obesity, Mol. Metab. 5 (2016) 699–708.
[65] I. Donkin, S. Versteyhe, L.R. Ingerslev, K. Qian, M. Mechta, L. Nordkap, B. Mortensen, E.V. Appel, N. Jorgensen, V.B. Kristiansen, T. Hansen, C.T. Workman, J.R. Zierath, R. Barres, Obesity and bariatric surgery drive epigenetic variation of spermatozoa in humans, Cell Metab. 23 (2016) 369–378.
[66] A. Soubry, Epigenetic inheritance and evolution: a paternal perspective on dietary influences, Prog. Biophys. Mol. Biol. 118 (2015) 79–85.
[67] A. Soubry, C. Hoyo, R.L. Jirtle, S.K. Murphy, A paternal environmental legacy: evidence for epigenetic inheritance through the male germ line, Bioessays 36 (2014) 359–371.
[68] N.A. Youngson, V. Lecomte, C.A. Maloney, P. Leung, J. Liu, L.B. Hesson, F. Luciani, L. Krause, M.J. Morris, Obesity-induced sperm DNA methylation changes at satellite repeats are reprogrammed in rat offspring, Asian J. Androl. 18 (2016) 930–936.
[69] J.R. Gannon, B.R. Emery, T.G. Jenkins, D.T. Carrell, The sperm epigenome: implications for the embryo, Adv. Exp. Med. Biol. 791 (2014) 53–66.
[70] H. Sasaki, Y. Matsui, Epigenetic events in mammalian germ-cell development: reprogramming and beyond, Nat. Rev. Genet. 9 (2008) 129–140.
[71] C. Krausz, J. Sandoval, S. Sayols, C. Chianese, C. Giachini, H. Heyn, M. Esteller, Novel insights into DNA methylation features in spermatozoa: stability and peculiarities, PLoS One 7 (2012) e44479.
[72] R.G. Urdinguio, G.F. Bayon, M. Dmitrijeva, E.G. Torano, C. Bravo, M.F. Fraga, L. Bassas, S. Larriba, A.F. Fernandez, Aberrant DNA methylation patterns of spermatozoa in men with unexplained infertility, Hum. Reprod. 30 (2015) 1014–1028.
[73] S.S. Hammoud, D.A. Nix, H. Zhang, J. Purwar, D.T. Carrell, B.R. Cairns, Distinctive chromatin in human sperm packages genes for embryo development, Nature 460 (2009) 473–478.
[74] A. Molaro, E. Hodges, F. Fang, Q. Song, W.R. McCombie, G.J. Hannon, A.D. Smith, Sperm methylation profiles reveal features of epigenetic inheritance and evolution in primates, Cell 146 (2011) 1029–1041.
[75] J. Li, R.A. Harris, S.W. Cheung, C. Coarfa, M. Jeong, M.A. Goodell, L.D. White, A. Patel, S.H. Kang, C. Shaw, A.C. Chinault, T. Gambin, A. Gambin, J.R. Lupski, A. Milosavljevic, Genomic hypomethylation in the human germline associates with selective structural mutability in the human genome, PLoS Genet. 8 (2012) e1002692.

[76] T. Eggermann, F. Brioude, S. Russo, M.P. Lombardi, J. Bliek, E.R. Maher, L. Larizza, D. Prawitt, I. Netchine, M. Gonzales, K. Gronskov, Z. Tumer, D. Monk, M. Mannens, K. Chrzanowska, M.K. Walasek, M. Begemann, L. Soellner, K. Eggermann, J. Tenorio, J. Nevado, G.E. Moore, D.J. Mackay, K. Temple, G. Gillessen-Kaesbach, T. Ogata, R. Weksberg, E. Algar, P. Lapunzina, Prenatal molecular testing for Beckwith-Wiedemann and Silver-Russell syndromes: a challenge for molecular analysis and genetic counseling, Eur. J. Hum. Genet. 24 (2016) 784–793.

[77] S. Atsem, J. Reichenbach, R. Potabattula, M. Dittrich, C. Nava, C. Depienne, L. Bohm, S. Rost, T. Hahn, M. Schorsch, T. Haaf, N. El Hajj, Paternal age effects on sperm FOXK1 and KCNA7 methylation and transmission into the next generation, Hum. Mol. Genet. 25 (2016) 4996–5005.

[78] T.G. Jenkins, K.I. Aston, C. Pflueger, B.R. Cairns, D.T. Carrell, Age-associated sperm DNA methylation alterations: possible implications in offspring disease susceptibility, PLoS Genet. 10 (2014) e1004458.

[79] C.J. Marques, F. Carvalho, M. Sousa, A. Barros, Genomic imprinting in disruptive spermatogenesis, Lancet 363 (2004) 1700–1702.

[80] C.J. Marques, P. Costa, B. Vaz, F. Carvalho, S. Fernandes, A. Barros, M. Sousa, Abnormal methylation of imprinted genes in human sperm is associated with oligozoospermia, Mol. Hum. Reprod. 14 (2008) 67–74.

[81] A. Poplinski, F. Tuttelmann, D. Kanber, B. Horsthemke, J. Gromoll, Idiopathic male infertility is strongly associated with aberrant methylation of MEST and IGF2/H19 ICR1, Int. J. Androl. 33 (2010) 642–649.

[82] N. El Hajj, U. Zechner, E. Schneider, A. Tresch, J. Gromoll, T. Hahn, M. Schorsch, T. Haaf, Methylation status of imprinted genes and repetitive elements in sperm DNA from infertile males, Sex. Dev. 5 (2011) 60–69.

[83] H. Kobayashi, A. Sato, E. Otsu, H. Hiura, C. Tomatsu, T. Utsunomiya, H. Sasaki, N. Yaegashi, T. Arima, Aberrant DNA methylation of imprinted loci in sperm from oligospermic patients, Hum. Mol. Genet. 16 (2007) 2542–2551.

[84] N. Khazamipour, M. Noruzinia, P. Fatehmanesh, M. Keyhanee, P. Pujol, MTHFR promoter hypermethylation in testicular biopsies of patients with non-obstructive azoospermia: the role of epigenetics in male infertility, Hum. Reprod. 24 (2009) 2361–2364.

[85] W. Wu, O. Shen, Y. Qin, X. Niu, C. Lu, Y. Xia, L. Song, S. Wang, X. Wang, Idiopathic male infertility is strongly associated with aberrant promoter methylation of methylenetetrahydrofolate reductase (MTHFR), PLoS One 5 (2010) e13884.

[86] L. Nanassy, L. Liu, J. Griffin, D.T. Carrell, The clinical utility of the protamine 1/protamine 2 ratio in sperm, Protein Pept. Lett. 18 (2011) 772–777.

[87] P. Navarro-Costa, P. Nogueira, M. Carvalho, F. Leal, I. Cordeiro, C. Calhaz-Jorge, J. Goncalves, C.E. Plancha, Incorrect DNA methylation of the DAZL promoter CpG island associates with defective human sperm, Hum. Reprod. 25 (2010) 2647–2654.

[88] H. Heyn, H.J. Ferreira, L. Bassas, S. Bonache, S. Sayols, J. Sandoval, M. Esteller, S. Larriba, Epigenetic disruption of the PIWI pathway in human spermatogenic disorders, PLoS One 7 (2012) e47892.

[89] S. Houshdaran, V.K. Cortessis, K. Siegmund, A. Yang, P.W. Laird, R.Z. Sokol, Widespread epigenetic abnormalities suggest a broad DNA methylation erasure defect in abnormal human sperm, PLoS One 2 (2007) e1289.

[90] M. Laqqan, E.F. Solomayer, M. Hammadeh, Aberrations in sperm DNA methylation patterns are associated with abnormalities in semen parameters of subfertile males, Reprod. Biol. (2017).

[91] M. Laqqan, S. Tierling, Y. Alkhaled, C. Lo Porto, E.F. Solomayer, M. Hammadeh, Spermatozoa from males with reduced fecundity exhibit differential DNA methylation patterns, Andrology 5 (2017) 971–978.

[92] S.E. Pacheco, E.A. Houseman, B.C. Christensen, C.J. Marsit, K.T. Kelsey, M. Sigman, K. Boekelheide, Integrative DNA methylation and gene expression analyses identify DNA packaging and epigenetic regulatory genes associated with low motility sperm, PLoS One 6 (2011) e20280.

[93] M. Bibikova, B. Barnes, C. Tsan, V. Ho, B. Klotzle, J.M. Le, D. Delano, L. Zhang, G.P. Schroth, K.L. Gunderson, J.B. Fan, R. Shen, High density DNA methylation array with single CpG site resolution, Genomics 98 (2011) 288–295.

[94] M. Bibikova, J. Le, B. Barnes, S. Saedinia-Melnyk, L. Zhou, R. Shen, K.L. Gunderson, Genome-wide DNA methylation profiling using Infinium(R) assay, Epigenomics 1 (2009) 177–200.

[95] J. Sandoval, H. Heyn, S. Moran, J. Serra-Musach, M.A. Pujana, M. Bibikova, M. Esteller, Validation of a DNA methylation microarray for 450,000 CpG sites in the human genome, Epigenetics 6 (2011) 692–702.

[96] K.I. Aston, P.J. Uren, T.G. Jenkins, A. Horsager, B.R. Cairns, A.D. Smith, D.T. Carrell, Aberrant sperm DNA methylation predicts male fertility status and embryo quality, Fertil. Steril. 104 (2015) 1388–1397. e1–e5.

[97] T. Doshi, C. D'Souza, G. Vanage, Aberrant DNA methylation at Igf2-H19 imprinting control region in spermatozoa upon neonatal exposure to bisphenol A and its association with post implantation loss, Mol. Biol. Rep. 40 (2013) 4747–4757.

[98] R. Lambrot, C. Xu, S. Saint-Phar, G. Chountalos, T. Cohen, M. Paquet, M. Suderman, M. Hallett, S. Kimmins, Low paternal dietary folate alters the mouse sperm epigenome and is associated with negative pregnancy outcomes, Nat. Commun. 4 (2013) 2889.

[99] L. Ly, D. Chan, M. Aarabi, M. Landry, N.A. Behan, A.J. MacFarlane, J. Trasler, Intergenerational impact of paternal lifetime exposures to both folic acid deficiency and supplementation on reproductive outcomes and imprinted gene methylation, Mol. Hum. Reprod. (2017).

[100] E.J. Radford, M. Ito, H. Shi, J.A. Corish, K. Yamazawa, E. Isganaitis, S. Seisenberger, T.A. Hore, W. Reik, S. Erkek, A.H. Peters, M.E. Patti, A.C. Ferguson-Smith, In utero effects. In utero undernourishment perturbs the adult sperm methylome and intergenerational metabolism, Science 345 (2014) 1255903.

[101] M. Manikkam, C. Guerrero-Bosagna, R. Tracey, M.M. Haque, M.K. Skinner, Transgenerational actions of environmental compounds on reproductive disease and identification of epigenetic biomarkers of ancestral exposures, PLoS One 7 (2012) e31901.

[102] M. Manikkam, R. Tracey, C. Guerrero-Bosagna, M.K. Skinner, Plastics derived endocrine disruptors (BPA, DEHP and DBP) induce epigenetic transgenerational inheritance of obesity, reproductive disease and sperm epimutations, PLoS One 8 (2013) e55387.

[103] M. Aarabi, M.C. San Gabriel, D. Chan, N.A. Behan, M. Caron, T. Pastinen, G. Bourque, A.J. MacFarlane, A. Zini, J. Trasler, High-dose folic acid supplementation alters the human sperm methylome and is influenced by the MTHFR C677T polymorphism, Hum. Mol. Genet. 24 (2015) 6301–6313.

[104] I. Berthaut, D. Montjean, L. Dessolle, K. Morcel, F. Deluen, C. Poirot, A. Bashamboo, K. McElreavey, C. Ravel, Effect of temozolomide on male gametes: an epigenetic risk to the offspring? J. Assist. Reprod. Genet. 30 (2013) 827–833.

[105] K. Northstone, J. Golding, G. Davey Smith, L.L. Miller, M. Pembrey, Prepubertal start of father's smoking and increased body fat in his sons: further characterisation of paternal transgenerational responses, Eur. J. Hum. Genet. 22 (2014) 1382–1386.

[106] M.E. Pembrey, L.O. Bygren, G. Kaati, S. Edvinsson, K. Northstone, M. Sjostrom, J. Golding, Sex-specific, male-line transgenerational responses in humans, Eur. J. Hum. Genet. 14 (2006) 159–166.

[107] A. Soubry, L. Guo, Z. Huang, C. Hoyo, S. Romanus, T. Price, S.K. Murphy, Obesity-related DNA methylation at imprinted genes in human sperm: Results from the TIEGER study, Clin. Epigenetics 8 (2016) 51.

[108] M. Laqqan, S. Tierling, Y. Alkhaled, C.L. Porto, E.F. Solomayer, M.E. Hammadeh, Aberrant DNA methylation patterns of human spermatozoa in current smoker males, Reprod. Toxicol. 71 (2017) 126–133.

CHAPTER

12

Noninvasive Methods of Embryo Selection

Sergio Oehninger

The Jones Institute for Reproductive Medicine, Eastern Virginia Medical School, Norfolk, VA, United States

BACKGROUND AND SIGNIFICANCE

Assisted reproductive technologies (ART) provide infertile couples with a unique means to achieve a pregnancy. Over 5 million children worldwide have been born as a result of ART since their inception, over three decades ago. It is estimated that 1.5 million in vitro fertilization (IVF) cycles are performed each year worldwide, resulting in 350,000 children born per year [1,2]. Statistics from the United States unequivocally demonstrate increasing pregnancy rates since inception of the techniques in the early 1980s [1,3]. As a result, more and more couples with a variety of diagnoses of female and/or male infertility are successfully being treated by IVF with uterine embryo transfer. Today, around 1.6% of all infants born in the United States and up to 5%–8% in some European countries every year are conceived using ART [4–8].

Although the use of ART is still relatively low as compared with the potential demand, its use has doubled over the past decade. In the United States, in 2015, more than 80% of clinics identified by the Centers for Disease Control (CDC) also reported to the Society for Assisted Reproductive Technologies (SART), which is the society that groups most ART clinics [1]. In that year, in women <35 years of age with good prognosis, the live birth per egg retrieval cycle was 43%, with an average of 1.6 embryos transferred per cycle (including 33% of elective single-embryo transfers), with a 44% implantation rate. Among this good prognosis group of patients, there was a 12% chance for twins per egg retrieval cycle and a 0.3% chance for a triplet pregnancy. However, the efficiency of IVF on a per-retrieved egg basis is still low. Jones and colleagues [9] reported on a large number of IVF cycles and estimated that after the recovery of 61,813 metaphase II (MII) oocytes, which lead to 24,066 transferred MII oocytes (as embryos), there were 2059 deliveries with an efficiency on the basis of recovered eggs of only 9%. Cobo et al. [10,11] recently reported that the egg efficiency with vitrification in a large oocyte donation program resulted in an oocyte-to-baby rate of 7%.

https://doi.org/10.1016/B978-0-12-812571-7.00013-7

Although the live birth rate in younger women (<35 years) has reached a remarkable rate of 50% per embryo transfer cycle, the implantation rate (which expresses the true efficiency of the technology on a per transferred embryo basis) does not typically exceed the mid-40s even in the best prognosis patients. This leads to many programs' clinical policy of transferring more than one embryo, resulting in the undesirable and high-incidence outcome of recently reported multiple pregnancies, not only twins (30%) but also triplets (2%) as expressed per established pregnancies [1,3].

Consequently, the high significance of the concept of developing "noninvasive methods of embryo selection" relies on the discussion and possibility of the discovery of a noninvasive method to identify specific biomarkers of embryo implantation competence during the in vitro culture period. The success of IVF is highly dependent on a precise selection of embryos. Due to the increased use of elective single-embryo transfer worldwide, the need for more cost-effective and noninvasive methods for embryo selection has become a critical challenge.

This discovery would significantly contribute to the optimization of ART by allowing (i) an improved selection of embryos with the highest implantation competence for uterine transfer (thereby increasing current implantation rates) and (ii) a concomitant reduction of the number of embryos transferred per attempt, with a significant impact on decreasing the medical complications and economic burden to the health system determined by the occurrence of multiple gestations. The extremely negative consequences of twins and triplets and even more for higher-order magnitude multiple pregnancies, include, among others, increased hospital costs, rearing burden, and developmental issues.

Based on data accumulated in the last two decades from the use of preimplantation genetic screening (PGS) as an "invasive" method of embryo selection, it has been established that a variable percentage of embryos are chromosomally abnormal, with a dramatically increased proportion observed as maternal age advances. The use of PGS following embryo biopsy as an invasive screening test has had controversial results and has increased the cost of treatment. Munne and collaborators [12] recently presented a large volume of data on PGS using array comparative genomic hybridization (aCGH). The data showed that implantation rates after the transfer of euploid embryos are independent of maternal age; in other words, the transfer of chromosomally normal embryos selected after PGS appears to eliminate the negative impact of advancing maternal age on implantation. It is interesting to see that the average implantation rate was maintained at around 66% across the age spectrum. While the data are reassuring, it is also apparent that even with chromosomally normal embryos, there is a 33% failure of embryonic implantation. This points to the presence of either endometrial deficiencies (window of receptivity or other uterine anomalies) or other "occult" embryonic dysfunction independent of the ploidy status (e.g., other undiagnosed chromosomal anomalies including mosaicism or other genetic, cytoplasmic, or metabolic factors). These observations highlight the need to continue our search for other biomarkers of implantation competence.

Current laboratory assessment of embryos is based on the examination of zygote/embryo morphology, cleavage status, and development to the blastocyst stage. However, his assessment is only weakly predictive of implantation [13,14]. Time-lapse cinematography (TLC) has been recently added to the armamentarium to enhance embryo selection [15–17]. This imaging system allows continuous embryo observation without the need to remove the embryo from culturing conditions [15]. In addition to embryo morphology and cleavage, as well as

TABLE 12.1 Methods of Embryo Selection in Human IVF

Noninvasive methods
• Pronuclear morphology
• Fast cleavage
• Day 3 and 5 scores
• Time-lapse cinematography (TLC)
• Identification of biomarkers through "omics" technologies
• Cell-free DNA
Invasive methods
• Chromosomal status (ploidy) through array-CGH, SNP-arrays, qPCR, or NGS
• Mitochondrial DNA content
Combined methods
• Proteomics and TLC
• Oxygen consumption and TLC
• Blastocoel fluid proteomics/metabolomics/cell-free DNA

morphokinetics by TLC analysis, other noninvasive methods of embryo selection have been attempted. These have been based on the measurement of metabolic determinants of embryo viability during culture (i.e., determination of pyruvate, glucose, amino acid, and oxygen modifications) [18,19]. More recently, the analysis of the proteome/secretome and metabolome of the developing embryo during days of in vitro culture have been introduced [20,21]. The latter approaches can be used in biomarker discovery. An ideal biomarker should be specific, easily detectable with minimum or no invasion, and it should be present at the earliest stages of the process under consideration. In the ART scenario, the biomarker(s) should have a positive impact on clinical management with robust statistical power for prediction of implantation.

Here, a comprehensive analysis of existing data for the current methods of embryo selection is presented, with emphasis on noninvasive techniques (Table 12.1) [15,16,22,23].

NONINVASIVE METHODS OF EMBRYO SELECTION

Morphological Methods

The selection of the highest-quality embryos to transfer is mainly based on morphological evaluations, but this approach has traditionally been observational and lacking a true origin on evidence-based medicine [24]. In the last three decades, numerous classification systems for embryo grading have been developed [25].

Zygote Pronuclear Scoring Systems

Scott and Smith [26] and Tesarik and Greco [27] developed scoring systems to assess zygotes at the pronuclear stage. The most commonly used systems assess the number and relative position of the nucleolar precursor bodies in each pronucleus, with evaluation of the pronucleus size and alignment, and take into consideration certain cytoplasmic features. The predictive value of pronuclear scoring is controversial, and most of the studies are retrospective [28].

Cleavage-Stage Scoring Systems

A few systems were developed based on blastomere numbers, embryo fragmentation (defined as anuclear, membrane-bound extracellular cytoplasmic structures), multinucleation, and blastomere symmetry [29].

Blastocyst Scoring Systems

The most commonly used scoring system for evaluating blastocysts is the one proposed by Gardner and collaborators [30,31]. This system takes into consideration the extent to which the volume of the embryo is occupied by the blastocoel, as well as the number and organization of cells in each inner cell mass and trophectoderm compartments.

The variety of grading systems has led to attempts to standardize scoring methods by the most renowned reproductive professional organizations. Standardized criteria for grading embryos have recently been proposed by two organizations, that is, SART (collected into the SART national registry—SART-CORS) [32] and the Alpha Executive and ESHRE Special Interest Group of Embryology [33].

Nonetheless, examination of embryo morphology is a subjective process associated with high and variable intraobserver and interobserver variability [34]. The evaluation time has to be as short as possible to prevent embryo exposure to pH and temperature shifts and excessive light exposure, which have adverse effects on embryo development and quality [25]. Morphological approaches have limited predictive value, with an average of approximately 20%–40% of the embryos identified using this method actually implanting [15].

Kinematic Methods: Time Lapse Cinematography (TLC)

Typically, embryos are removed from the incubator once a day to assess cleavage and morphology, but this type of monitoring only gives a basic transitory view of a more widely dynamic process. The embryos do not tolerate prolonged or frequent removal from optimal culturing conditions, which limits the number of observations that can be made. A major advantage of TLC is that the embryos can be continuously evaluated without removing them from the incubator. A camera is built into the incubator and takes pictures of the embryos at preset intervals. Relying on proper software, a video can be made that represents their development. There are a few systems commercially available [15,16]. Hierarchical models (with algorithms created based on different kinetic markers) have been proposed and tested in retrospective analyses [35,36], but the predictive ability of these markers has yet to be tested prospectively in large studies and using clinically meaningful end points.

Payne and colleagues [37] reported earlier on the kinetic events of human embryonic development, whereas Mio and Meada [38] described the kinetics of the events up to the blastocyst stage [39]. Their work was followed by observations made by several other groups that tried to correlate these kinetic and morphological markers with embryo development, implantation potential, pregnancy rate, and chromosomal status [39–43].

As with any novel clinical technology, it has been important to establish the safety of the methodology. Distinct and periodic light exposure, heat due to motion, friction of moving parts, the presence of magnetic fields, sheer stress of moving culture dishes, and fumes from lubricants related to TLC systems may bring harmful effect(s) on embryo quality. To date, no negative impact regarding any of these parameters has been reported [15,44–46].

The benefits of TLC versus conventional morphology evaluation and grading may therefore include a more stable culture environment (incubator effect), less handling (typically by using a single culture medium), less risk as it does not rely on the limited information from a few static observations, ability to allow for computer annotations of developmental milestones, and ability to decrease intra- and interobserver variability. In addition, it has been proposed that TLC can aid in the prediction of blastocyst formation, implantation, and even chromosomal status. However, the literature shows limited retrospective trials, few prospective studies, and the confounding factor of a still heterogeneous and sometimes unclear nomenclature [15,17].

Several meta-analyses of time-lapse studies have been published. Kaser and Racowsky [16] assessed the following morphokinetic parameters: pronuclear dynamics and morphology; duration of first cytokinesis and reappearance of nuclei after cleavage; time of various cleavage stages; duration of various cleavage stages; duration of cleavage cycles and mitotic synchronicity; and time to morula, blastocyst, and hatching. In this review, five studies used combined parameter grading to generate a cumulative score, and two studies retrospectively compared implantation rates following embryo selection by conventional morphology alone or with the addition of a hierarchal time-lapse classification. The authors concluded that while several studies suggest higher implantation rates for fast-cleaving embryos and those with a timely duration of the two-cell and three-cell stages, no single morphokinetic parameter has been consistently shown to predict implantation potential.

A more recent, larger and comprehensive meta-analysis was performed by Chen et al. [47]. These authors included only randomized controlled trials that compared TLC with conventional methods. Ten studies were included, four studies that randomized oocytes and six that randomized women. For the oocyte-based review (two studies, including 1154 embryos), the pooled analysis observed no significant difference between the TLC group and the control group for blastocyst rate (relative risk 1.08, 95% CI 0.94–1.25). The quality of evidence was moderate for all outcomes in oocyte-based review. For the woman-based review, only one study (842 women) provided a live birth rate (relative risk 1.23, 95% CI 1.06–1.44), and the pooled result showed no significant difference in ongoing pregnancy rate (relative risk 1.04, 95% CI 0.80–1.36, four studies, including 1403 women) between two groups. The quality of the evidence was low or very low for all outcomes in the woman-based review. The authors concluded that there is insufficient evidence to support that TLC is superior to conventional methods for human embryo incubation and selection. The authors speculated that clinical

TLC may have the potential to improve IVF outcomes, but currently there is insufficient evidence to support regular TLC use when considering ongoing pregnancy rates and blastocyst formation rates.

It has been documented that embryo morphology on day 3 or 5 does not predict aneuploidy [48]. Delayed blastulation is not associated with increased aneuploidy rates, but the absence of blastulation is associated with increased aneuploidy. Nonetheless, dynamic blastomere behavior reflects human embryo ploidy by the four-cell stage [49]. Campbell et al. [50] provided evidence that TLC using defined morphokinetic data and a specific algorithm can be used to classify human preimplantation embryos according to their risk of aneuploidy, without performing biopsy and PGS, and that this correlates well with clinical outcomes.

Other studies have shown that morphology by TLC does not predict ploidy, with no differences observed according to slow or fast dividing embryos [51]. These authors retrospectively applied key time-lapse imaging events to stratify embryos into three groups: low, medium, and high risk of aneuploidy. The actual ploidy results (from aCGH) were compared with expectations. The model failed to segregate euploid embryos from aneuploid embryos. Further analysis indicated that the variability of embryos among patients was too great to allow the selection of euploid embryos based on simple morphokinetic thresholds. The authors concluded that the selection of embryos based on morphokinetics alone is unlikely to identify euploid embryos accurately for transfer or yield higher rates of live delivery.

Although promising results are available, which support the use of TLC in clinical practice, definitely, more well- and prospectively designed studies with specific end points need to be carried out to prove the optimization of embryo-blastocyst selection and correlations with implantation and ploidy status.

"Omics"

The integration of "omics" techniques is called "system biology" and provides a vast amount of information with a high potential to unravel the complex interaction of molecular networks underlying the function of any organism in health and disease. Despite some advances, none of the "omics" techniques introduced so far has been unequivocally validated in the IVF clinical arena.

Due to ethical reasons, studies focusing on early human embryo development in vitro have been largely descriptive. Normal, IVF-derived embryos donated for research are not readily available; as a consequence, only limited research has been performed with "discarded" embryos that not necessarily provide physiological information. The ability to screen the embryos in the IVF laboratory using inverted light microscopy to ascertain which embryo has the highest chance of implantation has received a lot of attention, but these features are only predictive to a limited extent. Furthermore, in some studies, morphology by newer TLC is not a good predictor of euploidy; as up to 53% of blastocysts selected for transfer were aneuploidy [52], and no differences in euploidy rates were seen between fast and slow growing blastocysts [53].

To date, there are only a few reports where proteomics and metabolomics have been applied to the study of early human embryos and the in vitro culture systems used [20,21]. Katz-Jaffe et al. [54] examined protein biomarkers of development in mouse and human embryos using surface-enhanced laser desorption and ionization time of flight (SELDI-TOF). Results suggested the presence of different proteomic profiles at various embryonic developmental stages. These results, however, have not been validated in further experiments, and questions have been raised about this primary identification [55]. Others have used proteomics in order to define the human embryonic secretome, with the report that the protein lipocalin-1 could be detected in the secretome of human blastocysts [56]. However, no predictive ability was documented regarding implantation potential by culture media from transferred embryos resulting in pregnancy versus those not resulting in pregnancy.

Other approaches for the analysis of the secretome have included the use of protein microarrays containing 120 antibody targets [57]. Similarly, a targeted proteomics platform, Luminex, has provided great insight into the proteome of the fluids of the female reproductive tract [58].

The first limitation of a proteomic analysis is the requirement of large amounts of material [21,55], and in IVF, the media droplets typically may range from only 20–50 μL. The second limitation is that the embryo culture media consist of a complex mixture of proteins (i.e., human-derived plasma proteins added as supplements). This mixture of proteins, peptides, metabolites, and other compounds, combined with very small amounts of secreted metabolites/proteins by an embryo, make the analytic sensitivity a major limiting factor [59]. Nonetheless, a few previous studies have detected proteins in conditioned culture media from early embryos grown in vitro in a variety of species. Some of these molecules were identified in humans by immunologic methods, including paf, IGF-II, HLA-G, hCG, apolipoprotein A1, and superoxide dismutase-1 [59–62]. However, there has been debate as to the reliability of some of these measurements. While it appears that the embryos may indeed release some of these signals, the sensitivity of the assays, the impact of IVF culture conditions (various media and interference of protein supplements used clinically), and the true levels of the secreted molecules (orders of magnitude of the concentrations) have been questioned and require more extensive verification [59,62]. Importantly, in the mouse system, the intact zona pellucida of eggs and zygotes were permeable to sugar-lectin markers up to a molecular weight range of 170 (eggs) and 110 kDa (zygotes), with no movement through the zona over these limits [63]. This third limitation of zona pellucida permeability highlights the significance of a metabolomics approach compared with a proteomic analysis.

Nyalwidhe et al. [60] published a proteomic analysis of IVF embryo conditioned media in the search for biomarkers of human embryo developmental potential. Peptides/proteins were affinity purified based on their physical-chemical properties and profiled by mass spectrometry (MS) for differential expression. Some of the identified proteins were further characterized by western blot and ELISA, and absolute quantification was achieved by multiple reaction monitoring. These results did not result in the identification of newly secreted peptides/proteins that could be used as biomarkers, or in protein profile changes that provided clinical application, highlighting the need for developing a new approach. Consequently, in spite of its promise, the analysis of the embryonic secretome has yet to be adopted clinically. Another determinant for this is the lack of uniform and universal MS platforms and its associated costs [19].

In addition to the use of proteomics, a few publications have reported on the use of metabolomics as an alternative noninvasive method. Metabolomics is the high-throughput identification of the general profile of metabolites in a system and shares with proteomics the use of the MS methodology. MS and proton nuclear magnetic resonance spectroscopy (^{1}H-NMR) have been the most commonly applied techniques in metabolic profiling. A few studies have used Raman and near-infrared (NIR) spectroscopies [64–66]. Hardarson et al. [67] concluded that adding NIR spectroscopy to embryo morphology does not improve the chance of a viable pregnancy and that NIR technology cannot be used as an objective measure of embryo quality. Sánchez-Ribas et al. [68] used liquid chromatography/gas chromatography coupled with MS to detect changes in the metabolic profiles of aneuploidies. They found two metabolites, caproate and rosterone sulfate, and two unknown compounds that were differentially expressed between normal and trisomy 21 day-3 embryos. However, these discoveries lack clinical application due to the resolution of the technique at the laboratory level. Cortezzi et al. [69] used mass spectroscopy fingerprinting, but clearly, the resolution and sensitivity of mass spectroscopy (ESI-Q-TOF-MS) limit the ability of identifying biomarkers with high precision. Vergouw et al. [70] concluded that there was no evidence that embryo selection by near-infrared spectroscopy in addition to morphology is able to improve live birth rates.

Others have published the use of "targeted metabolomics," where one or several known metabolites used by the embryo can be quantified noninvasively [19]. Examples of this include glucose and one of its metabolites lactate, as well as amino acids together with ammonium. Embryo metabolic activity has been associated with improved embryo quality, as well as improved implantation and pregnancy rates. The analysis of the embryo metabolism involves either the detection of specific nutrients that have been consumed from the surrounding culture medium or the detection of metabolic factors that are secreted from the embryo into the culture medium. Specifically, the embryo's glycolytic activity, pyruvate consumption, and changes in various amino acid concentrations in the spent culture medium have been strongly correlated with embryo developmental competence and pregnancy outcome [71–75]. However, randomized control trials are still missing to confirm the value of these measurements to select the best embryo for transfer. Moreover, it is difficult to standardize these methods so that they can be used routinely in the clinical setting [23].

Despite these considerations, the investigation of the changes in the metabolome of the conditioned culture media appears to be an ideal and new noninvasive approach, as these variations reflect and amplify the activities at a functional level, that is, the true embryonic phenotype. Most of the techniques published so far lack resolution to identify the vast number, chemical complexity, heterogeneity, and changes of proteins and metabolites. It is possible that the very recent introduction of Fourier transform ion cyclotron resonance mass spectrometry (FTICR-MS) may overcome many of these limitations [76,77]. Its ultrahigh resolution and high mass accuracy grant the FTICR-MS superiority relative to the other MS methodologies and allow for the simultaneous detection of thousands of chemical compounds per single sample at nanomolar to femtomolar concentrations. Moreover, the ability to couple FTICR-MS with various soft ionization methods allows for subsequent detection of a wide range of compound classes from different types of organic mixtures [78,79]. It is expected that this technique and the development of improved mass spectrometry methods may enhance detection capabilities of biomarkers of implantation.

In this context, embryos with optimal developmental abilities will exhibit defined changes in the culture milieu including (i) a unique metabolic fingerprinting, such as the presence of defined metabolites not previously present in the medium, and/or (ii) modifications of the culture medium resulting in reproducible profiles associated with depletion or augmentation of specific metabolites. These profiles can then be tested in algorithms to statistically distinguish embryos with the highest implantation competence with high accuracy and reproducibility. Thereafter, the predictive metabolites can be identified and precisely characterized in mass and chemical structure.

However, according to current trials in women undergoing ART, there is insufficient evidence to show that the metabolomic assessment of embryos before implantation has any meaningful effect on rates of live birth, ongoing pregnancy, or miscarriage rates. Siristatidis et al. reported on a Cochrane meta-analysis [80] and found that the existing evidence varied from very low to low quality. Therefore, it was concluded that there is no evidence to support or refute the use of this technique for subfertile women undergoing ART. Robust evidence is needed from further prospective and randomized trials, which study the effects on live birth and miscarriage rates for the metabolomic assessment of embryo viability.

Cell-Free DNA in Conditioned Media

Free fetal DNA is present in maternal blood due to cell turnover as the placenta and fetus develop. It is unknown when the free DNA is released, that is, whether it starts as soon as an embryo begins to divide or at a later time point in embryonic development. In an elegant study, Shamonki et al. [81] hypothesized that free embryonic DNA is released early in embryonic development, soon after an embryo begins to divide. As such, their proof-of-concept study aimed to demonstrate that free embryonic DNA is present in spent IVF culture media. The greatest advantage of embryonic screening of spent (conditioned) IVF media would be the ability to evaluate an embryo without the potential for damage from biopsy as a noninvasive method of selection based on chromosomal status. These authors presented two novel findings: the presence of free embryonic DNA in embryo conditioned media and a chromosomal result that was consistent with trophectoderm biopsy. The authors amplified free DNA by whole genomic amplification (WGA), and after amplification, the DNA was labeled, purified, and hybridized for aCGH. The authors speculated that improvements in DNA collection, amplification, and testing may allow for PGS without biopsy in the future.

The study demonstrated several strengths. The authors showed that "control" media used in the IVF laboratory setting do not have identifiable DNA. They further showed that media exposed to developing embryos acquires free DNA and that a genetic screening result can be rendered similar to the corresponding trophectoderm biopsy. However, the study had significant and acknowledged weaknesses. The number of samples tested was small. Furthermore, most samples demonstrated quantity and/or quality insufficient to reach a reliable screening DNA result. Only two samples were of reliable quality to yield a screening result, albeit both were identical to the corresponding embryo biopsy. Even if this technique becomes reliable, there are still controversial data demonstrating that PGS improves live birth rate on a per retrieval basis for all patient populations [82,83].

INVASIVE METHODS OF EMBRYO SELECTION

These methods require micromanipulation and a microsurgical approach to obtain whole blastomeres or trophectoderm cells via embryo biopsy.

PGS With Array-CGH, qPCR, or Next Generation Sequencing (NGS)

Among the proposed invasive methods, PGS is based on analyzing chromosomal normalcy but requires the use of micromanipulation techniques with embryo biopsy of one-two blastomeres (at the cleavage stage) or several trophectoderm cells (at the blastocyst stage) [82–84]. Nowadays, there is general agreement that the advantage of a biopsy at the blastocyst stage involves a more differentiated embryo and possibility to extract 5–10 trophectoderm cells while not perturbing the inner cell mass, thereby reducing risk to the embryo and increasing accuracy of diagnosis [85]. The technique for DNA analysis is still evolving at a fast pace [86].

PGS has now been shown to significantly improve implantation rates and ongoing pregnancy rates in at least three randomized clinical trials as well as in several case-controlled prospective studies and two meta-analyses [12,87]. Additionally, there is also strong evidence that the incidence of miscarriage is reduced. A publication from the US Centers for Disease Control and Prevention has provided further insight into the value of PGS for patients of advanced maternal age [88]. The investigators analyzed IVF data from the United States between 2011 and 2012 and compared the results for treatment cycles with or without PGS. They concluded that a statistically significant decrease in miscarriage rates and increased odds of live birth were apparent in cycles of women over 37 years old when PGS had been used. In addition, as mentioned earlier, it has been demonstrated that the transfer of a chromosomally normal blastocyst results in an implantation rate that remains constant irrespective of maternal age [12].

Furthermore, Rubio et al. [89] recently reported on a multicenter, randomized trial that determined the clinical value of PGS for aneuploidy screening in women of advanced maternal age (between 38 and 41 years). Interventions were day-3 embryo biopsy, aCGH, blastocyst transfer, and vitrification, and the main outcome measures were delivery and live birth rates in the first transfer and cumulative outcome rates. Results demonstrated that PGS for aneuploidy screening is superior compared with controls, not only in clinical outcomes at the first embryo transfer but also in dramatically decreasing miscarriage rates and shortening the time to pregnancy. There was no benefit of PGS for cumulative delivery rates, as might be expected, but no detrimental effects were observed. For those patients without any euploid embryos (31%), application of PGS may avoid a high risk of miscarriage (39%) and unnecessary extra transfer cycles, whereas for those with euploid embryos, it reduces the time to a successful pregnancy with minimal risk of loss [90].

Three major controversial aspects of PGS remain unresolved: one is the fact that PGS has not been shown to increase cumulative pregnancy rates; the second one is that the error rate of PGS is still not zero so some euploid embryos, which might have produced viable pregnancies, might be incorrectly discarded, while other aneuploid ones may be inadvertently replaced; and finally, the significance and true incidence of embryo mosaicism continues to be an important clinical challenge [12,91].

Mitochondrial DNA Content

Fragouli et al. [92] examined biological and clinical relevance of the quantity of mitochondrial DNA (mtDNA) in human embryos. These were examined via a combination of aCGH, quantitative PCR (qPCR), and next-generation sequencing (NGS), providing information on chromosomal status, amount of mtDNA, and the presence of mutations in the mitochondrial genome. The results of this study suggested that increased mtDNA may be related to elevated metabolism and are associated with reduced viability, a possibility consistent with the "quiet embryo" hypothesis [73,75]. Of clinical significance, these authors proposed that mtDNA content represents a novel biomarker with potential value for IVF treatment, revealing chromosomally normal blastocysts incapable of producing a viable pregnancy. In a subsequent publication, it was communicated that elevated mtDNA levels, above a previously defined threshold, are strongly associated with blastocyst implantation failure and may represent an independent biomarker of embryo viability [93]. More data are needed to confirm these findings.

COMBINED METHODS OF EMBRYO SELECTION

A few initial reports have combined application of noninvasive methods. Dominguez at el. [94] proposed a new strategy for diagnosing embryo implantation potential by combining proteomics and TLC. A hierarchical classification of embryos was developed based on the presence of interleukin-6 (Il-6) in the conditioned culture media (measured by immunoassay) and the second duration of second cell cycle (i.e., the time from division to a 2-blastomere until division to a 3-blastomere embryo). This classification generated four embryo categories with increasing expected implantation potential.

In another report, Tejera et al. [95] used a combination of metabolic measurement and TLC. The authors examined embryo oxygen consumption with microsensors and performed measurements from pronucleus formation to the eight-cell cleavage. They observed differences between embryos that achieved an ongoing pregnancy and those that did not and proposed that this combined method provides an embryo selection method based on oxygen uptake and chronology of cytokinesis timing.

Others have recently introduced application of a combined approach using invasive and noninvasive methods. It was recently reported that the blastocyst fluid contains DNA, possibly representing another source of DNA for genetic analysis [96]. Should these findings be confirmed, the aspiration of the blastocoel fluid, a procedure termed blastocentesis, could easily become the preferred means of blastocyst biopsy. D'Alessandro et al. [97] reported on the metabolomic profile of this fluid with high-performance liquid chromatography-electrospray ionization-mass spectrometry (HPLC-ESI-MS). Palini et al. [98] performed a study that aimed to (1) verify the presence of DNA in blastocoel fluid; (2) estimate whether the chromosomal status predicted by its analysis corresponds with the ploidy condition of polar bodies, blastomeres, and trophectoderm cells; and (3) estimate whether the chromosomal status predicted by polar bodies, blastomeres, and trophectoderm cells analysis corresponds with the ploidy condition of the whole embryo. Blastocyst micropuncture and aspiration of blastocoel fluid was carried out followed by a real-time PCR-based approach with amplification of the multicopy genes of TSPY1 (on the Y chromosome) and TBC1D3 (on chromosome 17). WGA technology was also attempted, and aCGH was performed to confirm the reliability of this approach. Reliable quantification was not possible due to sensitivity. The median genomic DNA content in blastocoel fluid was estimated to be 9.9 pg per sample. The data indicated that amplifiable DNA exists in the blastocoel fluid. The suitability of this DNA for PCR, WGA, and aCGH is demonstrated. However, its usefulness for diagnostic purposes still needs to be proven. If this approach turns out to be feasible, potential risks associated with embryo biopsy could be reduced [99].

Hammond et al. [100] cautioned about these types of testing, as they found that there is DNA contamination within the culture media after manufacturing (contrary to previous work) [81,98]. Results also suggested that cumulus cells may also contribute to the additional DNA detected in spent embryo media. Any DNA originating from apoptotic cells of the embryo may not accurately represent the embryo. While the presence of this genetic material is an exciting discovery, the DNA in the blastocoel fluid and embryo culture medium appears to be of low yield and integrity, which makes it challenging to study. Further research aimed at assessing the methodologies used for both isolating and analyzing this genetic material, as well as establishing its origin, is needed in order to evaluate its potential for clinical use. Should such methodologies prove to be routinely successful and the DNA recovered demonstrated to be embryonic in origin, then they may be used in a minimally invasive and less technical methodology for genetic analysis and embryo viability assessment than those currently available.

SUMMARY AND CONCLUSIONS

The selection of the embryos with highest implantation competence constitutes the key goal of a state-of-the-art IVF lab. Morphology is still the standard reference method for the evaluation of embryo development and quality. Considering the limitations and flaws of studies reported so far, better designed prospective and randomized trials are needed to comprehensively evaluate the effectiveness of clinical TLC. Although present evidence suggests that clinical TLC may have the potential to improve IVF outcomes. There is insufficient evidence to support regular clinical use of TLC when considering ongoing pregnancy rates, blastocyst formation rates, or embryo ploidy. Nevertheless, recent data indicate that the combination of morphology and TLC and possibly the addition of optimized proteomic/metabolomic approaches may provide information of superior clinical value. These noninvasive methods of embryo selection will undoubtedly be further developed into useful clinical tools.

Potential links between morphokinetic data obtained by TLC and embryo ploidy (PGS) are being elucidated. Current invasive methods of embryo selection are growing in use in the clinical arena, with the introduction of new molecular methods of determining embryo chromosomal complement that are proving to be accurate and reproducible, with the future trending toward aCGH arrays or NGS as a rapid and reliable means of analysis. Combined data from embryo metabolism, morphokinetics, and ploidy (results obtained by biopsy or cell-free in conditioned media), and/or other candidate markers such as microRNAs [101] may add to establish high predictive value for the most relevant clinical outcomes.

References

[1] Centers for Disease Control (CDC), 2015 Assisted Reproductive Technologies Report.

[2] European IVF-Monitoring Consortium (EIM) for the European Society of Human Reproduction and Embryology (ESHRE), C. Calhaz-Jorge, C. de Geyter, M.S. Kupka, J. de Mouzon, K. Erb, E. Mocanu, T. Motrenko, G. Scaravelli, C. Wyns, Assisted reproductive technology in Europe, 2012: results generated from European registers by ESHRE, Hum. Reprod. 31 (2016) 1638–1652.

[3] Society for Assisted Reproductive Technologies (SART), SART CORS, ClinicSummaryReporthttps://www.sartcorsonline.com/rptCSR_PublicMultYear.aspx?ClinicPKID=0 154000, 2015.
[4] K.G. Nygren, A.N. Andersen, Assisted reproductive technology in Europe, 1999. Results generated from European registers by ESHRE, Hum. Reprod. 17 (2002) 3260–3274.
[5] A.P. Ferraretti, V. Goossens, J. de Mouzon, S. Bhattacharya, J.A. Castilla, V. Korsak, M. Kupka, K.G. Nygren, A. Nyboe Andersen, European IVF-monitoring (EIM), Consortium for European Society of Human Reproduction and Embryology (ESHRE), Assisted reproductive technology in Europe, 2008: results generated from European registers by ESHRE, Hum. Reprod. 27 (2012) 2571–2584.
[6] S. Sunderam, D.M. Kissin, L. Flowers, J.E. Anderson, S.G. Folger, D.J. Jamieson, W.D. Barfield, Centers for Disease Control and Prevention (CDC), Assisted reproductive technology surveillance—United States, 2009, MMWR Surveill. Summ. 61 (2012) 1–23.
[7] L.A. Schieve, O. Devine, C.A. Boyle, J.R. Petrini, L. Warner, Estimation of the contribution of non-assisted reproductive technology ovulation stimulation fertility treatments to US singleton and multiple births, Am. J. Epidemiol. 170 (2009) 1396–1407.
[8] European Society of Human Reproduction and Embryology (ESHRE), 2013 (ART Fact Sheet).
[9] H.W. Jones Jr., S. Oehninger, S. Bocca, L. Stadtmauer, J. Mayer, Reproductive efficiency of human oocytes fertilized in vitro, Facts Views Vis. Obgyn 2 (2010) 169–171.
[10] A. Cobo, N. Garrido, A. Pellicer, J. Remohí, Six years' experience in ovum donation using vitrified oocytes: report of cumulative outcomes, impact of storage time, and development of a predictive model for oocyte survival rate, Fertil. Steril. 104 (2015) 1426–1434.
[11] A. Cobo, J.A. García-Velasco, A. Coello, J. Domingo, A. Pellicer, J. Remohí, Oocyte vitrification as an efficient option for elective fertility preservation, Fertil. Steril. 105 (2016) 755–764.
[12] S. Munné, J. Grifo, D. Wells, Mosaicism: "survival of the fittest" versus "no embryo left behind", Fertil. Steril. 105 (2016) 1146–1149.
[13] Z.P. Nagy, D. Sakkas, B. Behr, Symposium: innovative techniques in human embryo viability assessment. Non-invasive assessment of embryo viability by metabolomic profiling of culture media ('metabolomics'), Reprod. Biomed. Online 17 (2008) 502–507.
[14] S. Munné, S. Chen, P. Colls, J. Garrisi, X. Zheng, N. Cekleniak, M. Lenzi, P. Hughes, J. Fischer, M. Garrisi, G. Tomkin, J. Cohen, Maternal age, morphology, development and chromosome abnormalities in over 6000 cleavage-stage embryos, Reprod. Biomed. Online 14 (2007) 628–634.
[15] P. Kovacs, Embryo selection: the role of time-lapse monitoring, Reprod. Biol. Endocrinol. 12 (2014) 124–135.
[16] D.J. Kaser, C. Racowsky, Clinical outcomes following selection of human preimplantation embryos with time-lapse monitoring: a systematic review, Hum. Reprod. Update 20 (2014) 617–631.
[17] C. Racowsky, P. Kovacs, W.P. Martins, A critical appraisal of time-lapse imaging for embryo selection: where are we and where do we need to go? J. Assist. Reprod. Genet. 32 (2015) 1025–1030.
[18] R.G. Sturmey, D.R. Brison, H.J. Leese, Symposium: innovative techniques in human embryo viability assessment. Assessing embryo viability by measurement of amino acid turnover, Reprod. Biomed. Online 17 (2008) 486–496.
[19] D.K. Gardner, M. Meseguer, C. Rubio, N.R. Treff, Diagnosis of human preimplantation embryo viability, Hum. Reprod. Update 21 (2015) 727–747.
[20] M.G. Katz-Jaffe, S. McReynolds, D.K. Gardner, W.B. Schoolcraft, The role of proteomics in defining the human embryonic secretome, Mol. Hum. Reprod. 15 (2009) 271–277.
[21] E. Seli, C. Robert, M.A. Sirard, OMICS in assisted reproduction: possibilities and pitfalls, Mol. Hum. Reprod. 16 (2010) 513–530.
[22] G.A. Sigalos, O. Triantafyllidou, N.F. Vlahos, Novel embryo selection techniques to increase embryo implantation in IVF attempts, Arch. Gynecol. Obstet. 294 (2016) 1117–1124.
[23] M. Omidi, A. Faramarzi, A. Agharahimi, M.A. Khalili, Noninvasive imaging systems for gametes and embryo selection in IVF programs: a review, J. Microsc. (2017), https://doi.org/10.1111/jmi.12573 (Epub ahead of print).
[24] J. Holte, L. Berglund, K. Milton, C. Garello, G. Gennarelli, A. Revelli, T. Bergh, Construction of an evidence-based integrated morphology cleavage embryo score for implantation potential of embryos scored and transferred on day 2 after oocyte retrieval, Hum. Reprod. 22 (2007) 548–557.
[25] R. Machtinger, C. Racowsky, Morphological systems of human embryo assessment and clinical evidence, Reprod. Biomed. Online 26 (2013) 210–221.

[26] L.A. Scott, S. Smith, The successful use of pronuclear embryo transfers the day following oocyte retrieval, Hum. Reprod. 13 (1998) 1003–1013.
[27] J. Tesarik, E. Greco, The probability of abnormal preimplantation development can be predicted by a single static observation on pronuclear stage morphology, Hum. Reprod. 14 (1999) 1318–1323.
[28] C.C. Skiadas, C. Racowsky, K. Elder, J. Cohen (Eds.), Development rate, cumulative scoring and embryonic viability, in: Human Preimplantation Embryo Selection, Taylor and Francis, Colchester, UK, 2007, pp. 101–121.
[29] M. Alikani, J. Cohen, G. Tomkin, G.J. Garrisi, C. Mack, R.T. Scott, Human embryo fragmentation in vitro and its implications for pregnancy and implantation, Fertil. Steril. 71 (1999) 836–842.
[30] D.K. Gardner, W.B. Schoolcraft, Culture and transfer of human blastocysts, Curr. Opin. Obstet. Gynecol. 11 (1999) 307–311.
[31] D.K. Gardner, M. Lane, J. Stevens, T. Schlenker, W.B. Schoolcraft, Blastocyst score affects implantation and pregnancy outcome: towards a single blastocyst transfer, Fertil. Steril. 3 (2000) 1155–1158.
[32] C. Racowsky, M. Vernon, J. Mayer, G.G. Ball, B. Behr, K.O. Pomeroy, D. Wininger, W.E. Gibbons, J. Conaghan, J.E. Stern, Standardization of grading embryo morphology, Fertil. Steril. 94 (2010) 1152–1153.
[33] Alpha Scientists in Reproductive Medicine, ESHRE Special Interest Group of Embryology, The Istanbul consensus workshop on embryo assessment: proceedings of an expert meeting, Reprod. Biomed. Online 22 (2011) 632–646.
[34] G. Paternot, J. Devroe, S. Debrock, T.M. D'Hooghe, C. Spiessens, Intra- and inter-observer analysis in the morphological assessment of early-stage embryos, Reprod. Biol. Endocrinol. 7 (2009) 105–111.
[35] C.C. Wong, K.E. Loewke, N.L. Bossert, B. Behr, C.J. De Jonge, T.M. Baer, R.A. Reijo Pera, Non-invasive imaging of human embryos before embryonic genome activation predicts development to the blastocyst stage, Nat. Biotechnol. 28 (2010) 1115–1121.
[36] M. Meseguer, J. Herrero, A. Tejera, K.M. Hilligsøe, N.B. Ramsing, J. Remohí, The use of morphokinetics as a predictor of embryo implantation, Hum. Reprod. 26 (2011) 2658–2671.
[37] A. Payne, S.P. Flaherty, M.F. Barry, C.D. Matthews, Preliminary observations on polar body extrusion and pronuclear formation in human oocytes using time-lapse video cinematography, Hum. Reprod. 12 (1997) 532–541.
[38] Y. Mio, K. Maeda, Time-lapse cinematography of dynamic changes occurring during in vitro development of human embryos, Am. J. Obstet. Gynecol. 199 (2008) 660–665.
[39] J.G. Lemmen, I. Agerholm, S. Ziebe, Kinetic markers of human embryo quality using time-lapse recordings of IVF/ICSI-fertilized oocytes, Reprod. Biomed. Online 17 (2008) 385–391.
[40] M. Cruz, B. Gadea, N. Garrido, K.S. Pedersen, M. Martínez, I. Pérez-Cano, M. Muñoz, M. Meseguer, Embryo quality, blastocyst and ongoing pregnancy rates in oocyte donation patients whose embryos were monitored by time-lapse imaging, J. Assist. Reprod. Genet. 28 (2011) 569–573.
[41] I. Rubio, A. Galán, Z. Larreategui, F. Ayerdi, J. Bellver, J. Herrero, M. Meseguer, Clinical validation of embryo culture and selection by morphokinetic analysis: a randomized, controlled trial of the EmbryoScope, Fertil. Steril. 102 (2014) 1287–1294.
[42] Y. Motato, M.J. de los Santos, M.J. Escriba, B.A. Ruiz, J. Remohí, M. Meseguer, Morphokinetic analysis and embryonic prediction for blastocyst formation through an integrated time-lapse system, Fertil. Steril. 105 (2016) 376–384.
[43] S. Armstrong, N. Arroll, L.M. Cree, V. Jordan, C. Farquhar, Time-lapse systems for embryo incubation and assessment in assisted reproduction, Cochrane Database Syst. Rev. (2) (2015) CD011320. https://doi.org/10.1002/14651858.CD011320.pub2.
[44] B. Källén, O. Finnström, A. Lindam, E. Nilsson, K.G. Nygren, P.O. Olausson, Blastocyst versus cleavage stage transfer in in vitro fertilization: differences in neonatal outcome? Fertil. Steril. 94 (2010) 1680–1683.
[45] M. Meseguer, I. Rubio, M. Cruz, N. Basile, J. Marcos, A. Requena, Embryo incubation and selection in a time-lapse system improves pregnancy outcome compared with standard incubator: a retrospective cohort study, Fertil. Steril. 98 (2012) 1481–1491.
[46] C. Wong, A.A. Chen, B. Behr, S. Shen, Time-lapse microscopy and image analysis in basic and clinical embryo development research, Reprod. Biomed. Online 26 (2013) 120–129.
[47] M. Chen, S. Wei, J. Hu, J. Yuan, F. Liu, Does time-lapse imaging have favorable results for embryo incubation and selection compared with conventional methods in clinical in vitro fertilization? A meta-analysis and systematic review of randomized controlled trials, PLoS One 12 (2017) e0178720.

[48] L. Kroener, G. Ambartsumyan, C. Briton-Jones, D. Dumesic, M. Surrey, S. Munné, D. Hill, The effect of timing of embryonic progression on chromosomal abnormality, Fertil. Steril. 98 (2012) 876–880.

[49] S.L. Chavez, K.E. Loewke, J. Han, F. Moussavi, P. Colls, S. Munne, B. Behr, R.A. Reijo Pera, Dynamic blastomere behaviour reflects human embryo ploidy by the four-cell stage, Nat. Commun. 3 (2012) 1251–1263.

[50] A. Campbell, S. Fishel, N. Bowman, S. Duffy, M. Sedler, C.F. Hickman, Modelling a risk classification of aneuploidy in human embryos using non-invasive morphokinetics, Reprod. Biomed. Online 26 (2013) 477–485.

[51] Y.G. Kramer, J.D. Kofinas, K. Melzer, N. Noyes, C. McCaffrey, J. Buldo-Licciardi, D.H. McCulloh, J.A. Grifo, Assessing morphokinetic parameters via TLC to predict euploidy: are aneuploidy risk classification models universal? J. Assist. Reprod. Genet. 31 (2014) 1231–1242.

[52] K.E. Melzer, C. McCaffrey, P. Colls, S. Munne, J. Grifo, Developmental morphology and continuous time-lapse microscopy (TLM) of human embryos: can we predict aneuploidy? Fertil. Steril. 98 (2012) S136.

[53] J. Stevens, M. Rawlins, A. Janesch, N. Treff, W.B. Schoolcraft, M.G. Katz-Jaffe, Time-lapse observation of embryo development identifies later stage morphology based parameters associated with blastocyst quality but no chromosomal constitution, Fertil. Steril. 98 (2012) S30.

[54] M.G. Katz-Jaffe, W.B. Schoolcraft, D.K. Gardner, Analysis of protein expression (secretome) by human and mouse preimplantation embryos, Fertil. Steril. 86 (2006) 678–685.

[55] C.M. Combelles, C. Racowsky, Protein expression profiles of early embryos-an important step in the right direction: just not quite ready for prime time, Fertil. Steril. 86 (2006) 493.

[56] S. McReynolds, L. Vanderlinden, J. Stevens, K. Hansen, W.B. Schoolcraft, M.G. Katz-Jaffe, Lipocalin-1: a potential marker for noninvasive aneuploidy screening, Fertil. Steril. 95 (2011) 2631–2633.

[57] A. Dominguez, B. Gadea, F.J. Esteban, J.A. Horcajadas, A. Pellicer, C. Simon, Comparative protein-profile analysis of implanted versus non-implanted human blastocysts, Hum. Reprod. 23 (2008) 1993–2000.

[58] N.J. Hannan, P. Paiva, K.L. Meehan, L.J. Rombauts, D.K. Gardner, L.A. Salamonsen, Analysis of fertility-related soluble mediators in human uterine fluid identifies VEGF as a key regulator of embryo implantation, Endocrinology 152 (2011) 4948–4956.

[59] Y. Ménézo, K. Elder, S. Viville, Soluble HLA-G release by the human embryo: an interesting artefact? Reprod. Biomed. Online 13 (2006) 763–764.

[60] J. Nyalwidhe, T. Burch, S. Bocca, L. Cazares, S. Green-Mitchell, M. Cooke, P. Birdsall, G. Basu, O.J. Semmes, S. Oehninger, The search for biomarkers of human embryo developmental potential in IVF: a comprehensive proteomic approach, Mol. Hum. Reprod. 19 (2013) 250–263.

[61] L. Yu, T. Burch, S. Bocca, J. Nyalwide, E. Jones, J.A. Horcajadas, S. Oehninger, Characterization of the secretome of 2-cell mouse embryos cultured in vitro to the blastocyst stage with and without protein supplementation, J. Assist. Reprod. Genet. 31 (2014) 757–765.

[62] I. Sargent, A. Swales, N. Ledee, N. Kozma, J. Tabiasco, P. Le Bouteiller, sHLA-G production by human IVF embryos: can it be measured reliably? J. Reprod. Immunol. 75 (2007) 128–132.

[63] M. Legge, Oocyte and zygote zona pellucida permeability to macromolecules, J. Exp. Zool. 271 (1995) 145–150.

[64] A. Seli, D. Sakkas, R. Scott, S.C. Kwok, S.M. Rosendahl, D.H. Burns, Non-invasive metabolomic profiling of embryo culture media using Raman and near-infrared spectroscopy correlates with reproductive potential of embryos in women undergoing in vitro fertilization, Fertil. Steril. 88 (2007) 1350–1357.

[65] E. Seli, L. Botros, D. Sakkas, D.H. Burns, Noninvasive metabolomic profiling of embryo culture media using proton nuclear magnetic resonance correlates with reproductive potential of embryos in women undergoing in vitro fertilization, Fertil. Steril. 90 (2008) 2183–2189.

[66] R. Scott, E. Seli, K. Miller, D. Sakkas, K. Scott, D.H. Burns, Noninvasive metabolomic profiling of human embryo culture media using Raman spectroscopy predicts embryonic reproductive potential: a prospective blinded pilot study, Fertil. Steril. 90 (2008) 77–83.

[67] T. Hardarson, A. Ahlström, L. Rogberg, L. Botros, T. Hillensjö, G. Westlander, D. Sakkas, M. Wikland, Non-invasive metabolomic profiling of day 2 and 5 embryo culture medium: a prospective randomized trial, Hum. Reprod. 27 (2012) 89–96.

[68] I. Sánchez-Ribas, M. Riqueros, P. Vime, L. Puchades-Carrasco, T. Jönsson, A. Pineda-Lucena, A. Ballesteros, F. Domínguez, C. Simón, Differential metabolic profiling of non-pure trisomy 21 human preimplantation embryos, Fertil. Steril. 98 (2012) 1157–1164.

[69] S.S. Cortezzi, E.C.C. SS, M.G. Trevisan, C.R. Ferreira, A.S. Setti, D.P. Braga, C. Figueira Rde, A. Iaconelli Jr., M.N. Eberlin, E. Borges Jr., Prediction of embryo implantation potential by mass spectrometry fingerprinting of the culture medium, Reproduction 145 (2013) 453–462.

[70] C.G. Vergouw, N.W. Heymans, T. Hardarson, I.A. Sfontouris, K.A. Economou, A. Ahlström, L. Rogberg, T.G. Lainas, D. Sakkas, D.C. Kieslinger, E.H. Kostelijk, P.G. Hompes, R. Schats, C.B. Lambalk, No evidence that embryo selection by near-infrared spectroscopy in addition to morphology is able to improve live birth rates: results from an individual patient data meta-analysis, Hum. Reprod. 29 (2014) 455–461.

[71] H.J. Leese, J. Conaghan, K.L. Martin, K. Hardy, Early human embryo metabolism, Bioessays 15 (1993) 259–264.

[72] M. Lane, D.K. Gardner, Differential regulation of mouse embryo development and viability by amino acids, J. Reprod. Fertil. 109 (1997) 153–164.

[73] H.J. Leese, Quiet please, do not disturb: a hypothesis of embryo metabolism and viability, Bioessays 24 (2002) 845–849.

[74] D.K. Gardner, M. Lane, W.B. Schoolcraft, Physiology and culture of the human blastocyst, J. Reprod. Immunol. 55 (2002) 85–100.

[75] C.G. Baumann, D.G. Morris, J.M. Sreenan, H.J. Leese, The quiet embryo hypothesis: molecular characteristics favoring viability, Mol. Reprod. Dev. 74 (2007) 1345–1353.

[76] A.G. Marshall, Milestones in Fourier transform ion cyclotron resonance mass spectrometry technique development, Int. J. Mass. Spectrom. 200 (2000) 331–356.

[77] N.K. Kaiser, J.P. Quinn, G.T. Blakney, C.L. Hendrickson, A.G. Marshall, A novel 9.4 tesla FTICR mass spectrometer with improved sensitivity, mass resolution, and mass range, J. Am. Soc. Mass. Spectrom. 22 (2011) 1343–1351.

[78] H.A. Abdulla, R.L. Sleighter, P.G. Hatcher, Two dimensional correlation analysis of fourier transform ion cyclotron resonance mass spectra of dissolved organic matter: a new graphical analysis of trends, Anal. Chem. 85 (2013) 3895–3902.

[79] H.A. Abdulla, E.C. Minor, P.G. Hatcher, Using two-dimensional correlations of ^{13}C NMR and FTIR to investigate changes in the chemical composition of dissolved organic matter along an estuarine transect, Environ. Sci. Technol. 44 (2010) 8044–8049.

[80] C.S. Siristatidis, E. Sertedaki, D. Vaidakis, Metabolomics for improving pregnancy outcomes in women undergoing assisted reproductive technologies, Cochrane Database Syst. Rev. 5 (2017) CD011872.

[81] M.I. Shamonki, H. Jin, Z. Haimowitz, L. Liu, Proof of concept: preimplantation genetic screening without embryo biopsy through analysis of cell-free DNA in spent embryo culture media, Fertil. Steril. 106 (2016) 1312–1318.

[82] S. Mastenbroek, S. Repping, Preimplantation genetic screening: back to the future, Hum. Reprod. 29 (2014) 1846–1850.

[83] A. Capalbo, F.M. Ubaldi, D. Cimadomo, R. Maggiulli, C. Patassini, L. Dusi, F. Sanges, L. Buffo, R. Venturella, L. Rienzi, Consistent and reproducible outcomes of blastocyst biopsy and aneuploidy screening across different biopsy practitioners: a multicentre study involving 2586 embryo biopsies, Hum. Reprod. 31 (2016) 199–208.

[84] P. Donoso, C. Staessen, B.C. Fauser, P. Devroey, Current value of preimplantation genetic aneuploidy screening in IVF, Hum. Reprod. Update 13 (2007) 15–25.

[85] D. Wells, K. Kaur, J. Grifo, M. Glassner, J.C. Taylor, E. Fragouli, S. Munne, Clinical utilisation of a rapid low-pass whole genome sequencing technique for the diagnosis of aneuploidy in human embryos prior to implantation, J. Med. Genet. 51 (2014) 553–562.

[86] D. Goodrich, X. Tao, C. Bohrer, A. Lonczak, T. Xing, R. Zimmerman, Y. Zhan, R.T. Scott Jr., N.R. Treff, A randomized and blinded comparison of qPCR and NGS-based detection of aneuploidy in a cell line mixture model of blastocyst biopsy mosaicism, J. Assist. Reprod. Genet. 33 (2016) 1473–1480.

[87] G.L. Harton, S. Munné, M. Surrey, J. Grifo, B. Kaplan, D.H. McCulloh, D.K. Griffin, D. Wells, PGD Practitioners Group, Diminished effect of maternal age on implantation after preimplantation genetic diagnosis with array comparative genomic hybridization, Fertil. Steril. 100 (2013) 1695–1703.

[88] J. Chang, S.L. Boulet, G. Jeng, L. Flowers, D.M. Kissin, Outcomes of in vitro fertilization with preimplantation genetic diagnosis: an analysis of the United States Assisted Reproductive Technology Surveillance Data, 2011–2012, Fertil. Steril. 105 (2016) 394–400.

[89] C. Rubio, J. Bellver, L. Rodrigo, G. Castillón, A. Guillén, C. Vidal, J. Giles, M. Ferrando, S. Cabanillas, J. Remohí, A. Pellicer, C. Simón, In vitro fertilization with preimplantation genetic diagnosis for aneuploidies in advanced maternal age: a randomized, controlled study, Fertil. Steril. 107 (2017) 1122–1129.

[90] S. Munné, J. Cohen, Advanced maternal age patients benefit from preimplantation genetic diagnosis of aneuploidy, Fertil, Steril. 107 (2017) 1145–1146.

[91] D. Marin, R.T. Scott Jr., N.R. Treff, Preimplantation embryonic mosaicism: origin, consequences and the reliability of comprehensive chromosome screening, Curr. Opin. Obstet. Gynecol. 29 (2017) 168–174.

[92] E. Fragouli, K. Spath, S. Alfarawati, F. Kaper, A. Craig, C.E. Michel, F. Kokocinski, J. Cohen, S. Munne, D. Wells, Altered levels of mitochondrial DNA are associated with female age, aneuploidy, and provide an independent measure of embryonic implantation potential, PLoS Genet. 11 (2015) e1005241.

[93] K. Ravichandran, C. McCaffrey, J. Grifo, A. Morales, M. Perloe, S. Munne, D. Wells, E. Fragouli, Mitochondrial DNA quantification as a tool for embryo viability assessment: retrospective analysis of data from single euploid blastocyst transfers, Hum. Reprod. 32 (2017) 1282–1292.

[94] A. Dominguez, M. Meseguer, B. Aparicio-Ruiz, P. Piqueras, A. Quiñonero, C. Simón, New strategy for diagnosing embryo implantation potential by combining proteomics and time-lapse technologies, Fertil. Steril. 104 (2015) 908–914.

[95] A. Tejera, D. Castelló, J.M. de Los Santos, A. Pellicer, J. Remohí, M. Meseguer, Combination of metabolism measurement and a time-lapse system provides an embryo selection method based on oxygen uptake and chronology of cytokinesis timing, Fertil. Steril. 106 (2016) 119–126.

[96] L. Gianaroli, M.C. Magli, A. Pomante, A.M. Crivello, G. Cafueri, M. Valerio, A.P. Ferraretti, Blastocentesis: a source of DNA for preimplantation genetic testing. Results from a pilot study, Fertil. Steril. 102 (2014) 1692–1699.

[97] A. D'Alessandro, G. Federica, S. Palini, C. Bulletti, L. Zolla, A mass spectrometry-based targeted metabolomics strategy of human blastocoele fluid: a promising tool in fertility research, Mol. Biosyst. 8 (2012) 953–958.

[98] S. Palini, L. Galluzzi, S. de Stefani, M. Bianchi, D. Wells, D. Magnani, C. Bulletti, Genomic DNA in human blastocoele fluid, Reprod. Biomed. Online 26 (2013) 603–610.

[99] M.L. LaBonte, An analysis of US fertility centre educational materials suggests that informed consent for preimplantation genetic diagnosis may be inadequate, J. Med. Ethics. 38 (2012) 479–484.

[100] E.R. Hammond, A.N. Shelling, L.M. Cree, Nuclear and mitochondrial DNA in blastocoele fluid and embryo culture medium: evidence and potential clinical use, Hum. Reprod. 31 (2016) 1653–1661.

[101] A. Capalbo, F.M. Ubaldi, D. Cimadomo, L. Noli, Y. Khalaf, A. Farcomeni, D. Ilic, L. Rienzi, MicroRNAs in spent blastocyst culture medium are derived from trophectoderm cells and can be explored for human embryo reproductive competence assessment, Fertil. Steril. 105 (2016) 225–235.

CHAPTER

13

Genetic Selection of the Human Embryos: From FISH to NGS, Past and Future

Sarthak Sawarkar, *Santiago Munne*[†]

[*]University of Kent, Canterbury, United Kingdom [†]Yale University, New Haven, CT, United States

As a technology, in vitro fertilization or IVF has completely transformed the field of human infertility. Since its introduction in 1978, IVF has captured widespread public attention. Assisted reproductive technologies (ART) are available throughout the civilized world with practices differing greatly compared with the early days. Scientific developments and refinements in clinical and laboratory technologies have led to IVF evolving into an efficient, safe, and accessible medical technique. By definition, IVF leads to the generation of embryos in vitro. Studies have shown that greater than 50% cleavage and blastocyst-stage embryos produced in vitro are found to be chromosomally abnormal or aneuploid, with the numbers reaching as high as 80% in women over 42 years of age [1]. While some aneuploid embryos arrest during culture, most do not. Nowadays, most embryo testing in the United States is performed on blastocyst-stage embryos. Within these tested embryos, about 50% are abnormal with a mean maternal age of about 38 years. Low success of artificial reproductive treatments is observed in embryos that are detected with numerical chromosomal abnormalities since they are usually not compatible with either implantation or birth. Aneuploidy has clearly shown to have detrimental effects on efficacy of ART. High numbers up to 70% of chromosomal abnormalities are often observed in cases that result in spontaneous abortions. Selection of normal (euploid) embryos to transfer during IVF to improve the rates of success for the procedure is necessary. This process of selection against aneuploidy is known as preimplantation genetic diagnosis (PGD) or preimplantation genetic screening (PGS).

Reproductomics
https://doi.org/10.1016/B978-0-12-812571-7.00014-9

PGS VERSION 1 AND LIMITATIONS

Fluorescent in situ hybridization (FISH) to analyze day 3 blastomere biopsy or polar bodies biopsied from oocytes and/or zygotes was one of the earliest strategies utilized for PGS [2]. FISH allowed the study and examination of anywhere between 5 and 12 chromosomes per oocyte or zygote. However, a complete evaluation of the chromosome complement of the 23 pairs was not possible using FISH. Despite the technical limitation of limited coverage, FISH was able to detect more than 80% of chromosomally abnormal embryos compared with the abnormalities detected by array technology, a later version of PGS. Largely conflicting results for the first version of PGS using FISH were published. Some studies showed a definite improvement in implantation rates, reduction in spontaneous abortions, and take-home-baby rates. However, some studies presented an opposing picture of no improvements in rates or even a detrimental effect of PGD for aneuploidy. The studies that demonstrated a positive effect were not performed in a randomized fashion. In contrast, the studies showing a detrimental effect were performed in a randomized fashion; however, they had several biological fallacies. It was primarily argued that the cause for the detrimental effect observed was that the cleavage-stage embryos used for analysis in these studies have shown to have high rates of chromosomal mosaicism. It was concluded that since the analysis is based upon a single cell, mosaicism would obfuscate the results. Although mosaicism is common in cleavage-stage embryos (according to FISH analyses), most chromosomal abnormalities observed in these embryos occur in every cell within them. In this case, however, the biopsied cell is chromosomally different compared with the remaining cells in the embryo. These remainder cells may contain errors affecting different chromosomes compared with the biopsied cell, but the clinical diagnosis of "abnormal" would still be valid.

The most likely cause explaining the intercenter differences in PGD outcomes is variations in the biopsy and genetic techniques employed. Embryo biopsy is probably the most pertinent variable in PGD. One of the studies that presented the detrimental effects of PGD used a two-cell biopsy from each cleavage-staged embryo for analysis. However, researchers from the same group in a later publication reported a detrimental effect of a two-celled biopsy compared with the one-celled biopsy from cleavage-stage embryos. Biopsying even a single cell in suboptimal laboratory conditions has been shown to be extremely damaging to the embryo potential. Mastenbroek et al. reported a very high rate of diagnostic failure (20%).This resulted in several embryos included in the study being transferred without a diagnosis. The implantation rate of these nondiagnosed embryos was observed to be 59% lower than the control. One should note that in this case, the only difference between the test and the control group appears to be the biopsy. This in turn suggests that embryo viability was drastically reduced by the possibility of incorrect biopsy procedures used in the clinics involved in the study.

The second factor in determining "good results" with FISH is low error rate. Low error rate is the rate attained by reanalyzing cells from nonreplaced embryos (arrested and abnormal embryos) and determining the correctness of the original diagnosis by correlating the results of the two experiments. Due to varying practices in PGD laboratories in the steps after the biopsy of the embryo, error rates have been shown to vary widely between PGD laboratories, from 2%–7% [3–5] to 40%–50% [6,7]. Higher error rates have been indicative of decreased implantation rates [8].

When FISH is performed using appropriate, well-validated methods, it has been shown to detect almost 90% of chromosome abnormalities detected by aCGH [9,10]; some PGD laboratories appear to have consistent results with cleavage-staged embryos and FISH. However, PGD/PGS as a field has definitely moved away from the biopsying embryos at the cleavage stage. Increasingly, embryo trophectoderm is being biopsied at the blastocyst stage of the embryo at day 5 or day 6. Alternatively, biopsies from polar bodies of oocytes or zygotes are also being used by some laboratories. These advanced stages of embryonic development are more resilient to technical manipulation. The first version of PGS (FISH) has been replaced by more comprehensive methods of DNA analysis. These methods detect almost 100% chromosomal abnormalities. Version 2 of PGS is extremely redundant and involves testing of every chromosome multiple times at different sites within each chromosome. The techniques in this version are less subjective, highly automated, and less prone to errors.

PGS VERSION 2

In this section, we shall take a look at four techniques that are a part of the second and a more updated version of PGS used to detect chromosome abnormalities. Comparative genome hybridization (CGH) was a revolutionary, molecular cytogenetic technique. CGH allowed for the performance of a global assay to determine chromosomal losses and gains in embryos [11]. CGH was initially used in day 3 cleavage-stage embryo biopsies. However, CGH as a technique has several drawbacks. CGH is extremely time-consuming and labor-intensive. This makes fresh transfers, that is, day 3 embryo biopsy and day 5 embryo transfers logistically infeasible. During the early days of use of CGH for PGD/PGS, embryo freezing was not very efficient, and the survival rates of embryos post thawing were low. Hence, CGH was temporarily abandoned and not reused till vitrification was a standard procedure. Since then, CGH has been clinically applied to both polar body biopsies and blastocyst biopsies. Vitrification along with CGH significantly improved the implantation rates from 46.5% to 72.2% in cycles with screening, with almost 100% of blastocysts surviving the embryo biopsy [12]. Recently, blastocyst culture, vitrification, and embryo biopsy have been widely used and in fact have come to be an essential aspect of ART as reviewed by Rienzi et al. in 2016 [13]. Frozen cycle transfers accounted for almost 30% of transfers in North America, while over 50% cycles in Nordic countries are frozen transfers and are routinely performed for cases undergoing PGS. However, please note that freezing the embryos adds extra cost to the cycle and some patients prefer to have a fresh cycle instead.

Two other techniques from the second version that are used for comprehensive chromosome analysis are microarray CGH (array CGH/aCGH) and single-nucleotide polymorphism (SNP) microarrays. One of the biggest advantages of these techniques is the short turnaround time required.

Various biopsy techniques have been applied in various IVF laboratories to enable PGS labs to run their diagnostic tests. In the below section, we shall take a look into what these techniques are and what are the advantages and pitfalls these techniques bring toward the final diagnosis.

BIOPSY TECHNIQUES

Currently, almost every embryo screened with PGS is from ART. During ART, patients undergo controlled ovarian stimulation. All oocytes obtained during ART are stripped from their surrounding cumulus cells using hyaluronidase enzyme. This stripping is performed because cumulus cells can potentially provide a source of nonembryonic DNA, which could cause "contamination" during PGD [14,15]. The goal of PGS is to perform genetic analysis on embryonic DNA. Intracytoplasmic sperm injection (ICSI) is used most routinely to fertilize the oocytes in order to reduce the rate of fertilization failures and also to prevent any contamination caused by DNA from residual sperm attached to the zona pellucida [15,16]. Postfertilization of the embryos can be biopsied at different stages of its development, and one or more cells can be retrieved for extracting DNA that will be used during PGS. Additionally, polar body biopsy can also be performed to obtain the first and second polar bodies extruded from the oocyte.

Polar Body Biopsy

Genetic material for PGD can be obtained with prefertilization by the removal of the first polar body from the mature (metaphase II) oocyte. The first polar body can be used to deduce the genotype of the oocyte prior to fertilization, while upon fertilization, the second polar body can also be obtained and tested [17]. Polar body PGD was first applied by Verlinsky and colleagues [18], and the accuracy of their method has been shown in a large number of PGD cases carried out for several genetic disorders [19]. It should be noted that PGD for single-gene disorders requires that both the first (prefertilization) and second (postfertilization) polar bodies are tested, since meiotic recombination may lead to unclear (heterozygous) results in the first polar body. Polar body biopsy is also more commonly used for aneuploidy screening of the oocytes. The ultimate benefit of this method as claimed by its proponents is that it does not involve biopsy of the embryo and therefore avoids any potential harm (even if minimal). Furthermore, in some countries (e.g., Germany), where testing of embryos is forbidden or highly restricted, it is the only choice for obtaining genetic material for PGD. However, polar body biopsy only provides information regarding the oocyte genotype, and therefore, it cannot be used for screening paternally derived genetic disorders [20]. Additionally, any chromosomal aneuploidies inherited from the sperm and any aneuploidies that arise because of mitotic errors after the zygote formation will not be detected [21].

Cleavage Stage Biopsy

One of the methods of obtaining genetic material for PGS is by performing biopsy on cleavage-stage embryos on day 3 postfertilization (Fig. 13.1). Cleavage-stage biopsy was the most popular form of biopsy used in PGS. However, recently, it has been replaced by day 5 blastocyst biopsy that we will discuss later in the chapter.

Embryos have been observed to grow in vitro until they reach day 7. In the cleavage-stage biopsy technique, the embryo biopsy is performed at day 3, by breaching the zona pellucida [22]. The embryo usually contains 6–12 cells at the cleavage stage (day 3), which are totipotent

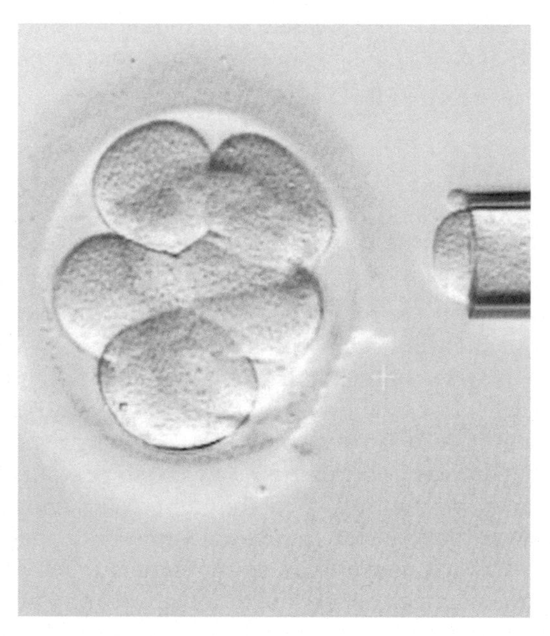

FIG. 13.1 Biopsy of a day 3 postfertilization cleavage-stage embryo to remove a single blastomere.

[17]. One or two blastomeres are extracted and used for PGD. The removal of two cells has been linked to decreased implantation potential and reduced rate of blastocyst formation that in turn led to fewer two-cell blastomere retrievals. Suitable embryos that are successful in attaining the blastocyst stage are selected and transferred to the uterus of the woman on day 4 or day 5 postfertilization. Using blastomeres from day 3 embryos offers the potential of checking the embryo for genetic disorders inherited from both parents and also leaves room to test for any meiotic or mitotic chromosomal imbalances. Hence, compared with polar body biopsies, cleavage-stage biopsies offer wider scope of information. Mosaicism is one of the typical problems that occur during the screening of cleavage-stage embryos for chromosomal anomalies. Mosaicism refers to the presence of two or more distinct cell lines with different chromosomal constitutions in the same embryo that occur due to errors that happen during mitosis postzygote formation—especially during the first three mitotic divisions [23–26]. The cause of these errors is probably the absence of cell-cycle checkpoints during early cleavage divisions. The incidence of mosaicism upon the analysis of cleavage-stage embryos utilizing CCS methods has reported to be ~66% [27,28]. Therefore, a single blastomere biopsied from a day 3 embryo might not always be representative of the rest of the embryo. This leads to an overall drop in the sensitivity of the test as more false negative and false positive results are observed. This could possibly lead to the exclusion from transfer of euploid embryos or even worse, transfer of an abnormal aneuploid embryo [29]. Postimplantation embryos have a lower degree of mosaicism as compared with preimplantation cleavage-stage embryos. Spontaneous miscarriage specimens exhibit mosaicism at a degree of <10% [30]. While for first trimester, the incidence of mosaicism in ongoing pregnancies is observed to be even lower (1%–2%) [31]. It therefore seems that mosaicism disappears prior to the period of first trimester. Mechanisms that

possibly play a role in the disappearance are the loss of mosaic embryos or a shift of the mosaic embryos toward normality (i.e., selection against abnormal cells in the embryo).

Blastocyst Biopsy

The most prevalent option nowadays for obtaining genetic material from the embryo is by blastocyst (day 5) biopsy. At the blastocyst stage, the inner cell mass that consists of the only pluripotent cell line inside the embryo that will eventually develop into definitive structures of the fetus has differentiated from the trophectoderm. The trophectoderm later develops into extraembryonic tissues [17]. Approximately 3–10 cells are removed during biopsy at the blastocyst stage. Biopsy of the trophectoderm is advantageous since no cells are extracted from the inner cell mass. Also in contrast to the other two biopsy strategies, trophectoderm biopsy obtains multiple cells for carrying out PGD/PGS. This leads to an overall improved accuracy of the test. Furthermore, blastocysts are robust compared with the earlier embryonic stages and tolerate insult of the biopsy better than cleavage-stage embryos. Two disadvantages of this method are that firstly, only about 50% of the embryos growing in vitro reach the blastocyst stage [32]. This means that the total number of embryos available for transfer reduces by half as compared with embryos that attain cleavage-stage (day 3) development. While this might not be a concern for younger patients with a high ovarian oocyte reserve, it is definitely a concern in patients of advanced maternal age. Secondly, the time left for diagnosis after blastocyst biopsy that is performed on day 5 or 6 is limited since the embryos need to be transferred by day 6/7 [14]. However, a solution has been seen for the second drawback of immediate transfer through the introduction of vitrification as a method to cryopreserve blastocysts [33]. Cryopreservation is the use of extremely low temperatures to preserve structurally intact living cells and tissues. Cryopreservation of blastocysts compounded with high survival rates obtained after biopsy [12] makes blastocyst biopsy increasingly attractive for PGS. Theoretically, cryopreservation of biopsied blastocysts grants for an unlimited amount of time for PGS to take place. Furthermore, vitrification of blastocysts and their transfer later on during a natural ovulatory cycle rather during a hormonally stimulated cycle has been found to result in a better clinical outcome [34]. Mosaicism has been found to exist in blastocysts too [23,30]; the incidence of mosaicism however is observed to be considerably lower (24%–32.4%) [23] as compared with cleavage-stage embryos. Through data collected from various studies, it appears that a proportion of mosaic embryos undergo developmental arrest even before they reach the blastocyst stage [30,35]. This explains the lower rates of mosaicism at the blastocyst stage when compared with day 3 embryos. It is important to note that as a corollary, it also suggests that culture to the blastocyst stage might be beneficial since it would identify embryos of increased developmental competence. Furthermore, the probability of misdiagnosis is believed to be considerably lower since only a small number of mosaic blastocysts are found to have a significant number of euploid cells. Compared with day 3 biopsy where 1–2 cells are retrieved, 5–10 cells are obtained in a day 5 biopsy for analysis, and thus, diagnosis is not dependent on a single cell. In fact, it was determined by different studies that array and NGS platforms commonly used in clinical practice for performance of CCS are able to detect a mosaicism rate of as low as 25% (array) and 10%–20% (NGS) in a trophectoderm biopsy [36–38], preventing a large degree of the transfer of mosaic embryos that would possibly have a negative effect on IVF outcome.

TECHNIQUES ROUTINELY USED IN PGS LABORATORIES

As mentioned above, several version 2 technologies are presently being used in the PGS laboratories throughout the world for comprehensive chromosome screening. In this part of the chapter, we will cover array CGH, SNP arrays, and next-generation sequencing (NGS) techniques that are routinely being used for testing embryos. The benefits and drawbacks for every technology will be discussed along with the technique itself. Microarray CGH (array CGH or aCGH), single-nucleotide polymorphism (SNP) microarrays, and NGS can be used for comprehensive chromosome analysis of single cells from day 3 biopsy and day 5 embryos.

aCGH

Array CGH (aCGH) is widely used for the cytogenetic analysis in both prenatal and post-natal samples [39–42] since it is cost-effective, rapid, and permits chromosomal regions to be screened at high resolution. CGH and aCGH provide a quantitative analysis based on comparing the relative amount of DNA from two different sources, one from the clinical sample (e.g., a cell from an embryo) and another from a chromosomally normal individual. DNA samples from the two sources are differentially labeled and hybridized to either metaphase chromosomes (CGH) or probes on a microarray (aCGH). In the case of aCGH, each probe reveals the relative amounts of these two DNAs at a single chromosomal site. Since multiple copies of each probe are placed on the microarray and each chromosome is tested at several distinct loci, the diagnosis is very accurate. A variety of aCGH platforms are available for the purpose of aneuploidy screening. The one most commonly used for the purpose of PGD utilizes bacterial artificial chromosome (BAC) probes, about 150,000bp in length, covering all chromosome bands and giving a 4MB or lower resolution. Even higher resolutions are achievable however not recommended since at that level, the difference between clinically significant deletions/duplications and normally occurring copy number variations becomes less clear. The microarray validated for PGD has 4000 probes and thus covers ~25% of the genome sequence [43]. Microarray CGH has a similar accuracy rate to conventional CGH. Microarray CGH is capable of producing similar results to those obtained in the promising CGH study performed by Schoolcraft et al. [12].

Chromosome imbalances (aneuploidies, deletions, duplications, and unbalanced translocations) are easily detected using CGH and aCGH. A major limitation of these approaches however is that diploidy cannot be distinguished from changes involving loss or gain of an entire set of chromosomes (e.g., haploidy, triploidy, and tetraploidy). A fair question to ask at this point is that how important is this? In a paper by Munne et al. (personal communication), about 7.7% ($n=91{,}073$) of the supposedly 2PN embryos tested were haploid or polyploid, but the majority of them had additional abnormalities detectable by aCGH or CGH, and only 1.8% of all embryos were homogeneously polyploid or haploid. Furthermore, of those embryos, the majority were arrested embryos, by day 4, leaving only 0.2% of the developing embryos uniformly polyploid or haploid. This suggests that failure to detect polyploid embryos definitely leaves a rare possibility for a misdiagnosis but is not likely to have a significant impact on the clinical efficiency of the screening using aCGH or CGH.

SNP Arrays

Single-nucleotide polymorphisms are areas of the genome where a single nucleotide in the DNA sequence varies within the population. Most SNPs are biallelic, existing in one of two forms, and are found scattered throughout the genome. A haplotype (a contiguous series of polymorphisms on the same chromosome) can be assembled by determining the genotype of multiple SNPs along the length of each chromosome. This ultimately allows the tracking of inheritance of individual chromosomes or pieces of chromosomes from parents to embryos. Current SNP microarrays with the aid of powerful computers and software simultaneously assay hundreds of thousands of SNPs, distinguishing how many copies of each chromosome was inherited by an embryo [43–45].

Chromosome screening methods (CGH, aCGH, SNP microarrays, and NGS) rely on whole genome amplification (WGA) to amplify DNA from the single cell or small number of cells removed from a developing embryo [46]. CGH can be performed in combination with a variety of WGA methods. SNP microarrays are especially sensitive to the type of amplification technique used and are not compatible with all WGA methods. WGA methods like multiple displacement amplification (MDA), PicoPlex, and GenomePlex are most commonly used for SNP microarrays. These amplification methods allow for better overall coverage of the genome compared with earlier WGA methods (e.g., degenerate oligonucleotide primed PCR). These WGA methods are less likely to preferentially amplify some parts of the genome while leaving others under amplified or altogether unamplified. While the technologies are quite different from one another, they all (arrays (CGH-based and SNP-based) and NGS) are trying to answer the same underlying question: how many copies of each chromosome are present in an embryo sample? The small size of the SNP array probes can lead to poor hybridization efficiencies and low signal intensities for individual probes. This factor when coupled with the failure of WGA methods to uniformly amplify the complete genome may lead to many probes yielding a "no result" (i.e., a low "call rate"). Additionally, preferential amplification (PA) and/or allele dropout of one SNP allele versus another can lead to a great deal of "noise" in the system, which requires sophisticated interpretation. To tackle this "noise," several methods for the cleaning up of data from SNP microarrays have been developed: (1) quantitative approaches, assessing only the intensity of SNP calls; (2) qualitative methods, looking only at the inheritance of specific SNPs and requiring comparison with parental DNA samples; and (3) techniques combining both the quantitative and qualitative methods, using both inheritance patterns and SNP intensity calls. A quantitative approach compares the intensity of each SNP against the other SNPs. A purely quantitative approach for aneuploidy screening may not require parental testing ahead of the cycle; however, this approach would not be compatible with combination testing of single-gene defects with aneuploidy screening (discussed below). This approach is currently in a very primitive stage of development. It is necessary to assess parental DNA prior to clinical embryo testing for qualitative approaches. The key requirement is the deduction of the four parental haplotypes for each chromosome. Embryo testing is then focused on detecting the individual parental haplotypes, revealing how many chromosomes were inherited from each parent, that is, karyomapping [43]. This approach has the disadvantage that mitotic abnormalities, in which only two haplotypes are present in a trisomy (i.e., caused by duplication of one of the two chromosomes in the embryo after fertilization), will not be detected. This can result in misdiagnosis of a substantial amount of embryos, since 30% of aneuploid embryos may contain mitotic abnormalities (mosaics). A qualitative/quantitative approach has also been applied clinically and probably can obviate the issues mentioned above for purely qualitative or quantitative approaches [44,45]. All of the analysis approaches still share one major limitation. It pertains to the diagnosis of tetraploidies. In a tetraploid cell, only two haplotypes are present (i.e., a postmeiotic duplication of a euploid cell); therefore, all SNPs will have the same intensity. SNP-based microarrays do offer a few benefits over aCGH: (a) if a qualitative analysis is used, SNP-based microarrays have the capability to detect the paternal origin of any chromosomal abnormalities. This may be valuable in certain rare instances of young couples producing many chromosome abnormalities. However, it is of little relevance to cases of advanced maternal age (AMA) where at least 90% of the aneuploidies will be maternal in origin and those of paternal origin are most likely mitotic errors where the paternal chromosome was randomly recruited as the extra chromosome. These errors offer no predictive value for other embryos within the cohort or even for future cycles. (b) SNP microarrays applied to PGD for chromosome rearrangements can differentiate between balanced (carrier) and normal embryos. However, because the rate of abnormalities in translocation cases is generally very high (>80%) [47], majority of PGD cycles do not have a surplus of embryos with a balanced chromosome constitution. In most cases, whatever balanced embryos are available are needed for transfer. (c) SNP arrays can directly produce a fingerprint of the embryo. This permits for assessment of which of the transferred embryos led to a pregnancy. However, if a laboratory is using aCGH, a similar test can be performed by utilizing a small aliquot of the DNA produced by WGA to perform conventional DNA fingerprinting. (d) Lastly, qualitative SNP arrays can also detect uniparental disomy (UDP); however, this is also a very rare event (e.g., UDP 15 occurs in 0.001% of newborns (OMIM)).

A major disadvantage of a qualitative or combination approach to SNP array analysis is the need to assess parental DNA before the PGD cycle. This complicates patient management that is already tricky. It also adds substantially to the cost of the test and precludes ad hoc decisions on biopsy for PGD. Approximately 20% of IVF cycles with planned PGD are canceled on day 3 due to low embryo numbers. Precycle parental testing that is necessary for an SNP analysis would force patients to spend money on testing that is ultimately unnecessary.

NGS

Massively parallel genome sequencing also known as next-generation sequencing (NGS) is the latest and currently the most popular technique being utilized in PGS laboratories throughout the United States. Based on internal data, as of December 2017, more than 90% PGS cases being performed in the two major PGS laboratories in the United States were being performed using NGS. This number is expected to only increase in the near future. Compared with other techniques like the SNP array (32,000 reads) and aCGH (2700 reads), high-resolution NGS (hr-NGS) that is currently being used in the laboratory derives about 700,000 reads per sample. hr-NGS utilizes similar upstream technology in terms of biopsy and WGA as aCGH and SNP arrays. The starting material for the testing is obtained through the biopsy of the trophectoderm from the embryos. Whole genome amplification is performed on the biopsy. At this stage, the protocol for hr-NGS deviates from aCGH and SNP. One of the major benefits of NGS is the ability to scale up and run a large number of samples together by multiplex sequencing, thereby reducing the cost per sample for every procedure. Multiplexing is the ability to run multiple samples on a single run on the sequencing machine. NGS makes this a possibility by the virtue of the addition of "barcodes." Barcodes are small (6–10bp) sequences that are unique oligonucleotides that ligate to the ends of the fragments of the DNA (derived from the blastocyst day 5 biopsy) being sequenced. Barcodes help differentiate between different samples when sequenced. The possibility to scale up as a result of multiplexing is unique to hr-NGS.

In most cases, DNA sequencing unlike aCGH and SNP array isn't limited in coverage to the areas of the genome that are attached to the chip. The process of DNA sequencing involves processing one base at a time. Sequencing is performed on DNA that has been randomly amplified during WGA stage. Limitations if any would be at the WGA stage emanating from amplification failure of certain regions. This would be more of a limitation of the WGA method and would be shared by all platforms (SNP, aCGH, and NGS) alike.

In NGS, sequences obtained post sequencing are aligned or laid out in a line and compared with a reference or a master sequence, which is a database of previously sequenced genomes. Post alignment in hr-NGS, the sequenced fragments are then mapped back to each chromosome.

Validation of aCGH, SNP Arrays, and NGS

Due to the intrinsic and often unanticipated problems with every new technology, any new technology should always be validated against other, more established, proved methods [48]. Evaluating a new approach against itself may prevent the detection of technique-related flaws. In PGS, validation of different platforms has been performed using cell lines, polar body biopsies compared, gametes, blastomere biopsies, and blastocyst biopsies. There is always a possibility of validation by inadequate or incorrect methods, and it may lead to false or incorrect assumptions when (1) the analysis of cell lines with well-defined chromosome abnormalities that cannot mimic mosaicism and other particularities of the cell are being tested; (2) analysis of eggs or embryos using one technique is compared with analysis of polar bodies or the remainder of the embryo by the same technique that precludes identifying abnormalities not detectable by that technique; and (3) undiagnosed embryos are blindly replaced (either by fingerprinting the embryo or single embryo transfer), and obstetric outcomes determine the "fate" of each tested embryo that does not account for the status of nonimplanted embryos. Additionally, if the tools used for analysis are qualitative in nature, they will miss the presence of two chromosomes of the same grandparental origin. The errors caused by mosaicism will also not be taken into account during this validation mode, hence giving us bogus 99.9% confidence results. In our opinion, the optimal method for validating any new technique or technology for PGS is to reanalyze embryos that were not transferred to the patient, either because they were diagnosed as chromosomally abnormal or because they underwent arrest. The reanalysis of these embryos should be done with another well-established technique, that is, "gold standard." This would discern deficiencies of the new method under evaluation and account for issues related to embryo biology, such as mosaicism. To simplify the comparison between studies, an error should be classified as diagnosing an embryo as euploid when reanalysis shows that it was abnormal or vice versa. Due to the extent of mosaicism, an error rate per chromosome has questionable relevance and no clinical importance compared with an error rate per embryo.

SNP microarrays have undergone some validation experiments, such as comparison of PGD results and analysis of babies born [45], SNP microarray reanalysis of embryos previously analyzed by SNP arrays [49], and using data from one set of SNPs as internal controls for another set of SNPs. To date, no studies have confirmed the original diagnosis by reanalyzing the remaining embryonic cells with a different technique.

Microarray CGH for PGD has been validated by the analysis of single cells from known cell lines (Dagan Wells, personal communication) and additionally by analyzing eggs with aCGH and correlating them to the results obtained using aCGH of the corresponding PBs (Montag and Gianaroli, personal communication). Further validation was performed on day 3 embryos analyzed by PGD with aCGH. The embryos that were not replaced because of chromosome or morphological abnormalities were reanalyzed in most of their remaining cells by FISH, as the gold standard comparative method, using 12 probes for the most common chromosome abnormalities. Plus probes for any chromosomes found abnormal according to aCGH were also used. Only 1.9% of embryos were found to be erroneously diagnosed [43]. This is even lower than the 7% error rate expected solely from mosaicism as calculated in FISH studies [4].

More recently, hr-NGS validation was performed by Fiorentino et al. in 2014 and Kung et al. in 2015 [50,51]. In these studies, dependability of NGS-based 24-chromosome copy number assignments was gauged with previously established array-CGH-based diagnoses (as the gold standard) of the same WGA products at the level of individual chromosome copy numbers for the entire 24 chromosomes of each embryo biopsy tested and also for the overall diagnosis of aneuploidy or euploidy. The sensitivity of NGS was observed to be 100%. Also NGS specificity for an aneuploidy call (consistency of chromosome copy number assignment) was observed to be a 100%. An overall error rate of 0% was observed during the validation of NGS.

Clinical Results

Of the techniques discussed, CGH is the technique with the largest PGS clinical dataset that is available [12,52–54]. Sher et al. [52] detected a 74% ongoing pregnancy rate per transfer and 63% per retrieval in women with an average age of 37.5 years. For patients of a similar age receiving a blastocyst transfer, Schoolcraft et al. [12] detected a significant increase in implantation rates, from 46.5% to 72.2% ($P<0.001$) following embryo selection using CGH. It is important to note that both aforementioned studies showing high implantation rates avoided cleavage-stage embryo biopsy and transferred embryos, which had earlier been cryopreserved in a later cycle. In addition to the possible benefits of transferring euploid embryos by selection, there may be additional benefits related with transfer in a nonstimulated cycle that is possible due to cryopreservation [55]. As observed by Schoolcraft et al., the loss of blastocyst-stage embryos post devitrification was minimal (0.7%) [12]. Certain datasets [56] show that only 118/151 PGD cycles had normal embryos for transfer in a population 38 years of age after a day 3 biopsy followed by aCGH and day 5 replacement. The pregnancy rate was significantly higher (~59%) per transfer when compared with the controls (~38%) with a transfer ($P<0.001$). The ongoing pregnancy rate for the PGD group was also found to be much higher (~54%) per transfer, as compared with the controls (~31.1%) with a transfer ($P<0.001$). Results from day 5 (blastocyst) biopsy have shown to be much more remarkable than the day 3 biopsies [57]. Day 5 biopsies are observed to have about 3–10 cells. Hence automatically, they have more DNA than day 3 biopsies, which consist of only single cells. Also at day 5, the embryo is much more resilient than it is at day 3 [57]. This resilience seems to increase the overall implantation rate by almost 20% when day 5 blastocyst biopsies are used instead of cleavage-stage biopsies. However, at this point, it is important to understand that biopsies are subject to differ greatly between operators. This in turn alters the success rates quite drastically.

As discussed by Greco et al. [58] and Tormasi et al. (2015) at ASRM, almost 17% of mosaics that were undiagnosed by aCGH were detected using NGS. Consequently, when losses from previously aCGH-analyzed euploid embryos were reanalyzed using NGS, almost 50% of the cases ended up being mosaic, as presented by Grifo and Munne at ASRM (2015). As per Frogouli et al. (PGDIS, 2016), mosaic embryos do in fact have a lower rate of implantation. Moreover, Friedenthal et al. presented at ESHRE (2017) that the implantation rate and the ongoing pregnancy rates of NGS were higher as compared with aCGH. However, it is important to note that mosaic embryos do implant and can result in a healthy pregnancy in about 30% ongoing pregnancies [58]. We will discuss the paradigm shift in handling mosaic embryos and the change in classification system of mosaics in the PGS field toward the end of this chapter.

An important factor to consider regarding different techniques used in CCS is that it is probable that the difference between clinical results is related to the stage at which biopsy was carried out in and the technique of the embryologist carrying out the biopsy, rather than to the differences in the actual method of chromosome screening. Moreover, additional variables like media differences from one manufacturer to another, differences in stimulation protocol from one IVF lab to another, temperature, and pH could also potentially be affecting the final outcomes. Such variables either need to be accounted for or controlled in future studies to improve the understanding that we have. Also obstetric outcome data are usually lacking in the PGS field in general. These data should be freely shared between the reference laboratories and the IVF centers to determine the clinical efficacy of the PGS technique at large.

In summary, data on the clinical application of comprehensive chromosome analysis techniques certainly suggest a better prognosis for patients undergoing ART with increased implantation rates, ongoing pregnancy rates, and decreased miscarriage rates.

NGS, Mosaicism and How We Should Deal With It

One of the key disadvantages of the FISH technique as mentioned previously is that even when performed with utmost optimality, it is unable to detect more than 90% of chromosomally abnormal embryos due to the inability to concurrently evaluate all 24 chromosomes. This is compared with the detection for the same using a relatively newer technique of array comparative genome hybridization (aCGH) [59]. Hence, the lack of coverage of the entire genome and the dearth of resolution was the issue faced by FISH that led to the increased use of aCGH. Despite the obvious advantages of aCGH, there was an initial resistance to adopt the better technique. Performing aCGH was tedious and more time-consuming as compared with FISH. This meant that embryos had to be cryopreserved, not a regular clinical practice then. Moreover, with aCGH, almost 98% abnormal embryos (many previously missed by FISH) were being detected that led to an overall increase in the number of abnormal embryos that IVF centers received reports for. The increase in the number of abnormal embryos within the cohort in turn meant fewer embryos available for transfer compared with when FISH was being used for PGS. hr-NGS has an increased dynamic range compared with even aCGH. This increased range enables the detection of mosaicism in multicellular samples [50]. As explained above, embryo mosaicism is the presence of two or more cells in a single embryo that have a different genetic makeup. Chromosomal mosaicism is an established and relatively a common phenomenon in the preimplantation embryo that has both diploid and aneuploid cells [60]. About 21% embryos are observed to be mosaic in the clinic [61]. Due to the mitotic nature of mosaicism, the phenomenon does not have an increased incidence in the advance maternal age group. Reanalysis of embryos with NGS (post analysis with aCGH) to detect mosaicism showed that the risk of miscarriage could potentially be reduced by half as compared with analysis by aCGH. Additionally, mosaic embryos diagnosed with NGS seem to significantly implant less than the embryos that were diagnosed euploid using NGS [61]. This reduces the chances of the observed mosaicism being artefactual in nature. Considering the compelling trends from data, one would be tempted to consider all mosaics as abnormal embryos. However, it is important that we pause before making such an assumption and question whether all mosaic embryos are the same. It also begs the question as to whether some mosaic embryos have a higher chance of success compared with other mosaics. Based on the clinical trends currently being observed, it is very likely that there is a range of abnormality observed within mosaics, and not all mosaic embryos are the same in their levels of abnormality. The "low-level" (up to 20% abnormal cells within the biopsy) mosaics transfer better than the "high-level" (80% abnormal cells within the biopsy) mosaics [61]. This could explain as to why we observe that some mosaic embryos implant successfully and others do not. Hence, considering the eventuality of mosaic embryos, what we need to be looking toward is a paradigm shift in the approach we take toward classifying embryos after PGS (seen in Fig. 13.2). Currently, embryos are classified as normal or abnormal, with an error rate of about 2%–10%. With the newer thought process, what we need to consider including additional categories of embryo classification. These categories could be defined as the mosaic category. All embryos that fall within this category could be further subclassified as "high mosaics" or "low mosaics," each having low or high chance of implantation, respectively.

Lastly, deprioritizing of mosaic embryos should be the norm. Prioritizing euploid embryos over even the low mosaics would also be desirable. Human primordial germ cells develop from pluripotent epiblast cells [62]. By day 24, they are segregated in the dorsal yolk sac [62,63]. Blastocysts are biopsied at the 5 day stage. The presence of mosaic cells at this stage could very well lead to the embryo developing into an individual possessing mosaic tissues, if the transfer of such a mosaic embryo is successful. According to Campbell et al., parents that have somatic mosaic tissues might also have germ-line mosaicism that could possibly cause intergenerational recurrence of disease [64]. This would be highly undesirable in the long term. Hopefully, adopting the above alternate paradigm to classify embryos would decrease rates of miscarriages and rates of errors, increase healthy births, and also reduce the likelihood of the individual or its future generations facing any trouble from the disease.

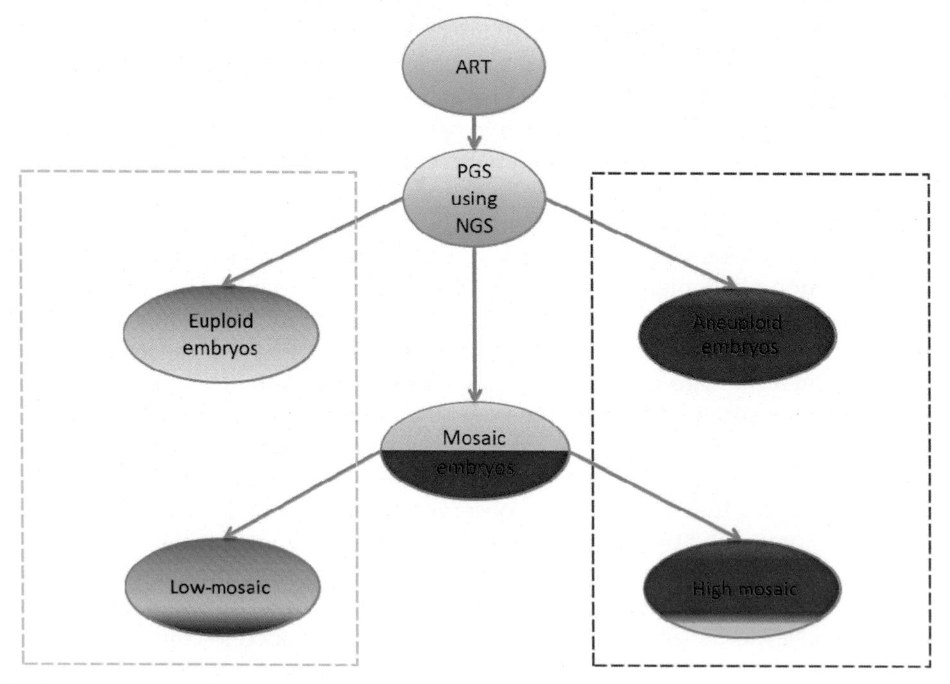

FIG. 13.2 An alternate paradigm to guide PGS classification of embryos while using NGS.

References

[1] S. Munne, Preimplantation genetic diagnosis for aneuploidy and translocations using array comparative genomic hybridization, Curr. Genomics 13 (6) (2012) 463–470.

[2] S. Munne, et al., Diagnosis of major chromosome aneuploidies in human preimplantation embryos, Hum. Reprod. 8 (12) (1993) 2185–2191.

[3] M.C. Magli, et al., Embryo morphology and development are dependent on the chromosomal complement, Fertil. Steril. 87 (3) (2007) 534–541.

[4] P. Colls, et al., Increased efficiency of preimplantation genetic diagnosis for infertility using "no result rescue", Fertil. Steril. 88 (1) (2007) 53–61.

[5] C. Gutierrez-Mateo, et al., Validation of microarray comparative genomic hybridization for comprehensive chromosome analysis of embryos, Fertil. Steril. 95 (3) (2011) 953–958.

[6] C.B. Coulam, et al., Discordance among blastomeres renders preimplantation genetic diagnosis for aneuploidy ineffective, J. Assist. Reprod. Genet. 24 (1) (2007) 37–41.

[7] E.B. Baart, et al., Fluorescence in situ hybridization analysis of two blastomeres from day 3 frozen-thawed embryos followed by analysis of the remaining embryo on day 5, Hum. Reprod. 19 (3) (2004) 685–693.

[8] S. Munne, D. Wells, J. Cohen, Technology requirements for preimplantation genetic diagnosis to improve assisted reproduction outcomes, Fertil. Steril. 94 (2) (2010) 408–430.

[9] P. Colls, et al., Increased efficiency of preimplantation genetic diagnosis for aneuploidy by testing 12 chromosomes, Reprod. Biomed. Online 19 (4) (2009) 532–538.

[10] S. Munne, et al., Improved detection of aneuploid blastocysts using a new 12-chromosome FISH test, Reprod. Biomed. Online 20 (1) (2010) 92–97.

[11] J. Houldsworth, R.S. Chaganti, Comparative genomic hybridization: an overview, Am. J. Pathol. 145 (6) (1994) 1253–1260.

[12] Schoolcraft, W.B., et al., Clinical application of comprehensive chromosomal screening at the blastocyst stage, Fertil. Steril. 94 (5) (2010) 1700–1706.

[13] L. Rienzi, et al., Oocyte, embryo and blastocyst cryopreservation in ART: systematic review and meta-analysis comparing slow-freezing versus vitrification to produce evidence for the development of global guidance, Hum. Reprod. Update 23 (2) (2017) 139–155. https://doi.org/10.1093/humupd/dmw038.
[14] C. Spits, K. Sermon, PGD for monogenic disorders: aspects of molecular biology, Prenat. Diagn. 29 (1) (2009) 50–56.
[15] A.R. Thornhill, et al., ESHRE PGD Consortium 'Best practice guidelines for clinical preimplantation genetic diagnosis (PGD) and preimplantation genetic screening (PGS), Hum. Reprod. 20 (1) (2005) 35–48.
[16] I. Liebaers, et al., Clinical experience with preimplantation genetic diagnosis and intracytoplasmic sperm injection, Hum. Reprod. 13 (Suppl. 1) (1998) 186–195.
[17] P. Braude, et al., Preimplantation genetic diagnosis, Nat. Rev. Genet. 3 (12) (2002) 941–953.
[18] Y. Verlinsky, et al., Analysis of the first polar body: preconception genetic diagnosis, Hum. Reprod. 5 (7) (1990) 826–829.
[19] S. Rechitsky, et al., Accuracy of preimplantation diagnosis of single-gene disorders by polar body analysis of oocytes, J. Assist. Reprod. Genet. 16 (4) (1999) 192–198.
[20] C.M. Ogilvie, P.R. Braude, P.N. Scriven, Preimplantation genetic diagnosis—an overview, J. Histochem. Cytochem. 53 (3) (2005) 255–260.
[21] D.K. Griffin, et al., Dual fluorescent in situ hybridisation for simultaneous detection of X and Y chromosome-specific probes for the sexing of human preimplantation embryonic nuclei, Hum. Genet. 89 (1) (1992) 18–22.
[22] K. Sermon, A. Van Steirteghem, I. Liebaers, Preimplantation genetic diagnosis, Lancet 363 (9421) (2004) 1633–1641.
[23] E. Fragouli, et al., Cytogenetic analysis of human blastocysts with the use of FISH, CGH and aCGH: scientific data and technical evaluation, Hum. Reprod. 26 (2) (2011) 480–490.
[24] J.D. Delhanty, et al., Multicolour FISH detects frequent chromosomal mosaicism and chaotic division in normal preimplantation embryos from fertile patients, Hum. Genet. 99 (6) (1997) 755–760.
[25] S. Munne, et al., Chromosome mosaicism in cleavage-stage human embryos: evidence of a maternal age effect, Reprod. Biomed. Online 4 (3) (2002) 223–232.
[26] M.G. Katz-Jaffe, A.O. Trounson, D.S. Cram, Mitotic errors in chromosome 21 of human preimplantation embryos are associated with non-viability, Mol. Hum. Reprod. 10 (2) (2004) 143–147.
[27] D. Wells, J.D. Delhanty, Comprehensive chromosomal analysis of human preimplantation embryos using whole genome amplification and single cell comparative genomic hybridization, Mol. Hum. Reprod. 6 (11) (2000) 1055–1062.
[28] L. Voullaire, et al., Chromosome analysis of blastomeres from human embryos by using comparative genomic hybridization, Hum. Genet. 106 (2) (2000) 210–217.
[29] M. Li, et al., Fluorescence in situ hybridization reanalysis of day-6 human blastocysts diagnosed with aneuploidy on day 3, Fertil. Steril. 84 (5) (2005) 1395–1400.
[30] M.A. Santos, et al., The fate of the mosaic embryo: chromosomal constitution and development of day 4, 5 and 8 human embryos, Hum. Reprod. 25 (8) (2010) 1916–1926.
[31] F.J. Los, D. Van Opstal, C. van den Berg, The development of cytogenetically normal, abnormal and mosaic embryos: a theoretical model, Hum. Reprod. Update 10 (1) (2004) 79–94.
[32] L. Van Landuyt, et al., Blastocyst formation in in vitro fertilization versus intracytoplasmic sperm injection cycles: influence of the fertilization procedure, Fertil. Steril. 83 (5) (2005) 1397–1403.
[33] M. Youssry, et al., Current aspects of blastocyst cryopreservation, Reprod. Biomed. Online 16 (2) (2008) 311–320.
[34] H.J. Chang, et al., Optimal condition of vitrification method for cryopreservation of human ovarian cortical tissues, J. Obstet. Gynaecol. Res. 37 (8) (2011) 1092–1101.
[35] M. Sandalinas, et al., Developmental ability of chromosomally abnormal human embryos to develop to the blastocyst stage, Hum. Reprod. 16 (9) (2001) 1954–1958.
[36] T. Mamas, et al., Detection of aneuploidy by array comparative genomic hybridization using cell lines to mimic a mosaic trophectoderm biopsy, Fertil. Steril. 97 (4) (2012) 943–947.
[37] L.E. Northrop, et al., SNP microarray-based 24 chromosome aneuploidy screening demonstrates that cleavage-stage FISH poorly predicts aneuploidy in embryos that develop to morphologically normal blastocysts, Mol. Hum. Reprod. 16 (8) (2010) 590–600.
[38] D.K. Kalousek, I.J. Barrett, A.B. Gartner, Spontaneous abortion and confined chromosomal mosaicism, Hum. Genet. 88 (6) (1992) 642–646.

[39] C. Sismani, et al., Cryptic genomic imbalances in patients with de novo or familial apparently balanced translocations and abnormal phenotype, Mol. Cytogenet. 1 (2008) 15.
[40] S. Goobie, et al., Molecular and clinical characterization of de novo and familial cases with microduplication 3q29: guidelines for copy number variation case reporting, Cytogenet. Genome Res. 123 (1-4) (2008) 65–78.
[41] A.L. Beaudet, J.W. Belmont, Array-based DNA diagnostics: let the revolution begin, Annu. Rev. Med. 59 (2008) 113–129.
[42] E. Stejskalova, et al., Cytogenetic and array comparative genomic hybridization analysis of a series of hepatoblastomas, Cancer Genet. Cytogenet. 194 (2) (2009) 82–87.
[43] A.H. Handyside, et al., Karyomapping: a universal method for genome wide analysis of genetic disease based on mapping crossovers between parental haplotypes, J. Med. Genet. 47 (10) (2010) 651–658.
[44] D.S. Johnson, et al., Preclinical validation of a microarray method for full molecular karyotyping of blastomeres in a 24-h protocol, Hum. Reprod. 25 (4) (2010) 1066–1075.
[45] N.R. Treff, et al., Robust embryo identification using first polar body single nucleotide polymorphism microarray-based DNA fingerprinting, Fertil. Steril. 93 (7) (2010) 2453–2455.
[46] E. Vanneste, et al., Chromosome instability is common in human cleavage-stage embryos, Nat. Med. 15 (5) (2009) 577–583.
[47] S. Munne, et al., Outcome of preimplantation genetic diagnosis of translocations, Fertil. Steril. 73 (6) (2000) 1209–1218.
[48] R.T. Scott Jr., N.R. Treff, Assessing the reproductive competence of individual embryos: a proposal for the validation of new "-omics" technologies, Fertil. Steril. 94 (3) (2010) 791–794.
[49] N.R. Treff, et al., A novel single-cell DNA fingerprinting method successfully distinguishes sibling human embryos, Fertil. Steril. 94 (2) (2010) 477–484.
[50] F. Fiorentino, et al., Development and validation of a next-generation sequencing-based protocol for 24-chromosome aneuploidy screening of embryos, Fertil. Steril. 101 (5) (2014) 1375–1382.
[51] A. Kung, et al., Validation of next-generation sequencing for comprehensive chromosome screening of embryos, Reprod. Biomed. Online 31 (6) (2015) 760–769.
[52] G. Sher, et al., Oocyte karyotyping by comparative genomic hybridization [correction of hybrydization] provides a highly reliable method for selecting "competent" embryos, markedly improving in vitro fertilization outcome: a multiphase study, Fertil. Steril. 87 (5) (2007) 1033–1040.
[53] E. Fragouli, et al., Comparative genomic hybridization of oocytes and first polar bodies from young donors, Reprod. Biomed. Online 19 (2) (2009) 228–237.
[54] E. Fragouli, et al., Comprehensive chromosome screening of polar bodies and blastocysts from couples experiencing repeated implantation failure, Fertil. Steril. 94 (3) (2010) 875–887.
[55] B.S. Shapiro, et al., High ongoing pregnancy rates after deferred transfer through bipronuclear oocyte cryopreservation and post-thaw extended culture, Fertil. Steril. 92 (5) (2009) 1594–1599.
[56] S. Munne, E.S. Surrey, J. Grifo, E. Marut, M. Opsahl, T.H. Taylor, Preimplantation genetic diagnosis using array CGH significantly increases ongoing pregnancy rates per transfer, Fertil. Steril. 94 (4) (2010) S81.
[57] R.T. Scott Jr., et al., Cleavage-stage biopsy significantly impairs human embryonic implantation potential while blastocyst biopsy does not: a randomized and paired clinical trial, Fertil. Steril. 100 (3) (2013) 624–630.
[58] E. Greco, M.G. Minasi, F. Fiorentino, Healthy babies after intrauterine transfer of mosaic aneuploid blastocysts, N. Engl. J. Med. 373 (21) (2015) 2089–2090.
[59] N.R. Treff, R.T. Scott Jr., Methods for comprehensive chromosome screening of oocytes and embryos: capabilities, limitations, and evidence of validity, J. Assist. Reprod. Genet. 29 (5) (2012) 381–390.
[60] J. van Echten-Arends, et al., Chromosomal mosaicism in human preimplantation embryos: a systematic review, Hum. Reprod. Update 17 (5) (2011) 620–627.
[61] S. Munne, D. Wells, Detection of mosaicism at blastocyst stage with the use of high-resolution next-generation sequencing, Fertil. Steril. 107 (5) (2017) 1085–1091.
[62] T. Fujimoto, Y. Miyayama, M. Fuyuta, The origin, migration and fine morphology of human primordial germ cells, Anat. Rec. 188 (3) (1977) 315–330.
[63] A.I. Marques-Mari, et al., Differentiation of germ cells and gametes from stem cells, Hum. Reprod. Update 15 (3) (2009) 379–390.
[64] I.M. Campbell, et al., Parental somatic mosaicism is underrecognized and influences recurrence risk of genomic disorders, Am. J. Hum. Genet. 95 (2) (2014) 173–182.

CHAPTER

14

Unraveling the Causes of Failed Fertilization After Intracytoplasmic Sperm Injection Due to Oocyte Activation Deficiency

Davina Bonte, Ramesh Reddy Guggilla, Panagiotis Stamatiadis, Petra De Sutter, Björn Heindryckx

Ghent-Fertility and Stem cell Team (G-FaST), Department for Reproductive Medicine, Ghent University Hospital, Ghent, Belgium

BACKGROUND

Recognized as a disorder by the WHO, infertility is defined as "a disease of the reproductive system defined by the failure to achieve a clinical pregnancy after 12 months or more of regular unprotected sexual intercourse" (WHO). Clinical data are showing that 10%–16% of all couples are suffering from infertility problems worldwide [1]. The start of the assisted reproductive technology (ART) field dates back to the 1890s when Prof Walter Heape (University of Cambridge, the United Kingdom) reported for the first time an embryo transplantation in rabbits [2]. Although much more research was performed in the meantime, it lasted almost one century until the world's first in vitro fertilization (IVF) baby, Louise Brown, was born in 1978 [3]. Since then, IVF has become a very common method to bypass some infertility problems and is, to date, often used on a routine basis in the IVF clinics worldwide. However, in 1992, Palermo et al. reported the birth of the first baby born after intracytoplasmic sperm injection (ICSI), mainly to treat male factor infertility by microinjection of one sperm into the oocyte [4]. In Europe, 6.5 million ART cycles have been performed between 1997 and 2002, resulting in almost 1.2 million successful live births [5]. Nowadays, ICSI accounts for the major part (70%–80%) of all cycles done worldwide [6]. Despite average fertilization rates of 70%, complete fertilization failure still occurs in 1%–5% of ICSI cycles, mainly attributed to an oocyte activation deficiency (OAD) that can be sperm- or oocyte-related [7–11].

https://doi.org/10.1016/B978-0-12-812571-7.00015-0

OOCYTE ACTIVATION

The Oocyte Activation Mechanism

Oocyte activation is an essential process induced by sperm entry, converting a mature arrested oocyte into a fertilized oocyte. In many vertebrate species, including human, oocytes remain arrested at the metaphase of the second meiotic division (metaphase II, MII) until the fertilizing sperm fuses with the oocyte [12]. Subsequently, the sperm not only introduces its genomic material into the oocyte but also delivers a sperm oocyte-activating factor, believed to be phospholipase C zeta (PLCζ), which will trigger the start of the so-called oocyte activation [13,14]. This oocyte activation process comprises different events, including the exocytosis of cortical granules (CGs), meiotic cell-cycle resumption, pronuclear formation, translation of maternal mRNAs, and meiosis-to-mitosis transition [15–20].

It has been shown in mouse that the sperm factor PLCζ targets membrane-associated PIP_2 molecules located on vesicles in the oocyte cytoplasm [21]. This sperm factor will hydrolyze phosphatidylinositol 4,5-bisphosphate (PIP_2) leading to the generation of two new molecules, inositol 1,4,5-trisphosphate (IP_3) and diacylglycerol (DAG) [1,22,23] (see Fig. 14.2).

Such as calcium (Ca^{2+}), IP_3 will bind to the IP_3 receptor (IP_3R), which is located on the smooth endoplasmic reticulum (ER) in the oocyte [22,24]. As a result, the IP_3R will undergo a conformational change, increasing the sensitivity of the IP_3R and resulting in Ca^{2+} release from the ER stores [14,25,26]. At low intracellular Ca^{2+} concentrations, the IP_3-induced Ca^{2+} release (IICR) is stimulated (positive feedback loop), where at high intracellular Ca^{2+} concentrations, this mechanism is inhibited and IP_3R channels will close again [27,28]. This dual regulation of the IP_3R makes it possible to support long-lasting Ca^{2+} oscillations without inducing Ca^{2+}-induced apoptosis. Indeed, in mammals, multiple Ca^{2+} rises, the so-called Ca^{2+} oscillations, are observed after sperm entry and will evoke different downstream events. Next to the IP_3R-induced Ca^{2+} release (IICR), studies in mouse have shown that ER-associated ryanodine receptors (RyRs) also contribute to the Ca^{2+} efflux to the oocyte cytoplasm. RyRs are functional and able in triggering CG exocytosis but not meiotic resumption. However, their contribution to oocyte activation is not essential [29].

In mouse oocytes, Ca^{2+} oscillations start about 1–2 min after sperm-egg fusion and last up to 5 or 6 h, stopping around the time of the pronucleus formation [30,31]. The frequency of Ca^{2+} oscillations is species-specific. In human and bovine eggs, one Ca^{2+} transient every 30–60 min is observed, while in mice, one Ca^{2+} transient every 10–20 min is seen [32–34] (see Fig. 14.1).

The increase in intracellular Ca^{2+} levels will activate Ca^{2+}/calmodulin-dependent protein kinase II (CaMKII) [35]. Together with Polo-like kinase 1 (PLK1), CaMKII phosphorylates its substrate, namely, the early mitotic inhibitor 2 (Emi2), which is a key component of the cytostatic factor (CSF) activity. In nonphosphorylated conditions, Emi2 will suppress the activity of the anaphase-promoting complex/cyclosome (APC/C). However, when Ca^{2+} oscillations occur during oocyte activation, Emi2 will get phosphorylated and will be recognized by the SKP2-cullin 1-F-box protein (SCF) ubiquitin-ligase complex. When targeted for destruction by this ubiquitin-ligase complex, Emi2 cannot longer exhibit its APC/C inhibitory function. Therefore, APC/C will become active and be able to degrade cyclin B1 (CNB1), which together with cyclin-dependent kinase 1 (CDK1) forms the maturation-promoting factor (MPF).

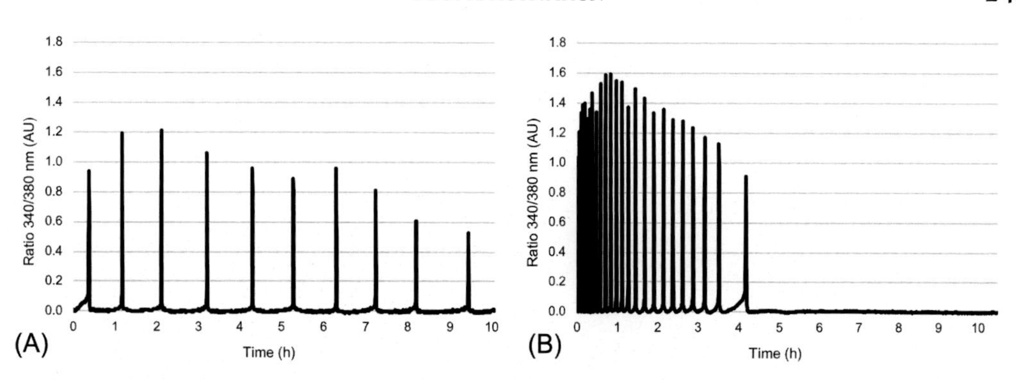

FIG. 14.1 Calcium oscillating pattern recorded following (A) human ICSI and (B) mouse ICSI. *AU*, arbitrary units. (Recorded at the Department for Reproductive Medicine—Ghent University.)

High MPF levels ensure an arrest of the oocyte at the metaphase stage of the second meiotic division (MII stage). However, CNB1 destruction will lower the MPF levels ensuring meiotic cell-cycle resumption [36] (see Fig. 14.2). In addition, APC/C activity will lead to the degradation of securin, a separase inhibitor, and consequently half of the sister chromatids will be expulsed into a second polar body providing a haploid set of mother chromosomes to the oocyte [18,36]. Besides, Ca^{2+} oscillations induce the inactivation of the MAPK pathway, which allows the formation of the pronuclei [1,37].

On the other hand, DAG together with the induced Ca^{2+} oscillations will target the protein kinase C (PKC) pathway [22,38]. Six PKC isoforms have been shown to be expressed in mouse and rat oocytes. Halet et al. suggested that certain PKC isoforms translocate to the oocyte cortex where they contribute to the induction of Ca^{2+} rises by phosphorylating membrane proteins involved in the store-operated Ca^{2+} entry (SOCE; see further) [39]. Indeed, it has been demonstrated that PKC isoforms α, β, and γ translocate to the oocyte cortex after sperm-oocyte interaction or Ca^{2+} ionophore-induced oocyte activation [40,41]. However, which Ca^{2+} pump is phosphorylated by the PKCs in mammalian oocytes is yet unknown. In addition, it was hypothesized that PKC phosphorylates possibly proteins from the CG exocytosis machinery. Indeed, upon fertilization, PKCα is recruited to the oocyte plasma membrane, where it will assist with the induction of the CG exocytosis [42]. In the oocyte cortex, PKCs will phosphorylate myristoylated alanine-rich C-kinase substrate (MARCKS) proteins. MARCKS proteins are known to be involved in the cytoskeletal actin network, preventing polyspermy [43]. Phosphorylation of MARCKS proteins will disassemble the actin network, allowing the CGs to reach the oocyte membrane and to exocytose [37,42] (see Fig. 14.2).

Zinc and Its Involvement in Oocyte Activation

Recently, special attention has been paid to the importance of zinc ions (Zn^{2+}) during mammalian oocyte activation. Inside the oocyte, Zn^{2+} molecules are stored in vesicles, and changes in their distribution and quantity during both oocyte maturation and activation correspond to those of CGs [44]. During fertilization or parthenogenetic activation in mouse, it has been shown that each intracellular Ca^{2+} transient is followed by the release of Zn^{2+} molecules to the

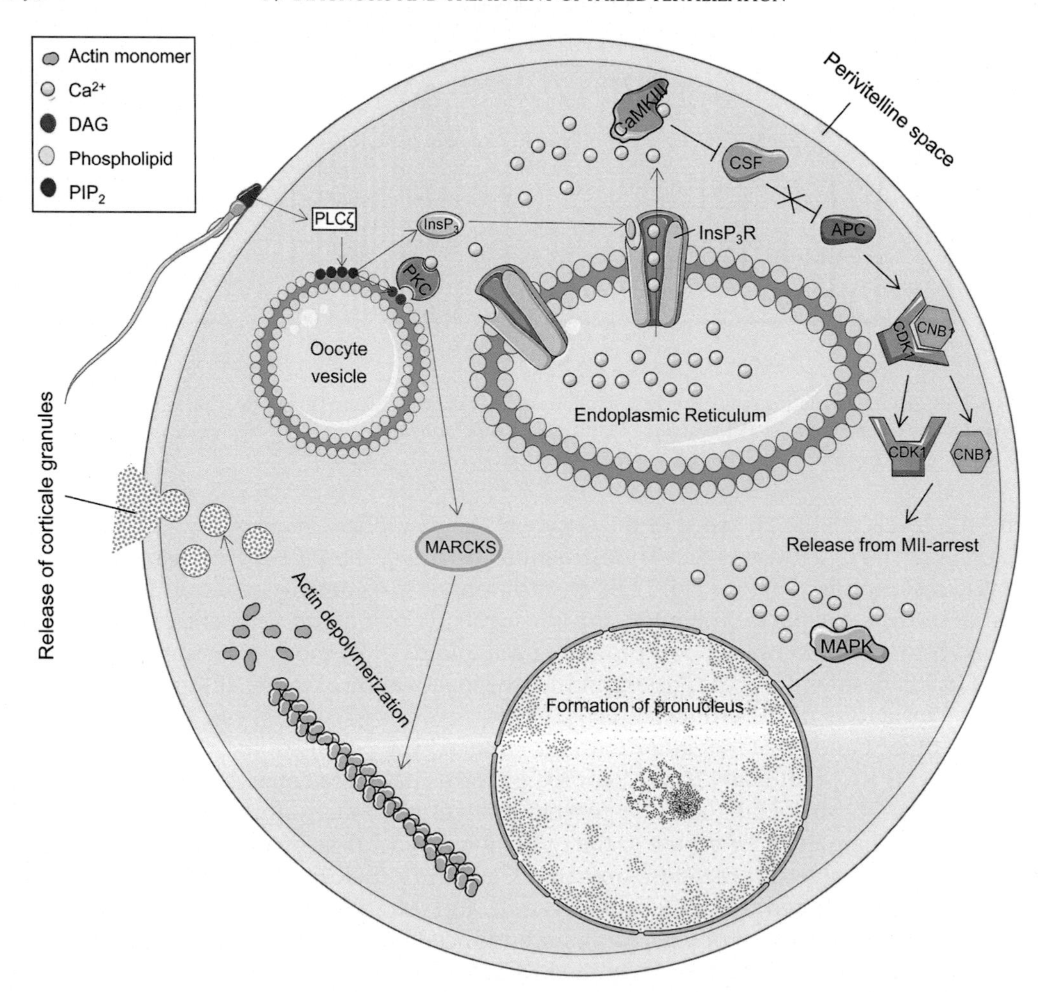

FIG. 14.2 Oocyte activation triggered by a sperm-specific factor, phospholipase C zeta (PLCζ). Upon sperm-oocyte fusion, PLCζ is released into the ooplasm and hydrolyzes membrane-associated PIP_2 molecules into IP_3 and DAG. Subsequently, IP_3 binds to its receptor (IP_3R), thereby releasing Ca^{2+} from intracellular stores. The intracellular Ca^{2+} rises mediate cortical granule exocytosis via the PKC pathway, formation of the pronuclei via MAPK inhibition, and meiotic resumption via CaMKII activation. *Reproduced with permission from M. Yeste, et al., Oocyte activation deficiency: a role for an oocyte contribution? Hum. Reprod. Update 22(1) (2016) 23–47.*

extracellular milieu of the oocyte, termed "zinc spark". This Zn^{2+} exocytosis is shown to be Ca^{2+}-dependent and allows mature oocytes to initiate meiotic resumption [45]. Artificial Zn^{2+} depletion by the Zn^{2+} chelator TPEN is sufficient to induce successful fertilization in both mouse and pig, without the manifestation of Ca^{2+} oscillations. In addition, parthenogenetic TPEN activation resulted in healthy viable mice and pig clones using somatic cell nuclear transfer, in the absence of Ca^{2+} transients [46,47]. On the contrary, Zn^{2+} overloading using the Zn^{2+} ionophore ZnPT maintains the oocyte meiotic arrest [48]. These findings suggest that Zn^{2+} sparks occur downstream the Ca^{2+} oscillations and they are sufficient and essential for successful oocyte activation and full-term embryonic development [45,49]. However, studies in human are very scarce. Recently, Duncan et al. activated human oocytes with calcium-ionomycin, ionomycin, or human PLCζ cRNA microinjection [50]. These oocyte-activating agents induced intracellular Ca^{2+} rises and Zn^{2+} sparks in the extracellular space. Treatment of human oocytes with TPEN was sufficient to induce meiotic resumption and transition from a meiotic to a mitotic cell. Together, these results demonstrate important functions for Zn^{2+} in oocyte activation and suggest Zn^{2+} as a biomarker for early human development [50].

The mechanism behind the alleviation of meiotic arrest by Zn^{2+} exocytosis remains unclear. However, Bernhardt et al. demonstrated that Zn^{2+} chelation induces meiotic resumption without Emi2 degradation and that Emi2 contains a Zn^{2+}-binding region [48]. Therefore, a model is proposed where Ca^{2+}-dependent Zn^{2+} exocytosis inhibits the Emi2 activity, resulting in CNB1 destruction, a decrease in MPF levels, and finally the continuation of the meiosis process. Furthermore, Sun et al. showed that Zn^{2+} molecules support the activity of phosphatase Cdc25 that prevents the phosphorylation of CDK1 [51]. This suggests that upon fertilization, the intracellular Zn^{2+} decrease reduces the Cdc25 activity, allowing CDK1 to become phosphorylated and inactivated, resulting in the alleviation of the meiotic arrest. Further investigation is of utmost importance to confirm this hypothesis and unravel the complete Zn^{2+} mechanism in the oocyte activation process [52].

The Regulation of Calcium Cycling During Oocyte Activation

The mechanism behind the Ca^{2+} oscillations, which demonstrates Ca^{2+} cycling, is made possible by the precise coordination between Ca^{2+} channels, Ca^{2+} pumps, and Ca^{2+} exchangers [53]. Although this mechanism for maintaining Ca^{2+} homeostasis is not completely understood, this section summarizes the current state of the art.

Upon fertilization, Ca^{2+} is released from the ER through the IP_3R into the cytoplasm of the oocyte, resulting in an intracellular Ca^{2+} increase. Subsequently, plasma membrane Ca^{2+} ATPases (PMCAs) and Na^+/Ca^{2+} exchangers export a big part of the cytoplasmic Ca^{2+} ions to the extracellular milieu. As a consequence, intracellular Ca^{2+} levels will be going back to baseline Ca^{2+} levels [54,55]. However, the intracellular Ca^{2+} stores in the ER are limited. Therefore, the need for Ca^{2+} ions migrating from the extracellular milieu to the oocyte cytoplasm is required to support the Ca^{2+} oscillations that are observed after fertilization in mammals. This Ca^{2+} import is believed to be achieved by both store-operated Ca^{2+} entry (SOCE) [54,55] and sarcoendoplasmic reticulum Ca^{2+} ATPases (SERCAs) that facilitate the influx of Ca^{2+} toward the oocyte cytoplasm and the ER of the oocyte, respectively [53].

In mammals, the two SOCE key players are the stromal interaction molecule (STIM) 1 and 2 and the Ca^{2+}-release-activated Ca^{2+} channel protein 1 (Orai1) [56,57]. Whereas Orai1 is a plasma membrane-associated Ca^{2+} channel, STIM 1 and 2 are ER membrane-associated Ca^{2+} sensors. STIM 2 is only activated when small decreases in ER-stored Ca^{2+} occur, whereas STIM1 is activated when the ER stores are drastically depleted of Ca^{2+} ions [58]. When the ER luminal part of STIM1 protein senses low Ca^{2+} levels, STIM1s cluster and migrate toward the plasma membrane of the oocyte. This allows interaction with and subsequent opening of the Orai1 channel, which in its turn results in Ca^{2+} entry into the oocyte. This Ca^{2+} is subsequently pumped back into the ER by the SERCAs. This Ca^{2+} cycling mechanism ensures the prolonged period of Ca^{2+} rises [56,58,59] (see Fig. 14.3).

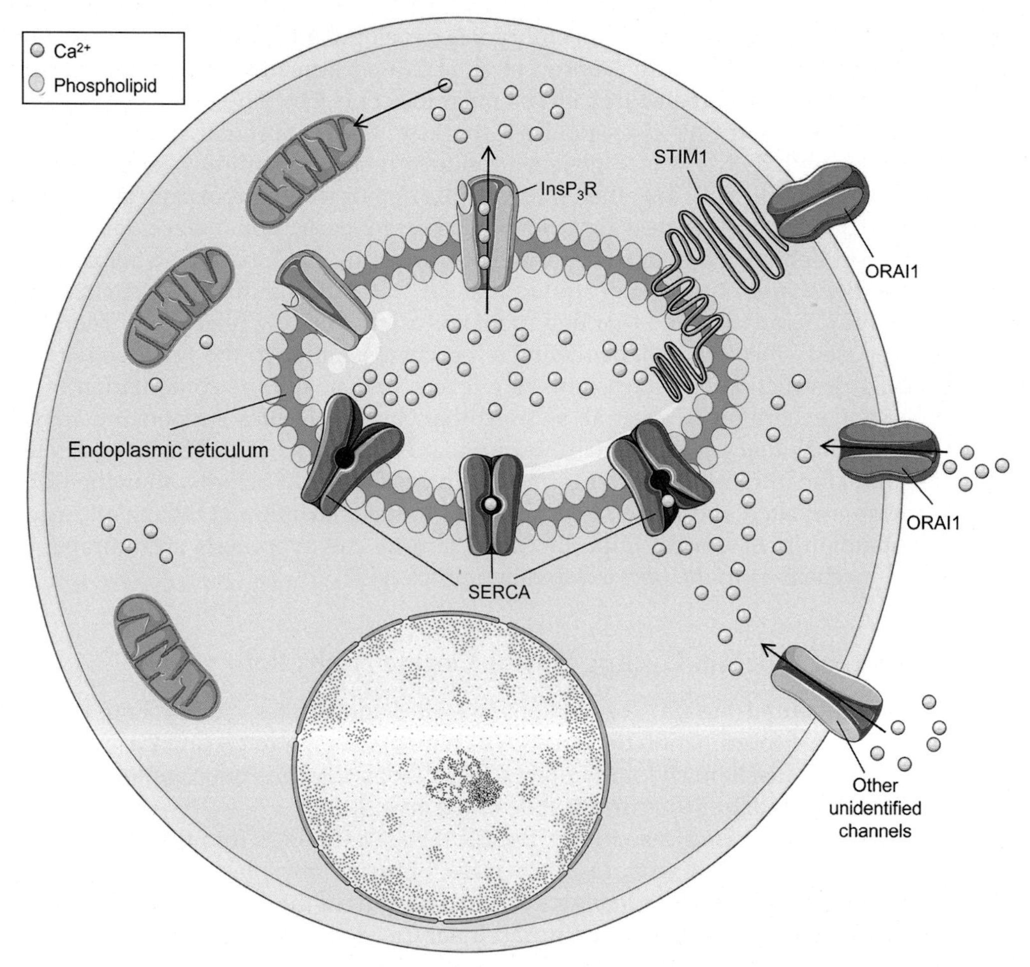

FIG. 14.3 Oocyte mechanisms involved in calcium cycling during oocyte activation. Upon fertilization, Ca^{2+} is released from the ER through the IP_3R into the cytoplasm of the oocyte, resulting in an intracellular Ca^{2+} increase. Since the intracellular Ca^{2+} stores in the ER are limited, the need for Ca^{2+} refilling of these stores is required to support the long-lasting Ca^{2+} oscillations that are observed after fertilization in mammals. This Ca^{2+} import is believed to be achieved by both store-operated Ca^{2+} entry (STIM1/2 and Orai1) and sarcoendoplasmic reticulum Ca^{2+} ATPases (SERCAs) that facilitate the influx of Ca^{2+} toward the oocyte cytoplasm and the ER of the oocyte, respectively. It is also important to note the role of mitochondria. *Reproduced with permission from M. Yeste, et al., Oocyte activation deficiency: a role for an oocyte contribution? Hum. Reprod. Update 22(1) (2016) 23–47.*

Although SOCE has been proved to support Ca^{2+} influx during oocyte activation in mouse and pig experiments [57,60–62], its importance and necessity are being questioned. In mouse, it has been shown that the interference of the STIM1/2-Orai system still resulted in Ca^{2+} oscillations [63,64]. This raises the possibility that SOCE-independent Ca^{2+} influx mechanisms are responsible for sustaining the long-lasting Ca^{2+} transients. Recently, the necessity of STIM1 and STIM2 was considered by creating oocyte-specific conditional knockout (cKO) mice for

STIM1 and STIM2 and STIM1/2 double-cKO mice. Results showed that the absence of STIM1 and/or STIM2 did not impair the ER Ca^{2+} stores, the Ca^{2+} influx following store depletion, or the normal pattern of Ca^{2+} oscillations. Similar results were seen in Orai1 knockout (KO) mice. Together, these data show that the STIM-Orai mechanism is dispensable to support the Ca^{2+} oscillations in mouse oocytes upon fertilization [65]. Mouse studies determined the involvement of different channels in a SOCE-independent way. Experiments using KO mice provided evidence for the essential role of the T-type Ca^{2+} channel CaV3.2 in Ca^{2+} influx to support ER Ca^{2+} stores and Ca^{2+} oscillations [66]. In addition, the transient receptor potential melastatin 7 (TRPM7) channel was found to be a promising candidate to contribute to the Ca^{2+} influx. As this hypothesis was based on results obtained from pharmacological inhibition experiments, a genetic approach will be required to confirm the contribution of TRPM7 or TRPM7-like channel [65]. These results postulate that, for example, porcine oocytes might rely more on SOCE than mouse oocytes do, which could explain the differences seen between species. However, further research is needed to unravel these mechanisms.

Besides the ER, mitochondria also play an important role in oocytes upon fertilization. After IP_3-induced Ca^{2+} release (IICR) from the ER into the oocyte cytoplasm, Ca^{2+} is taken up by mitochondria via the mitochondrial Ca^{2+} uptake 1 (MICU1) channel [67] (see Fig. 14.3). Within the mitochondrial matrix, the presence of Ca^{2+} ions will stimulate the production of energy (ATP). In its turn, mitochondrial ATP is released outside the mitochondria and will be consumed by SERCA pumps to restore the Ca^{2+} levels in the ER, which is then ready to induce a new Ca^{2+} rise in the oocyte cytoplasm. Therefore, mitochondria are also important Ca^{2+} stores that contribute to the Ca^{2+} homeostasis and the induction of Ca^{2+} rises [68,69].

Calcium: The Key Player in Oocyte Activation and Further Development

Regarding the oocyte activation mechanism explained above, there is general consensus that Ca^{2+} fulfills a key role in the oocyte activation process. Upon fertilization in mammals, the Ca^{2+} pattern consists of a specific signature of repetitive Ca^{2+} transients or Ca^{2+} oscillations, which last for several hours. Ducibella et al. found that, in mouse, different oocyte activation events require different numbers of Ca^{2+} oscillations for their initiation and finalization [70]. For example, a single Ca^{2+} pulse is sufficient to start CG exocytosis, whereas meiotic resumption and pronucleus formation require at least four and eight Ca^{2+} pulses, respectively. Besides, the completion of these oocyte activation events requires a greater number of Ca^{2+} transients compared with their initiation [70]. However, other evidence demonstrates that if the total Ca^{2+} signal input reaches a certain threshold, regardless of the number of Ca^{2+} transients, this would be enough to induce both oocyte activation and pre- and postimplantation development [71]. This has led to the hypothesis that the summation of all Ca^{2+} rises and not the specific oscillatory Ca^{2+} pattern is essential to trigger and successfully complete oocyte activation [71,72]. Nonetheless, one has to be careful with extrapolating these results to further development to term. Indeed, data in mammals show that small changes in the oscillatory nature of the Ca^{2+} signal can have a negative impact on both oocyte activation and gene expression and longer-term effects, such as reduced embryonic development, even at the postimplantation level [73–75]. It has to be noted that both oscillatory and monotonic Ca^{2+} stimulation protocols exist, which is paradoxical, and this raises many questions about the necessity and sufficiency of a specific Ca^{2+} pattern to induce normal embryonic development

[72,76]. However, since the oscillatory Ca^{2+} pattern is the normal trigger seen during mammalian fertilization, it is obvious to assume that prolonged Ca^{2+} transients would be of an advantage to cause normal oocyte activation. In conclusion, it is clear that Ca^{2+} plays a key role during the start of the oocyte activation mechanism, which will trigger several downstream Ca^{2+}-dependent pathways in order to achieve successful development.

THE SPERM FACTOR PLCζ

The reproductive biology field community has spent many years of investigating how the sperm could initiate intracellular Ca^{2+} release in the oocyte. Nowadays, it has been generally accepted that in mammalian species, a soluble protein factor is introduced by the sperm into the oocyte cytosol after sperm-oocyte fusion [77,78]. The first evidence for this sperm factor hypothesis was delivered by experiments in sea urchins. It was shown that microinjection of sea urchin cytosolic sperm extracts was able to trigger fertilization in sea urchin oocytes [79]. Besides, sperm extracts from various mammalian species are also able to activate oocytes from different species, meaning that the sperm factor causing Ca^{2+} release is not species-specific [80].

Despite the recent controversies over the last decade, the gamete-specific phospholipase C ζ (PLCζ) is widely considered to be the sole physiological stimulus, which induces Ca^{2+} oscillations and subsequent oocyte activation [14,81–85]. Chronologically, the PLCζ sperm protein was first identified in mouse [14] and subsequently in cynomolgus monkey, human, pig, and rat [85–87]. Later studies have also identified PLCζ orthologs in nonmammalian species, such as the medaka fish and the domestic chicken [87,88]. Although PLCζ is not species-specific for mammals, its intrinsic activity differs between different species. Studies found that the intrinsic activity of PLCζ RNA to induce Ca^{2+} oscillations varies between species: rat $<$ mouse $<$ monkey $<$ human $<$ horse (ranked according to increasing intrinsic activity in mouse oocytes) [85,87,89–91].

Saunders et al. showed that the amount of PLCζ required in order to induce Ca^{2+} oscillations in oocytes corresponds to the amount contained in a single sperm [14]. This study also demonstrated that immunodepletion of PLCζ from sperm extracts abolishes their ability to induce Ca^{2+} oscillatory activity in mouse oocytes [14]. The necessity and importance of PLCζ in mammalian fertilization have been emphasized by a vast amount of biochemical and clinical data supporting PLCζ as the sole sperm factor and by its association with clinical cases of male factor infertility [1,92–96]. Recently, a study demonstrated that sperm derived from PLCζ KO mice fail to trigger Ca^{2+} oscillations in mouse oocytes, resulting in polyspermy. These data provide even more substantial evidence that PLCζ is the key oocyte-activating factor [97].

PLCζ Structure

PLCζ has a typical PLC domain and is characterized by the most basic domain organization among all PLC isoforms, consisting of four EF-hand domains at the N-terminus, the characteristic X and Y catalytic domains, flanked by a C-terminal C2 domain [14,26]. Recently, an unstructured region between the X and Y domains, referred as the X-Y linker, was discovered

[98]. In contrast to the other PLC isoforms, PLCζ lacks a pleckstrin homology (PH) domain and Src homology (SH) domains. This makes PLCζ the smallest of all 13 known PLC isozymes with a molecular weight of 70 kDa in humans and 74 kDa in mice [14,85]. PLCζ shares the highest homology with phospholipase C delta 1 (PLCδ1) (47% similarity and 33% identity). Hence, a plethora of data regarding the role of PLCζ are based on knowledge acquired by structural studies on PLCδ1 [14,99].

The X and Y catalytic domains of PLCζ form the active site of the enzyme, which is conserved among and characteristic to all PLCs [17,100]. Opposite to the autoinhibitory role of the X-Y linker region in the majority of the PLCs [101,102], the X-Y linker in PLCζ is suggested to be essential for binding to its substrate, namely, PIP_2 [98]. The different function of the X-Y linker region of PLCζ compared with other PLCs can be explained on a molecular level. The PLCζ X-Y linker region is longer and contains a cluster of basic residues, not found in other PLC isoforms [14,25,85]. Nomikos et al. showed that the positively charged residues within this X-Y linker domain help in targeting PLCζ to the negatively charged PIP_2 via electrostatic interactions [98,103,104]. However, removal of the X-Y linker does not completely abolish the enzymatic activity of PLCζ [98]. Furthermore, it has been suggested that the high Ca^{2+} sensitivity of PLCζ can be attributed to the EF-hand domains, which is a major determinant of the ability of the enzyme to effectively trigger high-frequency Ca^{2+} oscillations compared with the rest of the PLC isoforms [105]. This is underlined by studies deleting both EF-hand domains, which dramatically increased the Ca^{2+} concentration for half-maximal PLCζ activity, EC_{50}, from 80 nM to 30 μM [106]. Substitution of PLCζ EF-hand domains with that from PLCδ1 led to a tenfold decrease in Ca^{2+} sensitivity, without affecting any other biochemical properties of the enzyme, such as oscillation-inducing activity or PIP_2 affinity [107]. In addition, it has been shown that the positively charged residues of the N-terminal lobe of the EF-hand domain, together with the X-Y linker domain, interact with the negatively charged PIP_2, making it susceptible to hydrolysis by the catalytic domains [105]. The exact role of the C2 domain in the PLCζ function still remains unclear. However, the importance of this domain is shown in cases where deletion of the PLCζ C2 domain or replacement by the PLCδ1 C2 domain abolished its Ca^{2+} oscillation-inducing activity, leaving the Ca^{2+} sensitivity of the enzyme unaffected [107]. There is also evidence that the PLCζ C2 domain is not important for binding to PIP_2 [104,108].

PLCζ and Its (Biochemical) Characteristics

Immunocytochemistry analysis examining PLCζ distribution revealed that PLCζ has the specific feature to target cytoplasmic PIP_2-containing vesicles in the oocyte cytoplasm [21]. Furthermore, experiments demonstrated that removal of phosphates from PIP_2, by a specific inositol phosphatase that targets only plasma membrane PIP_2 molecules, did not inhibit PLCζ-mediated Ca^{2+} oscillations. However, directing the phosphatase to the cytosolic small vesicles significantly blocked the PLCζ-mediated Ca^{2+} oscillations [21].

Another important characteristic of PLCζ is its inability to induce Ca^{2+} oscillatory activity in somatic cells. Studies induced PLCζ expression in Chinese hamster ovary (CHO) cells to levels ~1000 times higher than the PLCζ concentration found in oocytes [109]. Nevertheless, no differences were seen in the PLCζ-transfected cells for resting Ca^{2+} levels or Ca^{2+} responses to extracellular ATP, compared with control cells. In addition, sperm extracts containing PLCζ failed to induce Ca^{2+} oscillations in CHO cells. Despite these observations, PLCζ-transfected CHO cell extracts displayed high recombinant protein expression and PLC enzymatic activity. In contrast, when extracts from PLCζ-transfected cells were injected into mouse oocytes, Ca^{2+} oscillations were observed [109]. These findings suggest that either an oocyte-specific factor provides the necessary interactions with PLCζ leading to the induction of Ca^{2+} oscillations or that somatic cells contain factors that inhibit PLCζ-mediated Ca^{2+} oscillations [25,109]. It is also possible that this specificity is related to the localization pattern of PLCζ in the oocyte [110].

PLCζ is very effective in generating Ca^{2+} oscillations in mammalian oocytes, because of its biochemical characteristics compared with other PLCs. PLCζ is effective at the very low concentration of ~40 fg. In contrast, PLCδ1-induced Ca^{2+} release requires concentrations of approximately 40 times more than PLCζ (>1 pg), while PLCβ1 or PLCγ are ineffective to induce Ca^{2+} oscillations in mammalian oocytes, even at concentrations reaching 5 pg [14,78,106]. PLCζ exhibits high Ca^{2+} sensitivity, with an EC_{50} of 80 nM, making PLCζ 100-fold more Ca^{2+} sensitive than the PLCδ1 isoform. At a resting Ca^{2+} level concentration of ~100 nM, PLCζ is half-maximally active. Small increases in intracellular Ca^{2+} will significantly affect PLCζ activity and subsequently cause the increase of IP_3 production [106], and hence, there is a positive feedback loop of Ca^{2+} and IP_3 response [106]. Furthermore, in mouse oocytes injected with PLCζ, IP_3 levels follow in synchrony the changes in Ca^{2+} concentration during their oscillatory period [111].

PLCζ Localization in Sperm Cells

Numerous issues arise when attempting to identify the PLCζ localization in mammalian sperm, particularly in human. PLCζ appears to be located in distinct regions within the sperm head in mammals, with a potential different role assigned to each population [90,112,113]. Three distinct populations have been identified in human, namely, acrosomal, equatorial, and postacrosomal regions of the sperm head, with a potential additional population in the sperm tail as indicated by immunofluorescence studies [84,95,99,112,114,115]. In mouse and porcine sperm, PLCζ is located in acrosomal and postacrosomal regions, with a tail population also identified in porcine sperm [84,113,116,117]. The postacrosomal localization of PLCζ enables the rapid diffusion into the oocyte cytoplasm, enabling Ca^{2+} oscillations to occur immediately after sperm-oocyte fusion [118]. So, the localization of PLCζ further supports the role of PLCζ as the oocyte-activating sperm factor.

PLCζ Deficiency and Male Factor Infertility

Male factor infertility accounts for 25%–30% of all infertility cases and might be caused by aberrations in various factors, including semen quality, sperm morphology, or other epigenetic modifications [119]. Based on the type of morphological defects, sperm samples are classified as oligozoospermia, teratozoospermia, asthenozoospermia, azoospermia, or multiple morphological abnormalities of sperm flagella (MMAF) [119–123]. There are several genes linked to be responsible for the causation of these phenotypes; however, 10%–25% of the male fertility cases still remain unexplained [93]. The advent of ICSI to overcome specific types of male infertility shed a light of hope for the patients suffering from male infertility.

However, in 1%–3% of the ICSI cycles, total fertilization failure occurs even after repeated ICSI cycles, and the underlying cause is believed to be the failure of the oocyte to activate (OAD) [124].

Research over the years has shown that PLCζ deficiency in infertile patients is associated with OAD. Yoon et al. reported the first link between PLCζ and OAD. In this study, sperm from patients that failed to activate human oocytes after ICSI could not initiate Ca^{2+} oscillations when injected into mouse oocytes [115]. Heytens et al. showed the first genetic link between PLCζ and OAD, with the identification of a point mutation in the PLCζ gene in an infertile male giving rise to failed fertilization after ICSI [95]. The mutation substitutes the amino acid histidine to proline at position 398 in the Y catalytic domain of the PLCζ active site (PLCζH398P) (see Fig. 14.4). The injection of PLCζH398P complementary RNA (cRNA) into mouse oocytes resulted in abnormal Ca^{2+} oscillations that were unable to activate oocytes. Protein modeling on this substitution revealed that the substitution would result in the disruption of the local folding of the active site of PLCζ [95]. The cRNA of the PLCζH398P equivalent mutation in mouse PLCζ, PLCζH435P, failed to induce Ca^{2+} oscillations in mouse oocytes [125]. However, the major question that arose was how a heterozygous PLCζH398P mutation could lead to infertility in a patient. One hypothesized that the infertility of the patient could be due to a dominant negative mechanism [95]. However, injection with both cRNA wild-type PLCζ and a 10 times more concentrated cRNA mutant PLCζH435P into mouse oocytes elicited a normal Ca^{2+} oscillation pattern. This finding indicates that PLCζH398P doesn't act in a dominant negative manner [125]. Kashir et al. demonstrated that the wild-type PLCζ (PLCζWT) protein could induce Ca^{2+} oscillations in mouse oocytes typical of those seen at fertilization, whereas the PLCζH398P recombinant protein failed to induce Ca^{2+} oscillations in mouse oocytes [126].

Furthermore, Kashir et al. identified a second point mutation in the same patient after analyzing his genomic DNA [96]. The mutation occurred in the X catalytic domain of the PLCζ active site resulting in the substitution of histidine to leucine at position 223 of the PLCζ open reading frame (ORF) (PLCζH223L) (see Fig. 14.4). The injection of PLCζH233L cRNA into mouse oocytes resulted in abnormal Ca^{2+} oscillation patterns. In this study, they also performed segregation analysis of the mutation by sequencing both parents and the half-brother and daughter, who was born after application of assisted oocyte activation (AOA) [96]. The segregation analysis revealed that the PLCζH398P mutation was inherited from the father and the PLCζH223L mutation was inherited from the mother [96]. The half-brother didn't inherit any mutation, and the daughter inherited the PLCζH223L mutation from the patient, her father [96]. The two mutations were also analyzed using custom-made single-nucleotide polymorphism assays in 100 fertile men and 8 infertile patients with OAD, but the mutations were not found in these fertile and infertile (OAD) men. However, it has to be noted that six out of the eight OAD patients were suffering from globozoospermia, a genetic syndrome that

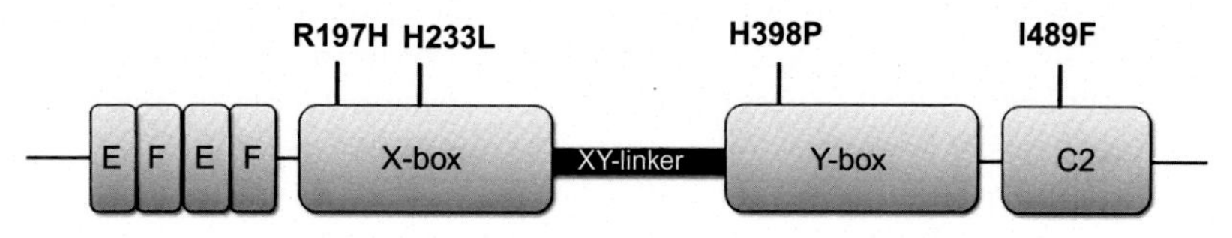

FIG. 14.4 Physical localization of mutations identified in PLCζ in different studies.

results in the formation of a sperm with abnormal morphology and that has genes other than PLCζ linked to this occurrence [96] (see "PLCζ Deficiency in Globozoospermic Individuals" section).

More genetic evidence for the involvement of PLCζ mutations in patients with failed fertilization after ICSI came from a study from Escoffier et al. [94]. In this study, they performed a whole exome sequencing (WES) on two infertile Tunisian brothers who had failed fertilization after ICSI. The WES analysis revealed four homozygous variants common in both infertile brothers. The variants in the EPS8, RP11-1021N1.1, and LKAAEAR1 gene were predicted not to be deleterious. The variant in the C2 domain of the PLCζ gene (c.1465A>T; I489F) was found homozygous in both the affected brothers and heterozygous in the third fertile brother [94] (see Fig. 14.4). Since they found that I489 is well conserved across species, the I489F mutation could be deleterious. The I489F mutation was not found in the 60,000 individuals described in the ExAC database (exac.broadinstitute.org). The found amino acid substitution was confirmed using Sanger sequencing. Besides, the PLCζ immunofluorescence analysis on the sperm of both patients showed no signal except faint punctuates staining over the acrosome; however, Western blot analysis didn't yield any signal. Following the injection of WT and I489F cRNA in mouse oocytes, they found that WT PLCζ showed homogenous distribution in the ooplasm and partly overlaps with the distribution of the ER, whereas the I489F PLCζ showed uneven distribution that didn't overlap with the ER. Altogether, these findings suggest that the mutation affects the targeting and/or anchoring properties of PLCζ, which probably hinders the retention of PLCζ in mature sperm. Moreover, the PLCζ protein quantification by Western blot analysis on the sperm cells of the patients showed no signal, whereas a clear band of approximately 70 kDa was observed in the protein extracts from fertile men [94].

Another study analyzed PLCζ in 15 patients who had low or failed fertilization. Sequencing analysis of PLCζ in those 15 patients resulted in the identification of a heterozygous missense variant, R197H, in one of the patients [127]. The R197H mutation is located on the X catalytic domain of the PLCζ protein (see Fig. 14.4). However, PLCζ protein levels, distribution, and localization in the sperm cells from the patient showed no significant difference with that of fertile donor sperm cells [127].

So, to date, there are four missense mutations reported in the PLCζ gene in four different patients experiencing low or complete failed fertilization after ICSI: H398P, H233L, I489F, and R197H [94–96,127] (see Fig. 14.4). The mutation prediction analysis (by Polyphen and SIFT) on the four mutations showed that all the mutations are damaging, except for the H233L mutation (see Table 14.1).

All the studies till now have been focusing on the coding region of PLCζ. However, sequencing of the intronic and promoter regions of PLCζ is very important. Mutations in these regulatory regions might affect the overall expression of the gene. Indeed, a study on Chinese Holstein bulls has shown that genetic variants in the promoter region of PLCζ affect the semen quality traits and thereby fertilization outcome [128]. Since PLCζ and another testis-specific gene called CAPZA3 contain a common bidirectional promoter, it is worth comparing both the PLCζ and CAPZA3 expression levels in fertile and infertile men. In a more recent study, a parallel assessment of PLCζ and CAPZA3 mRNA levels was performed on patients with failed fertilization after ICSI and on fertile controls [129]. The results showed a correlation between the mRNA expression levels of both PLCζ and

TABLE 14.1 List of Mutations Identified in the PLCζ Gene in Different Studies

Author	Coding Region Change	Amino Acid Change	ExAC Browser (Allele Frequency)	Polyphen	SIFT
Heytens et al. (2009) [95]	c.1193A>C	p.His398Pro	No frequency	Damaging	Damaging
Kashir et al. (2012) [96]	c.698A>T	p.His233Leu	0,0007	Benign	Tolerated
Escoffier et al. (2016) [94]	c.1465A>T	p.lle489Phe	0,000008	Damaging	Damaging
Ferrer-Vaquer et al. (2016) [127]	c.590G>A	p.Arg197His	0,000016	Damaging	Damaging

To date, four missense mutations are reported in the PLCζ gene from four different patients experiencing low or complete failed fertilization after ICSI, namely, H398P, H233L, I489F, and R197H. H398P and H233L mutations were found in the same patient in a compound heterozygous state, located on the X and Y catalytic domain of PLCζ, respectively. The I489F mutation was found in a homozygous state in two infertile brothers and is located on the C2 domain of PLCζ. The R197H mutation was found in heterozygous state in a patient with low fertilization failure and is located on the X catalytic domain of the protein.

CAPZA3. Furthermore, men experiencing low fertilization rates appeared to have low expression of both genes. Sequencing one of the patients with low expression helped in identifying a mutation in the promoter region of CAPZA3; however, this mutation is located outside but close to the PLCζ promoter region. Since reduced PLCζ levels were observed in this patient, further studies are required to know if this region is also part of the PLCζ promoter and if this mutation is the cause for the low PLCζ expression [129]. Therefore, including the regulatory and intronic regions of the PLCζ gene in the mutation screening tests are of utmost importance.

PLCζ Deficiency in Globozoospermic Individuals

Globozoospermia is a condition, which is characterized by the presence of round spermatozoa lacking the acrosome [130,131]. Globozoospermia was first reported in 1965 [132], and Schirren et al. later described the structural details of globozoospermia using electron microscopy [133]. Globozoospermia can be of two types: (i) total globozoospermia, where the ejaculate contains 100% of round-headed sperm cells without an acrosome, and (ii) partial globozoospermia, where the ejaculate comprises less than 100% of acrosomeless round-headed sperm cells [134].

Few genes have been related with globozoospermia, among others DPY19L2 [135–140], which is the major gene causing globozoospermia, and SPATA16 [141,142]. PLCζ is likely to be associated with the inner acrosomal membrane (IAM) and the perinuclear theca facing the acrosome [84,143]. After the acrosome reaction, both the IAM and the perinuclear theca remain, which enables the associated PLCζ to be delivered into the oocyte with sperm-oocyte fusion. However, in DPY19L2-associated globozoospermia, the acrosome and the acrosome-associated perinuclear theca and outer nuclear envelope are discarded from the sperm cells [144,145]. Besides, the SPATA16 protein is localized to the Golgi apparatus and to the proacrosomal vesicles, which are transported to the acrosome during spermatogenesis, which suggests its role in acrosome formation [141,146]. So, this can explain the reason for infertility in patients with globozoospermia.

The identification of many familial cases of globozoospermia and also the KO mouse models showing the teratozoospermic phenotype suggest a strong genetic contribution for the occurrence of this phenotype [147–151]. To date, there are approximately 63 genetically modified mice reported in the Mouse Genome Informatics database (http://www.informatics.jax.org) with a phenotype related to globozoospermia. However, only a few mutations have been identified in their human orthologs.

Globozoospermic patients have varied PLCζ expression levels in their sperm cells, going from no expression to very low PLCζ expression [95,152]. Since PLCζ is required for normal oocyte activation, sperm lacking the PLCζ protein usually cannot initiate oocyte activation [14]. Besides, sperm cells with low PLCζ protein levels result in low-frequency Ca^{2+} oscillations in mouse oocytes [95]. Although studies demonstrated successful activation of the oocyte resulting in live births in some men with globozoospermia [147,153–160], Ca^{2+} ionophore treatment is generally used to overcome fertilization failure after ICSI in most cases of globozoospermia [142,161–166]. So, in some cases, a single sperm analysis might be sufficient to diagnose globozoospermic patients and thus unmask their OAD.

The Genetic Link With Globozoospermia

A homozygosity mapping using genome-wide scan analysis from an Ashkenazi Jewish family with three globozoospermic brothers revealed a homozygous mutation (c.848G>A) in the SPATA16 gene [141]. Mutations in the SPATA16 gene are rare; however, a new SPATA16 mutation was recently reported in a single man from a large cohort of globozoospermic men [142].

In 2011, an SNP-based genetic linkage analysis study was performed on a cohort of 20 Tunisian patients with total globozoospermia. This analysis identified a common homozygosity region on 12q14.2, which includes DPY19L2, a gene strongly expressed in the testis. Fifteen out of the 20 patients analyzed carried a 200 kb deletion in the DPY19L2 gene [135]. Several other studies on globozoospermic patients identified a DPY19L2 deletion in different frequencies based on the ethnic background of the patients [136–140]. Homozygous and compound heterozygous point mutations in DPY19L2 further broaden the mutational spectrum of DPY19L2-dependent globozoospermia [137–139]. This is making DPY19L2 mutations a major cause of globozoospermia. Conventional ICSI in nine globozoospermic patients with complete deletion of the DPY19L2 gene surprisingly resulted in some fertilization in some of those couples. These outcomes can be explained by the fact that still, 10% of human sperm showed traces of PLCζ associated with small remnants of the acrosome [143].

Mutations in two genes, ZPBP1 and PICK1, are reported to be associated with a globozoospermia-like phenotype, although a stronger evidence is required for their involvement in the causation of the disease. A heterozygous missense mutation and a splice mutation were reported in ZPBP1 in patients with abnormal sperm head morphology, but their involvement in the cause of the disease is not well established [167]. Similarly, in a Chinese family affected with globozoospermia, a homozygous missense mutation (G198A) in exon 13 was identified in the PICK1 gene [168]. However, the identification of a relatively mild mutation without any functional validation in a single familial case makes this gene not a very convincing candidate. Identification of more mutations in the PICK1 gene in globozoospermic patients and further functional analysis of the mutations found are required to make PICK1 as a candidate causative gene for globozoospermia [168].

THE OOCYTE FACTOR IP_3R

To date, it has been clear that PLCζ deficiency caused by abnormal localization, reduced activity/expression, or genetic mutations in PLCζ has been associated with OAD and ICSI failure. However, not all cases of OADs can be attributed to deficiencies in this sperm factor, supporting also the pivotal role of the oocyte in the activation mechanism. Indeed, PLCζ can be expressed in somatic cells, but cannot elicit Ca^{2+} oscillations in these cells, suggesting an essential role for oocyte-related factors in eliciting Ca^{2+} oscillations [109]. So, although the sperm factor PLCζ is able to elicit Ca^{2+} oscillations in the oocyte, suggesting that the oocyte calcium-releasing machinery is functional, it does not guarantee successful fertilization.

The ability of an oocyte to correctly respond to an activating factor (e.g., PLCζ and AOA agents) develops during the meiotic maturation of the oocyte. Meiotic maturation events starting from the first meiotic division (prophase I) to the metaphase of the second meiotic division (metaphase II, MII) render the mammalian oocyte able to generate the necessary sperm-induced long-lasting Ca^{2+} oscillations. It is known that several important cytoplasmic changes take place during final oocyte maturation, including reorganization of ER with an increase of Ca^{2+} concentration in the ER lumen and an increase in the number and sensitivity of the IP_3 receptors. All these events are required in order to result in the correct Ca^{2+} pattern needed for successful embryo development [66,169,170]. Specific attention has to be paid to the IP_3R, since it is a very important oocyte factor during oocyte activation. Multiple IP_3R modifications occur during oocyte maturation affecting its sensitivity and ensuring that the Ca^{2+} machinery of the oocyte is optimized. As shown in animal models, these IP_3R modifications include its redistribution, elevated levels, and IP_3R posttranslational modifications, in particular phosphorylation [171–175].

During mouse oocyte maturation, the IP_3Rs reorganize. Whereas small IP_3R clusters (<1 μM) are located around the entire oocyte cortex in immature oocytes, large IP_3R clusters (1–2 μM diameter) are located in the cortex adjacent to the meiotic spindle in mature oocytes [169]. This latter region corresponds to the location where sperm-egg fusion occurs [176], the CGs are located [177,178], and the Ca^{2+} rises are generated [179]. Although one study showed similar observations in human oocytes [180], another study could not confirm these findings, although the same polyclonal antibody was used [181]. Also, the number of IP_3Rs increases during oocyte maturation resulting in an almost twofold increase in mature oocytes compared with immature oocytes [169,182,183]. The importance of IP_3R is also underlined in studies, where RNA interference (RNAi) was used to reduce the expression of IP_3R in mouse oocytes, which caused the reduction of the IP_3R and subsequently a significant decrease in the frequency of Ca^{2+} oscillations [184]. Also, posttranslational modifications are important for the regulation of the IP_3R function, with an important role for the cell cycle kinases (M-phase kinases). Polo-like kinase 1 (Plk1), mitogen-activated protein kinase (MAPK), and CDK1 are the most important players in modulating the sensitivity of the IP_3R [19]. Plk1 is known to be important during the early stage of the oocyte maturation [175]. Studies in mouse and *Xenopus* showed that Plk1 phosphorylates an MPM-2 epitope present on the IP_3R1 [172,175]. However, it is suggested that Plk1 is not the only cell cycle kinase involved in this IP_3R phosphorylation process [172]. Indeed, MAPK is one of the other kinases involved in IP_3R phosphorylation, also targeting MPM-2 epitopes during mouse and *Xenopus* oocyte maturation. However, the site of phosphorylation between both species is located on a different site on the IP_3R [185,186]. A third kinase important in increasing the IP_3R sensitivity is CDK1, the subunit of MPF. Data in *Xenopus* showed that both CDK1 and MAPK need to be simultaneously activated in order to achieve full sensitization of the IP_3R in the oocyte [186]. In addition, other wide-ranging kinases are also able to increase IP_3R-mediated Ca^{2+} release, such as PKA, PKC, and CaMKII, which all have an important role in oocytes [37]. Furthermore, the role of the IP_3R during fertilization has been shown by several studies. When IP_3R inhibition is induced by a monoclonal antibody targeting the receptor, sperm-induced Ca^{2+} transients remain absent [187]. Oppositely, microinjection of IP_3 or adenophostin (an IP_3 analog) exposure is able to induce Ca^{2+} oscillations in mammalian oocytes [80,188]. Unfortunately, the role of the IP_3R has not been fully elucidated in human. In conclusion, the IP_3R is one of the main factors in the oocyte, located upstream in the oocyte activation process, which requires definitely more attention.

POSSIBLE OTHER SPERM-RELATED OR OOCYTE-RELATED FACTORS ESSENTIAL FOR OOCYTE ACTIVATION

Other Sperm Factors

Studies with the aim to identify novel sperm factors have been at the forefront of developmental research. Although the evidence supporting PLCζ as the sperm factor is getting stronger through the years, several other sperm factor candidates have been proposed such as citrate synthase [189], the truncated form of the KIT receptor (TR-KIT) [190], and the most controversial factor postacrosomal sheath WW domain-binding protein (PAWP) [191,192].

Citrate synthase is a 45kDa protein that is present in sperm extracts of newt (*Cynops pyrrhogaster*) [189]. Harada et al. reported that citrate synthase is able to induce egg activation by an intracellular Ca^{2+} increase when microinjected into unfertilized newt eggs. The findings in this study suggest that citrate synthase is one of the main components of the sperm factor for egg activation in newt [189]. However, its importance in other species, like mammals, has not been shown.

TR-KIT becomes expressed in the postmeiotic stages of spermatogenesis and is accumulated in mature sperm cells. In spermatozoa, TR-KIT is expressed in the postacrosomal region of the sperm head and in the sperm midpiece [190]. Sette et al. reported that mature mouse oocytes underwent complete parthenogenetic activation after microinjection of synthetic TR-KIT mRNA or recombinant TR-KIT [190].

PAWP is an alkaline extractable protein with a proline-rich C-terminus that is localized to the postacrosomal sheath-perinuclear theca (PAS-PT) of the sperm head [191,192]. The structure of PAWP consists of an N-terminal that exhibits sequence homology to WW domain-binding protein 2, a C-terminal comprising a PPXY consensus binding site and an unknown repeating motif (YGXPPXG) [191,193]. It has been shown that microinjection of recombinant PAWP into porcine, bovine, and *Xenopus* oocytes promotes meiotic resumption and pronucleus formation [191] and its injection causes Ca^{2+} oscillations in *Xenopus* oocytes [193]. Furthermore, mouse and human oocytes microinjected with PAWP cRNA or recombinant PAWP induced Ca^{2+} oscillations and oocyte activation similar to those after ICSI [194]. On the contrary, a competitive inhibitor, PAWP-derived *PPGY* peptide, blocked the sperm-induced Ca^{2+} oscillations [194]. Moreover, they found correlations between PAWP expression levels and fertilization and embryonic development, with high PAWP levels being associated with higher fertilization rates and lower number of arrested embryos post-ICSI [195]. However, one other group was not able to replicate the same results since they found that recombinant mouse PAWP protein did not show any in vitro PIP_2 hydrolyzing activity and both PAWP cRNA and recombinant protein were unable to elicit Ca^{2+} oscillations in mouse oocytes [196]. Satouh et al. created PAWP KO mice of which the sperm cells showed no defects in spermatogenesis and did elicit normal Ca^{2+} oscillation patterns and normal embryo development in mouse. These findings showed that PAWP is not a prerequisite for male fertility and spermatogenesis in mouse [197]. In human, it has been shown that the I489F mutation led to infertility in one patient (see "PLCζ Deficiency and Male Factor Infertility" section); however, PAWP showed normal expression and localization [94]. Another study reported that both PLCζ, PAWP, and TR-KIT mRNA and protein expression levels were significantly lower in globozoospermic individuals ($n=12$) compared with fertile men ($n=12$) [198]. Extended analysis reinforced these findings, with significantly reduced mRNA and protein levels of both PLCζ and PAWP in patients with globozoospermia ($n=21$) compared with fertile men ($n=25$) [199]. Anyhow, further research is required to eliminate the controversies about PAWP and its involvement in oocyte activation.

Other Oocyte Factors

During oocyte activation, various oocyte-specific factors in the oocyte participate in order to complete the fertilization process successfully. Aberrations in one of these factors can prevent proper activation of the oocyte and further embryonic development [1]. Research about the importance of all these factors will be shortly described in this section.

As we mentioned before, the sperm factor PLCζ specifically targets PIP_2-containing cytoplasmic vesicles in the oocyte and hydrolyzes PIP_2 into IP_3 and DAG [21]. It has been shown that a reduced number of PIP_2-containing vesicles or vesicles with depleted PIP_2, induced by inositol phosphatases, led to the abolishment of Ca^{2+} oscillations during fertilization [200]. In addition, deficits or abnormal localization of PIP_2 in the cytoplasmic vesicles results in OAD [21]. Once the PIP_2 is hydrolyzed, the formed IP_3 molecules will bind to the ER-located IP_3R that will trigger a downstream cascade [109,201]. The function and importance of the IP_3R has been described above (see "The Oocyte Factor IP_3R" section).

Besides, sperm entry also triggers the PKC pathway, which is targeted by DAG [38]. PKC signaling has been proposed to be involved in CG exocytosis and the generation of the second polar body [202,203]. It has been shown that PKC activity can affect the Ca^{2+} pattern at fertilization. When triggering the PKC signaling by PMA or PKCα overexpression, Ca^{2+} oscillations were induced, whereas PKC inhibition by bisindolylmaleimide did prevent the occurrence of Ca^{2+} transients [39]. However, there has been some controversy about the importance of PKC in oocyte activation. While some report PKC to be essential upon fertilization [200], others indicate that the activation of CaMKII is sufficient to support all oocyte activation events [204,205]. Therefore, further research is required to reveal if PKC abnormalities might be the cause for some OADs.

Since the necessity of Ca^{2+} influx into the oocyte to induce long-lasting Ca^{2+} oscillations (see "The Regulation of Calcium Cycling During Oocyte Activation" section), aberrations in the elements of the SOCE system might be harmful for the activation of the oocyte. Therefore, loss of function due to genetic mutations or abnormal trafficking of STIM1/2, ORAI1, SERCA, and other membrane channel proteins could lead to Ca^{2+} depletion in the ER, which prevents further Ca^{2+} oscillations, oocyte activation events, and embryo developments due to the lack of Ca^{2+} in the Ca^{2+} stores [206].

Also, abnormalities in other possible oocyte factors downstream of the Ca^{2+} oscillations, such as CaMKII and MAPK, can result in oocyte activation failure. Genetic mutations in CaMKII and/or MAPK might prevent the completion of the oocyte activation, despite the presence of Ca^{2+} oscillations in the oocyte [207].

From the above, it is clear that there are several key oocyte proteins involved in the cascade triggered by the sperm factor upon fertilization. Far less attention has been paid to the potential involvement of any specific oocyte factor in OAD, but more research on the identification of oocyte-related factors, especially in human, is required to unravel the oocyte activation mechanism and to generate new targets for clinical diagnostic assays or therapeutic interventions.

DIAGNOSTIC TOOLS TO REVEAL THE OOCYTE ACTIVATION POTENTIAL IN HUMAN SPERM

Following low or failed fertilization after ICSI, the question arises whether this outcome is due to a sperm-related or an oocyte-related factor. Determining the underlying cause of this oocyte activation problem is of utmost importance, in order to give the patients the right prognosis and to define their future treatment options. Also, we should be aware of the possible genetic origin of this fertilization failure, as shown in some found mutations in PLCζ, which can be transmitted to their progeny (see "PLCζ Deficiency and Male Factor Infertility" section). Nowadays, some appropriate and reliable sperm-specific diagnostic tests are available. These diagnostic approaches include sperm morphology evaluations, quantification, and localization of PLCζ; PLCζ genetic screening; and the determination of the activation and Ca^{2+} oscillating capacity of the spermatozoa.

Immunofluorescent Staining/Western Blotting/qPCR

Numerous studies report that the assessment of PLCζ in human sperm can be used to develop a useful diagnostic tool. Indeed, different PLCζ diagnostic approaches (such as immunofluorescent staining, Western blot, and qPCR) do already exist for diagnosing patients with OAD, but with low predictive potential.

Immunofluorescent analyses on the sperm of individuals with low or failed fertilization showed reduced/absent levels of PLCζ compared with fertile control individuals [94–96,115,126,152,208]. This reduced expression is possibly caused not only by an outright absence of expression but also by an incorrect pattern of localization. However, few studies did not detect any correlation between the PLCζ levels and the fertilizing capacity of the sperm [127]. Therefore, the expression of PLCζ might be more complex than was expected. Kashir et al. have discovered a significant variance in the localization pattern and total levels of PLCζ within the sperm samples of both fertile men and OAD patients when performing immunofluorescent analysis [209]. These results correspond to the findings of Grasa et al.

who also detected variability in PLCζ localization among individuals, although that study analyzed sperm from control subjects only [112]. Therefore, quantitative immunofluorescence analysis might not be eligible for serving as a clinically prognostic indicator of the oocyte activation capability. On the contrary, a recent study including more patient samples showed that the total levels of PLCζ did correlate with the fertilization outcome after ICSI, although some cases experiencing total fertilization failure couldn't be explained by reduced/absent PLCζ expression levels [210]. All of the analyses solely depend on the use of polyclonal antibodies, so one should be cautious in interpreting these results as polyclonal antibodies can be unspecific. In conclusion, there is no clear association between the total levels of PLCζ and fertility status, thus requiring a need for a more robust and sensitive assay to determine PLCζ in human sperm. A more extended and detailed analysis, using monoclonal antibodies, is recommended to verify its diagnostic and predictive potential. Furthermore, studies examining the expression levels of human PLCζ fail to exhibit a consistent Western blot profile, even in cases where the same antibody was used [94,95,115,143]. The variance observed in antibody specificity following immunoblotting directly affects the credibility of these findings, reinforcing the need for an epitope-specific monoclonal antibody for PLCζ.

Using quantitative real-time PCR (qPCR), Aghajanpour et al. evaluated the expression of PLCζ mRNA in the semen of globozoospermic men or those with previously low or failed fertilization relative to fertile individuals. The study found that the mRNA expression levels of PLCζ in men with globozoospermia or with previous low or total fertilization failure were significantly lower compared with fertile men (with high fertilization rates). In addition, there was a significant difference in the PLCζ levels between the sperm from globozoospermic patients and patients experiencing low or failed fertilization [211]. However, it is unknown how closely the mRNA expression levels of PLCζ reflect the levels of their corresponding protein.

So, despite the attempts to assess the PLCζ expression levels using immunofluorescent or quantitative analysis, the predictive potential is rather low.

Heterologous ICSI Tests to Reveal Oocyte Activating Capacity of Spermatozoa

Mouse Oocyte Activation Test (MOAT)

Daily, several couples experience low or failed fertilization following standard ICSI treatment. In 1995, a method was developed to assess the sperm-activating capacity of human spermatozoa, called the mouse oocyte activation test (MOAT) [212]. The MOAT compares the activation percentage (two-cell formation) of mouse oocytes after piezo-driven heterologous ICSI using human sperm from the patient (test sperm sample) to control sperm with proven activation capacity (positive control sample). Injection of culture medium (sham injection) and exposure to medium only (medium control) serve as negative controls [213] (see Fig. 14.5).

The MOAT enables classification into three groups depending on the mouse oocyte activation percentage: (i) if 20% or fewer oocytes are activated, patients are classified in MOAT group 1 (low activating group); (ii) patients with a MOAT result between 21% and 84% are classified as MOAT group 2 (intermediate activating group); and (iii) if the mouse oocyte activation percentage is 85% or more, patients belong to the MOAT group 3 (normal/high activating group) [214] (see Fig. 14.5). The origin of previous fertilization problems in MOAT group 1 patients is most presumably a sperm-related deficiency, whereas the cause in MOAT group

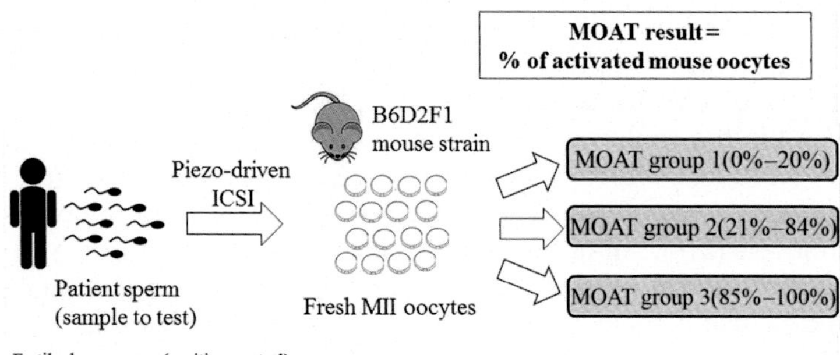

FIG. 14.5 Schematic overview of the mouse oocyte activation test (MOAT). The MOAT is a diagnostic test that evaluates the oocyte activation capacity of human spermatozoa. The MOAT compares the activation percentage of mouse oocytes after heterologous ICSI using sperm from the patient with that of control sperm with proved activation capacity (positive control). Injection of culture medium (sham injection) and exposure of mouse oocytes to the medium only serve as negative controls.

3 patients is most probably an oocyte-related deficiency. For MOAT group 2 patients, experiencing a reduced activation potential of their spermatozoa, it could initially not be clearly demonstrated if the cause of the underlying fertilization problem is oocyte- or sperm-related [214]. Further Ca^{2+} pattern analysis (see "Mouse Oocyte Calcium Analysis (MOCA)" section) revealed a sperm-related origin of OAD in this MOAT group 2.

Although MOAT can reveal sperm deficiencies in some cases, one has to take into account that human PLCζ exhibits greater potency compared with its mouse counterpart [85,215]. Therefore, the use of mouse oocytes for determining the activating capacity of the sperm (by MOAT and MOCA; see "Mouse Oocyte Calcium Analysis (MOCA)" section) might reveal only extreme cases of activation deficiency, while cases where only a subtle reduction of activation capacity is present remain undiagnosed. Furthermore, it has been reported that the oocyte activation rate can double when two instead of one human spermatozoon are injected into a single mouse oocyte, which points toward a quantitative-related activation deficiency [216]. In addition to its relatively low sensitivity, MOAT presents other constraints such as the need for mouse housing facilities and ethical issues regarding the heterologous ICSI procedure. Therefore, only a handful of other IVF laboratories worldwide carried out MOAT to assess the activation potential of spermatozoa of globozoospermic patients or patients with a history of failed fertilization after ICSI [212,213,217,218]. This implies that more easy and user-friendly diagnostic tools are warranted.

Other Heterologous ICSI Tests

Also, some other heterologous ICSI tests have been performed using bovine, hamster, and rabbit oocytes.

After microinjecting hamster oocytes with sperm from patients, the ability of the human spermatozoa to induce sperm head decondensation and male pronucleus formation, which are crucial events during fertilization, was verified. In these studies, sperm samples were used from patients with severe male factor infertility (e.g., abnormal sperm morphology) [219,220] or patients scheduled for IVF/ICSI [221]. Only in one case report, they injected hamster oocytes with sperm from a patient experiencing fertilization failure after IVF and very low fertilization after ICSI. The test revealed that the sperm capacity to activate hamster oocytes was significantly diminished compared to a fertile donor [222].

Since the sperm centrosome is important during fertilization, fertilization failure can be the cause of sperm centrosome dysfunction in the oocyte [223]. Since the centrosome is paternally derived in rabbit oocytes, rabbit oocytes have been used as a model, in particular, to assess the human sperm centrosomal function. Sperm samples from 15 IVF patients were compared with two fertile donor samples. They reported significantly higher sperm aster formation rates in fertile donor sperm ($35.0 \pm 1.5\%$) compared with infertile sperm ($3.6 \pm 2.9\%$). Moreover, sperm aster formation rates in rabbit oocytes were significantly higher in patients experiencing high cleavage rates (>70%) after human IVF ($34.5 \pm 7.1\%$) compared with patients with low cleavage rates (<70%) after IVF treatment ($3.6 \pm 3.2\%$). Although sperm aster formation rates correlated with cleavage rates, the sperm aster formation rates after microinjecting fertile donor sperm into rabbit oocytes were low [224].

Also, bovine oocytes have been used to assess the human sperm centrosomal function since the centrosome is paternally derived into the bovine oocyte [225]. Hereby, human spermatozoa are injected by piezo-driven ICSI in bovine oocytes [226]. Since the sperm aster formation rates with fertile donor sperm were 60%, bovine oocytes are a better approach to assess the human sperm centrosomal function [226]. The centrosomal function was examined in some cases experiencing infertility. The rates of sperm aster formation were 15.8% in a case of globozoospermia [227], 16% in a case of dysplasia of the fibrous sheath (severe teratozoospermia and total sperm immotility) [228], and 47% (average rate) in 15 infertile men undergoing IVF treatment [229]. These rates were significantly lower than sperm aster formation rates of the fertile donor sperm samples, which were, respectively, 57.9%, 68%, and 66.1% [227–229]. In conclusion, this test is suitable to detect human centrosomal dysfunctions and shows that semen characteristics appear to be independent of sperm centrosomal function. Mouse and hamster oocytes cannot be used for this assay since the centrosome is maternally derived in these species [230,231].

Mouse Oocyte Calcium Analysis (MOCA)

Nowadays, the MOAT and other heterologous tests are one of the few diagnostic tools that are carried out to determine the oocyte-activating capacity of human spermatozoa. However, both sperm-related and oocyte-related deficiencies may still be the underlying problem in MOAT group 2 patients, since the result only showed a reduced activation capacity in some sperm samples [214,232].

One of the reasons of very low or failed fertilization after ICSI may be a deficiency in the oocyte to respond to the sperm factor and/or to induce several downstream pathways. On the other hand, failure of the sperm to induce a normal pattern of Ca^{2+} oscillations in the oocyte, which is essential for fertilization and further development, may also be the cause. Nowadays, the latter can be confirmed by the mouse oocyte Ca^{2+} analysis (MOCA). MOCA is a method that detects Ca^{2+} changes in the oocyte after microinjection of the patient's sperm into mouse oocytes and can be of added value [232]. The piezo-driven heterologous ICSI is preceded by exposure of mouse oocytes to a

Ca^{2+}-sensitive fluorescent dye (e.g., Fura-2 AM) and a plasma membrane-disrupting agent, L-α-lysophosphatidylcholine (Lyso-PC). The ICSI procedure is followed by a 2h measurement with an inverted epifluorescence microscope, providing excitation alternations between 340 and 380nm for Fura-2 AM. The intracellular Ca^{2+} levels are recorded as the ratio of fluorescence (340/380nm), which increase with rising intracellular Ca^{2+} concentration. Each oocyte is scored depending on the frequency of the Ca^{2+} rises: no Ca^{2+} spike (score "0"), 1–2 spikes (score "+"), 3–10 spikes (score "++"), and more than 10 spikes (score "+++") [232]. Next, the mean frequency (*F*) and the mean relative amplitude (*A*) of all Ca^{2+} patterns are calculated for one patient. It has been shown that there is a hyperbolic relationship between the MOAT result and the $A \times F$ value: low MOAT results show a low $A \times F$ value, which increases when the MOAT result is higher [232]. The $A \times F$ threshold value is 9, with lower values indicating a diminished Ca^{2+} inducing capacity of the patient's sperm and higher values pointing to an oocyte-related activation deficiency. This diagnostic test revealed a sperm-related activation problem in nearly all MOAT 2 patients. The Ca^{2+} pattern analysis showed a correlation with the MOAT result and additionally allows a more fine-tuned classification of the patients, revealing a more clear distinction between a sperm-related and oocyte-related activation deficiency [232].

It is apparent that the sperm plasma membrane has to disintegrate in order to release the sperm-activating factor into the oocyte and subsequently trigger oocyte activation. Since the plasma membrane of human sperm is physically and biochemically more stable than that of mouse sperm, the disintegration of the human sperm plasma membrane will take longer when injected into mouse oocytes compared with mouse sperm plasma membrane [233]. Therefore, heterologous ICSI of human sperm into mouse oocytes might lead to a delayed Ca^{2+} release. This would make the Ca^{2+} analysis a less reproducible and very time-consuming method. To circumvent this problem, human sperm are exposed to the membrane-disrupting agent Lyso-PC before the heterologous ICSI procedure.

Human Oocyte Calcium Analysis (HOCA)

The most recent and ideally developed diagnostic test of detecting the sperm-activating capacity is the human oocyte Ca^{2+} analysis (HOCA). For this, donated human in vitro matured oocytes are incubated with the Ca^{2+}-sensitive fluorescent dye PE3-AM. In analogy with the MOCA, each oocyte is injected with a single patient's sperm, followed by Ca^{2+} pattern measurement for 10h.

Since human oocytes are bigger in volume and human PLCζ exhibits greater potency compared with their mouse counterparts [85,215], more human PLCζ is necessary to activate human oocytes compared with mouse oocytes. Furthermore, human oocytes might respond differently to PLCζ compared with mouse oocytes. Therefore, although human PLCζ may provoke some Ca^{2+} rises in mouse oocytes, that amount may be insufficient to instigate Ca^{2+} oscillations when injected into human oocytes. Indeed, in one patient, we observed an abnormal Ca^{2+} pattern after HOCA, although the MOCA was normal [234]. More Ca^{2+} pattern analyses using human oocytes of other MOAT patients have to be performed to confirm our preliminary findings. Altogether, we hypothesize that HOCA would show a different Ca^{2+} pattern compared with MOCA. Thus, we cannot extrapolate results acquired from mouse oocytes to human oocytes. Ideally, a test using human oocytes (e.g., HOCA) will be able to detect the underlying cause with great certainty.

One has to remark that human oocytes are scarcely available for research purposes, thus limiting the execution of HOCA. Nikiforaki et al. showed that in vitro matured (IVM) human oocytes can be efficiently used for HOCA, taking into account the somewhat lower frequency and shorter duration of the provoked Ca^{2+} transients (in GV to MII and MI to MII stage oocytes) and the prolonged period of Ca^{2+} oscillations (in MI oocytes reaching the MII stage within 3h) compared with in vivo matured oocytes [235].

Recently, a very promising, noninvasive, and quantitative method to determine the occurrence and frequency of Ca^{2+} oscillations in oocytes has been developed [75]. This approach combines rapid time-lapse imaging of fertilized mouse oocytes with a computer analysis based on particle image velocimetry (PIV). This application allows the detection of rhythmic cytoplasmic movements in the oocyte. These movements are the result of contractions of the actomyosin cytoskeleton that are triggered by Ca^{2+} oscillations upon fertilization. The relationship between these cytoplasmic movements and Ca^{2+} oscillations allows the prediction of the developmental potential of the zygote [75]. Therefore, this method is able to predict the viability of in vitro fertilized oocytes. These findings were found using mouse oocytes; however, similar results were found when donated unfertilized human oocytes were activated [236]. Future studies are required to see if this approach is safe, useful, and reliable during IVF and ICSI cycles in human.

Genetic Screening for PLCζ Mutations

Owing to the fact that mutations in PLCζ lead to abnormal or no Ca^{2+} oscillations [94–96,126,127] and subsequently lead to oocyte activation deficiency (OAD), genetic screening of patients with OAD for the PLCζ gene is required. To date, there are four mutations reported in PLCζ, and all the mutations have shown low or no Ca^{2+} oscillations [94–96,126,127] (see "PLCζ Deficiency and Male Factor Infertility" section). Furthermore, in a recent study, it is shown that the sperm derived from a PLCζ KO mice failed to induce Ca^{2+} oscillations in oocytes suggesting PLCζ as the physiological trigger to induce Ca^{2+} oscillations [97]. Based on the observations from all these studies, it is advisable for patients suffering from OAD to undergo genetic screening of PLCζ to see if they have a mutation in this gene.

CLINICAL APPLICATION TO OVERCOME FERTILIZATION FAILURE AFTER ICSI

Assisted Oocyte Activation (AOA) in Human to Overcome OAD

Since fertilization failure still occurs in 1%–5% of all ICSI cycles, a lot of research has been done to overcome this problem, which is mainly caused by OADs. The oocyte activation process is one of the earliest events during human development, where Ca^{2+} plays an essential role in triggering various downstream pathways (see "Oocyte Activation" section). A lot of protocols have been tested to assist oocyte activation, and they can be categorized into three main groups: mechanical, physical, and chemical agents.

Mechanical oocyte activation is accomplished by vigorous aspiration of the oocyte cytoplasm during ICSI, which results in a major Ca^{2+} concentration increase in the oocyte. It has been shown that mechanical oocyte activation can overcome sperm-related and oocyte-related activation deficiencies in human [218]. This effect can be explained by (1) the Ca^{2+} influx coming from the external Ca^{2+}-containing medium; (2) the disruption of the ER during needle movements and aspiration; and/or (3) the injection of the sperm in the cortical region, where oocyte activation normally is triggered and where the responsiveness of intracellular Ca^{2+} stores is different [218,237]. Mechanical activation enabled normal fertilization and embryo development in three cases of sperm-related and three cases of oocyte-related activation deficiencies, resulting in five live births [218]. Furthermore, Ebner et al. reported that 14 patients experiencing total fertilization failure in 17 previous ICSI cycles were treated with the mechanical activation method. A mean fertilization rate of 53.6% and pregnancy rate of 33.3% were obtained by this oocyte manipulation; however, one live birth was reported [238].

A method of physical stimulation is electrical activation. The purpose of electrostimulation is to alter the activation potential of ion channels present in the oocyte plasma membrane. By the application of electrical pulses to the oocyte, Ca^{2+} coming from the extracellular milieu will enter the oocyte through the ion channels [216,239]. Some studies have been reporting electric activation as a successful procedure in patients with previous fertilization failure [216,240–242]. This method has also been shown to be effective in a man with globozoospermia, resulting in normal fertilization of the oocytes and the birth of a healthy girl [243].

To date, chemical agents are the most commonly used and most efficient oocyte-activating agents in practice. The purpose is to mimic as closely as possible the Ca^{2+} oscillations seen during natural fertilization in mammals. Several chemicals have been proposed to activate human oocytes. Depending on the mechanism they use to trigger human oocyte activation, these chemicals can be subdivided into Ca^{2+} transient-inducing agents, inducing one (e.g., ionomycin, calcimycin, and ethanol [244–248]) or more (e.g., strontium [247,249]) Ca^{2+} rises in the oocyte, and agents that do not trigger Ca^{2+} transients (e.g., puromycin and TPEN [50,250]).

Over the years, the most frequently used and investigated chemical agents are Ca^{2+} ionophores. These molecules are lipid soluble and able to transport Ca^{2+} ions across the oocyte plasma membrane, inducing one single Ca^{2+} transient in the oocyte [251]. Ionomycin and calcimycin (A23187 or GM508) are two Ca^{2+} ionophores frequently used to assist oocyte activation. Several articles have been published about the use of ICSI-AOA treatments to overcome low or failed fertilization after ICSI [214,245,252,253]. In many studies, Ca^{2+} ionophore treatment in globozoospermic patients also has shown its positive effect on the pregnancy outcomes [142,161–166]. Chemical activation by ionomycin exposure was also successful in a case of recurrent molar pregnancies where the sperm induced aberrant Ca^{2+} patterns upon fertilization [234]. Besides, Ca^{2+} ionophore treatment has shown its efficiency for both sperm-related and oocyte-related OADs [213,214,245]. However, the efficiency of AOA in oocyte-related activation failures remains to be further explored [245].

To date, none of these AOA protocols are routinely used. They are still experimental methods because studies show no sufficient clinical evidence. The lack of a standardized AOA protocol is the main causative factor for this. The variability between protocols complicates the comparison of the efficiency of these methods [76]. Indeed, it has been shown recently that ionomycin induced greater Ca^{2+} rises leading to higher activation and blastocyst formation rates in mouse oocytes compared with the commercially available calcimycin solution (GM508 Cult-Active). In human, also, higher activation rates were observed using ionomycin [246].

Furthermore, the oocytes are exposed to the AOA agents for a short time; however, the chance remains that these molecules may exert mutagenic, cytotoxic, or epigenetic side effects upon the oocytes and embryos. Given that the Ca^{2+} oscillation pattern plays an essential role in coordinating the oocyte activation process, AOA application might be a threat to the development of the embryo [253]. To date, sample sizes in studies are still too low, and follow-up studies of children born following AOA remain very scarce, but no adverse effects have been reported so far [254–257]. Also, the exact mechanism of action triggered by these AOA agents remains to be elucidated. So, more research is of utmost importance.

Recombinant PLCζ as a Potential Therapeutic Agent for OAD

PLCζ cRNA has already shown its efficiency in inducing Ca^{2+} oscillations and forming parthenogenetic blastocysts in human [258]; however, the use should not be recommended for clinical applications. Two reasons are the risk for uncontrollable transcription of the RNA and the hazard of RNA transcription into cDNA that can subsequently incorporate into the embryonic genome [259]. Therefore, the development of a recombinant human PLCζ (rhPLCζ) protein holds the promise for a safer and more effective treatment in cases of total fertilization failure due to a PLCζ deficiency [17,124]. In concrete terms, this means that the use of rhPLCζ protein would mimic the Ca^{2+} pattern induced during fertilization instead of only one single Ca^{2+} transient when exposed to a Ca^{2+} ionophore. These artificially induced Ca^{2+} oscillations will be more effective in triggering oocyte activation and subsequent development.

Kashir et al. were the first to report the production of an rhPLCζ protein, which upon microinjection in mouse oocytes induced Ca^{2+} oscillations as seen during normal fertilization. However, this protein was unpurified because of its production in mammalian cell lines [126]. The first purified rhPLCζ protein, expressed in a bacterial cell line, was reported the following year and was also able to induce Ca^{2+} oscillations in mouse oocytes. The major disadvantages were that this protein caused abnormal Ca^{2+} patterns in human oocytes and excessively high concentrations were needed to achieve these Ca^{2+} oscillations [260]. Further research provided an active, purified rhPLCζ protein that evoked a Ca^{2+} pattern that was similar to that observed during normal oocyte activation; however, no results on embryo development were reported [99]. Until a recombinant human PLCζ protein can be used in a clinical setting, AOA by chemical agents remains the best option for patients experiencing OAD.

References

[1] M. Yeste, et al., Oocyte activation deficiency: a role for an oocyte contribution? Hum. Reprod. Update 22 (1) (2016) 23–47.

[2] R.M. Kamel, Assisted reproductive technology after the birth of louise brown, J. Reprod. Infertil. 14 (3) (2013) 96–109.

[3] P.C. Steptoe, R.G. Edwards, Birth after the reimplantation of a human embryo, Lancet 2 (8085) (1978) 366.

[4] G. Palermo, et al., Pregnancies after intracytoplasmic injection of single spermatozoon into an oocyte, Lancet 340 (8810) (1992) 17–18.

[5] European IVF-Monitoring Consortium (EIM) for the European Society of Human Reproduction and Embryology (ESHRE), et al., Assisted reproductive technology in Europe, 2012: results generated from European registers by ESHRE, Hum. Reprod. 31 (8) (2016) 1638–1652.

[6] G.D. Palermo, et al., ICSI: where we have been and where we are going, Semin. Reprod. Med. 27 (2) (2009) 191–201.
[7] S.P. Flaherty, et al., Assessment of fertilization failure and abnormal fertilization after intracytoplasmic sperm injection (ICSI), Reprod. Fertil. Dev. 7 (2) (1995) 197–210.
[8] J. Tesarik, M. Sousa, More than 90% fertilization rates after intracytoplasmic sperm injection and artificial induction of oocyte activation with calcium ionophore, Fertil. Steril. 63 (2) (1995) 343–349.
[9] M. Sousa, J. Tesarik, Ultrastructural analysis of fertilization failure after intracytoplasmic sperm injection, Hum. Reprod. 9 (12) (1994) 2374–2380.
[10] V.Y. Rawe, et al., Cytoskeletal organization defects and abortive activation in human oocytes after IVF and ICSI failure, Mol. Hum. Reprod. 6 (6) (2000) 510–516.
[11] N.G. Mahutte, A. Arici, Failed fertilization: is it predictable? Curr. Opin. Obstet. Gynecol. 15 (3) (2003) 211–218.
[12] S.A. Stricker, Comparative biology of calcium signaling during fertilization and egg activation in animals, Dev. Biol. 211 (2) (1999) 157–176.
[13] D. Dozortsev, et al., Human oocyte activation following intracytoplasmic injection: the role of the sperm cell, Hum. Reprod. 10 (2) (1995) 403–407.
[14] C.M. Saunders, et al., PLC zeta: a sperm-specific trigger of Ca(2+) oscillations in eggs and embryo development, Development 129 (15) (2002) 3533–3544.
[15] D. Kline, J.T. Kline, Repetitive calcium transients and the role of calcium in exocytosis and cell cycle activation in the mouse egg, Dev. Biol. 149 (1) (1992) 80–89.
[16] K. Swann, Y. Yu, The dynamics of calcium oscillations that activate mammalian eggs, Int. J. Dev. Biol. 52 (5-6) (2008) 585–594.
[17] M. Nomikos, K. Swann, F.A. Lai, Starting a new life: sperm PLC-zeta mobilizes the Ca2+ signal that induces egg activation and embryo development: an essential phospholipase C with implications for male infertility, Bioessays 34 (2) (2012) 126–134.
[18] K.T. Jones, Mammalian egg activation: from Ca2+ spiking to cell cycle progression, Reproduction 130 (6) (2005) 813–823.
[19] T. Wakai, V. Vanderheyden, R.A. Fissore, Ca2+ signaling during mammalian fertilization: requirements, players, and adaptations, Cold Spring Harb. Perspect. Biol. 3 (4) (2011).
[20] R.M. Schultz, G.S. Kopf, Molecular basis of mammalian egg activation, Curr. Top. Dev. Biol. 30 (1995) 21–62.
[21] Y. Yu, et al., PLCzeta causes Ca(2+) oscillations in mouse eggs by targeting intracellular and not plasma membrane PI(4,5)P(2), Mol. Biol. Cell 23 (2) (2012) 371–380.
[22] P.G. Suh, et al., Multiple roles of phosphoinositide-specific phospholipase C isozymes, BMB Rep. 41 (6) (2008) 415–434.
[23] K. Swann, F.A. Lai, Egg activation at fertilization by a soluble sperm protein, Physiol. Rev. 96 (1) (2016) 127–149.
[24] S. Miyazaki, IP3 receptor-mediated spatial and temporal Ca2+ signaling of the cell, Jpn. J. Physiol. 43 (4) (1993) 409–434.
[25] M. Nomikos, et al., Sperm PLCzeta: from structure to Ca2+ oscillations, egg activation and therapeutic potential, FEBS Lett. 587 (22) (2013) 3609–3616.
[26] M. Nomikos, Novel signalling mechanism and clinical applications of sperm-specific PLCzeta, Biochem. Soc. Trans. 43 (3) (2015) 371–376.
[27] C.W. Taylor, S.C. Tovey, IP(3) receptors: toward understanding their activation, Cold Spring Harb. Perspect. Biol. 2 (12) (2010) a004010.
[28] C.J. Williams, Signalling mechanisms of mammalian oocyte activation, Hum. Reprod. Update 8 (4) (2002) 313–321.
[29] T. Ayabe, G.S. Kopf, R.M. Schultz, Regulation of mouse egg activation: presence of ryanodine receptors and effects of microinjected ryanodine and cyclic ADP ribose on uninseminated and inseminated eggs, Development 121 (7) (1995) 2233–2244.
[30] Y. Lawrence, M. Whitaker, K. Swann, Sperm-egg fusion is the prelude to the initial Ca2+ increase at fertilization in the mouse, Development 124 (1) (1997) 233–241.
[31] G. Halet, et al., Ca2+ oscillations at fertilization in mammals, Biochem. Soc. Trans. 31 (Pt 5) (2003) 907–911.
[32] C.T. Taylor, et al., Oscillations in intracellular free calcium induced by spermatozoa in human oocytes at fertilization, Hum. Reprod. 8 (12) (1993) 2174–2179.

[33] R.A. Fissore, C. Pinto-Correia, J.M. Robl, Inositol trisphosphate-induced calcium release in the generation of calcium oscillations in bovine eggs, Biol. Reprod. 53 (4) (1995) 766–774.
[34] R. Deguchi, et al., Spatiotemporal analysis of Ca(2+) waves in relation to the sperm entry site and animal-vegetal axis during Ca(2+) oscillations in fertilized mouse eggs, Dev. Biol. 218 (2) (2000) 299–313.
[35] G. Dupont, Link between fertilization-induced Ca2+ oscillations and relief from metaphase II arrest in mammalian eggs: a model based on calmodulin-dependent kinase II activation, Biophys. Chem. 72 (1-2) (1998) 153–167.
[36] S. Madgwick, K.T. Jones, How eggs arrest at metaphase II: MPF stabilisation plus APC/C inhibition equals Cytostatic Factor, Cell Div. 2 (2007) 4.
[37] T. Ducibella, R. Fissore, The roles of Ca2+, downstream protein kinases, and oscillatory signaling in regulating fertilization and the activation of development, Dev. Biol. 315 (2) (2008) 257–279.
[38] Y. Nishizuka, Intracellular signaling by hydrolysis of phospholipids and activation of protein kinase C, Science 258 (5082) (1992) 607–614.
[39] G. Halet, et al., Conventional PKCs regulate the temporal pattern of Ca2+ oscillations at fertilization in mouse eggs, J. Cell Biol. 164 (7) (2004) 1033–1044.
[40] A. Luria, et al., Differential localization of conventional protein kinase C isoforms during mouse oocyte development, Biol. Reprod. 62 (6) (2000) 1564–1570.
[41] G. Halet, PKC signaling at fertilization in mammalian eggs, Biochim. Biophys. Acta 1742 (1-3) (2004) 185–189.
[42] L. Tsaadon, R. Kaplan-Kraicer, R. Shalgi, Myristoylated alanine-rich C kinase substrate, but not Ca2+/calmodulin-dependent protein kinase II, is the mediator in cortical granules exocytosis, Reproduction 135 (5) (2008) 613–624.
[43] E. Eliyahu, et al., Association between myristoylated alanin-rich C kinase substrate (MARCKS) translocation and cortical granule exocytosis in rat eggs, Reproduction 131 (2) (2006) 221–231.
[44] E.L. Que, et al., Quantitative mapping of zinc fluxes in the mammalian egg reveals the origin of fertilization-induced zinc sparks, Nat. Chem. 7 (2) (2015) 130–139.
[45] A.M. Kim, et al., Zinc sparks are triggered by fertilization and facilitate cell cycle resumption in mammalian eggs, ACS Chem. Biol. 6 (7) (2011) 716–723.
[46] T. Suzuki, et al., Full-term mouse development by abolishing Zn2+-dependent metaphase II arrest without Ca2+ release, Development 137 (16) (2010) 2659–2669.
[47] K. Lee, et al., Pig oocyte activation using a Zn(2)(+) chelator, TPEN, Theriogenology 84 (6) (2015) 1024–1032.
[48] M.L. Bernhardt, et al., A zinc-dependent mechanism regulates meiotic progression in mammalian oocytes, Biol. Reprod. 86 (4) (2012) 114.
[49] A.M. Kim, et al., Zinc availability regulates exit from meiosis in maturing mammalian oocytes, Nat. Chem. Biol. 6 (9) (2010) 674–681.
[50] F.E. Duncan, et al., The zinc spark is an inorganic signature of human egg activation, Sci. Rep. 6 (2016) 24737.
[51] L. Sun, et al., Zinc regulates the ability of Cdc25C to activate MPF/cdk1, J. Cell. Physiol. 213 (1) (2007) 98–104.
[52] A.R. Krauchunas, M.F. Wolfner, Molecular changes during egg activation, Curr. Top. Dev. Biol. 102 (2013) 267–292.
[53] T. Wakai, R.A. Fissore, Ca(2+) homeostasis and regulation of ER Ca(2+) in mammalian oocytes/eggs, Cell Calcium 53 (1) (2013) 63–67.
[54] M.J. Berridge, P. Lipp, M.D. Bootman, The versatility and universality of calcium signalling, Nat. Rev. Mol. Cell Biol. 1 (1) (2000) 11–21.
[55] M.D. Bootman, P. Lipp, M.J. Berridge, The organisation and functions of local Ca(2+) signals, J. Cell Sci. 114 (Pt 12) (2001) 2213–2222.
[56] M.D. Cahalan, STIMulating store-operated Ca(2+) entry, Nat. Cell Biol. 11 (6) (2009) 669–677.
[57] C. Wang, et al., Orai1 mediates store-operated Ca2+ entry during fertilization in mammalian oocytes, Dev. Biol. 365 (2) (2012) 414–423.
[58] O. Brandman, et al., STIM2 is a feedback regulator that stabilizes basal cytosolic and endoplasmic reticulum Ca2+ levels, Cell 131 (7) (2007) 1327–1339.
[59] Y.L. Miao, C.J. Williams, Calcium signaling in mammalian egg activation and embryo development: the influence of subcellular localization, Mol. Reprod. Dev. 79 (11) (2012) 742–756.
[60] C. Gomez-Fernandez, et al., Relocalization of STIM1 in mouse oocytes at fertilization: early involvement of store-operated calcium entry, Reproduction 138 (2) (2009) 211–221.

[61] K. Lee, C. Wang, Z. Machaty, STIM1 is required for Ca2+ signaling during mammalian fertilization, Dev. Biol. 367 (2) (2012) 154–162.
[62] C. Wang, et al., Store-operated Ca2+ entry sustains the fertilization Ca2+ signal in pig eggs, Biol. Reprod. 93 (1) (2015) 25.
[63] Y.L. Miao, et al., Calcium influx-mediated signaling is required for complete mouse egg activation, Proc. Natl. Acad. Sci. U. S. A. 109 (11) (2012) 4169–4174.
[64] T. Takahashi, et al., Ca(2)(+) influx-dependent refilling of intracellular Ca(2)(+) stores determines the frequency of Ca(2)(+) oscillations in fertilized mouse eggs, Biochem. Biophys. Res. Commun. 430 (1) (2013) 60–65.
[65] M.L. Bernhardt, et al., Store-operated Ca2+ entry is not required for fertilization-induced Ca2+ signaling in mouse eggs, Cell Calcium 65 (2017) 63–72.
[66] M.L. Bernhardt, et al., CaV3.2 T-type channels mediate Ca(2)(+) entry during oocyte maturation and following fertilization, J. Cell Sci. 128 (23) (2015) 4442–4452.
[67] F. Perocchi, et al., MICU1 encodes a mitochondrial EF hand protein required for Ca(2+) uptake, Nature 467 (7313) (2010) 291–296.
[68] M.R. Duchen, Mitochondria and calcium: from cell signalling to cell death, J. Physiol. 529 (Pt 1) (2000) 57–68.
[69] R. Rizzuto, P. Bernardi, T. Pozzan, Mitochondria as all-round players of the calcium game, J. Physiol. 529 (Pt 1) (2000) 37–47.
[70] T. Ducibella, et al., Egg-to-embryo transition is driven by differential responses to Ca(2+) oscillation number, Dev. Biol. 250 (2) (2002) 280–291.
[71] S. Toth, et al., Egg activation is the result of calcium signal summation in the mouse, Reproduction 131 (1) (2006) 27–34.
[72] T. Ducibella, R.M. Schultz, J.P. Ozil, Role of calcium signals in early development, Semin. Cell Dev. Biol. 17 (2) (2006) 324–332.
[73] J.P. Ozil, et al., Ca2+ oscillatory pattern in fertilized mouse eggs affects gene expression and development to term, Dev. Biol. 300 (2) (2006) 534–544.
[74] J.P. Ozil, D. Huneau, Activation of rabbit oocytes: the impact of the Ca2+ signal regime on development, Development 128 (6) (2001) 917–928.
[75] A. Ajduk, et al., Rhythmic actomyosin-driven contractions induced by sperm entry predict mammalian embryo viability, Nat. Commun. 2 (2011) 417.
[76] F. Vanden Meerschaut, et al., Assisted oocyte activation following ICSI fertilization failure, Reprod. Biomed. Online 28 (5) (2014) 560–571.
[77] J. Kashir, et al., Comparative biology of sperm factors and fertilization-induced calcium signals across the animal kingdom, Mol. Reprod. Dev. 80 (10) (2013) 787–815.
[78] K. Swann, F.A. Lai, PLCzeta and the initiation of Ca(2+) oscillations in fertilizing mammalian eggs, Cell Calcium 53 (1) (2013) 55–62.
[79] B. Dale, L.J. DeFelice, G. Ehrenstein, Injection of a soluble sperm fraction into sea-urchin eggs triggers the cortical reaction, Experientia 41 (8) (1985) 1068–1070.
[80] K. Swann, Ca2+ oscillations and sensitization of Ca2+ release in unfertilized mouse eggs injected with a sperm factor, Cell Calcium 15 (4) (1994) 331–339.
[81] K.T. Jones, et al., A mammalian sperm cytosolic phospholipase C activity generates inositol trisphosphate and causes Ca2+ release in sea urchin egg homogenates, FEBS Lett. 437 (3) (1998) 297–300.
[82] K.T. Jones, et al., Different Ca2+-releasing abilities of sperm extracts compared with tissue extracts and phospholipase C isoforms in sea urchin egg homogenate and mouse eggs, Biochem. J. 346 (Pt 3) (2000) 743–749.
[83] A. Rice, et al., Mammalian sperm contain a Ca(2+)-sensitive phospholipase C activity that can generate InsP(3) from PIP(2) associated with intracellular organelles, Dev. Biol. 228 (1) (2000) 125–135.
[84] S. Fujimoto, et al., Mammalian phospholipase Czeta induces oocyte activation from the sperm perinuclear matrix, Dev. Biol. 274 (2) (2004) 370–383.
[85] L.J. Cox, et al., Sperm phospholipase Czeta from humans and cynomolgus monkeys triggers Ca2+ oscillations, activation and development of mouse oocytes, Reproduction 124 (5) (2002) 611–623.
[86] A. Yoneda, et al., Molecular cloning, testicular postnatal expression, and oocyte-activating potential of porcine phospholipase Czeta, Reproduction 132 (3) (2006) 393–401.
[87] M. Ito, et al., Difference in Ca2+ oscillation-inducing activity and nuclear translocation ability of PLCZ1, an egg-activating sperm factor candidate, between mouse, rat, human, and medaka fish, Biol. Reprod. 78 (6) (2008) 1081–1090.

[88] K. Coward, et al., Phospholipase Czeta, the trigger of egg activation in mammals, is present in a non-mammalian species, Reproduction 130 (2) (2005) 157–163.

[89] C. Malcuit, M. Kurokawa, R.A. Fissore, Calcium oscillations and mammalian egg activation, J. Cell. Physiol. 206 (3) (2006) 565–573.

[90] J. Kashir, et al., Sperm-induced Ca2+ release during egg activation in mammals, Biochem. Biophys. Res. Commun. 450 (3) (2014) 1204–1211.

[91] K. Sato, et al., Molecular characteristics of horse phospholipase C zeta (PLCzeta), Anim. Sci. J. 84 (4) (2013) 359–368.

[92] S.N. Amdani, et al., Sperm factors and oocyte activation: current controversies and considerations, Biol. Reprod. 93 (2) (2015) 50.

[93] S.N. Amdani, et al., Phospholipase C zeta (PLCzeta) and male infertility: clinical update and topical developments, Adv. Biol. Regul. 61 (2016) 58–67.

[94] J. Escoffier, et al., Homozygous mutation of PLCZ1 leads to defective human oocyte activation and infertility that is not rescued by the WW-binding protein PAWP, Hum. Mol. Genet. 25 (5) (2016) 878–891.

[95] E. Heytens, et al., Reduced amounts and abnormal forms of phospholipase C zeta (PLCzeta) in spermatozoa from infertile men, Hum. Reprod. 24 (10) (2009) 2417–2428.

[96] J. Kashir, et al., A maternally inherited autosomal point mutation in human phospholipase C zeta (PLCzeta) leads to male infertility, Hum. Reprod. 27 (1) (2012) 222–231.

[97] A. Hachem, et al., PLCzeta is the physiological trigger of the Ca2+ oscillations that induce embryogenesis in mammals but offspring can be conceived in its absence, Development 144 (16) (2017) 2914–2924.

[98] M. Nomikos, et al., Novel regulation of PLCzeta activity via its XY-linker, Biochem. J. 438 (3) (2011) 427–432.

[99] M. Nomikos, et al., Phospholipase Czeta rescues failed oocyte activation in a prototype of male factor infertility, Fertil. Steril. 99 (1) (2013) 76–85.

[100] K. Swann, et al., PLCzeta(zeta): a sperm protein that triggers Ca2+ oscillations and egg activation in mammals, Semin. Cell Dev. Biol. 17 (2) (2006) 264–273.

[101] S.N. Hicks, et al., General and versatile autoinhibition of PLC isozymes, Mol. Cell 31 (3) (2008) 383–394.

[102] A. Gresset, et al., Mechanism of phosphorylation-induced activation of phospholipase C-gamma isozymes, J. Biol. Chem. 285 (46) (2010) 35836–35847.

[103] M. Nomikos, et al., Binding of phosphoinositide-specific phospholipase C-zeta (PLC-zeta) to phospholipid membranes: potential role of an unstructured cluster of basic residues, J. Biol. Chem. 282 (22) (2007) 16644–16653.

[104] M. Nomikos, et al., Phospholipase Czeta binding to PtdIns(4,5)P2 requires the XY-linker region, J. Cell Sci. 124 (Pt 15) (2011) 2582–2590.

[105] M. Nomikos, et al., Essential role of the EF-hand domain in targeting sperm phospholipase Czeta to membrane phosphatidylinositol 4,5-bisphosphate (PIP2), J. Biol. Chem. 290 (49) (2015) 29519–29530.

[106] M. Nomikos, et al., Role of phospholipase C-zeta domains in Ca2+-dependent phosphatidylinositol 4,5-bisphosphate hydrolysis and cytoplasmic Ca2+ oscillations, J. Biol. Chem. 280 (35) (2005) 31011–31018.

[107] M. Theodoridou, et al., Chimeras of sperm PLCzeta reveal disparate protein domain functions in the generation of intracellular Ca2+ oscillations in mammalian eggs at fertilization, Mol. Hum. Reprod. 19 (12) (2013) 852–864.

[108] Z. Kouchi, et al., The role of EF-hand domains and C2 domain in regulation of enzymatic activity of phospholipase Czeta, J. Biol. Chem. 280 (22) (2005) 21015–21021.

[109] S.V. Phillips, et al., Divergent effect of mammalian PLCzeta in generating Ca(2)(+) oscillations in somatic cells compared with eggs, Biochem. J. 438 (3) (2011) 545–553.

[110] N. Ogonuki, et al., Activity of a sperm-borne oocyte-activating factor in spermatozoa and spermatogenic cells from cynomolgus monkeys and its localization after oocyte activation, Biol. Reprod. 65 (2) (2001) 351–357.

[111] H. Shirakawa, et al., Measurement of intracellular IP3 during Ca2+ oscillations in mouse eggs with GFP-based FRET probe, Biochem. Biophys. Res. Commun. 345 (2) (2006) 781–788.

[112] P. Grasa, et al., The pattern of localization of the putative oocyte activation factor, phospholipase Czeta, in uncapacitated, capacitated, and ionophore-treated human spermatozoa, Hum. Reprod. 23 (11) (2008) 2513–2522.

[113] C. Young, et al., Phospholipase C zeta undergoes dynamic changes in its pattern of localization in sperm during capacitation and the acrosome reaction, Fertil. Steril. 91(5 Suppl.) (2009) 2230–2242.

[114] M. Kurokawa, et al., Functional, biochemical, and chromatographic characterization of the complete [Ca2+]i oscillation-inducing activity of porcine sperm, Dev. Biol. 285 (2) (2005) 376–392.
[115] S.Y. Yoon, et al., Human sperm devoid of PLC, zeta 1 fail to induce Ca(2+) release and are unable to initiate the first step of embryo development, J. Clin. Invest. 118 (11) (2008) 3671–3681.
[116] M. Nakai, et al., Pre-treatment of sperm reduces success of ICSI in the pig, Reproduction 142 (2) (2011) 285–293.
[117] K. Kaewmala, et al., Investigation into association and expression of PLCz and COX-2 as candidate genes for boar sperm quality and fertility, Reprod. Domest. Anim. 47 (2) (2012) 213–223.
[118] K. Swann, Soluble sperm factors and Ca2+ release in eggs at fertilization, Rev. Reprod. 1 (1) (1996) 33–39.
[119] M.M. Matzuk, D.J. Lamb, The biology of infertility: research advances and clinical challenges, Nat. Med. 14 (11) (2008) 1197–1213.
[120] A.N. Yatsenko, et al., X-linked TEX11 mutations, meiotic arrest, and azoospermia in infertile men, N. Engl. J. Med. 372 (22) (2015) 2097–2107.
[121] O. Okutman, et al., Exome sequencing reveals a nonsense mutation in TEX15 causing spermatogenic failure in a Turkish family, Hum. Mol. Genet. 24 (19) (2015) 5581–5588.
[122] Z.E. Kherraf, et al., SPINK2 deficiency causes infertility by inducing sperm defects in heterozygotes and azoospermia in homozygotes, EMBO Mol. Med. 9 (8) (2017) 1132–1149.
[123] M. Ben Khelifa, et al., Mutations in DNAH1, which encodes an inner arm heavy chain dynein, lead to male infertility from multiple morphological abnormalities of the sperm flagella, Am. J. Hum. Genet. 94 (1) (2014) 95–104.
[124] J. Kashir, et al., Oocyte activation, phospholipase C zeta and human infertility, Hum. Reprod. Update 16 (6) (2010) 690–703.
[125] M. Nomikos, et al., Male infertility-linked point mutation disrupts the Ca2+ oscillation-inducing and PIP(2) hydrolysis activity of sperm PLCzeta, Biochem. J. 434 (2) (2011) 211–217.
[126] J. Kashir, et al., Loss of activity mutations in phospholipase C zeta (PLCzeta) abolishes calcium oscillatory ability of human recombinant protein in mouse oocytes, Hum. Reprod. 26 (12) (2011) 3372–3387.
[127] A. Ferrer-Vaquer, et al., PLCzeta sequence, protein levels, and distribution in human sperm do not correlate with semen characteristics and fertilization rates after ICSI, J. Assist. Reprod. Genet. 33 (6) (2016) 747–756.
[128] Q. Pan, et al., PLCz functional haplotypes modulating promoter transcriptional activity are associated with semen quality traits in Chinese Holstein bulls, PLoS One 8 (3) (2013) e58795.
[129] S. Javadian-Elyaderani, et al., Diagnosis of genetic defects through parallel assessment of PLCzeta and CAPZA3 in infertile men with history of failed oocyte activation, Iran J. Basic Med. Sci. 19 (3) (2016) 281–289.
[130] A.F. Holstein, C. Schirren, C.G. Schirren, Human spermatids and spermatozoa lacking acrosomes, J. Reprod. Fertil. 35 (3) (1973) 489–491.
[131] A.H. Dam, et al., Globozoospermia revisited, Hum. Reprod. Update 13 (1) (2007) 63–75.
[132] W. Meyhofer, Contribution to the cytophotometric evaluation of pathologically changed spermatozoa with special reference to round-shaped spermatozoa following Feuglen and fast green staining, Z. Haut Geschlechtskr. 39 (4) (1965) 174–182.
[133] C.G. Schirren, A.F. Holstein, C. Schirren, Uber die morphogenese rundkopfiger spermatozoen des menschen, Andrologia 3 (1971) 117–125.
[134] L. Chansel-Debordeaux, et al., Reproductive outcome in globozoospermic men: update and prospects, Andrology 3 (6) (2015) 1022–1034.
[135] R. Harbuz, et al., A recurrent deletion of DPY19L2 causes infertility in man by blocking sperm head elongation and acrosome formation, Am. J. Hum. Genet. 88 (3) (2011) 351–361.
[136] I. Koscinski, et al., DPY19L2 deletion as a major cause of globozoospermia, Am. J. Hum. Genet. 88 (3) (2011) 344–350.
[137] C. Coutton, et al., Fine characterisation of a recombination hotspot at the DPY19L2 locus and resolution of the paradoxical excess of duplications over deletions in the general population, PLoS Genet. 9 (3) (2013) e1003363.
[138] E. Elinati, et al., Globozoospermia is mainly due to DPY19L2 deletion via non-allelic homologous recombination involving two recombination hotspots, Hum. Mol. Genet. 21 (16) (2012) 3695–3702.
[139] F. Zhu, et al., DPY19L2 gene mutations are a major cause of globozoospermia: identification of three novel point mutations, Mol. Hum. Reprod. 19 (6) (2013) 395–404.
[140] P. Noveski, et al., A homozygous deletion of the DPY19l2 gene is a cause of globozoospermia in men from the Republic of Macedonia, Balkan J. Med. Genet. 16 (1) (2013) 73–76.

[141] A.H. Dam, et al., Homozygous mutation in SPATA16 is associated with male infertility in human globozoospermia, Am. J. Hum. Genet. 81 (4) (2007) 813–820.
[142] N. Karaca, et al., First successful pregnancy in a globozoospermic patient having homozygous mutation in SPATA16, Fertil. Steril. 102 (1) (2014) 103–107.
[143] J. Escoffier, et al., Subcellular localization of phospholipase Czeta in human sperm and its absence in DPY19L2-deficient sperm are consistent with its role in oocyte activation, Mol. Hum. Reprod. 21 (2) (2015) 157–168.
[144] D. Escalier, Failure of differentiation of the nuclear-perinuclear skeletal complex in the round-headed human spermatozoa, Int. J. Dev. Biol. 34 (2) (1990) 287–297.
[145] V. Pierre, et al., Absence of Dpy19l2, a new inner nuclear membrane protein, causes globozoospermia in mice by preventing the anchoring of the acrosome to the nucleus, Development 139 (16) (2012) 2955–2965.
[146] L. Lu, et al., Gene functional research using polyethylenimine-mediated in vivo gene transfection into mouse spermatogenic cells, Asian J. Androl. 8 (1) (2006) 53–59.
[147] E.K. Dirican, et al., Clinical pregnancies and livebirths achieved by intracytoplasmic injection of round headed acrosomeless spermatozoa with and without oocyte activation in familial globozoospermia: case report, Asian J. Androl. 10 (2) (2008) 332–336.
[148] Z. Kilani, et al., Evaluation and treatment of familial globozoospermia in five brothers, Fertil. Steril. 82 (5) (2004) 1436–1439.
[149] B. Dale, et al., A morphological and functional study of fusibility in round-headed spermatozoa in the human, Fertil. Steril. 61 (2) (1994) 336–340.
[150] S. Florke-Gerloff, et al., Biochemical and genetic investigation of round-headed spermatozoa in infertile men including two brothers and their father, Andrologia 16 (3) (1984) 187–202.
[151] C. Coutton, et al., Teratozoospermia: spotlight on the main genetic actors in the human, Hum. Reprod. Update 21 (4) (2015) 455–485.
[152] S.L. Taylor, et al., Complete globozoospermia associated with PLCzeta deficiency treated with calcium ionophore and ICSI results in pregnancy, Reprod. Biomed. Online 20 (4) (2010) 559–564.
[153] K. Lundin, et al., Fertilization and pregnancy after intracytoplasmic microinjection of acrosomeless spermatozoa, Fertil. Steril. 62 (6) (1994) 1266–1267.
[154] Z.M. Kilani, et al., Triplet pregnancy and delivery after intracytoplasmic injection of round-headed spermatozoa, Hum. Reprod. 13 (8) (1998) 2177–2179.
[155] S. Stone, et al., A normal livebirth after intracytoplasmic sperm injection for globozoospermia without assisted oocyte activation: case report, Hum. Reprod. 15 (1) (2000) 139–141.
[156] L.G. Nardo, et al., Ultrastructural features and ICSI treatment of severe teratozoospermia: report of two human cases of globozoospermia, Eur. J. Obstet. Gynecol. Reprod. Biol. 104 (1) (2002) 40–42.
[157] H.B. Zeyneloglu, et al., Achievement of pregnancy in globozoospermia with Y chromosome microdeletion after ICSI, Hum. Reprod. 17 (7) (2002) 1833–1836.
[158] M.R. Banker, et al., Successful pregnancies and a live birth after intracytoplasmic sperm injection in globozoospermia, J. Hum. Reprod. Sci. 2 (2) (2009) 81–82.
[159] S. Bechoua, et al., Fertilisation and pregnancy outcome after ICSI in globozoospermic patients without assisted oocyte activation, Andrologia 41 (1) (2009) 55–58.
[160] B. Sahu, O. Ozturk, P. Serhal, Successful pregnancy in globozoospermia with severe oligoasthenospermia after ICSI, J. Obstet. Gynaecol. 30 (8) (2010) 869–870.
[161] A.V. Rybouchkin, et al., Fertilization and pregnancy after assisted oocyte activation and intracytoplasmic sperm injection in a case of round-headed sperm associated with deficient oocyte activation capacity, Fertil. Steril. 68 (6) (1997) 1144–1147.
[162] S.T. Kim, et al., Successful pregnancy and delivery from frozen-thawed embryos after intracytoplasmic sperm injection using round-headed spermatozoa and assisted oocyte activation in a globozoospermic patient with mosaic Down syndrome, Fertil. Steril. 75 (2) (2001) 445–447.
[163] A. Tejera, et al., Successful pregnancy and childbirth after intracytoplasmic sperm injection with calcium ionophore oocyte activation in a globozoospermic patient, Fertil. Steril. 90 (4) (2008) 1202 (e1–e5).
[164] K. Kyono, et al., A birth from the transfer of a single vitrified-warmed blastocyst using intracytoplasmic sperm injection with calcium ionophore oocyte activation in a globozoospermic patient, Fertil. Steril. 91 (3) (2009) 931 (e7-e11).
[165] H. Kamiyama, et al., Successful delivery following intracytoplasmic sperm injection with calcium ionophore A23187 oocyte activation in a partially globozoospermic patient, Reprod. Med. Biol. 11 (2012) 159–164.

[166] N. Karaca, et al., A successful healthy childbirth in a case of total globozoospermia with oocyte activation by calcium ionophore, J. Reprod. Infertil. 16 (2) (2015) 116–120.

[167] A.N. Yatsenko, et al., Association of mutations in the zona pellucida binding protein 1 (ZPBP1) gene with abnormal sperm head morphology in infertile men, Mol. Hum. Reprod. 18 (1) (2012) 14–21.

[168] G. Liu, Q.W. Shi, G.X. Lu, A newly discovered mutation in PICK1 in a human with globozoospermia, Asian J. Androl. 12 (4) (2010) 556–560.

[169] L.M. Mehlmann, K. Mikoshiba, D. Kline, Redistribution and increase in cortical inositol 1,4,5-trisphosphate receptors after meiotic maturation of the mouse oocyte, Dev. Biol. 180 (2) (1996) 489–498.

[170] A. Jedrusik, et al., Mouse oocytes fertilised by ICSI during in vitro maturation retain the ability to be activated after refertilisation in metaphase II and can generate Ca2+ oscillations, BMC Dev. Biol. 7 (2007) 72.

[171] V. Vanderheyden, et al., Regulation of inositol 1,4,5-trisphosphate-induced Ca2+ release by reversible phosphorylation and dephosphorylation, Biochim. Biophys. Acta 1793 (6) (2009) 959–970.

[172] V. Vanderheyden, et al., Regulation of inositol 1,4,5-trisphosphate receptor type 1 function during oocyte maturation by MPM-2 phosphorylation, Cell Calcium 46 (1) (2009) 56–64.

[173] T. Wakai, et al., Regulation of inositol 1,4,5-trisphosphate receptor function during mouse oocyte maturation, J. Cell. Physiol. 227 (2) (2012) 705–717.

[174] A. Ajduk, A. Malagocki, M. Maleszewski, Cytoplasmic maturation of mammalian oocytes: development of a mechanism responsible for sperm-induced Ca2+ oscillations, Reprod. Biol. 8 (1) (2008) 3–22.

[175] J. Ito, et al., Inositol 1,4,5-trisphosphate receptor 1, a widespread Ca2+ channel, is a novel substrate of polo-like kinase 1 in eggs, Dev. Biol. 320 (2) (2008) 402–413.

[176] B.E. Talansky, H.E. Malter, J. Cohen, A preferential site for sperm-egg fusion in mammals, Mol. Reprod. Dev. 28 (2) (1991) 183–188.

[177] T. Ducibella, et al., Quantitative studies of changes in cortical granule number and distribution in the mouse oocyte during meiotic maturation, Dev. Biol. 130 (1) (1988) 184–197.

[178] F.J. Longo, D.Y. Chen, Development of cortical polarity in mouse eggs: involvement of the meiotic apparatus, Dev. Biol. 107 (2) (1985) 382–394.

[179] D. Kline, et al., The cortical endoplasmic reticulum (ER) of the mouse egg: localization of ER clusters in relation to the generation of repetitive calcium waves, Dev. Biol. 215 (2) (1999) 431–442.

[180] P.T. Goud, et al., Presence and dynamic redistribution of type I inositol 1,4,5-trisphosphate receptors in human oocytes and embryos during in-vitro maturation, fertilization and early cleavage divisions, Mol. Hum. Reprod. 5 (5) (1999) 441–451.

[181] J.S. Mann, K.M. Lowther, L.M. Mehlmann, Reorganization of the endoplasmic reticulum and development of Ca2+ release mechanisms during meiotic maturation of human oocytes, Biol. Reprod. 83 (4) (2010) 578–583.

[182] J. Parrington, et al., Expression of inositol 1,4,5-trisphosphate receptors in mouse oocytes and early embryos: the type I isoform is upregulated in oocytes and downregulated after fertilization, Dev. Biol. 203 (2) (1998) 451–461.

[183] R.A. Fissore, et al., Differential distribution of inositol trisphosphate receptor isoforms in mouse oocytes, Biol. Reprod. 60 (1) (1999) 49–57.

[184] Z. Xu, et al., Maturation-associated increase in IP3 receptor type 1: role in conferring increased IP3 sensitivity and Ca2+ oscillatory behavior in mouse eggs, Dev. Biol. 254 (2) (2003) 163–171.

[185] B. Lee, et al., Phosphorylation of IP3R1 and the regulation of [Ca2+]i responses at fertilization: a role for the MAP kinase pathway, Development 133 (21) (2006) 4355–4365.

[186] L. Sun, et al., Kinase-dependent regulation of inositol 1,4,5-trisphosphate-dependent Ca2+ release during oocyte maturation, J. Biol. Chem. 284 (30) (2009) 20184–20196.

[187] S. Miyazaki, et al., Block of Ca2+ wave and Ca2+ oscillation by antibody to the inositol 1,4,5-trisphosphate receptor in fertilized hamster eggs, Science 257 (5067) (1992) 251–255.

[188] K.T. Jones, V.L. Nixon, Sperm-induced Ca(2+) oscillations in mouse oocytes and eggs can be mimicked by photolysis of caged inositol 1,4,5-trisphosphate: evidence to support a continuous low level production of inositol 1, 4,5-trisphosphate during mammalian fertilization, Dev. Biol. 225 (1) (2000) 1–12.

[189] Y. Harada, et al., Characterization of a sperm factor for egg activation at fertilization of the newt Cynops pyrrhogaster, Dev. Biol. 306 (2) (2007) 797–808.

[190] C. Sette, et al., Parthenogenetic activation of mouse eggs by microinjection of a truncated c-kit tyrosine kinase present in spermatozoa, Development 124 (11) (1997) 2267–2274.

[191] A.T. Wu, et al., PAWP, a sperm-specific WW domain-binding protein, promotes meiotic resumption and pronuclear development during fertilization, J. Biol. Chem. 282 (16) (2007) 12164–12175.
[192] A.T. Wu, et al., The postacrosomal assembly of sperm head protein, PAWP, is independent of acrosome formation and dependent on microtubular manchette transport, Dev. Biol. 312 (2) (2007) 471–483.
[193] M. Aarabi, et al., Sperm-borne protein, PAWP, initiates zygotic development in *Xenopus laevis* by eliciting intracellular calcium release, Mol. Reprod. Dev. 77 (3) (2010) 249–256.
[194] M. Aarabi, et al., Sperm-derived WW domain-binding protein, PAWP, elicits calcium oscillations and oocyte activation in humans and mice, FASEB J. 28 (10) (2014) 4434–4440.
[195] M. Aarabi, et al., Sperm content of postacrosomal WW binding protein is related to fertilization outcomes in patients undergoing assisted reproductive technology, Fertil. Steril. 102 (2) (2014) 440–447.
[196] M. Nomikos, et al., Sperm-specific post-acrosomal WW-domain binding protein (PAWP) does not cause Ca2+ release in mouse oocytes, Mol. Hum. Reprod. 20 (10) (2014) 938–947.
[197] Y. Satouh, K. Nozawa, M. Ikawa, Sperm postacrosomal WW domain-binding protein is not required for mouse egg activation, Biol. Reprod. 93 (4) (2015) 94.
[198] M. Tavalaee, M.H. Nasr-Esfahani, Expression profile of PLCzeta, PAWP, and TR-KIT in association with fertilization potential, embryo development, and pregnancy outcomes in globozoospermic candidates for intracytoplasmic sperm injection and artificial oocyte activation, Andrology 4 (5) (2016) 850–856.
[199] M. Kamali-Dolat Abadi, et al., Evaluation of PLCzeta and PAWP Expression in globozoospermic individuals, Cell J. 18 (3) (2016) 438–445.
[200] Y. Yu, et al., Regulation of diacylglycerol production and protein kinase C stimulation during sperm- and PLCzeta-mediated mouse egg activation, Biol. Cell 100 (11) (2008) 633–643.
[201] H. Wu, et al., Sperm factor induces intracellular free calcium oscillations by stimulating the phosphoinositide pathway, Biol. Reprod. 64 (5) (2001) 1338–1349.
[202] G.I. Gallicano, R.W. McGaughey, D.G. Capco, Activation of protein kinase C after fertilization is required for remodeling the mouse egg into the zygote, Mol. Reprod. Dev. 46 (4) (1997) 587–601.
[203] E. Eliyahu, R. Shalgi, A role for protein kinase C during rat egg activation, Biol. Reprod. 67 (1) (2002) 189–195.
[204] S. Madgwick, M. Levasseur, K.T. Jones, Calmodulin-dependent protein kinase II, and not protein kinase C, is sufficient for triggering cell-cycle resumption in mammalian eggs, J. Cell Sci. 118 (Pt 17) (2005) 3849–3859.
[205] J.G. Knott, et al., Calmodulin-dependent protein kinase II triggers mouse egg activation and embryo development in the absence of Ca2+ oscillations, Dev. Biol. 296 (2) (2006) 388–395.
[206] C. Wang, Z. Machaty, Calcium influx in mammalian eggs, Reproduction 145 (4) (2013) R97–R105.
[207] J.R. Von Stetina, T.L. Orr-Weaver, Developmental control of oocyte maturation and egg activation in metazoan models, Cold Spring Harb. Perspect. Biol. 3 (10) (2011) a005553.
[208] M. Nomikos, et al., Male infertility-linked point mutation reveals a vital binding role for the C2 domain of sperm PLCzeta, Biochem. J. 474 (6) (2017) 1003–1016.
[209] J. Kashir, et al., Variance in total levels of phospholipase C zeta (PLC-zeta) in human sperm may limit the applicability of quantitative immunofluorescent analysis as a diagnostic indicator of oocyte activation capability, Fertil. Steril. 99 (1) (2013) 107–117.
[210] S. Yelumalai, et al., Total levels, localization patterns, and proportions of sperm exhibiting phospholipase C zeta are significantly correlated with fertilization rates after intracytoplasmic sperm injection, Fertil. Steril. 104 (3) (2015) 561–568 (e4).
[211] S. Aghajanpour, et al., Quantitative expression of phospholipase C zeta, as an index to assess fertilization potential of a semen sample, Hum. Reprod. 26 (11) (2011) 2950–2956.
[212] A. Rybouchkin, et al., Intracytoplasmic injection of human spermatozoa into mouse oocytes: a useful model to investigate the oocyte-activating capacity and the karyotype of human spermatozoa, Hum. Reprod. 10 (5) (1995) 1130–1135.
[213] B. Heindryckx, et al., Treatment option for sperm- or oocyte-related fertilization failure: assisted oocyte activation following diagnostic heterologous ICSI, Hum. Reprod. 20 (8) (2005) 2237–2241.
[214] B. Heindryckx, et al., Efficiency of assisted oocyte activation as a solution for failed intracytoplasmic sperm injection, Reprod. Biomed. Online 17 (5) (2008) 662–668.
[215] M. Nomikos, et al., Human PLCzeta exhibits superior fertilization potency over mouse PLCzeta in triggering the Ca(2+) oscillations required for mammalian oocyte activation, Mol. Hum. Reprod. 20 (6) (2014) 489–498.
[216] K. Yanagida, et al., Successful fertilization and pregnancy following ICSI and electrical oocyte activation, Hum. Reprod. 14 (5) (1999) 1307–1311.

[217] Y. Araki, et al., Use of mouse oocytes to evaluate the ability of human sperm to activate oocytes after failure of activation by intracytoplasmic sperm injection, Zygote 12 (2) (2004) 111–116.
[218] J. Tesarik, et al., Use of a modified intracytoplasmic sperm injection technique to overcome sperm-borne and oocyte-borne oocyte activation failures, Fertil. Steril. 78 (3) (2002) 619–624.
[219] Y. Terada, et al., Use of mammalian eggs for assessment of human sperm function: molecular and cellular analyses of fertilization by intracytoplasmic sperm injection, Am. J. Reprod. Immunol. 51 (4) (2004) 290–293.
[220] A. Ahmadi, A. Bongso, S.C. Ng, Intracytoplasmic injection of human sperm into the hamster oocyte (hamster ICSI assay) as a test for fertilizing capacity of the severe male-factor sperm, J. Assist. Reprod. Genet. 13 (8) (1996) 647–651.
[221] P.T. Goud, et al., Chromatin decondensation, pronucleus formation, metaphase entry and chromosome complements of human spermatozoa after intracytoplasmic sperm injection into hamster oocytes, Hum. Reprod. 13 (5) (1998) 1336–1345.
[222] H.J. Chi, et al., Successful fertilization and pregnancy after intracytoplasmic sperm injection and oocyte activation with calcium ionophore in a normozoospermic patient with extremely low fertilization rates in intracytoplasmic sperm injection cycles, Fertil. Steril. 82 (2) (2004) 475–477.
[223] R. Asch, et al., The stages at which human fertilization arrests: microtubule and chromosome configurations in inseminated oocytes which failed to complete fertilization and development in humans, Hum. Reprod. 10 (7) (1995) 1897–1906.
[224] Y. Terada, et al., Centrosomal function assessment in human sperm using heterologous ICSI with rabbit eggs: a new male factor infertility assay, Mol. Reprod. Dev. 67 (3) (2004) 360–365.
[225] C.S. Navara, N.L. First, G. Schatten, Microtubule organization in the cow during fertilization, polyspermy, parthenogenesis, and nuclear transfer: the role of the sperm aster, Dev. Biol. 162 (1) (1994) 29–40.
[226] S. Nakamura, et al., Human sperm aster formation and pronuclear decondensation in bovine eggs following intracytoplasmic sperm injection using a Piezo-driven pipette: a novel assay for human sperm centrosomal function, Biol. Reprod. 65 (5) (2001) 1359–1363.
[227] S. Nakamura, et al., Analysis of the human sperm centrosomal function and the oocyte activation ability in a case of globozoospermia, by ICSI into bovine oocytes, Hum. Reprod. 17 (11) (2002) 2930–2934.
[228] V.Y. Rawe, et al., A pathology of the sperm centriole responsible for defective sperm aster formation, syngamy and cleavage, Hum. Reprod. 17 (9) (2002) 2344–2349.
[229] T. Yoshimoto-Kakoi, et al., Assessing centrosomal function of infertile males using heterologous ICSI, Syst. Biol. Reprod. Med. 54 (3) (2008) 135–142.
[230] G. Schatten, C. Simerly, H. Schatten, Microtubule configurations during fertilization, mitosis, and early development in the mouse and the requirement for egg microtubule-mediated motility during mammalian fertilization, Proc. Natl. Acad. Sci. U. S. A. 82 (12) (1985) 4152–4156.
[231] L. Hewitson, et al., Microtubule organization and chromatin configurations in hamster oocytes during fertilization and parthenogenetic activation, and after insemination with human sperm, Biol. Reprod. 57 (5) (1997) 967–975.
[232] F. Vanden Meerschaut, et al., Diagnostic and prognostic value of calcium oscillatory pattern analysis for patients with ICSI fertilization failure, Hum. Reprod. 28 (1) (2013) 87–98.
[233] K. Morozumi, et al., Simultaneous removal of sperm plasma membrane and acrosome before intracytoplasmic sperm injection improves oocyte activation/embryonic development, Proc. Natl. Acad. Sci. U. S. A. 103 (47) (2006) 17661–17666.
[234] D. Nikiforaki, et al., Sperm involved in recurrent partial hydatidiform moles cannot induce the normal pattern of calcium oscillations, Fertil. Steril. 102 (2) (2014) 581–588 (e1).
[235] D. Nikiforaki, et al., Oocyte cryopreservation and in vitro culture affect calcium signalling during human fertilization, Hum. Reprod. 29 (1) (2014) 29–40.
[236] K. Swann, et al., Phospholipase C-zeta-induced Ca2+ oscillations cause coincident cytoplasmic movements in human oocytes that failed to fertilize after intracytoplasmic sperm injection, Fertil. Steril. 97 (3) (2012) 742–747.
[237] J. Tesarik, M. Sousa, C. Mendoza, Sperm-induced calcium oscillations of human oocytes show distinct features in oocyte center and periphery, Mol. Reprod. Dev. 41 (2) (1995) 257–263.
[238] T. Ebner, et al., Complete oocyte activation failure after ICSI can be overcome by a modified injection technique, Hum. Reprod. 19 (8) (2004) 1837–1841.
[239] K. Versieren, et al., Developmental competence of parthenogenetic mouse and human embryos after chemical or electrical activation, Reprod. Biomed. Online 21 (6) (2010) 769–775.

[240] J. Zhang, et al., Electrical activation and in vitro development of human oocytes that fail to fertilize after intracytoplasmic sperm injection, Fertil. Steril. 72 (3) (1999) 509–512.
[241] R. Mansour, et al., Electrical activation of oocytes after intracytoplasmic sperm injection: a controlled randomized study, Fertil. Steril. 91 (1) (2009) 133–139.
[242] V. Baltaci, et al., The effectiveness of intracytoplasmic sperm injection combined with piezoelectric stimulation in infertile couples with total fertilization failure, Fertil. Steril. 94 (3) (2010) 900–904.
[243] A. Egashira, et al., A successful pregnancy and live birth after intracytoplasmic sperm injection with globozoospermic sperm and electrical oocyte activation, Fertil. Steril. 92 (6) (2009) 2037 (e5–e9).
[244] E.P. de Fried, et al., Human parthenogenetic blastocysts derived from noninseminated cryopreserved human oocytes, Fertil. Steril. 89 (4) (2008) 943–947.
[245] F. Vanden Meerschaut, et al., Assisted oocyte activation is not beneficial for all patients with a suspected oocyte-related activation deficiency, Hum. Reprod. 27 (7) (2012) 1977–1984.
[246] D. Nikiforaki, et al., Effect of two assisted oocyte activation protocols used to overcome fertilization failure on the activation potential and calcium releasing pattern, Fertil. Steril. 105 (3) (2016) 798–806 (e2).
[247] Y. Liu, et al., Three-day-old human unfertilized oocytes after in vitro fertilization/intracytoplasmic sperm injection can be activated by calcium ionophore a23187 or strontium chloride and develop to blastocysts, Cell. Reprogram. 16 (4) (2014) 276–280.
[248] Y. Liu, et al., Artificial oocyte activation and human failed-matured oocyte vitrification followed by in vitro maturation, Zygote 21 (1) (2013) 71–76.
[249] J.W. Kim, et al., Successful pregnancy after SrCl2 oocyte activation in couples with repeated low fertilization rates following calcium ionophore treatment, Syst. Biol. Reprod. Med. 60 (3) (2014) 177–182.
[250] P. De Sutter, et al., Parthenogenetic activation of human oocytes by puromycin, J. Assist. Reprod. Genet. 9 (4) (1992) 328–337.
[251] K. Swann, J.P. Ozil, Dynamics of the calcium signal that triggers mammalian egg activation, Int. Rev. Cytol. 152 (1994) 183–222.
[252] M. Montag, et al., The benefit of artificial oocyte activation is dependent on the fertilization rate in a previous treatment cycle, Reprod. Biomed. Online 24 (5) (2012) 521–526.
[253] M.H. Nasr-Esfahani, M.R. Deemeh, M. Tavalaee, Artificial oocyte activation and intracytoplasmic sperm injection, Fertil. Steril. 94 (2) (2010) 520–526.
[254] Y. Sato, et al., Follow up of children following new technology: TESE, IVM, and oocyte activation, Hum. Reprod. 26 (2011) I274.
[255] T. Takisawa, et al., Effect of oocyte activation by calcium ionophore A23187 or strontium chloride in patients with low fertilization rates and follow-up of babies, Hum. Reprod. 26 (2011) I178–79.
[256] T. Takisawa, et al., Effect of oocyte activation by calcium ionophore A23187 or strontium chloride in patients with low fertilization rates and follow-up of babies, Fertil. Steril. 96 (2011) S162.
[257] F. Vanden Meerschaut, et al., Neonatal and neurodevelopmental outcome of children aged 3-10 years born following assisted oocyte activation, Reprod. Biomed. Online 28 (1) (2014) 54–63.
[258] N.T. Rogers, et al., Phospholipase Czeta causes Ca2+ oscillations and parthenogenetic activation of human oocytes, Reproduction 128 (6) (2004) 697–702.
[259] C. Spadafora, Endogenous reverse transcriptase: a mediator of cell proliferation and differentiation, Cytogenet. Genome Res. 105 (2–4) (2004) 346–350.
[260] S.Y. Yoon, et al., Recombinant human phospholipase C zeta 1 induces intracellular calcium oscillations and oocyte activation in mouse and human oocytes, Hum. Reprod. 27 (6) (2012) 1768–1780.

CHAPTER

15

The Molecular Signature of the Endometrial Receptivity: Research and Clinical Application

*José P. Carrascosa**, *José A. Horcajadas**,†, *Juan M. Moreno-Moya*†

*Pablo de Olavide University, Sevilla, Spain †SINAE, Sevilla, Spain

THE ENDOMETRIUM

Definition

The human endometrium is the mucous membrane that coats the inner part of the uterus in mammals (Fig. 15.1). In humans and higher primates, it is hormonally regulated and changes dynamically during the menstrual cycle. These changes are necessary for the preparation of the receptive stage that is essential for embryo adhesion, implantation, and gestational development to keep nurturing and protecting the allogeneic fetus. No primates or mammalians have an estrous cycle. The key differences between an estrous cycle and a menstrual cycle are that the endometrium does not detach, but resorbs it if implantation does not occur and it is not hormonally regulated [1].

Anatomy

The uterus consists of two anatomically different sections: the uterine body (corpus uteri with the uterine cavity), which contains a smooth muscle layer also called the myometrium coated by a tunica mucosa named endometrium, and the cervix, which is also coated with the endocervix (Fig. 15.1).

The human endometrium consists of the epithelial, stromal, and vascular compartments and also immune-resident cells. All four of these compartments are located in two regions named "functionalis" and "basalis." The "functionalis" is regenerated each

https://doi.org/10.1016/B978-0-12-812571-7.00016-2

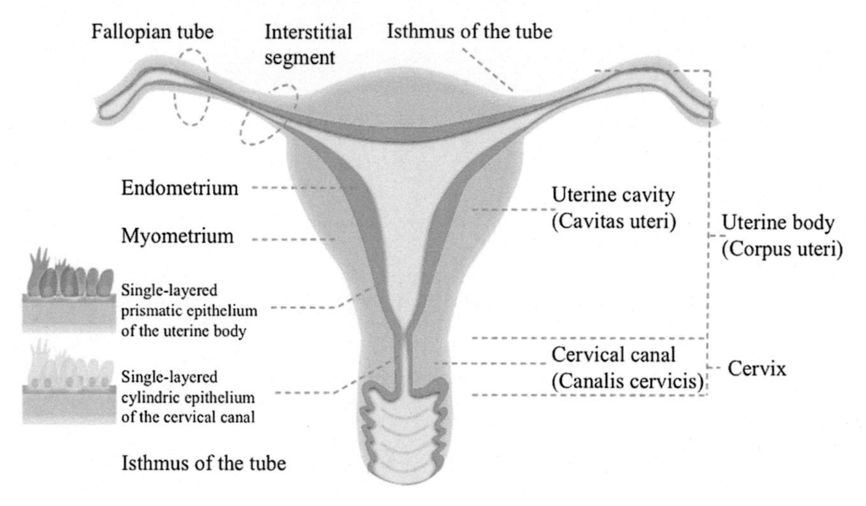

FIG. 15.1 Schematic representation of the uterus.

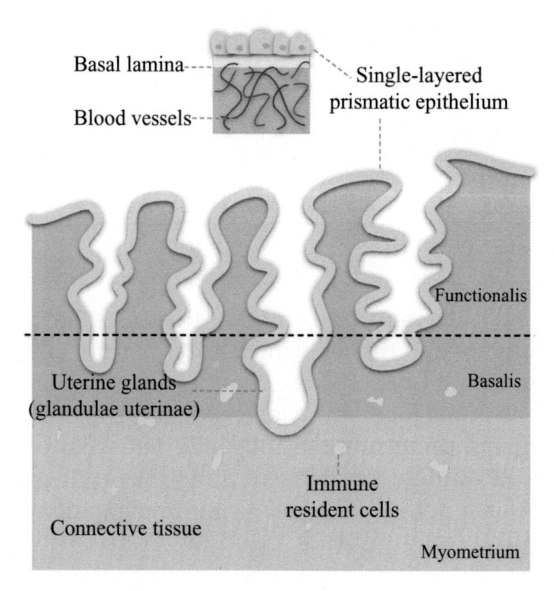

FIG. 15.2 Schematic representation of the endometrium.

month, while the "basalis" remains and is the resource for the cyclic regeneration of the endometrium (Fig. 15.2).

The epithelial compartment is a monolayer of polarized cuboid cells oriented toward the lumen of the uterus, which consists of two components, the luminal and the glandular epithelia. The luminal epithelium modifies its morphology across the menstrual cycle [2]. These changes are induced in response to estrogens and affect the plasma membrane, cytoskeleton, tight junctions, and microvilli, which are reduced during the secretory stage, while the apical protuberances originated by the endometrial fluid endocytosis [3] also called "pinopodes" increase. The luminal epithelium consists of cilium and noncilium cells, and their numbers increase during the proliferative phase of the menstrual cycle. In this phase, the rate of noncilium/cilium cells goes from 30:1 to 15:1, with a drop to 50:1 after 20 days of the menstrual cycle [4,5]. The main functions of the epithelium are the defense against external pathogens and embryo implantation. During the WOI, the epithelium undergoes multiple changes to facilitate the embryo attachment and allows for implantation to occur. During the proliferative phase, the glandular epithelium is covered by microvilli and presents different unions between cells, such as desmosomes, tight junctions, and gap junctions. At this stage, there are a large number of mitochondria closely related with no association with the rough endoplasmic reticulum and with poorly developed secretory apparatus [6,7]. The glandular epithelium contains epithelial cells that proliferate during the secretory phase forming large glands that produce and secrete molecules to nurture the implanting blastocyst. Some cellular characteristics are glycogen accumulation in the subnuclear cytoplasm and the development of the nuclear channel system and giant mitochondria [8,9]. The origin of the giant mitochondria could be due to the fusion of normal and small mitochondria, as a response to the action of progesterone in the DNA [10]. The function of these giant mitochondria may be to provide energy for the dynamic changes that these cells undergo.

The stromal compartment is a connective tissue formed by cells and the extracellular matrix. The principal cell type is the fibroblast, which is involved in matrix remodeling across the menstrual cycle, which principally occurs during the so-called "decidualization" process in the midluteal phase. Several morphological and biochemical changes characterize this process in response to estrogen and progesterone exposure. For example, morphologically, the elongated fibroblast-like ESCs transform into enlarged round cells with specific ultrastructural modifications. This is accompanied by the secretion of specific markers such as prolactin (PRL) and insulin-like growth factor-binding protein-1 (IGFBP-1) [11] and extracellular matrix proteins such as laminin, type IV collagen, fibronectin, and heparin sulfate proteoglycan as part of their differentiation program [12]. The onset of this process in the ESCs surrounding the terminal spiral arteries marks the end of the window of implantation (WOI).

The vascular compartment constitutes a complex network starting with the myometrium. The uterine arteries form the arcuate arteries that originate at the radial arteries, which cross the myometrium and reach the endometrial-myometrial junction where they differentiate into the basal arteries that give rise to the spiral arteries that support the basalis region. Basalis arteries ramify in the functional layer, and each one can support a 4–8 mm^2 of endometrial surface [13]. In the human endometrium, there are three main angiogenesis events: (i) during menstruation to repair the vascular compartment, (ii) during the fast growth in the proliferative phase, and (iii) during the secretory phase when the number of spiral arteries increases [14].

The immune-resident cells consist of uterine natural killer cells (uNKs), macrophages, and T cells, in which their main function is to protect the genital tract from infections and avoid immune rejection during embryo implantation. The leucocyte population found in a normal endometrium accounts for 10%–15% of the total stromal cell population, which is highest during late secretory and premenstrual stages [15]. Each immune cell type has essential functions that contribute to a successful embryo implantation and avoid a possible rejection. Regulatory T cells and regulatory uNKs have a key role in immunosuppression in order to avoid a fetus rejection [16]. Granulated lymphocytes are related to trophoblast invasion and the secretion of transforming growth factor (TGF) [17]. Macrophages have various functions such as the enrichment of the extracellular matrix, the regeneration of tissues and cytokines, and the production of proteases to rid the endometrium of cell debris after menstruation [18].

Menstrual Cycle

The menstrual cycle is exclusive to primates and humans, with other mammals having an estrous cycle characterized by the endometrial reabsorption. In the menstrual cycle, the endometrium is renewed by expelling it in each menstruation [19]. The cyclic regulation of the human endometrium is due to the effects of the ovarian steroids, estrogen and progesterone (E2 and P4, respectively), allowing for the coordination between the menstrual and ovarian cycles [20]. There are three main phases during the menstrual cycle: (i) the menstrual phase, (ii) the proliferative phase, and (iii) the secretory phase [21] (Fig. 15.3).

Menstruation phase: This phase begins with the initiation of menses (day 0) and takes 3–5 days. The drop of E2 and P4 induces the detachment of the functionalis layer and causes endometrial shedding. After this phase, the endometrium is thin, and the basalis layer is all that remains.

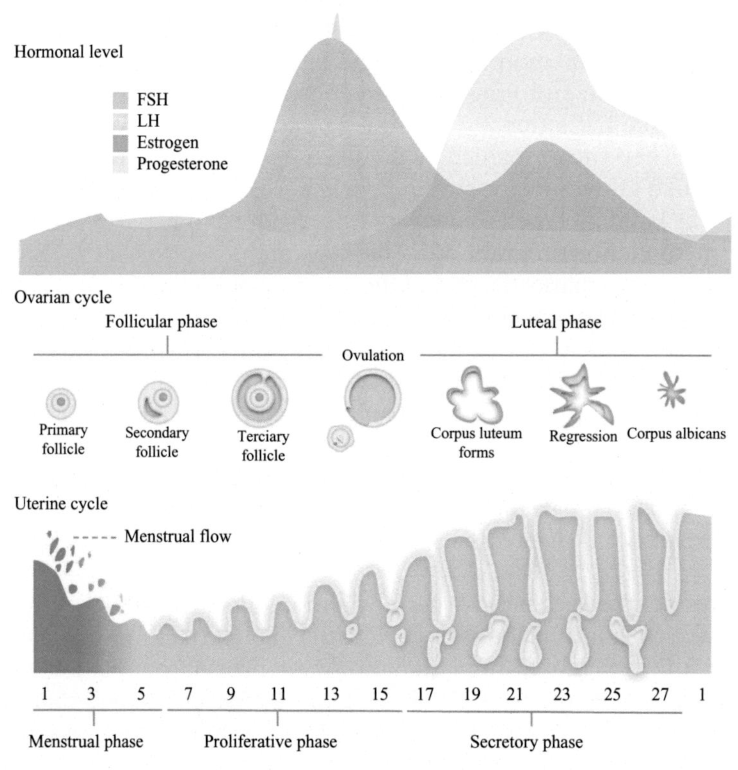

FIG. 15.3 Schematic view of the human menstrual and ovarian cycles.

Proliferative phase: This phase starts at the end of menstruation and lasts until the day of ovulation (day 14). The stromal and epithelial cells from the functionalis layer proliferate and regenerate in response to increasing levels of estrogens, secreted from the ovary. The glandular epithelium acquires linear shapes and vascularization. The endometrial thickness increases approximately from 4 to 7mm.

Secretory phase: This phase begins at ovulation (day 15) and lasts until menstruation (day 28). After ovulation, the corpus luteum starts to secrete high amounts of E2 and P4. The estrogens induce a slight proliferation, and the progesterone promotes decidualization. Between days 19 and 21, the endometrial epithelium becomes receptive [22]. However, if there is no implantation, the corpus luteum degenerates, E2 and P4 levels decrease, the endometrium becomes ischemic, and glandular secretion stops. The final result is the elimination of the functionalis layer and the start of a new cycle [23].

ENDOMETRIAL RECEPTIVITY

Definition

The human endometrium has two main functions: (i) the temporal acquisition of an adhesive phenotype to allow the embryo to implant, which is also called "endometrial receptivity," and (ii) its active participation in the initial dialogue with the embryo that will direct the invasion, placentation, fetal development, and finally parturition [24].

The human endometrium undergoes cyclic changes during the menstrual cycle in response to steroid hormones. The final goal of these changes is to reach the endometrial receptivity for the embryo implantation. The phase in which the endometrium is receptive is known as WOI. It corresponds to the period of time in which the endometrium remains receptive for an embryo and occurs in response to the presence of endogenous or exogenous progesterone and after appropriate stimulation with 17β-estradiol. These hormones bind to their receptors activating a signaling pathway that alters the endometrial signature [25]. In natural cycles, the WOI takes place between day 19 and 21 of the menstrual cycle and 7days after the surge in luteinizing hormone (LH) [26–28]. On any other day of the menstrual cycle, the endometrium is nonreceptive, and embryo implantation does not occur [29]. During the WOI, the endometrial epithelium acquires a dependent steroid hormone status allowing for the adhesion of the blastocyst [30]. When the luminal endometrial epithelium acquires a receptive status, it undergoes several changes such as "plasma membrane transformations," pinopode formation, and tight junctions [2].

Morphological Evaluation

Morphological evaluation is based on the cyclic histological changes reported by Noyes [21], in which 8000 endometrial biopsies across the menstrual cycle were sectioned and stained with hematoxylin-eosine. Noyes' criteria have been considered as the gold standard in endometrial evaluation and are based upon eight basic histological features. These include glandular mitosis, nucleus pseudostratification, basal vacuoles, secretion, stromal edema, pseudodecidual reaction, stromal mitosis, and leucocyte infiltration (Fig. 15.4). Randomized

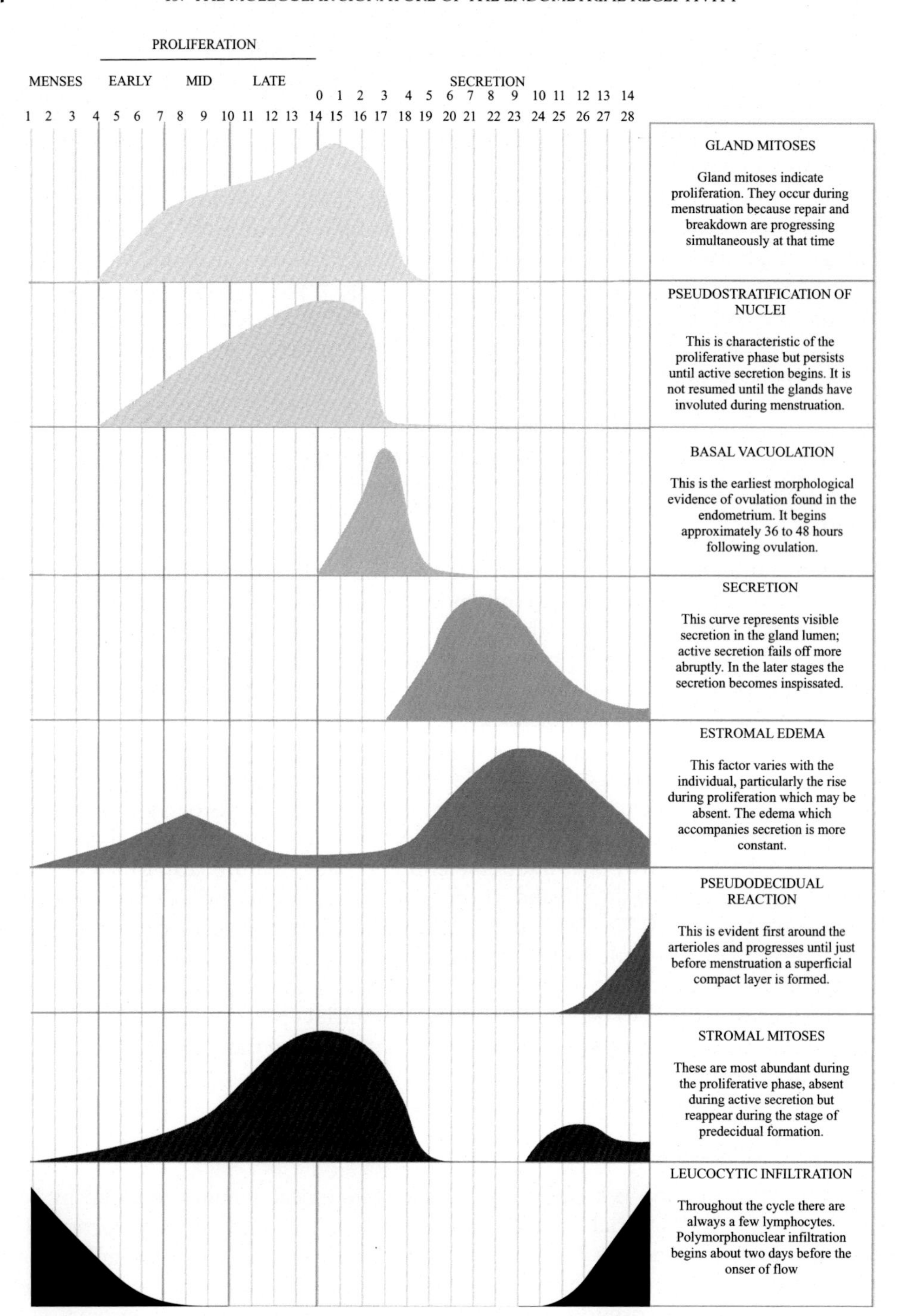

FIG. 15.4 Noyes histological criteria.

controlled trials have shown that these criteria cannot predict the endometrial receptivity state accurately [31,32]. The Practice Committee of the American Society for Reproductive Medicine (ASRM) advised against the use of endometrial biopsies for endometrial dating, as the results are based upon prospective, blind, and randomized controlled trials. Furthermore, endometrial biopsies cannot differentiate between fertile and infertile women and are inefficient in diagnosing and treating the luteal phase deficiency [33]. The accuracy in endometrial dating depends on several factors such as ovulation day, time of the biopsy, quality of the biopsy, the absence of endometrial lesions, fixative techniques, and pathologist's interpretation. Therefore, ovarian stimulation alters the endometrial maturation process. As a consequence, the endometrial status is not investigated morphologically in the standard workup performed in fertility clinics worldwide due to the absence of reliable histological tests.

In the implantation process, the embryo interacts in the beginning with the endometrial surface; therefore, the luminal epithelium was one of the first structures analyzed to discover markers of endometrial receptivity. The best morphological biomarker identified in the luminal epithelium associated with endometrial receptivity is the pinopodes. These structural adaptations are smooth mushroomlike protrusions at the surface of the luminal epithelium and have been proposed as a reliable marker for endometrial receptivity [34,35]. A controversial issue regarding pinopodes is their duration on the surface of the endometrium, with some studies reporting less than 48 h [36,37] and others showing that these structures remain in the postreceptive endometrium [38]. Additionally, their visualization requires scanning electron microscopy (SEM), which is too complex of a technique to be utilized routinely in clinical practice. The variability in the duration of pinopodes among different patients is the main reason against its use as a biomarker for endometrial receptivity.

A third noninvasive morphological evaluation of receptivity is the transvaginal ultrasound. Several parameters have been related to a healthy endometrium, such as the size, thickness, and vascular perfusion of the endometrium. However, they have not been well correlated with endometrial receptivity or gestation prediction after assisted reproductive technology [39–41].

Searching for an Ideal Marker of Endometrial Receptivity

The ideal marker of endometrial receptivity would be a gene with cellular specificity, by being able to show consistent expression and being able to differentiate among all the states of receptivity, thus allowing practitioners to choose the right moment for the embryo transfer. Based on several in vitro and in vivo human observations, numerous molecules have been suggested in the endometrial implantation process as biomarkers for endometrial receptivity. For instance, such biomarkers include leukemia inhibitory factor (LIF), a member of the interleukin-6 family, which binds to the LIF receptor (LIFR), and gp130 that regulates various cellular functions. The first evidence of LIF in embryo implantation came from a LIF-deficient female mouse that failed to implant. However, supplementation with LIF restored the implantation rate in this mouse [42]. Endometrial epithelial cells under the influence of progesterone begin to secret different molecules as LIF, which are released into the uterine lumen prior to the attachment of the blastocyst [43]. It has been proved that LIF enhances the adhesion of the trophoblast to the endometrium by upregulating the expression of the integrin heterodimers $\alpha V\beta 3$ and $\alpha V\beta 5$ [44]. High levels of LIFR and gp130 during immunostaining were correlated to low levels of SOCS1 and positively correlated to the formation of pinopodes. Therefore, LIF pathway seems to have disturbances in some cases of infertility, but not in all women with unexplained infertility or implantation failure [45]. As a result, LIF gene expression probably plays an important role in embryo implantation, as it is upregulated during the WOI in fertile women [46–48] and is downregulated on the mRNA level and protein level in infertile women [49]. Women with high LIF immunoreactivity during the WOI have greater chances of pregnancy than those with weaker expression; therefore, it may have an important role for IVF clinics [50]. However, preclinical and clinical trials have shown that treatment with a recombinant human LIF during the luteal phase after embryo transfer failed to improve the implantation rate in women with recurrent implantation failure (RIF) [51].

Other molecules proved to be implicated in the endometrium to reach a receptive status are leptin and leptin receptor (LEPR). Throughout the menstrual cycle, the presence of both leptin and LEPR has been reported in the endometrium [52]. Protein and mRNA of leptin and LEPR are expressed in the secretory endometrium and in cultured endometrial epithelial cells, with the secretion being regulated by the human blastocyst in vitro. Therefore, it has a possible physiological role in the implantation of a fertilized egg [53]. Several studies have reported an increase of serum leptin levels in women with unexplained infertility [54] and in women with immunologic recurrent abortion or unexplained recurrent abortion [55]. In another study, higher leptin concentrations in the follicular fluid were associated with higher oocyte fertilization rates in normogonadotropic patients in assisted reproductive technologies [56].

Integrins are transmembrane receptors with important roles in cell signaling and cell attachment to the cell's extracellular matrix, and they have also been related to endometrial receptivity. The appearance and distribution of some integrins are linked to the endometrial cycle, and the disturbance of integrin expression may be related with reduced uterine receptivity and infertility [57]. Abnormal protein β-3 integrin expression has been related with unexplained infertility in women. This protein has a key role in embryo implantation with a positive association to embryo implantation [58,59]. During the WOI, fertile women express the protein β-3 integrin; meanwhile, the lack of this protein is correlated to unexplained infertility, endometriosis, and hydrosalpinx [60,61]. This result has not been replicated by independent studies, that is, other studies observed no differences in the expression profiles between fertile and unfertile women [62] or in women with endometriosis [63]; therefore, its efficacy has been questioned. The αvβ3 integrin has a ligand called osteopontin that is coexpressed with the αvβ3 integrin across the menstrual cycle, with a peak at the WOI [64]. These results have been recently investigated in women with elevated serum progesterone and/or estradiol, and its expression/coexpression is not impaired during the WOI [65].The clinical value of osteopontin and integrin αvβ3 to asses endometrial receptivity seems to be uncertain.

The expression in the endometrium of a pleiotropic protein, prokineticin (PROK1), has a peak during the secretory phase of the menstrual cycle [66]. PROK1 upregulated the expression of LIF via the induction of intracellular calcium and the activation of intracellular phosphorylation cascades, thus suggesting its implication in enhancing endometrial receptivity including the regulation of endometrial cell proliferation and decidualization [67,68]. Recently, a new study has showed that an increase in mRNA levels of PROK1 and LIF in the endometrium is linked to recurrent pregnancy lost [69].

Heparin-binding epidermal growth factor-like growth factor (HB-EGF) is a member of the EGF family that binds to heparan sulfate proteoglycans and EGF receptors, which affects cell-to-cell interactions [70]. HB-EGF has pleiotropic biological functions, and it plays a role in endometrial decidualization and reduces TNFα and TNFβ that induce apoptosis of endometrial stromal cells [71]. It is synchronized with the endometrial cycle and coincides with a receptive phenotype. HB-EGF has a low mRNA expression during the proliferative stage, increasing in the secretory stage, reaching its highest expression immediately prior to the implantation window (day 19–21), and decreasing thereafter [72,73]. It has been proposed as a marker for the classification of the ultrasonic patterns and for local injury in the endometrium related with ultrasonic evaluation of endometrial receptivity [74]. In a healthy human endometrium, HB-EGF is expressed inside the luminal epithelial cells and on the surface of pinopodes, when the highest expression of pinopodes is present [75]. The dysregulation in the expression of HB-EGF in the endometrium has been associated with unexplained infertility and unexplained miscarriages [76,77]. It seems like HB-EGF is needed for normal decidualization of the endometrial stromal cells to reach a receptive state in the endometrium and for the initiation of implantation. Therefore, HB-EGF has a potential to be a biomarker of endometrial receptivity when combined with other biomarkers or techniques.

The two forms of IL1 agonists (IL1α and IL1β) have a similar affinity to the Il1 receptor (IL1R1) and, consequently, also show comparable, if not identical, biological activities. Both interleukins are present at high concentrations in the culture medium of human embryos with successful implantation rates [78]. Throughout the menstrual cycle of fertile women, IL1R1 displays a triphasic expression at the mRNA and protein levels, that is, it was low in the proliferative phase until day 13 (period 1), moderate during ovulation and implantation (period 2), and intense from day 22 (period 3) in both epithelial and stromal cells. However, there was no evidence of differential expression patterns between fertile women and women with unexplained infertility [79].

HOX genes are highly evolutionary conserved and codified transcription factors that are crucial for embryonic morphogenesis and differentiation [80]. The HOX genes that have been studied the most are HOXA10 and HOXA11. Both of these genes are regulated by sex steroids with a peak during the implantation window, suggesting a possible role in implantation [81,82]. Other HOX genes implicated in the regulation of endometrial receptivity have been shown to have an increased expression during the midsecretory phase of the menstrual cycle. These genes were HOXA9, HOXA11, and HOXA10D [83]. Comparison studies have been conducted to establish if there were differences at the mRNA and/or protein levels for HOXA10 and HOXA11 between fertile and infertile women. These results found no differences in the amount of protein or mRNA in patients with idiopathic infertility to those who were fertile [84].

Glycodelin is a glycoprotein of the lipocalin superfamily, is secreted by endometrial glands during the secretory phase, and is known by several names such as placental protein 14 (PP14) [85], progestogen-dependent endometrial protein (PEP) [86], or progesterone-associated endometrial protein (PAEP) [87]. It has a cyclic pattern suggesting it likely has steroid regulated and has a role in endometrial receptivity [88]. Glycodelin expression is synchronized with LIFR and the formation of pinopodes [89]. It has an immunosuppressive function in the endometrium during the WOI, allowing the possible maternal response to the implantation of the incoming embryo [90]. Women with unexplained infertility have lower concentrations of glycodelin in uterine fluids but not in plasma than normal fertile women, suggesting a possible role at the local level of glycodelin in endometrial receptivity [91].

Several groups have studied the expression of specific molecules in the human endometrium, in the different cell compartments during the different stages of the menstrual cycle [92]. For example, integrins are a family of cell adhesion receptors that interact with the extracellular matrix components to trigger specific cell signaling. Some authors have determined that the β3, β4, and β1 integrins were indicators of endometrial receptivity [93]. Mucins are another type of molecule in which researchers have focused upon. One of the best known mucins is mucin 1, which acts as a barrier for adhesion until it is specifically cleaved by the embryo [94,95]. Additionally, receptors for integrins such as osteopontin have been found to be increased in endometrial glands and secreted during receptivity [96–98]. However, alone, none of these signaling molecules have been shown to be clinical predictors of human endometrial receptivity.

Novel Molecular Characterization: "Omics"

The new technologies that have allowed the consecution of the human genome project [99] and recently the Encyclopedia of DNA Elements (ENCODE) project have completely changed the possibilities of research and clinical diagnostics in the field of human reproduction, from single molecules to the whole genome of a cell in a single experiment [100].

Transcriptomics

The term "genomic" consists of the study of DNA and chromosomes, while "transcriptomics" or "functional genomics" studies the global mRNA gene expression. The basis of functional genomics resides in the fact that all cells contain the same genetic information, but depending on the tissue, cell type, or biological process that is active, the cells selectively express the genetic material differently. For transcriptomics, the most common evaluation techniques are gene expression microarrays, microfluidic multiplexed PCR, and RNA sequencing (RNA-seq). Microarrays have been widely used in human endometrial studies for global gene expression in all the phases of the menstrual cycle, thus concluding that it is possible to classify human endometrial samples through their transcriptomic profile, establishing different clusters of genes for each phase of the endometrial cycle [101]. It has also been used in the human endometrium from the prereceptive states (LH+/LH+5) to the receptive states (LH+7/LH+9) in natural cycles [28,96,102–105]. Despite using the same technology, there are broad differences between studies due to dissimilarities in experimental design and data analysis. The main differences found were the day of the menstrual cycle for endometrial biopsy, phases of the menstrual cycle compared, genomic variation between patients, number of endometrial biopsies used, and pooling or not pooling the isolated RNA. Additionally, statistical methods and data analysis to accept a regulated gene varied among studies. Four of the studies fixed a minimal twofold increase to accept the gene regulation, where others considered a more stringent criterion of a threefold increase, and the samples were acquired from the same patient on two different days of the menstrual cycle, thus reducing the genomic variation between patients [106]. Mirkin compared the results of microarray analysis for five studies [107] finding that only one gene, the osteopontin, was presented in all of the studies. This glycoprotein is linked to the cellular adhesion process and embryo implantation [58].

The birth of IVF/ICSI techniques brought a new pathophysiological state that is characterized by the failure to achieve a clinical pregnancy after numerous failures (at least four good-quality embryos in three cycles). In order to identify possible markers of infertility, the endometrial gene expression profile of women with unexplained infertility was compared with fertile women in a few studies published in the past 10 years. Tapia and colleagues found 63 differentially expressed genes, and among these dysregulated genes were matrix metalloproteinase-7, CXC chemokine receptor 4, glycodelin, and C4b-binding protein [108]. Koler and coworkers showed 313 genes with modified expression levels in RIF patients with functions in the cell cycle, more specifically Wnt signaling and cellular adhesion pathways [109]. Altmäe and collaborators identified 145 upregulated and 115 downregulated genes in women with RIF versus the control women and found a dysregulation of gene pathways through leukocyte extravasation signaling, lipid metabolism, and detoxification [110].

Others studies have focused on the effects of gene expression profiles during natural and stimulated cycles, with agonist and different doses of antagonist [104,105,111–113], in the refractory endometrium by the presence of an intrauterine device (IUD) [114] or in the presence of uterine intramural leiomyomas [115].

Based on these studies that focused on endometrial receptivity between the prereceptive and the receptive stages for natural cycles, different predictive tests have been developed for the assessment of endometrial receptivity using different subsets of these differentially expressed genes. The first of these tools was a customized microarray named "endometrial receptivity array (ERA)" (commercialized by iGenomix SL) [116–118] that analyzes a total of 238 gene transcripts. This tool allows for the assessment of the endometrial receptivity status and pinpoints the ideal day to conduct the embryo transfer, which has been named personalized embryo transfer (pET) [119]. It has been employed in patients with endometriosis, with no changes in the endometrial signature; in fact, none of the 238 genes were significantly up- or downregulated [120].The second product named endometrial receptivity map (ER Map) [121] has been recently launched and composed of a curated list of 192 gene transcripts using fluidigm technology, a microfluidic multiplexed real-time PCR approach (commercialized by IGLS SL). This gene list was created using the data of 71 scientific papers published between 2005 and 2015, in which just 53 genes were in common with ERA and only 7 out of 25 genes were in common with the golden list proposed by Horcajadas in 2007 [106]. A simple comparison of this 192 WOI gene list with seven transcriptomic studies using microarrays [96,98,102,103,107,122] or RNA-seq [123] in natural cycles in healthy women showed large differences in the selected genes. One hundred and twenty-five out of the 192 selected genes in the ER Map study were not present in any of these studies, with just one gene being common between all of the studies, which was osteopontin. Four genes were shared by the ER Map and six out of seven of the research studies (claudin 4, growth arrest and DNA-damage-inducible protein, interleukin-15, and monoamine oxidase A). Both of these products, ERA and ER Map, are capable to classify the status of the endometrial sample as prereceptive, receptive, or postreceptive. This helps the clinician to plan the ideal day for the patient's embryo transfer according to their results, so the synchrony between the embryo and the receptive status is optimal. For example, a significant improvement has been found in aiding patients with repeated implantation failure predicted as "prereceptive" in ERA. These patients, at a "prereceptive" state, were replaced 2 days posterior to the usual day of embryo transfer in the next cycle, and this resulted in higher percentages of pregnancies and babies delivered [124]. A major disadvantage of these tests is that they require an endometrial biopsy, therefore making it practically impossible to perform the embryo transfer in that cycle.

Nowadays, microarrays have almost been replaced by a new technology called RNA sequencing (RNA-seq), which is based on next-generation sequencing (NGS). This technique has been widely used in transcriptomics to show the presence of RNA at a given time and to quantify the amount of RNA present. One of the advantages over microarrays is the power to detect even the genes with low expression that are undetectable with microarrays [125]. In fact, NGS has replaced microarrays in the ERA, by using it with a computational predictor and algorithm in order to be able to classify an endometrial sample based on their receptivity genes [126]. Recently, a new method to obtain endometrial stromal and epithelial cells and to study their transcriptome at a single-cell level by single-cell RNA-seq has been reported [127]. This approach could be interesting in new research in endometrial receptivity and embryo implantation with clinical application at the single-cell level.

Proteomics

The proteomics explores the existence of translated proteins and their relative quantity. The most commonly used techniques are the "two-dimensional differential gel electrophoresis" (2-D-DIGE), the "isotope-coded affinity tag" (ICAT), and the "isobaric tag for relative and absolute quantitation" (iTRAQ) that are based on liquid chromatography. The principal disadvantage of proteomics in general is that it requires very high protein concentrations and a mass spectrometry system in order to identify each peptide [128]. Dominguez's group has studied the proteome of the receptive versus the nonreceptive endometrium of healthy donor patients by 2-D-DIGE and found several molecules upregulated such as annexin A2 and stathmin 1 [129]. Other studies that have focused on the proliferative and secretory phases have identified 76 differentially expressed proteins, with 34 protein isoforms, thus indicating the importance of posttranslational modifications and processing [130]. Epithelial cells and stromal compartments have been studied separately by performing global quantitative MS during the proliferative and secretory phases to identify 318 and 19 overexpressed proteins between these phases, respectively [131]. To know more about the proteomics derived from the transcriptomics existing during the WOI, six endometrial biopsies diagnosed as receptive and six as nonreceptive were analyzed by DIGE and matrix-assisted laser desorption/ionization (MALDI). This study only revealed 24 proteins that were differentially expressed between receptive and nonreceptive endometrial samples affecting two pathways significantly (the carbohydrate biosynthetic process and nuclear mRNA splicing via spliceosome) [132]. A different group performed ICAT between the proliferative and the secretory endometrium and found five proteins with a consistent differential expression [133], with examples being the NMDA receptor subunit zeta 1 precursor and FRAT1. It has also been described in proteomics in relation to the in vitro decidualization process [12].

The main problem of using biopsies for proteomic analyses is the large amount of functional and structural proteins that are in the way of identifying other proteins present in lower levels. Another important issue found in the gel techniques is the difficulty to separate proteins of low molecular weight. Two possible solutions have been proposed, sample prefractionation and gel-free techniques, in order to identify more biological markers in protein studies [134].

The discovery of biomarkers in the uterine fluid is a less invasive approach used in numerous proteomic studies. The endometrial secretome between the prereceptive and receptive phases has been analyzed by DIGE, which has recognized 82 proteins that are differentially expressed. Posttranscriptional modifications were found in 14 proteins [135]. Soluble mediators such as vascular endothelial growth factor and placental growth factor (PlGF) have been identified in the uterine fluid, with potential functions on the peri-implantation embryo and the endometrial epithelium [136]. Analysis of the human uterine fluid of women at the prereceptive and receptive stages and women with unexplained infertility at the receptive phase has revealed 42 cytokines, chemokines, and growth factors that play important roles during implantation [137]. The main advantage of using uterine fluids for the discovery of biomarkers is the higher levels of secreted proteins that could also be discovered in other analyzable samples such as blood or urine. In transcriptomics, it seems more likely to obtain a protein fingerprint of multiple molecules than a single molecule able to identify the receptivity status in both fertile and infertile women.

In Table 15.1, which is adapted from Haouzi's review of endometrial receptivity studies in 2012 [138], it summarizes the most common molecules identified during receptivity between the different studies at the transcriptomic and proteomic levels. This scheme exemplifies that mRNA transcription and protein translation levels are not always correlated. Fig. 15.5 also shows that between the six mentioned transcriptomic studies, the results differ. This observation can also be explained by the patients' characteristics and date of biopsy.

Lipidomics

Lipidomics is a novel term assigned to the study of the global lipid profile in different types of samples. Some recent reports have applied lipidomics to the study of reproductive

TABLE 15.1 Relation Between Transcriptomic and Proteomic Results for Endometrial Receptivity

	Transcriptomic Studies						**Proteomic Studies**	
Name	**Carson et al. [97]**	**Riesewijk et al. [28]**	**Mirkin et al. [107]**	**Talbi (2006)**	**Haouzi et al. [105]**	**Díaz-Gimeno et al. [116]**	**Li et al. (2006)**	**Domínguez et al. [129]**
ANXA4	–	4	6.5	4.9	2.6	4.7	2.1	1.9
ANXA2	–	4	5.6	2	–	–	–	2.1
MAOA	–	15	–	–	9.9	8.4	–	3.4
TAGLN	–	6	–	–	5.9	–	–	1.7
LCP1	–	–	2.6	1.6	–	–	–	1.6
PGRMC1	–	–	–	−1.8	–	–	–	−2.4
STMN1	–	–	−3.2	–	–	–	–	−2.2
APOL2	–	–	–	–	2.4	–	–	3.7
ALDH1A3	–	–	–	–	16.5	–	–	1.8
S100A10	–	–	–	–	3.5	–	–	4.8

Modified from Haouzi.

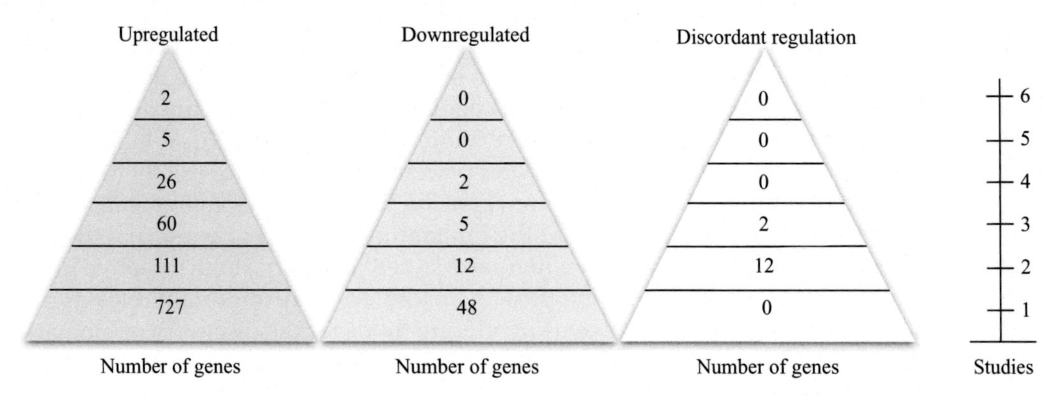

FIG. 15.5 Number of genes involved in endometrial receptivity common to the microarrays studies. *Adapted from Haouzi.*

function by assessing the lipid content of endometrial biopsies during pregnancy [139]. The technique used for this kind of metabolite evaluation is high-performance liquid chromatography (HPLC) that is used to fractionate, separate, and identify the composition and expression levels of lipids at single level [140]. Vilella and colleagues found that prostaglandins PGE2 and PGF2a are produced by synthesis located in the luminal epithelium and that these prostaglandins are secreted toward the endometrial fluid and increase during the WOI.

miRNomics

miRomics is a novel concept for the -omics that have been assigned to the study of global profiles of miRNAs (miRNAs). miRNAs are short noncoding RNAs that act as negative regulators of mRNAs by complementary binding to them and triggering the RNA-induced silencing complex (RISC). Thousands of miRNAs have been identified, and they can be quantified using the same strategies as for mRNA, that is, RNA-seq, PCR, and microarrays. In the endometrium, they have been proposed as novel biomarkers to study receptivity [141,142].

The traditional approach to obtain important biomarkers of endometrial receptivity was applied by comparing miRNA signatures in the receptive versus prereceptive endometrium from healthy women. The results displayed hsa-miR-30b and hsa-miR-30d to be significantly upregulated and hsa-miR-494 and hsa-miR-923 to be downregulated in the receptive endometrium. This mRNA had important functions in transcription, cell proliferation, and apoptosis and had significant participation in some relevant pathways, such as axon guidance, Wnt/β-catenin, ERK/MAPK, transforming growth factor-β (TGF-β), p53, and leukocyte extravasation [143].

To date, there have been few endometrial studies using RNA-seq, but it is expected to increase in the following years. It has been used to identify miRNA related to endometrial receptivity in both natural and stimulated cycles using endometrial biopsies or villus [123,144,145], to analyze the exosome carriers of miRNA in the uterine fluid, to investigate the environment in which a molecular dialogue happens between the embryo and the endometrium [146], and to study and quantify the miRNA existing throughout the female reproductive tract [147]. Several studies have identified miRNA profiles in endometrial biopsies

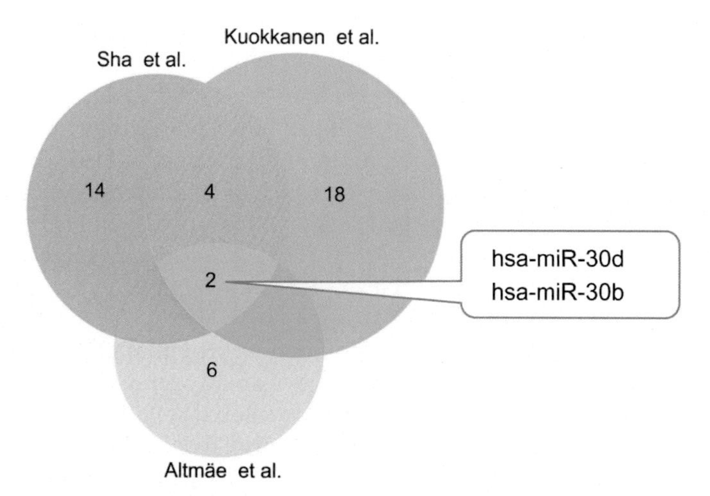

FIG. 15.6 Venn's diagram showing the number shared miRNAs in the three studies of endometrial receptivity.

in women with RIF. This miRNA signature may lead to implantation failure [148]. Recently, a meta-analysis of human endometrial receptivity to clarify that mRNA and miRNA are important has been published. It was based on the low number of genes overlapping between transcriptomic studies and had the main purpose of finding new potential diagnostic biomarkers. The analysis revealed 52 upregulated genes and 5 downregulated, with the upregulated transcripts with the highest scores being PAEP, SPP1, GPX3, MAOA, and GADD45A and the downregulated transcripts being SFRP4, EDN3, OLFM1, CRABP2, and MMP7. Bioinformatic prediction recognizes 348 microRNAs that could regulate 30 endometrial receptivity genes, and it was found that some miRNAs were dysregulated [149]. In Fig. 15.6, the Venn diagram shows how miRNAs were correlated between the different studies in regard to endometrial receptivity.

NONINVASIVE APPROACHES OF "OMICS"

There is also a trend toward the noninvasive omics in the study of endometrial receptivity, by not damaging the uterine context or interfering with conception in the same menstrual cycle. For example, the "secretomics," or the study of uterine secretions, would potentially replace biopsies as a nondisruptive technique. This concept uses the same technology as transcriptomics, proteomics, miRomics, and lipidomics. It has been shown that aspiration or flushing of the endometrial fluid may be performed using an embryo transfer catheter immediately before embryo transfer in IVF cycles without negatively affecting implantation rates [150].

The endometrial secretome is constituted by the secreted mediators that can modulate endometrial receptivity, the maintenance and nurturing of ascending spermatozoa, and the early development of the preimplantation embryo. The components of uterine fluid are derived from the luminal epithelium and glands, cellular debris, proteins selectively transudated from the blood, and contributions from the tubal fluid. The uterine cavity is very tight, and therefore, the volume of uterine fluid is low (it is difficult to retrieve more than 10 μL from a patient) [151].

The primary components of endometrial secretions are proteins, amino acids, electrolytes, glucose, urea, cytokines, growth factors, metalloproteinases and their inhibitors, immunoglobulins, α-1 antitrypsin precursor, haptoglobin, and transferrin [152]. The lipidomics of endometrial secretions is characterized by the presence of triglycerides, eicosanoids (prostaglandins (PGs), thromboxane, and leukotriene), endocannabinoids, and sphingolipids, which play a central role in the biology of reproduction [153]. A significant increase in the concentration of two specific lipids, PGE2 and PGF2α, was found between days 19 and 21 of the menstrual cycle, coinciding with the WOI [140].

The miRomics of endometrial fluid has also been approximated in two recent studies. Salamonsen's group observed in vitro the existence of small secreted vesicles that contained miRNAs in the conditioned media of endometrial epithelial cell lines, thus proposing them as a source of biomarkers [146]. A second study found that miRNAs in the endometrial fluid have a specific signature across the menstrual cycle, especially around the WOI [154]. They also observed that the murine embryo is able to incorporate miRNAs from culture medium, which could explain potential cross communication between mother and embryo during the WOI.

CONCLUSION

The endometrial receptivity is a key process for the success in assisted reproductive technology. It is of enormous interest for clinicians to explain the cause of repetitive implantation failure. This has increased even more in the past few years due to the fact of improvement in the implantation rates when the optimal receptivity days in the endometrium are found to be displaced in these patients. Omics studies have become the gold standard for assessing the endometrial status. However, there is still a long way to go in order to be able to decipher the functions and mechanisms of the different proteins, RNAs, and metabolites involved in this process.

References

[1] N.A. Campbell, R.J. Jane, Reproducción animal, in: Panamericana, seventh ed., 2005, pp. 964–986.
[2] C.R. Murphy, T.J. Shaw, Plasma membrane transformation: a common response of uterine epithelial cells during the peri-implantation period, Cell Biol. Int. 18 (1994) 1115–1128.
[3] M. Kabir-Salmani, S. Shiokawa, Y. Akimoto, K. Sakai, K. Sakai, M. Iwashita, Tissue transglutaminase at embryo-maternal interface, J. Clin. Endocrinol. Metab. 90 (2005) 4694–4702.
[4] R. Masterton, E.M. Armstrong, I.A. More, The cyclic variation in the percentage of ciliated cells in the normal human endometrium, J. Reprod. Fertil. 42 (1975) 537–540.
[5] A. Ferenczy, Surface ultrastructural response of the human uterine lining epithelium to hormonal environment, Acta Cytol. 21 (1977) 566–572.
[6] R.M. Wynn, The human endometrium: cyclic and gestational changes, in: R.M. Wynn, W.P. Jollie (Eds.), Biology of the Uterus, second ed., Plenum, New York, 1989, pp. 289–332.
[7] F.J. Cornillie, J.M. Laweryns, I.A. Brosens, Normal human endometrium, Gynecol. Obstet. Investig. 20 (1985) 113–129.

[8] I.A. More, D. McSeveney, The three dimensional structure of the nucleolar channel system in the endometrial glandular cell: Serial sectioning and high voltage electron microscopic studies, J. Anat. 130 (1980) 673–682.
[9] P. Dockery, T.C. Li, A.W. Rogers, I.D. Cooke, E.A. Lenton, The ultrastructure of the glandular epithelium in the timed endometrial biopsy, Hum. Reprod. 3 (1988) 826–834.
[10] T. Coaker, T. Downie, I.A. More, Complex giant mitochondria in the human endometrial glandular cell: serial sectioning, high-voltage electron microscopic, and three-dimensional reconstruction studios, J. Ultrastruct. Res. 78 (1982) 283–291.
[11] T. Wahlstrom, M. Seppala, Placental protein 12 (PP12) is induced in the endometrium by progesterone, Fertil. Steril. 41 (1984) 781–784.
[12] T. Garrido-Gomez, F. Dominguez, J.A. Lopez, E. Camafeita, A. Quinonero, J.A. Martinez-Conejero, et al., Modeling human endometrial decidualization from the interaction between proteome and secretome, J. Clin. Endocrinol. Metab. 96 (2011) 706–716.
[13] H.G. Augustin, M.L. Iruela-Arispe, P.A.W. Rogers, S.K. Smithe, Vascular morphogenesis in the female reproductive system, in: Cardiovascular Molecular Morphogenesis, third ed., Elsevier, 2006.
[14] C.E. Gargett, P.A. Rogers, Human endometrial angiogenesis, Reproduction 121 (2001) 181–186.
[15] J.N. Bulmer, P.M. Johnson, Immunohistological characterization of the decidual leucocytic infiltrate related to endometrial gland epithelium in early human pregnancy, Immunology 55 (1985) 35–44.
[16] S. Saito, A. Shiozaki, Y. Sasaki, A. Nakashima, T. Shima, M. Ito, Regulatory T cells and regulatory natural killer (NK) cells play important roles in feto-maternal tolerance, Semin. Immunopathol. 29 (2007) 115–122.
[17] J.N. Bulmer, L. Morrison, M. Longfellow, A. Ritson, D. Pace, Granulated lymphocytes in human endometrium: histochemical and immunohistochemical studies, Hum. Reprod. 6 (1991) 791–798.
[18] J.N. Bulmer, G.E. Lash, The role of uterine NK cells in normal reproduction and reproductive disorders, Adv. Exp. Med. Biol. 868 (2015) 95–126.
[19] H.N. Jabbour, R.W. Kelly, H.M. Fraser, H.O. Critchley, Endocrine regulation of menstruation, Endocr. Rev. 27 (2006) 17–46.
[20] H.O. Critchley, K.A. Robertson, T. Forster, T.A. Henderson, A.R. Williams, P. Ghazal, Gene expression profiling of mid to late secretory phase endometrial biopsies from women with menstrual complaint, Am. J. Obstet. Gynecol. 195 (2006). 406.e1–16.
[21] R.W. Noyes, A.T. Hertig, J. Rock, Dating the endometrial biopsy, Am. J. Obstet. Gynecol. 122 (1975) 262–263.
[22] J.D. Aplin, The cell biological basis of human implantation, Baillieres Best Pract. Res. Clin. Obstet. Gynaecol. 14 (2000) 757–764.
[23] S.M. Hawkins, M.M. Matzuk, The menstrual cycle: basic biology, Ann. N. Y. Acad. Sci. 1135 (2008) 10–18.
[24] C.A. Finn, L. Martin, The control of implantation, J. Reprod. Fertil. 39 (1974) 195–206.
[25] B.W. O'Malley, M.J. Tsai, Molecular pathways of steroid receptor action, Biol. Reprod. 46 (1992) 163–167.
[26] D. Navot, R.T. Scott, K. Droesch, L.L. Veeck, H.C. Liu, Z. Rosenwaks, The window of embryo transfer and the efficiency of human conception in vitro, Fertil. Steril. 55 (1991) 114–118.
[27] Y. Prapas, N. Prapas, E.E. Jones, A.J. Duleba, D.L. Olive, A. Chatziparasidou, et al., The window for embryo transfer in oocyte donation cycles depends on the duration of progesterone therapy, Hum. Reprod. 13 (3) (1998) 720.
[28] A. Riesewijk, J. Martín, R. van Os, J.A. Horcajadas, J. Polman, A. Pellicer, et al., Gene expression profiling of human endometrial receptivity on days LH+2 versus LH+7 by microarray technology, Mol. Hum. Reprod. 9 (2003) 253–264.
[29] T. Garrido-Gómez, M. Ruiz-Alonso, D. Blesa, P. Diaz-Gimeno, F. Vilella, C. Simón, Profiling the gene signature of endometrial receptivity: clinical results, Fertil. Steril. 99 (2013) 1078–1085.
[30] T. Tamaya, T. Murakami, H. Okada, Concentrations of steroid receptors in normal human endometrium in relation to the day of the menstrual cycle, Acta Obstet. Gynecol. Scand. 65 (1986) 195–198.
[31] C. Coutifaris, E.R. Myers, D.S. Guzick, M.P. Diamond, S.A. Carson, R.S. Legro, et al., Histological dating of timed endometrial biopsy tissue is not related to fertility status, Fertil. Steril. 82 (2004) 1264–1272.
[32] M.J. Murray, W.R. Meyer, R.J. Zaino, B.A. Lessey, D.B. Novotny, K. Ireland, et al., A critical analysis of the accuracy, reproducibility, and clinical utility of histologic endometrial dating in fertile women, Fertil. Steril. 81 (2004) 1333–1343.
[33] Practice Committee of the American Society for Reproductive Medicine, Current clinical irrelevance of luteal phase deficiency: a committee opinion, Fertil. Steril. 103 (2015) e27–e32.

[34] G. Nikas, P. Drakakis, D. Loutradis, C. Mara-Skoufari, E. Koumantakis, S. Michalas, et al., Uterine pinopodes as markers of the 'nidation window' in cycling women receiving exogenous oestradiol and progesterone, Hum. Reprod. 10 (1995) 1208–1213.
[35] G. Nikas, A. Psychoyos, Uterine pinopodes in peri-implantation human endometrium. Clinical relevance, Ann. N. Y. Acad. Sci. 816 (1997) 129–142.
[36] G. Nikas, A. Makrigiannakis, O. Hovatta, H.W. Jones Jr., Surface morphology of the human endometrium. Basic and clinical aspects, Ann. N. Y. Acad. Sci. 900 (2000) 316–324.
[37] L. Aghajanova, A. Stavreus-Evers, Y. Nikas, O. Hovatta, B.M. Landgren, Coexpression of pinopodes and leukemia inhibitory factor, as well as its receptor, in human endometrium, Fertil. Steril. 79 (2003) 808–814.
[38] C.E. Quinn, R.F. Casper, Pinopodes: a questionable role in endometrial receptivity, Hum. Reprod. Update 15 (2009) 229–236.
[39] R.P. Dickey, T.T. Olar, D.N. Curole, S.N. Taylor, E.M. Matulich, Relationship of first-trimester subchorionic bleeding detected by color Doppler ultrasound to subchorionic fluid, clinical bleeding, and pregnancy outcome, Obstet. Gynecol. 80 (1992) 415–420.
[40] J. Remohi, G. Ardiles, J.A. Garcia-Velasco, P. Gaitan, C. Simon, A. Pellicer, Endometrial thickness and serum oestradiol concentrations as predictors of outcome in oocyte donation, Hum. Reprod. 12 (1997) 2271–2276.
[41] J.A. Garcia-Velasco, V. Isaza, C. Caligara, A. Pellicer, J. Remohi, C. Simon, Factors that determine discordant outcome from shared oocytes, Fertil. Steril. 80 (2003) 54–60.
[42] C.L. Stewart, The role of leukemia inhibitory factor (LIF) and other cytokines in regulating implantation in mammals, Ann. N. Y. Acad. Sci. 734 (1994) 157–165.
[43] R.L. Jones, N.J. Hannan, T.J. Kaitu'u, J. Zhang, L.A. Salamonsen, Identification of chemokines important for leukocyte recruitment to the human endometrium at the times of embryo implantation and menstruation, J. Clin. Endocrinol. Metab. 89 (2004) 6155–6167.
[44] T.W. Chung, M.J. Park, H.S. Kim, H.J. Choi, K.T. Ha, Integrin αVβ3 and αVβ5 are required for leukemia inhibitory factor-mediated the adhesion of trophoblast cells to the endometrial cells, Biochem. Biophys. Res. Commun. 469 (2016) 936–940.
[45] L. Aghajanova, S. Altmae, K. Bjuresten, et al., Disturbances in the LIF pathway in the endometrium among women with unexplained infertility, Fertil. Steril. 91 (2009) 2602–2610.
[46] J.M. Borthwick, D.S. Charnock-Jones, B.D. Tom, M.L. Hull, R. Teirney, S.C. Phillips, et al., Determination of the transcript profile of human endometrium, Mol. Hum. Reprod. 9 (2003) 19–33.
[47] A. Arici, O. Engin, E. Attar, D.L. Olive, Modulation of leukemia inhibitory factor gene expression and protein biosynthesis in human endometrium, J. Clin. Endocrinol. Metab. 80 (1995) 1908–1915.
[48] K. Kojima, H. Kanzaki, M. Iwai, H. Hatayama, M. Fujimoto, T. Inoue, et al., Expression of leukemia inhibitory factor in human endometrium and placenta, Biol. Reprod. 50 (1994) 882–887.
[49] M. Mikołajczyk, J. Skrzypczak, K. Szymanowski, P. Wirstlein, The assessment of LIF in uterine flushing—a possible new diagnostic tool in states of impaired fertility, Reprod. Biol. 3 (2003) 259–270.
[50] P.C. Serafini, I.D. Silva, G.D. Smith, E.L. Motta, A.M. Rocha, E.C. Baracat, Endometrial claudin-4 and leukemia inhibitory factor are associated with assisted reproduction outcome, Reprod. Biol. Endocrinol. 7 (2009) 30.
[51] P.R. Brinsden, V. Alam, B. de Moustier, P. Engrand, Recombinant human leukemia inhibitory factor does not improve implantation and pregnancy outcomes after assisted reproductive techniques in women with recurrent unexplained implantation failure, Fertil. Steril. 91 (2009) 1445–1447.
[52] J. Kitawaki, H. Koshiba, H. Ishihara, I. Kusuki, K. Tsukamoto, H. Honjo, Expression of leptin receptor in human endometrium and fluctuation during the menstrual cycle, J. Clin. Endocrinol. Metab. 85 (2000) 1946–1950.
[53] R.R. González, P. Caballero-Campo, M. Jasper, A. Mercader, L. Devoto, A. Pellicer, et al., Leptin and leptin receptor are expressed in the human endometrium and endometrial leptin secretion is regulated by the human blastocyst, J. Clin. Endocrinol. Metab. 85 (2000) 4883–4888.
[54] B. Demir, S. Guven, E.S. Guven, Y. Atamer, G.S. Gunalp, T. Gul, Serum leptin level in women with unexplained infertility, J. Reprod. Immunol. 75 (2007) 145–149.
[55] Z. Saeed, S. Haleh, M. Afsaneh, A. Soheila, Z. Amir Hassan, I. Farah, et al., Serum leptin levels in women with immunological recurrent abortion, J. Reprod. Infertil. 11 (2010) 47–52.
[56] G. De Placido, C. Alviggi, R. Clarizia, A. Mollo, E. Alviggi, I. Strina, et al., Intra-follicular leptin concentration as a predictive factor for in vitro oocyte fertilization in assisted reproductive techniques, J. Endocrinol. Investig. 29 (2006) 719–726.

[57] B.A. Lessey, L. Damjanovich, C. Coutifaris, A. Castelbaum, S.M. Albelda, C.A. Buck, Integrin adhesion molecules in the human endometrium. Correlation with the normal and abnormal menstrual cycle, J. Clin. Invest. 90 (1992) 188–195.

[58] Y.J. Kang, K. Forbes, J. Carver, J.D. Aplin, The role of the osteopontin-integrin αvβ3 interaction at implantation: Functional analysis using three different in vitro models, Hum. Reprod. 29 (2014) 739–749.

[59] C. Schmitz, L. Yu, S. Bocca, S. Anderson, J.S. Cunha-Filho, B.S. Rhavi, et al., Role for the endometrial epithelial protein MFG-E8 and its receptor integrin αvβ3 in human implantation: Results of an in vitro trophoblast attachment study using established human cell lines, Fertil. Steril. 101 (2014) 874–882.

[60] B.A. Lessey, A.J. Castelbaum, J. Harris, J. Sun, S.L. Young, L. Wolf, Improvement in pregnancy rates with GnRH agonist in women with infertility, minimal or mild endometriosis and aberrant vß3 expression, Am. Soc. Reprod. Med. Ann. Mtg. 0165 (1996) S82.

[61] W.R. Meyer, A.J. Castelbaum, S. Somkuti, A.W. Sagoskin, M. Doyle, J.E. Harris, et al., Hydrosalpinges adversely affect markers of endometrial receptivity, Hum. Reprod. 12 (1997) 1393–1398.

[62] M.G. Boroujerdnia, R. Nikbakht, Beta3 integrin expression within uterine endometrium and its relationship with unexplained infertility, Pak. J. Biol. Sci. 11 (2008) 2495–2499.

[63] G. Casals, J. Ordi, M. Creus, F. Fábregues, F. Carmona, R. Casamitjana, et al., Expression pattern of osteopontin and αvβ3 integrin during the implantation window in infertile patients with early stages of endometriosis, Hum. Reprod. 27 (2012) 805–813.

[64] K.B. Apparao, M.J. Murray, M.A. Fritz, W.R. Meyer, A.F. Chambers, P.R. Truong, et al., Osteopontin and its receptor αvβ3 integrin are coexpressed in the human endometrium during the menstrual cycle but regulated differentially, J. Clin. Endocrinol. Metab. 86 (2001) 4991–5000.

[65] Z. He, Y. Ma, L. Li, J. Liu, H. Yang, C. Chen, et al., Osteopontin and integrin αvβ3 expression during the implantation window in IVF patients with elevated serum progesterone and Oestradiol level, Geburtshilfe Frauenheilkd. 76 (2016) 709–717.

[66] S. Battersby, H.O. Critchley, K. Morgan, R.P. Millar, H.N. Jabbour, Expression and regulation of the prokineticins (endocrine gland-derived vascular endothelial growth factor and Bv8) and their receptors in the human endometrium across the menstrual cycle, J. Clin. Endocrinol. Metab. 89 (2004) 2463–2469.

[67] J. Evans, R.D. Catalano, P. Brown, R. Sherwin, H.O. Critchley, A.T. Fazleabas, et al., Prokineticin 1 mediates fetal-maternal dialogue regulating endometrial leukemia inhibitory factor, FASEB J. 23 (2009) 2165–2175.

[68] L.J. Macdonald, K.J. Sales, V. Grant, P. Brown, H.N. Jabbour, R.D. Catalano, Prokineticin 1 induces Dickkopf 1 expression and regulates cell proliferation and decidualization in the human endometrium, Mol. Hum. Reprod. 17 (2011) 626–636.

[69] A. Karaer, Y. Cigremis, E. Celik, R. Urhan Gonullu, Prokineticin 1 and leukemia inhibitory factor mRNA expression in the endometrium of women with idiopathic recurrent pregnancy loss, Fertil. Steril. 102 (2014). 1091–95.e1.

[70] S.K. Das, X.N. Wang, B.C. Paria, D. Damm, J.A. Abraham, M. Klagsbrun, et al., Heparin-binding EGF-like growth factor gene is induced in the mouse uterus temporally by the blastocyst solely at the site of its apposition: a possible ligand for interaction with blastocyst EGF-receptor in implantation, Development 120 (1994) 1071–1083.

[71] K. Chobotova, N. Karpovich, J. Carver, S. Manek, W.J. Gullick, D.H. Barlow, et al., Heparin-binding epidermal growth factor and its receptors mediate decidualization and potentiate survival of human endometrial stromal cells, J. Clin. Endocrinol. Metab. 90 (2005) 913–919.

[72] H.J. Yoo, D.H. Barlow, H.J. Mardon, Temporal and spatial regulation of expression of heparin-binding epidermal growth factor-like growth factor in the human endometrium: a possible role in blastocyst implantation, Dev. Genet. 21 (1) (1997) 102–108.

[73] R.E. Leach, R. Khalifa, N.D. Ramirez, S.K. Das, J. Wang, S.K. Dey, et al., Multiple roles for heparin-binding epidermal growth factor-like growth factor are suggested by its cell-specific expression during the human endometrial cycle and early placentation, J. Clin. Endocrinol. Metab. 84 (9) (1999) 3355–3363.

[74] N. Wang, L. Geng, S. Zhang, B. He, J. Wang, Expression of PRB, FKBP52 and HB-EGF relating with ultrasonic evaluation of endometrial receptivity, PLoS One 7 (2012) e34010.

[75] A. Stavreus-Evers, L. Aghajanova, H. Brismar, H. Eriksson, B.M. Landgren, O. Hovatta, Co-existence of heparin-binding epidermal growth factor-like growth factor and pinopodes in human endometrium at the time of implantation, Mol. Hum. Reprod. 8 (2002) 765–769.

[76] L. Aghajanova, K. Bjuresten, S. Altmäe, B.M. Landgren, A. Stavreus-Evers, HB-EGF but not amphiregulin or their receptors HER1 and HER4 is altered in endometrium of women with unexplained infertility, Reprod. Sci. 15 (2008) 484–492.

[77] P.K. Wirstlein, M. Mikołajczyk, J. Skrzypczak, Correlation of the expression of heparanase and heparin-binding EGF-like growth factor in the implantation window of nonconceptual cycle endometrium, Folia Histochem. Cytobiol. 51 (2013) 127–134.

[78] E.E. Karagouni, A. Chryssikopoulos, T. Mantzavinos, N. Kanakas, E.N. Dotsika, Interleukin-1beta and interleukin-1alpha may affect the implantation rate of patients undergoing in vitro fertilization-embryo transfer, Fertil. Steril. 70 (1998) 553–559.

[79] F. Bigonnesse, Y. Labelle, A. Akoum, Triphasic expression of interleukin-1 receptor type I in human endometrium throughout the menstrual cycle of fertile women and women with unexplained infertility, Fertil. Steril. 75 (2001) 79–87.

[80] S.B. Carroll, Homeotic genes and the evolution of arthropods and chordates, Nature 376 (1995) 479–485.

[81] H.S. Taylor, A. Arici, D. Olive, P. Igarashi, HOXA 10 is expressed in response to sex steroids at the time of implantation in the human endometrium, J. Clin. Invest. 101 (1998) 1379–1384.

[82] H.S. Taylor, P. Igarashi, D.L. Olive, A. Arici, Sex steroids mediate HOXA11 expression in the human peri-implantation endometrium, J. Clin. Endocrinol. Metab. 84 (1999) 1129–1135.

[83] B. Xu, D. Geerts, Z. Bu, J. Ai, L. Jin, Y. Li, H. Zhang, G. Zhu, Regulation of endometrial receptivity by the highly expressed HOXA9, HOXA11 and HOXD10 HOX-class homeobox genes, Hum. Reprod. 29 (2014) 781–790.

[84] M. Szczepańska, P. Wirstlein, M. Luczak, P. Jagodzinski, J. Skrzypczak, Expression of HOXA-10 and HOXA-11 in the endometria of women with idiopathic infertility, Folia Histochem. Cytobiol. 49 (2011) 111–118.

[85] H. Bohn, W. Kraus, W. Winckler, New soluble placental tissue proteins: their isolation, characterization, localization and quantification, Placenta. Suppl. 4 (1982) 67–81.

[86] S.G. Joshi, K.M. Ebert, D.P. Swartz, Detection and synthesis of a progestagen-dependent protein in human endometrium, J. Reprod. Fertil. 59 (1980) 273–285.

[87] K. M1, M. Julkunen, M. Seppälä, HinfI polymorphism in the human progesterone associated endometrial protein (PAEP) gene, Nucleic Acids Res. 19 (1991) 5092.

[88] M. Amir, S. Romano, S. Goldman, E. Shalev, Plexin-B1, glycodelin and MMP7 expression in the human fallopian tube and in the endometrium, Reprod. Biol. Endocrinol. 7 (2009) 152.

[89] A. Stavreus-Evers, E. Mandelin, R. Koistinen, L. Aghajanova, O. Hovatta, M. Seppälä, Glycodelin is present in pinopodes of receptive-phase human endometrium and is associated with down-regulation of progesterone receptor B, Fertil. Steril. 85 (2006) 1803–1811.

[90] L. Aghajanova, C. Simón, J.A. Horcajadas, Are favorite molecules of endometrial receptivity still in favor? Expert. Rev. Obstetrics. Gynecol. 3 (2008) 487–501.

[91] A. Mackenna, T.C. Li, C. Dalton, A. Bolton, I. Cooke, Placental protein 14 levels in uterine flushing and plasma of women with unexplained infertility, Fertil. Steril. 59 (1993) 577–582.

[92] L. Aghajanova, A.E. Hamilton, L.C. Giudice, Uterine receptivity to human embryonic implantation: histology, biomarkers, and transcriptomics, Semin. Cell Dev. Biol. 19 (2008) 204–211.

[93] B.A. Lessey, Endometrial responsiveness to steroid hormones: a moving target, J. Soc. Gynecol. Investig. 11 (2) (2004) 61.

[94] M. Meseguer, J.D. Aplin, P. Caballero-Campo, J.E. O'Connor, J.C. Martin, J. Remohi, et al., Human endometrial mucin MUC1 is up-regulated by progesterone and down-regulated in vitro by the human blastocyst, Biol. Reprod. 64 (2001) 590–601.

[95] M. Meseguer, A. Pellicer, C. Simon, MUC1 and endometrial receptivity, Mol. Hum. Reprod. 4 (1998) 1089–1098.

[96] J.M. Borthwick, D.S. Charnock-Jones, B.D. Tom, M.L. Hull, R. Teirney, S.C. Phillips, et al., Determination of the transcript profile of human endometrium, Mol. Hum. Reprod. 9 (2003) 19–33.

[97] D.D. Carson, E. Lagow, A. Thathiah, R. Al-Shami, M.C. Farach-Carson, M. Vernon, et al., Changes in gene expression during the early to mid-luteal (receptive phase) transition in human endometrium detected by high-density microarray screening, Mol. Hum. Reprod. 8 (9) (2002) 871.

[98] A. Riesewijk, J. Martin, R. van Os, J.A. Horcajadas, J. Polman, A. Pellicer, et al., Gene expression profiling of human endometrial receptivity on days LH+2 versus LH+7 by microarray technology, Mol. Hum. Reprod. 9 (2003) 253–264.

[99] E.S. Lander, L.M. Linton, B. Birren, C. Nusbaum, M.C. Zody, J. Baldwin, et al., Initial sequencing and analysis of the human genome, Nature 409 (2001) 860–921.

[100] ENCODE Project Consortium, B.E. Bernstein, E. Birney, I. Dunham, E.D. Green, C. Gunter, et al., An integrated encyclopedia of DNA elements in the human genome, Nature 489 (2012) 57–74.

[101] A.P. Ponnampalam, G.C. Weston, A.C. Trajstman, B. Susil, P.A. Rogers, Molecular classification of human endometrial cycle stages by transcriptional profiling, Mol. Hum. Reprod. 10 (2004) 879–893.

[102] D.D. Carson, E. Lagow, A. Thathiah, R. Al-Shami, M.C. Farach-Carson, M. Vernon, et al., Changes in gene expression during the early to mid-luteal (receptive phase) transition in human endometrium detected by high-density microarray screening, Mol. Hum. Reprod. 8 (2002) 871–879.

[103] L.C. Kao, S. Tulac, S. Lobo, B. Imani, P. Yang, A. Germeyer, et al., Global gene profiling in human endometrium during the window of implantation, Endocrinology 143 (2002) 2119–2138.

[104] S. Mirkin, G. Nikas, J.G. Hsiu, J. Diaz, S. Oehninger, Gene expression profiles and structural/functional features of the peri-implantation endometrium in natural and gonadotropin-stimulated cycles, J. Clin. Endocrinol. Metab. 89 (2004) 5742–5752.

[105] D. Haouzi, S. Assou, K. Mahmoud, S. Tondeur, T. Rème, B. Hedon, et al., Gene expression profile of human endometrial receptivity: comparison between natural and stimulated cycles for the same patients, Hum. Reprod. 24 (2009) 1436–1445.

[106] J.A. Horcajadas, A. Pellicer, C. Simón, Wide genomic analysis of human endometrial receptivity: new times, new opportunities, Hum. Reprod. Update 13 (2007) 77–86.

[107] S. Mirkin, M. Arslan, D. Churikov, A. Corica, J.I. Diaz, S. Williams, et al., In search of candidate genes critically expressed in the human endometrium during the window of implantation, Hum. Reprod. 20 (2005) 2104–2117.

[108] A. Tapia, L.M. Gangi, F. Zegers-Hochschild, J. Balmaceda, R. Pommer, L. Trejo, et al., Differences in the endometrial transcript profile during the receptive period between women who were refractory to implantation and those who achieved pregnancy, Hum. Reprod. 23 (2008) 340–351.

[109] M. Koler, H. Achache, A. Tsafrir, Y. Smith, A. Revel, R. Reich, Disrupted gene pattern in patients with repeated in vitro fertilization (IVF) failure, Hum. Reprod. 24 (2009) 2541–2548.

[110] S. Altmäe, J.A. Martínez-Conejero, A. Salumets, C. Simón, J.A. Horcajadas, A. Stavreus-Evers, Endometrial gene expression analysis at the time of embryo implantation in women with unexplained infertility, Mol. Hum. Reprod. 16 (2010) 178–187.

[111] J.A. Horcajadas, A. Riesewijk, J. Polman, R. van Os, A. Pellicer, S. Mosselman, et al., Effect of controlled ovarian hyperstimulation in IVF on endometrial gene expression profiles, Mol. Hum. Reprod. 11 (2005) 195–205.

[112] C. Simon, J. Oberyé, J. Bellver, C. Vidal, E. Bosch, J.A. Horcajadas, et al., Similar endometrial development in oocyte donors treated with either high- or standard-dose GnRH antagonist compared to treatment with a GnRH agonist or in natural cycles, Hum. Reprod. 20 (2005) 3318–3327.

[113] Y. Liu, K.F. Lee, E.H. Ng, W.S. Yeung, P.C. Ho, Gene expression profiling of human peri-implantation endometria between natural and stimulated cycles, Fertil. Steril. 90 (2008) 2152–2164.

[114] J.A. Horcajadas, A.M. Sharkey, R.D. Catalano, J.R. Sherwin, F. Domínguez, L.A. Burgos, et al., Effect of an intrauterine device on the gene expression profile of the endometrium, J. Clin. Endocrinol. Metab. 91 (2006) 3199–3207.

[115] J.A. Horcajadas, E. Goyri, M.A. Higón, J.A. Martínez-Conejero, P. Gambadauro, G. García, et al., Endometrial receptivity and implantation are not affected by the presence of uterine intramural leiomyomas: a clinical and functional genomics analysis, J. Clin. Endocrinol. Metab. 93 (2008) 3490–3498.

[116] P. Diaz-Gimeno, J.A. Horcajadas, J.A. Martinez-Conejero, F.J. Esteban, P. Alama, A. Pellicer, et al., A genomic diagnostic tool for human endometrial receptivity based on the transcriptomic signature, Fertil Steril 95 (2011) 50–60. 60.e1–15.

[117] P. Diaz-Gimeno, M. Ruiz-Alonso, D. Blesa, N. Bosch, J.A. Martinez-Conejero, P. Alama, et al., The accuracy and reproducibility of the endometrial receptivity array is superior to histology as a diagnostic method for endometrial receptivity, Fertil. Steril. 99 (2012) 508–517.

[118] M. Ruiz-Alonso, D. Blesa, C. Simon, The genomics of the human endometrium, Biochim. Biophys. Acta 1822 (2012) 1931–1942.

[119] M. Ruiz-Alonso, D. Blesa, P. Díaz-Gimeno, E. Gómez, M. Fernández-Sánchez, F. Carranza, et al., The endometrial receptivity array for diagnosis and personalized embryo transfer as a treatment for patients with repeated implantation failure, Fertil. Steril. 100 (2013) 818–824.

[120] J.A. Garcia-Velasco, A. Fassbender, M. Ruiz-Alonso, D. Blesa, T. D'Hooghe, C. Simon, Is endometrial receptivity transcriptomics affected in women with endometriosis? A pilot study, Reprod. BioMed. Online 31 (2015) 647–654.
[121] J.P. Carrascosa, D. Cotán, I. Jurado, M. Oropesa-Ávila, P. Sánchez-Martín, R.F. Savaris, et al., The effect of copper on endometrial receptivity and induction of apoptosis on decidualized human endometrial stromal cells. Reprod. Sci. (2017), https://doi.org/10.1177/1933719117732165.
[122] P. Díaz-Gimeno, J.A. Horcajadas, J.A. Martínez-Conejero, F.J. Esteban, P. Alamá, A. Pellicer, et al., A genomic diagnostic tool for human endometrial receptivity based on the transcriptomic signature, Fertil Steril 95 (2011) 50–60. 60.e1–15.
[123] S. Hu, G. Yao, Y. Wang, H. Xu, X. Ji, Y. He, et al., Transcriptomic changes during the pre-receptive to receptive transition in human endometrium detected by RNA-Seq, J. Clin. Endocrinol. Metab. 99 (2014) E2744–E2753.
[124] M. Ruiz-Alonso, N. Galindo, A. Pellicer, C. Simon, What a difference two days make: "personalized" embryo transfer (pET) paradigm: a case report and pilot study, Hum. Reprod. 29 (2014) 1244–1247.
[125] E. Gómez, M. Ruíz-Alonso, J. Miravet, C. Simón, Human endometrial transcriptomics: implications for embryonic implantation, Cold Spring Harb. Perspect. Med. 5 (2015) a022996.
[126] C.T. Valdes, A. Schutt, C. Simon, Implantation failure of endometrial origin: it is not pathology, but our failure to synchronize the developing embryo with a receptive endometrium, Fertil. Steril. 108 (2017) 15–18.
[127] K. Krjutškov, S. Katayama, M. Saare, M. Vera-Rodriguez, D. Lubenets, K. Samuel, et al., Single-cell transcriptome analysis of endometrial tissue, Hum. Reprod. 31 (2016) 844–853.
[128] W.W. Wu, G. Wang, S.J. Baek, R.F. Shen, Comparative study of three proteomic quantitative methods, DIGE, cICAT, and iTRAQ, using 2D gel- or LC-MALDI TOF/TOF, J. Proteome Res. 5 (2006) 651–658.
[129] F. Dominguez, T. Garrido-Gomez, J.A. Lopez, E. Camafeita, A. Quinonero, A. Pellicer, et al., Proteomic analysis of the human receptive versus non-receptive endometrium using differential in-gel electrophoresis and MALDI-MS unveils stathmin 1 and annexin A2 as differentially regulated, Hum. Reprod. 24 (2009) 2607–2617.
[130] J.I.C. Chen, N.J. Hannan, Y. Mak, P.K. Nicholls, J. Zhang, A. Rainczuk, et al., Proteomic characterization of midproliferative and midsecretory human endometrium, J. Proteome Res. 8 (2009) 2032–2044.
[131] B.L. Hood, B. Liu, A. Alkhas, Y. Shoji, R. Challa, G. Wang, et al., Proteomics of the human endometrial glandular epithelium and stroma from the proliferative and secretory phases of the menstrual cycle, Biol. Reprod. 92 (2015) 106.
[132] T. Garrido-Gómez, A. Quiñonero, O. Antúnez, P. Díaz-Gimeno, J. Bellver, C. Simón, et al., Deciphering the proteomic signature of human endometrial receptivity, Hum. Reprod. 29 (2014) 1957–1967.
[133] L. DeSouza, G. Diehl, E.C. Yang, J. Guo, M.J. Rodrigues, A.D. Romaschin, et al., Proteomic analysis of the proliferative and secretory phases of the human endometrium: protein identification and differential protein expression, Proteomics 5 (2005) 270–281.
[134] T.A. Edgell, L.J. Rombauts, L.A. Salamonsen, Assessing receptivity in the endometrium: the need for a rapid, non-invasive test, Reprod. BioMed. Online 27 (2013) 486–496.
[135] J.G. Scotchie, M.A. Fritz, M. Mocanu, B.A. Lessey, S.L. Young, Proteomic analysis of the luteal endometrial secretome, Reprod. Sci. 16 (2009) 883–893.
[136] N.K. Binder, J. Evans, L.A. Salamonsen, D.K. Gardner, T.J. Kaitu'u-Lino, N.J. Hannan, Placental growth factor is secreted by the human endometrium and has potential important functions during embryo development and implantation, PLoS One 11 (2016) e0163096.
[137] N.J. Hannan, P. Paiva, K.L. Meehan, L.J. Rombauts, D.K. Gardner, L.A. Salamonsen, Analysis of fertility-related soluble mediators in human uterine fluid identifies VEGF as a key regulator of embryo implantation, Endocrinology 152 (2011) 4948–4956.
[138] D. Haouzi, H. Dechaud, S. Assou, J. De Vos, S. Hamamah, Insights into human endometrial receptivity from transcriptomic and proteomic data, Reprod. BioMed. Online 24 (2012) 23–34.
[139] J.H. Durn, K.M. Marshall, D. Farrar, P. O'Donovan, A.J. Scally, D.F. Woodward, et al., Lipidomic analysis reveals prostanoid profiles in human term pregnant myometrium, Prostaglandins, Leukotrienes Essent. Fatty Acids 82 (2010) 21–26.
[140] F. Vilella, L. Ramirez, O. Berlanga, S. Martinez, P. Alama, M. Meseguer, et al., PGE2 and PGF2alpha concentrations in human endometrial fluid as biomarkers for embryonic implantation, J. Clin. Endocrinol. Metab. 98 (2013) 4123–4132.

[141] S. Kuokkanen, B. Chen, L. Ojalvo, L. Benard, N. Santoro, J.W. Pollard, Genomic profiling of microRNAs and messenger RNAs reveals hormonal regulation in microRNA expression in human endometrium, Biol. Reprod. 82 (2010) 791–801.

[142] S. Altmae, J.A. Martinez-Conejero, F.J. Esteban, M. Ruiz-Alonso, A. Stavreus-Evers, J.A. Horcajadas, et al., MicroRNAs miR-30b, miR-30d, and miR-494 regulate human endometrial receptivity, Reprod. Sci. 20 (2013) 308–317.

[143] S. Altmäe, J.A. Martinez-Conejero, F.J. Esteban, M. Ruiz-Alonso, A. Stavreus-Evers, J.A. Horcajadas, et al., MicroRNAs miR-30b, miR-30d, and miR-494 regulate human endometrial receptivity, Reprod. Sci. 20 (2013) 308–317.

[144] A.G. Sha, J.L. Liu, X.M. Jiang, J.Z. Ren, C.H. Ma, W. Lei, et al., Genome-wide identification of micro-ribonucleic acids associated with human endometrial receptivity in natural and stimulated cycles by deep sequencing, Fertil. Steril. 96 (2011). 150–155.e155.

[145] J.M. Wang, Y. Gu, Y. Zhang, Q. Yang, X. Zhang, L. Yin, et al., Deep-sequencing identification of differentially expressed miRNAs in decidua and villus of recurrent miscarriage patients, Arch. Gynecol. Obstet. 293 (2016) 1125–1135.

[146] Y.H. Ng, S. Rome, A. Jalabert, A. Forterre, H. Singh, C.L. Hincks, et al., Endometrial exosomes/microvesicles in the uterine microenvironment: a new paradigm for embryo-endometrial cross talk at implantation, PLoS One 8 (2013) e58502.

[147] C.J. Creighton, A.L. Benham, H. Zhu, M.F. Khan, J.G. Reid, A.K. Nagaraja, et al., Discovery of novel microRNAs in female reproductive tract using next generation sequencing, PLoS One 5 (2010) e9637.

[148] C. Shi, H. Shen, L.J. Fan, J. Guan, X.B. Zheng, X. Chen, et al., Endometrial MicroRNA signature during the window of implantation changed in patients with repeated implantation failure, Chin. Med. J. 130 (2017) 566–573.

[149] S. Altmäe, M. Koel, U. Võsa, P. Adler, M. Suhorutšenko, T. Laisk-Podar, et al., Meta-signature of human endometrial receptivity: a meta-analysis and validation study of transcriptomic biomarkers, Sci. Rep. 7 (2017) 10077.

[150] M.H. van der Gaast, K. Beier-Hellwig, B.C. Fauser, H.M. Beier, N.S. Macklon, Endometrial secretion aspiration prior to embryo transfer does not reduce implantation rates, Reprod. BioMed. Online 7 (2003) 105–109.

[151] L.A. Salamonsen, T. Edgell, L.J. Rombauts, A.N. Stephens, D.M. Robertson, A. Rainczuk, et al., Proteomics of the human endometrium and uterine fluid: a pathway to biomarker discovery, Fertil. Steril. 99 (2013) 1086–1092.

[152] C.M. Boomsma, A. Kavelaars, M.J. Eijkemans, K. Amarouchi, G. Teklenburg, D. Gutknecht, et al., Cytokine profiling in endometrial secretions: a non-invasive window on endometrial receptivity, Reprod. BioMed. Online 18 (2009) 85–94.

[153] O. Berlanga, H.B. Bradshaw, F. Vilella-Mitjana, T. Garrido-Gomez, C. Simon, How endometrial secretomics can help in predicting implantation, Placenta 32 (Suppl 3) (2011) S271–S275.

[154] F. Vilella, J.M. Moreno-Moya, N. Balaguer, A. Grasso, M. Herrero, S. Martinez, et al., Hsa-miR-30d, secreted by the human endometrium, is taken up by the pre-implantation embryo and might modify its transcriptome, Development 142 (2015) 3210–3221.

CHAPTER

16

The Acquisition of the Human Endometrial Receptivity Phenotype: Lessons From Proteomic Studies

Delphine Haouzi, Samir Hamamah

ART/PGD Department, Arnaud de Villeneuve Hospital, INSERM Unit 1203, Montpellier, France

INTRODUCTION

Despite numerous advances in assisted reproductive technologies (ART), only 10%–15% of transferred embryos give birth to a child. A part of the failures is directly attributable to the embryo itself, but the limiting majority factor is the embryonic implantation failure due to uterine abnormalities and/or a default of the fetomaternal dialogue [1,2]. The embryonic implantation is a complex biological process that requires a receptive endometrium, a competent embryo, and a synchronized dialogue between the embryo and the maternal tissue. The endometrium is said receptive during a very short period called the implantation windows, of which the duration is generally around 2 days, from day 20 to day 24 of a normal menstrual cycle. To date, the medical tools provided to the clinician including medical imaging or histological analyses do not allow to appreciate the endometrial receptivity status. Yet, it is estimated that more than two-thirds of unsuccessful IVF/ICSI attempts were due to implantation failures. The necessity to develop reliable tools allowing to diagnose the endometrial receptivity status remains a major issue in ART. In the objective, numerous gene array studies have determined the mRNA changes during the menstrual cycle allowing to define a transcriptomic signature during the implantation window [3,4]. However, biological processes and molecular mechanisms that govern the acquisition of the endometrial receptivity phenotype are ultimately controlled by proteins, and therefore, analysis of the changes in the protein profile is required to acquire a complete overview. In addition, the determination of

https://doi.org/10.1016/B978-0-12-812571-7.00017-4

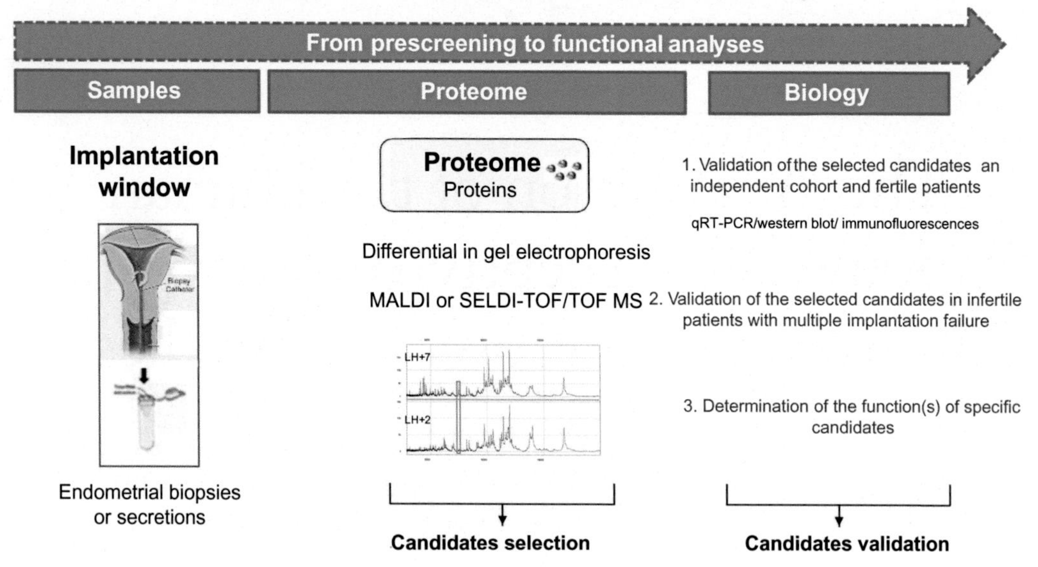

FIG. 16.1 Study workflow from proteomic prescreening up to investigation of the function(s) of candidate proteins for sufficiently powered studies.

the functions of identified proteins remains indispensable for understanding the signaling pathways controlling human endometrial receptivity and for sufficiently powered studies [5,6] (Fig. 16.1).

PROTEIN EXPRESSION PROFILE OF HUMAN ENDOMETRIAL RECEPTIVITY UNDER NATURAL CYCLE

Proteomic Analyses From Endometrial Biopsies

Studies on endometrium receptivity using high-throughput mass spectrometry-based proteomic approaches are still scarce (Table 16.1) [5,7–12].

So far, only four studies compared the endometrial protein expression profiles between the mid- or late-proliferative and mid-secretory stages [7–10]. Of these reports, one focused on comparison between proliferative and secretory phase endometrium using the isotope-coded affinity tag (ICAT) technology [7], whereas the three others used the two-dimensional DIGE (differential in gel electrophoresis) method followed by MALDI-TOF/TOF MS (mass spectrometry) to compare either late-proliferative and mid-secretory [9] or mid-proliferative and mid-secretory phase endometrium [8,10]. According to the different endometrial sampling

TABLE 16.1 Comparison of the Study Design of the Mass Spectrometry-Based Proteomic Analyses of Endometrial Samples at Different Phases of the Menstrual Cycle

Study	Number of Samples Used for Proteomic Analysis	Patient Population	Compared Samples (Number of Samples)	Proteomic Approach	Number of Proteins Significantly Differentially Expressed	Validation in Independent Cohort of Patients
[7]	6	Hysterectomy specimens	Proliferative vs mid-secretory ($n=3$) ($n=3$)	Isotope-coded affinity tags + multidimensional liquid chromatography + nanobore LC-MS/MS	5	No
[8]	6	Normal fertile women	Mid-proliferative (days 8–10) vs mid-secretory (days 19–23) ($n=3$) ($n=3$)	DIGE + MALDI-TOF/TOF MS	41	Validation of three candidates by immunohistochemistry
[9]	11	Healthy women	Late-proliferative vs mid-secretory (LH + 6) ($n=5$) ($n=6$)	DIGE + MALDI-TOF-TOF	8	Validation of two candidates by immunohistochemistry
[10]	12	Normal menstrual cycle and free of uterine abnormalities	Mid-proliferative (days 7–10) vs mid-secretory (days 20–24) ($n=6$) ($n=6$)	DIGE + MALDI-MS-MS	45	Validation of six candidates by Western blot
[11]	16	Ovum donors	Early-secretory (LH + 2) vs mid-secretory (LH + 7) ($n=8$) ($n=8$)	DIGE + MALDI-MS	2	Validation by immunohistochemistry and Western blot
[12]	8	Fertile women	Early-secretory (LH + 2) vs mid-secretory (LH + 7) ($n=4$) ($n=4$)	DIGE + MALDI-TOF/MS	31	No
[5]	18	Patients referred for ICSI for male infertility	Early-secretory (LH + 2) vs mid-secretory (LH + 7) ($n=9$) ($n=9$)	DIGE + SELDI-TOF/MS-MS	216	Validation of two candidates by immunohistochemistry and identification of the functions of one candidate

phases analyzed, this is not surprising to notice that the reported protein profiles were completely different with no candidates in common between these four studies. Many factors could justify these disparities (different proteomic approaches, recruited patients, number of samples, and statistical methodologies) [4,13]. Particularly, the endometrial sample size was generally low (Table 16.1), and the clinical populations were mostly not well defined. However, when comparing similar endometrial samples (mid-proliferative vs mid-secretory or early-secretory vs mid-secretory), specific protein signatures in common between studies were identified (Table 16.2). In particular, two members of the annexin family of calcium-dependent phospholipid-binding proteins, annexin A4 and A5 (ANXA4 and ANXA5), have been identified as overexpressed during the mid-secretory phase compared with the mid-proliferative phase in the studies of Chen et al. [8] and Rai et al. [10].

Although their specific functions in the acquisition of the endometrial receptivity phenotype remain unknown, several members of the annexin family have been involved in membrane-related events along exocytotic and endocytotic pathways. Using a human endometrial explant model, it has been previously reported that ANXA4 protein was exclusively expressed in epithelial cells, localized to both glandular and luminal epithelium, and

TABLE 16.2 List of Proteins That Are Differentially Expressed According to the Cycle Phases and in Common Between Several Studies

	Mid-Proliferative vs Mid-Secretory Phases		Early-Secretory vs Mid-Secretory Phases		
Protein name	**[8]**	**[10]**	**[11]**	**[12]**	**[5]**
ACTG1	√	√			
HSPA9	√	√			
KRT8	√	√			
ACTB	√	√			
MVP	√	√			
ARHGDIB	√	√			
RAB11B	√	√			
NPM1	√	√			
ANXA5	√	√			
ANXA4	√	√	√	√	
ANXA2			√	√	
F13A1			√	√	
CTSB				√	√
FGB				√	√
VIM			√	√	√
PARK7			√	√	√

was present in high levels throughout the menstrual cycle except during early-secretory phase [14]. In addition, progesterone significantly increased the ANXA4 mRNA and protein after 48 h in culture of proliferative explants, suggesting a key role of this member of the annexin family in the acquisition of the endometrial receptivity phenotype [14]. Moreover, the overexpression of the ANXA4 mRNA during the mid-secretory phase was associated with a successful pregnancy as well spontaneously as after ICSI cycle compared with patients who did not achieve a pregnancy [15]. Accordingly, ANXA4 may be a potential prognostic biomarker of successful pregnancy. The role and localization of ANXA5 in endometrium remain always unknown. However, mechanism study in renal cell carcinoma demonstrated that ANXA5 promotes migration, invasion, epithelial-mesenchymal transition, and matrix metallopeptidase expression, some major biological functions involved in the implantation process [16].

Similarly, three other studies compared the endometrium proteome during the transition from the early-secretory to the mid-secretory stage and highlighted shared proteins (Table 16.2) [5,11,12]. Two of them targeted the two-dimensional differential in gel electrophoresis (DIGE) method followed by MALDI-TOF/TOF MS [11,12], and only one used the SELDI-TOF technique [5]. Differently from other mass spectrometry approaches, the SELDI-TOF technique allows retaining proteins of a sample in a selective way on various chemical surfaces. This selective fixation, combined with the modulation of the stringency of the washes, gives the possibility to detect and analyze minority proteins or proteins with low molecular weight (<20 kDa) that would otherwise have been masked by the presence of the most abundant proteins in the sample. Due to its high sensibility, the SELDI-TOF technology allows the analysis of proteins for which the classic techniques do not give conclusive results. Moreover, few studies analyzed the protein profile changes between early-secretory and mid-secretory endometrium in the same patients and in a large cohort of patients, to minimize the interpatient variability. In these conditions, the study of Bissonnette et al. [5] identified 216 proteins as differentially expressed between the early- and mid-secretory phases, including several S100 family members (S100A4, S100A6, S100A10, and S100A11). The S100 family includes a group of low-molecular-weight acidic proteins that contain two EF-hand calcium-binding motifs. Binding to Ca^{2+} results in a conformational change of S100 proteins that exposes the hydrophobic residues of the C-terminal extension, thus enabling their interaction with various target proteins. The only exception is S100A10 that is permanently locked in an active conformation [17]. They are expressed in many cell types and have a broad range of extracellular and intracellular functions, including cell growth, proliferation, differentiation, apoptosis, inflammation, motility, migration, and invasion [18]. Elevated expression of S100 proteins is associated with several cancers. They are involved in tumor progression because of their roles in cell proliferation, metastasis, angiogenesis, and immune evasion [19]. S100A10 and S100A11 also have roles in fertility. Several studies have reported S100A10 upregulation during the implantation windows in humans and rhesus monkey and specifically at the implantation site in rhesus monkey [5,11,20]. Moreover, downregulation of S100A11 during the implantation window has been related to pregnancy failure in humans [21]. In addition, S100A10 and S100A11 mRNA levels were significantly reduced in nonreceptive endometrium compared with receptive endometrium in patients with repeated implantation failures [5]. These findings suggest a key role of these two proteins in the acquisition of the endometrial receptivity phenotype.

Immunofluorescence analysis of endometrial tissue sections showed that S100A10 was expressed in epithelial luminal and glandular endometrial epithelial cells and in endometrial stromal cells, mainly localized in the cytoplasm and cytomembrane of luminal and glandular epithelia [5,22]. S100A10 silencing in primary human endometrial stromal and epithelial cells resulted in migration inhibition [5]. A similar effect was previously reported in human macrophages, and S100A10 role in cell migration seems to be plasmin-dependent [23,24]. Plasmin contribution to cell migration is dependent on its ability to degrade extracellular matrix (ECM) proteins and activate other proteinases with matrix-degrading activity called metalloproteinases [24,25]. Transcriptome analysis of human endometrial epithelial and stromal cells where S100A10 was silenced by shRNA revealed the deregulation of 37 and 256 genes, respectively, related to the components of the extracellular matrix and intercellular connections. Functional annotations of these deregulated genes highlighted alterations of the leukocyte extravasation signaling and angiogenesis pathways that play a crucial role during implantation [5,26]. In addition, S100A10 silencing also affected decidualization and secretory transformation of primary endometrial stromal cells and epithelial cells, respectively, and promoted apoptosis in serum-starved endometrial epithelial cells. Therefore, by affecting migration, decidualization, and apoptosis, some major biological functions involved in the implantation process, S100A10 might play a key role during implantation. These findings suggest that S100A10 is a relevant candidate biomarker for predicting implantation failure, particularly due to inadequate endometrial receptivity (Fig. 16.1).

Finally, to gain insights into the proteomic signature of endometrial receptivity, Garrido-Gómez et al. [27] analyzed the difference in protein expression profile between endometria said receptive and nonreceptive resulting in a pregnancy and a nonpregnancy, respectively, using 2D-DIGE followed by MALDI-TOF/TOF MS. They identified 24 proteins as differentially expressed between nonreceptive vs receptive endometria, including the annexin A6 (ANXA6) and the progesterone receptor membrane component 1 (PGRMC1) that are underexpressed in receptive samples, suggesting a key role of these proteins in the acquisition of the endometrial receptivity phenotype. However, functional analyses must be performed to identify their(s) specific role(s) in this process.

Proteomic Analyses From Endometrial Secretions

The viscous fluid secreted by the endometrium during the mid-secretory phase, reflecting endometrial function and the embryo-endometrial dialogue prior to implantation, is also considered an important store of biomarkers for minimally invasive assessment of endometrial receptivity, avoiding to practice an endometrial biopsy. Two approaches have been proposed to obtain endometrial secretions: either by washing the endometrial cavity or by direct aspiration [28–32]. Several studies have demonstrated quantitative and qualitative changes of the protein patterns of endometrial secretion during the menstrual cycle [32–34]. Particular attention has been paid to the leukemia inhibitory factor (LIF) and the progestagen-associated endometrial protein (PAEP), an interleukin-6 family cytokine and a member of the kernel lipocalin superfamily, respectively, which have been shown to be crucial for successful implantation. These proteins vary during the menstrual cycle and appear to be differentially expressed between fertile and subfertile women [33,34]. Indeed, LIF concentrations in flushings from women with unexplained infertility were significantly lower than those in flushings from normal fertile women although the LIF content in endometrial secretion fluids was not correlated with endometrial maturity. These findings suggest that LIF plays an important role in human embryo implantation, rather than in the acquisition of the endometrial receptivity phenotype.

The application of a broad-based proteomic analysis (2D-DIGE MS/MS) to endometrial secretion during the early- and mid-secretory stages of the menstrual cycle revealed 82 proteins differentially expressed between these two stages with at least a twofold change difference [32]. These proteins were involved in numerous signaling pathways including coagulation, immune/inflammatory response, transport, transcription regulation, stress response, cell structure, metabolism, and complement as previously reported in functional annotations of transcriptomic studies [35]. On the other hand, using 2D-DIGE MS/MS to assess the differential in proteome between the mid-proliferative and mid-secretory phase of fertile patients revealed significant changes according to the cycle stage and the fertility status [36]. A number of proteins were identified by mass spectrometry, including antithrombin III (ANT3) and alpha-2-macroglobulin (A2M), whose production was confirmed in endometrial epithelium. A2M is overexpressed in both endometrial tissue and uterine fluid during the implantation windows with a specific localization in glandular and luminal endometrial epithelium and decreased in uterine infertile women, suggesting a role in the acquisition of the endometrial receptivity phenotype. In the mouse model, it has been previously shown that A2M controls trophoblast positioning in implantation sites [37], probably by its action as protease inhibitor on matrix metalloproteinase remodeling. A comprehensive catalogue of proteins of the endometrial secretions during the secretory phase of the menstrual cycle using different proteomic strategies, combining gel-based and chromatography-based methods coupled with MS, has been reported providing the basis for a better understanding of embryo implantation process [38]. This strategy allowed to identify only 56 proteins in common to the three proteomic approaches. Genes ontology based on biological processes revealed that majority of identified proteins in endometrial fluid were involved in metabolism, regulation of biological process, response to stimulus, cell organization, and biogenesis and transport. However, to date, the use of these proteins as biomarkers of endometrial receptivity needs to be tested.

PROTEIN EXPRESSION PROFILE OF HUMAN ENDOMETRIAL RECEPTIVITY UNDER STIMULATED CYCLE

Many studies have provided evidence for controlled ovarian stimulation (COS) effects on periovulatory endometrial histology, luteal function, and implantation under in vitro fertilization conditions [4]. Evidence for a direct effect of supraphysiological levels of sex steroid concentrations on the endometrium comes mainly from histological, histochemical, and gene expression studies. The COS protocol has been shown to be associated with an advancement of endometrial maturation during the periovulatory period whatever the protocol applied (GnRH agonists or antagonists) with a lack of clinical pregnancies if exceeding 3 days [4,39]. Although numerous studies reported a deleterious effect of high progesterone level on the day of human chorionic gonadotropin (hCG) administration on clinical pregnancies when day-3 embryos transfer during IVF/ICSI procedure were performed, no detrimental effect was reported on the pregnancy outcome when blastocyst transfer was performed [40–43]. These findings suggest that endometrial receptivity was not necessarily affected by a periovulatory endometrial maturation advancement and that the endometrium can probably recover from exposure to supraphysiological steroid concentrations. These concepts are reinforced by previous studies that have demonstrated the absence of histological advancement and no modification of gene expression profile during the implantation window under COS protocols [44–46]. However, the impact of ovarian stimulation on the endometrial receptivity remains questionable with conflicting findings [26,47–49]. Despite these conflicting results, there are still considerable efforts made for performing proteomic studies using high-throughput mass spectrometry-based proteomic approaches to better understand molecular mechanisms governing the acquisition of the endometrial receptivity phenotype. Majority of studies analyzed protein patterns in endometrial secretions rather than endometrial tissues as an alternative approach to avoid to perform an invasive biopsy that excludes its use during the luteal phase of cycles in which embryo transfers must be performed during IVF/ICSI procedures. In this objective, endometrial secretions removed immediately prior to embryo transfer provide a relevant noninvasive alternative avoiding to interfere with embryo implantation [30,31]. Endometrial secretions contain a variety of molecules secreted by endometrial epithelial cells that are presumed to support embryo implantation during the implantation window. Majority of studies in this field focused on growth factors and cytokines that are known to be involved in the implantation process such as LIF and PAEP. While it has been reported that ovarian stimulation has little impact on LIF and PEAP expression in the uterine secretion during the implantation window compared with natural cycles, others demonstrated that LIF assessment can be used as a predictor of reproductive success [29,50–52]. Boomsma et al. [53] analyzed the protein expression patterns of 17 cytokines and growth factors by multiplex immunoassay in endometrial secretions in a natural and a subsequent IVF cycle during the implantation windows in the same patients. After multivariable logistic regression, significantly higher concentrations of IL-1β, IL-5, IL-10, IL-12, IL-17, TNF-α, HB-EGF, CCL11, and DKK1 were reported in endometrial secretions from stimulated compared with natural cycles, supporting an altered protein expression profile under COS protocols. In addition, the vascular endothelial growth factor A, VEGFA, was the only candidate that demonstrated a decrease in its concentration after COS although no difference was reported in endometrial biopsies after COS compared with natural cycles [54]. VEGFA plays a central role in endometrial angiogenesis, a physiological process involved in the acquisition of the endometrial receptivity phenotype and the implantation process. The decrease in VEGFA in endometrial secretions under COS was consistent with clinical studies showing that subendometrial blood flows were lower in COS cycles compared with natural cycles [55]. However, in which measure these changes in the intrauterine medium affect endometrial receptivity and implantation process remain unknown. Boomsma et al., [56] investigated also in which extent these 17 soluble regulators of implantation can be predictive of implantation and clinical pregnancies in endometrial secretions prior to embryo transfer following COS in a cohort of 210 women. They identified CCL2 and CXCL10 as negatively and positively associated with implantation, respectively. Chemokines and chemokine receptors play an important role in reproductive immunology by recruiting leukocytes during early pregnancy. Recently reported was an overexpression of CCL2 in the glandular and luminal epithelium of the endometrium during early pregnancy compared with that in nonpregnant pigs. In addition, in vitro experiments demonstrated that CCL2 induces proliferation and cell cycle progression of porcine luminal epithelial cells via the activation of the PI3K/MAPK pathways and reduces endoplasmic reticulum stress regulatory gene expression [57]. In the same way, it has been previously reported a potential role of CXCL10 in the recruitment of immune cells into the endometrium during the implantation period in pigs [58]. All these findings suggest a key role of CCL2 and CXCL10 in maternal-fetal interaction by improving uterine receptivity by recruitment of immune cells during early gestational period in pigs.

On the other hand, after multivariable logistic regression, Boomsma et al. [53] identified TNF-α and IL-1β as positively and negatively associated with clinical pregnancy, respectively [56]. TNF-α is a pro-inflammatory cytokine that has been reported to be an important regulator of trophoblastic matrix metalloproteinases and, therefore, of implantation process [59]. The downexpression of IL-1β in endometrial secretions associated with clinical pregnancy was consistent with a previous report demonstrating higher IL-1β levels in endometrial flushings from women with implantation failure [60].

To date, no large-scale proteomic expression studies have been conducted to identify the potential impact of COS in endometrial secretions during the implantation window, allowing to obtain a global and complete overview.

CONCLUSIONS AND PERSPECTIVES

Proteomic studies using large-scale approaches to understand mechanisms associated to the acquisition of the human endometrial receptivity phenotype and the impact of controlled ovarian stimulation during IVF/ICSI cycles are small number and recent. In addition, there are few studies that have attempted to identify biological functions of identified proteins that are necessary steps to move forward in our knowledge on endometrial receptivity. Analysis of protein patterns in endometrial secretion fluid could give a new insight to understand the mechanisms involved in the process of embryo implantation and may offer a relatively noninvasive means of assessing endometrial receptivity during fertility treatment cycles. However, to date, the clinical value of endometrial secretion analysis in terms of predicting endometrial receptivity and consequently fertility is limited. Although it is a promising technique, endometrial secretion analysis has a number of limitations susceptible to be exceeded in synergy with the technological progress.

References

[1] C. Coughlan, W. Ledger, Q. Wang, F. Liu, A. Demirol, T. Gurgan, R. Cutting, K. Ong, H. Sallam, T.C. Li, Recurrent implantation failure: definition and management, Reprod. Biomed. Online 28 (2014) 14–38.
[2] C.T. Valdes, A. Schutt, C. Simon, Implantation failure of endometrial origin: it is not pathology, but our failure to synchronize the developing embryo with a receptive endometrium, Fertil. Steril. 108 (2017) 15–18.
[3] D. Haouzi, K. Mahmoud, M. Fourar, K. Bendhaou, H. Dechaud, J. De Vos, T. Rème, D. Dewailly, S. Hamamah, Identification of new biomarkers of human endometrial receptivity in the natural cycle, Hum. Reprod. 24 (2009) 198–205.
[4] D. Haouzi, H. Dechaud, S. Assou, J. De Vos, S. Hamamah, Insights into human endometrial receptivity from transcriptomic and proteomic data, Reprod. Biomed. Online 24 (2012) 23–34.
[5] L. Bissonnette, L. Drissennek, Y. Antoine, L. Tiers, C. Hirtz, S. Lehmann, H. Perrochia, F. Bissonnette, I.J. Kadoch, D. Haouzi, S. Hamamah, Human S100A10 plays a crucial role in the acquisition of the endometrial receptivity phenotype, Cell Adh. Migr. 10 (2016) 282–298.

[6] E.W. Deutsch, C.M. Overall, J.E. Van Eyk, M.S. Baker, Y.K. Paik, S.T. Weintraub, L. Lane, L. Martens, Y. Vandenbrouck, U. Kusebauch, W.S. Hancock, H. Hermjakob, R. Aebersold, R.L. Moritz, G.S. Omenn, Human Proteome Project mass spectrometry data interpretation guidelines 2.1, J. Proteome Res. 15 (2016) 3961–3970.
[7] L. DeSouza, G. Diehl, E.C. Yang, J. Guo, M.J. Rodrigues, A.D. Romaschin, T.J. Colgan, K.W. Siu, Proteomic analysis of the proliferative and secretory phases of the human endometrium: protein identification and differential protein expression, Proteomics 5 (2005) 270–281.
[8] J.I. Chen, N.J. Hannan, Y. Mak, P.K. Nicholls, J. Zhang, A. Rainczuk, P.G. Stanton, D.M. Robertson, L.A. Salamonsen, A.N. Stephens, Proteomic characterization of midproliferative and midsecretory human endometrium, J. Proteome Res. 8 (2009) 2032–2044.
[9] T. Parmar, S. Gadkar-Sable, L. Savardekar, R. Katkam, S. Dharma, P. Meherji, C.P. Puri, G. Sachdeva, Protein profiling of human endometrial tissues in the midsecretory and proliferative phases of the menstrual cycle, Fertil. Steril. 92 (2009) 1091–2003.
[10] P. Rai, V. Kota, C.S. Sundaram, M. Deendayal, S. Shivaji, Proteome of human endometrium: identification of differentially expressed proteins in proliferative and secretory phase endometrium, Proteomics Clin. Appl. 4 (2010) 48–59.
[11] E. Domínguez, T. Garrido-Gómez, J.A. López, E. Camafeita, A. Quiñonero, A. Pellicer, C. Simón, Proteomic analysis of the human receptive versus non-receptive endometrium using differential in-gel electrophoresis and MALDI-MS unveils stathmin 1 and annexin A2 as differentially regulated, Hum. Reprod. 24 (2009) 2607–2617.
[12] J. Li, Z. Tan, M. Li, T. Xia, P. Liu, W. Yu, Proteomic analysis of endometrium in fertile women during the prereceptive and receptive phases after luteinizing hormone surge, Fertil. Steril. 95 (2011) 1161–1163.
[13] S. Altmäe, F.J. Esteban, A. Stavreus-Evers, C. Simón, L. Giudice, B.A. Lessey, J.A. Horcajadas, N.S. Macklon, T. D'Hooghe, C. Campoy, B.C. Fauser, L.A. Salamonsen, A. Salumets, Guidelines for the design, analysis and interpretation of 'omics' data: focus on human endometrium, Hum. Reprod. Update 20 (2014) 12–28.
[14] A.P. Ponnampalam, P.A. Rogers, Cyclic changes and hormonal regulation of annexin IV mRNA and protein in human endometrium, Mol. Hum. Reprod. 12 (2006) 661–669.
[15] A. Allegra, A. Marino, P.C. Peregrin, A. Lama, A. García-Segovia, G.I. Forte, R. Núñez-Calonge, C. Agueli, S. Mazzola, A. Volpes, Endometrial expression of selected genes in patients achieving pregnancy spontaneously or after ICSI and patients failing at least two ICSI cycles, Reprod. Biomed. Online 25 (2012) 481–491.
[16] J. Tang, Z. Qin, P. Han, W. Wang, C. Yang, Z. Xu, R. Li, B. Liu, C. Qin, Z. Wang, M. Tang, W. Zhang, High Annexin A5 expression promotes tumor progression and poor prognosis in renal cell carcinoma, Int. J. Oncol. 50 (2017) 1839–1847.
[17] L. Santamaria-Kisiel, A.C. Rintala-Dempsey, G.S. Shaw, Calcium-dependent and -independent interactions of the S100 protein family, Biochem. J. 396 (2006) 201–214.
[18] R. Donato, B.R. Cannon, G. Sorci, F. Riuzzi, K. Hsu, D.J. Weber, C.L. Geczy, Functions of S100 proteins, Curr. Mol. Med. 13 (2013) 24–57.
[19] A.R. Bresnick, D.J. Weber, D.B. Zimmer, S100 proteins in cancer, Nat. Rev. Cancer 15 (2015) 96–109.
[20] X.Y. Sun, F.X. Li, J. Li, Y.F. Tan, Y.S. Piao, S. Tang, Y.L. Wang, Determination of genes involved in the early process of embryonic implantation in rhesus monkey (Macaca mulatta) by suppression subtractive hybridization, Biol. Reprod. 70 (2004) 1365–1373.
[21] X.M. Liu, G.L. Ding, Y. Jiang, H.J. Pan, D. Zhang, T.T. Wang, R.J. Zhang, J. Shu, J.Z. Sheng, H.F. Huang, Down-regulation of S100A11, a calcium-binding protein, in human endometrium may cause reproductive failure, J. Clin. Endocrinol. Metab. 97 (2012) 3672–3683.
[22] X. Wei, S. Tong, Q. Yan, Cyclic changes of S100A10 expression in human endometrium, Zhonghua Yi Xue Za Zhi 94 (2014) 2152–2155. (in Chinese).
[23] P.A. O'Connell, A.P. Surette, R.S. Liwski, P. Svenningsson, D.M. Waisman, S100A10 regulates plasminogen-dependent macrophage invasion, Blood 116 (2010) 1136–1146.
[24] K.D. Phipps, A.P. Surette, P.A. O'Connell, D.M. Waisman, Plasminogen receptor S100A10 is essential for the migration of tumor-promoting macrophages into tumor sites, Cancer Res. 71 (2011) 6676–6683.
[25] S.R. Gross, C.G. Sin, R. Barraclough, P.S. Rudland, Joining S100 proteins and migration: for better or for worse, in sickness and in health, Cell. Mol. Life Sci. 71 (2014) 1551–1579.
[26] D. Haouzi, S. Assou, K. Mahmoud, S. Tondeur, T. Rème, B. Hedon, J. De Vos, S. Hamamah, Gene expression profile of human endometrial receptivity: comparison between natural and stimulated cycles for the same patients, Hum. Reprod. 24 (2009) 1436–1445.

[27] T. Garrido-Gómez, A. Quiñonero, O. Antúnez, P. Díaz-Gimeno, J. Bellver, C. Simón, F. Domínguez, Deciphering the proteomic signature of human endometrial receptivity, Hum. Reprod. 29 (2014) 1957–1967.

[28] A.R. Gargiulo, R.N. Fichorova, J.A. Politch, J.A. Hill, D.J. Anderson, Detection of implantation-related cytokines in cervicovaginal secretions and peripheral blood of fertile women during ovulatory menstrual cycles, Fertil. Steril. 82 (2004) 1226–1234.

[29] M.H. van der Gaast, I. Classen-Linke, C.A. Krusche, K. Beier-Hellwig, B.C. Fauser, H.M. Beier, N.S. Macklon, Impact of ovarian stimulation on mid-luteal endometrial tissue and secretion markers of receptivity, Reprod. Biomed. Online 17 (2008) 553–563.

[30] C.M. Boomsma, A. Kavelaars, M.J. Eijkemans, K. Amarouchi, G. Teklenburg, D. Gutknecht, B.J. Fauser, C.J. Heijnen, N.S. Macklon, Cytokine profiling in endometrial secretions: a non-invasive window on endometrial receptivity, Reprod. Biomed. Online 18 (2009) 85–94.

[31] M.H. van der Gaast, N.S. Macklon, K. Beier-Hellwig, C.A. Krusche, B.C. Fauser, H.M. Beier, I. Classen-Linke, The feasibility of a less invasive method to assess endometrial maturation-comparison of simultaneously obtained uterine secretion and tissue biopsy, BJOG 116 (2009) 304–312.

[32] J.G. Scotchie, M.A. Fritz, M. Mocanu, B.A. Lessey, S.L. Young, Proteomic analysis of the luteal endometrial secretome, Reprod. Sci. 16 (2009) 883–893.

[33] S.M. Laird, E.M. Tuckerman, C.F. Dalton, B.C. Dunphy, T.C. Li, X. Zhang, The production of leukaemia inhibitory factor by human endometrium: presence in uterine flushings and production by cells in culture, Hum. Reprod. 12 (1997) 569–574.

[34] T.C. Li, E. Ling, C. Dalton, A.E. Bolton, I.D. Cooke, Concentration of endometrial protein PP14 in uterine flushings throughout the menstrual cycle in normal, fertile women, Br. J. Obstet. Gynaecol. 100 (1993) 460–464.

[35] D. Haouzi, H. Dechaud, S. Hamamah, Réceptivité de l'endomètre et impact des protocoles de stimulation ovarienne à la lumière des analyses transcriptomiques et protéomiques, mt Médecine de la Reproduction, Gynécol. Endocrinol. 14 (2012) 135–143.

[36] N.J. Hannan, A.N. Stephens, A. Rainczuk, C. Hincks, L.J.F. Rombauts, L.A. Salamonsen, 2D-DiGE analysis of the human endometrial secretome reveals differences between receptive and nonreceptive states in fertile and infertile women, J. Proteome Res. 9 (2010) 6256–6264.

[37] S. Esadeg, H. He, R. Pijnenborg, F. Van Leuven, B.A. Croy, Alpha-2 macroglobulin controls trophoblast positioning in mouse implantation sites, Placenta 24 (2003) 912–921.

[38] J. Casado-Vela, E. Rodriguez-Suarez, I. Iloro, A. Ametzazurra, N. Alkorta, J.A. García-Velasco, R. Matorras, B. Prieto, S. González, D. Nagore, L. Simón, F. Elortza, Comprehensive proteomic analysis of human endometrial fluid aspirate, J. Proteome Res. 8 (2009) 4622–4632.

[39] E.M. Kolibianakis, C. Bourgain, C. Albano, E. Osmanagaoglu, J. Smitz, A. Van Steirteghem, P. Devroey, Effect of ovarian stimulation with recombinant follicle-stimulating hormone, gonadotropin releasing hormone antagonists, and human chorionic gonadotropin on endometrial maturation on the day of oocyte pick-up, Fertil. Steril. 78 (2002) 1025–1029.

[40] J.N. Hugues, E. Massé-Laroche, J. Reboul-Marty, O. Boîko, C. Meynant, I. Cédrin-Durnerin, Impact of endogenous luteinizing hormone serum levels on progesterone elevation on the day of human chorionic gonadotropin administration, Fertil. Steril. 96 (2011) 600–604.

[41] C. Sonigo, G. Dray, C. Roche, I. Cédrin-Durnerin, J.N. Hugues, Impact of high serum progesterone during the late follicular phase on IVF outcome, Reprod. Biomed. Online 29 (2014) 177–186.

[42] E.G. Papanikolaou, E.M. Kolibianakis, C. Pozzobon, P. Tank, H. Tournaye, C. Bourgain, A. Van Steirteghem, P. Devroey, Progesterone rise on the day of human chorionic gonadotropin administration impairs pregnancy outcome in day 3 single embryo transfer, while has no effect on day 5 single blastocyst transfer, Fertil. Steril. 91 (2009) 949–952.

[43] R. Ochsenkühn, A. Arzberger, V. von Schonfeldt, J. Gallwas, N. Rogenhofer, A. Crispin, C.J. Thaler, U. Noss, Subtle progesterone rise on the day of human chorionic gonadotropin administration is associated with lower live birth rates in women undergoing assisted reproductive technology: a retrospective study with 2,555 fresh embryo transfers, Fertil. Steril. 98 (2012) 347–354.

[44] E. Labarta, J.A. Martínez-Conejero, P. Alamá, J.A. Horcajadas, A. Pellicer, C. Simón, E. Bosch, Endometrial receptivity is affected in women with high circulating progesterone levels at the end of the follicular phase: a functional genomics analysis, Hum. Reprod. 26 (2011) 1813–1825.

[45] D. Haouzi, L. Bissonnette, A. Gala, S. Assou, F. Entezami, H. Perrochia, H. Dechaud, J.N. Hugues, S. Hamamah, Endometrial receptivity profile in patients with premature progesterone elevation on the day of HCG administration, Biomed. Res. Int. 2014 (2014) 951937.

[46] R.W. Noyes, A.T. Hertig, J. Rock, Dating the endometrial biopsy, Fertil. Steril. 1 (2012) 3–25.
[47] J.A. Horcajadas, P. Mínguez, J. Dopazo, F.J. Esteban, F. Domínguez, L.C. Giudice, A. Pellicer, C. Simón, Controlled ovarian stimulation induces a functional genomic delay of the endometrium with potential clinical implications, J. Clin. Endocrinol. Metab. 93 (2008) (2008) 4500–4510.
[48] S. Mirkin, G. Nikas, J.G. Hsiu, J. Díaz, S. Oehninger, Gene expression profiles and structural/functional features of the peri-implantation endometrium in natural and gonadotropin-stimulated cycles, J. Clin. Endocrinol. Metab. 89 (2004) 5742–5752.
[49] D. Haouzi, S. Assou, C. Dechanet, T. Anahory, H. Dechaud, J. De Vos, S. Hamamah, Controlled ovarian hyperstimulation for in vitro fertilization alters endometrial receptivity in humans: protocol effects, Biol. Reprod. 82 (2010) 679–686.
[50] F. Olivennes, N. Lédée-Bataille, M. Samama, J. Kadoch, J.L. Taupin, S. Dubanchet, G. Chaouat, R. Frydman, Assessment of leukemia inhibitory factor levels by uterine flushing at the time of egg retrieval does not adversely affect pregnancy rates with in vitro fertilization, Fertil. Steril. 79 (2003) 900–904.
[51] M. Mikolajczyk, P. Wirstlein, J. Skrzypczak, The impact of leukemia inhibitory factor in uterine flushing on the reproductive potential of infertile women--a prospective study, Am. J. Reprod. Immunol. 58 (2007) 65–74.
[52] L. Aghajanova, Update on the role of leukemia inhibitory factor in assisted reproduction, Curr. Opin. Obstet. Gynecol. 22 (2010) 213–219.
[53] C.M. Boomsma, A. Kavelaars, M.J. Eijkemans, B.C. Fauser, C.J. Heijnen, N.S. Macklon, Ovarian stimulation for in vitro fertilization alters the intrauterine cytokine, chemokine, and growth factor milieu encountered by the embryo, Fertil. Steril. 94 (2010) 1764–1768.
[54] Y.L. Lee, Y. Liu, P.Y. Ng, K.F. Lee, C.L. Au, E.H. Ng, P.C. Ho, W.S. Yeung, Aberrant expression of angiopoietins-1 and -2 and vascular endothelial growth factor-A in peri-implantation endometrium after gonadotrophin stimulation, Hum. Reprod. 23 (2008) 894–903.
[55] E.H. Ng, C.C. Chan, O.S. Tang, W.S. Yeung, P.C. Ho, Comparison of endometrial and subendometrial blood flow measured by three-dimensional power Doppler ultrasound between stimulated and natural cycles in the same patients, Hum. Reprod. 19 (2004) 2385–2390.
[56] C.M. Boomsma, A. Kavelaars, M.J. Eijkemans, E.G. Lentjes, B.C. Fauser, C.J. Heijnen, N.S. Macklon, Endometrial secretion analysis identifies a cytokine profile predictive of pregnancy in IVF, Hum. Reprod. 24 (2009) 1427–1435.
[57] W. Lim, H. Bae, F.W. Bazer, G. Song, Cell-specific expression and signal transduction of C-C motif chemokine ligand 2 and atypical chemokine receptors in the porcine endometrium during early pregnancy, Dev. Comp. Immunol. 81 (2018) 312–323.
[58] J. Han, M.J. Gu, I. Yoo, Y. Choi, H. Jang, M. Kim, C.H. Yun, H. Ka, Analysis of cysteine-X-cysteine motif chemokine ligands 9, 10, and 11, their receptor CXCR3, and their possible role on the recruitment of immune cells at the maternal-conceptus interface in pigs, Biol. Reprod. 97 (2017) 69–80.
[59] M. Cohen, A. Meisser, L. Haenggeli, P. Bischof, Involvement of MAPK pathway in TNF-alpha-induced MMP-9 expression in human trophoblastic cells, Mol. Hum. Reprod. (2006) 225–232.
[60] N. Inagaki, C. Stern, J. McBain, A. Lopata, L. Kornman, D. Wilkinson, Analysis of intra-uterine cytokine concentration and matrix-metalloproteinase activity in women with recurrent failed embryo transfer, Hum. Reprod. 18 (2003) 608–615.

CHAPTER

17

Stem Cell-Derived Spermatozoa

Jasin Taelman, Swati Mishra, Margot Van der Jeught, Björn Heindryckx

Ghent-Fertility and Stem Cell Team (G-FAST), Department for Reproductive Medicine, Ghent University Hospital, Ghent, Belgium

INTRODUCTION

Stem cells have three basic properties: prolonged proliferation capacity, they are unspecialized and have the potential to differentiate into cells of any lineage. These "unspecialized" cells can give rise to "specialized" cell types, such as the cells of the heart and nervous system and, in interest of this chapter, sperm cells. Stem cells are therefore pluripotent, providing the opportunity for regenerative medicine, potentially allowing the development of therapies to modify, repair or replace malfunctioning, injured or missing cells in the body. Pluripotent stem cells therefore may serve as progenitors for gametes or germ cells, in cases where they are absent or abnormal.

Existing interventions in assisted reproduction technology (ART) cannot fully benefit patients devoid of viable gametes or diagnosed with idiopathic infertility, ovarian or testicular failure or aging, some genetic diseases, or repeated incidence of poor quality embryos or IVF failure [1–3]. For such patients wishing to have their own genetic offspring, the prospect of stem cell-derived (SCD) gametes provides hope. These gametes would be adult germ cells (sperm and oocytes), matured in vitro, either from isolated germ cell precursors (primordial germ cells, PGCs), or by differentiation of pluripotent stem cells toward the germ cell lineage.

The generation of SCD gametes may allow same-sex couples and postmenopausal women to have genetic descendants [4]. As stem cells can be differentiated to form gametes of either gender, it might be possible to reprogram somatic cells obtained from same-sex couples. The same can be achieved for women of postreproductive age with diminished ovarian reserves.

Male factor infertility is a growing concern in the developing world. Many research groups have reported a consistent decline of sperm quality over the last 50 years, a further challenge for couples who wish to have biological children [5–7]. While the cause of

Reproductomics
https://doi.org/10.1016/B978-0-12-812571-7.00018-6

male factor infertility is multifactorial, there are many cases of unknown origin, congenital and acquired. Many of these cases can be helped by the direct injection of sperm into the oocyte (intracytoplasmic sperm injection, ICSI) in conjunction with ART. However, when this is unsuccessful, our knowledge of early human fetal germline development is currently too inadequate to provide further possibilities for genetic parenthood. At present, ethical considerations and the unavailability of human embryo material for research have limited our understanding. Therefore, SCD sperm cells not only provide hope as a therapy for male infertility but may also broaden our understanding of human germline development.

Primarily, the process of obtaining an in vitro source of gametes may provide a useful model for molecular studies of human germline formation [4]. In addition, the technology may provide a supply of gametes for toxicity testing in the context of human fertility, the correction of genetic diseases, the development of novel contraceptives and for the study of genetic factors relating to infertility [4,8].

IN VIVO DEVELOPMENT OF SPERMATOZOA

Our current understanding of in vivo germline development heavily relies on animal models, especially mice. Although species-specific differences in developmental pathways exist, such models have provided us with a valuable foundation for unraveling spermatogenesis in human.

The Concept of Germ Cells

Germ cells are distinct from somatic cells and are responsible for transmission of information to the next generation ensuring continuation of the species [9]. During embryonic development in multicellular organisms, the appearance of primordial germ cells (PGCs) marks the initiation of the germ cell lineage. The stable differentiation and development of PGCs and the diploid embryonic precursors of male and female gametes assure normal fertility of the developing individual and correct transmission of genetic information [10]. Studies into the timing and mechanisms of germline specification help to understand the selective pressures that act on precursor cells prior to gametogenesis, as well as the factors influencing heritable variation [11]. While histological evidence of the existence and development of PGCs date back to the early twentieth century, their detailed developmental trajectories in humans remain largely unknown [10].

Mammalian Germline Development

Data derived from animal models show two distinct pathways of germ cell lineage specification. One is through maternal inheritance of localized determinants in the germ plasm (collection of cytoplasmic RNAs, RNA-binding proteins and organelles within the mature egg) before or immediately following fertilization called "preformation," as observed in organisms such as *Caenorhabditis elegans* and *Drosophila melanogaster* [10,12]. The second is through induction from surrounding pluripotent embryonic cells called "epigenesis," occurring relatively later in development, as seen in mammals [11,12]. Epigenesis is regarded to be the more frequent mode of germ cell specification [11].

The mammalian germ cell lineage originates from a cluster of cells in the postimplantation embryo and potentially gives rise to PGCs. In mice, germ cell lineage commitment occurs in the epiblast around embryonic day (E) 6.0, induced by members of the transforming growth factor β1 super family, such as bone morphogenetic protein 4 (BMP4), which is secreted by the surrounding extraembryonic ectoderm [13,14] (Fig. 17.1). BMP4 induces the expression of other germ cell specification marker genes. Firstly, PR domain-containing 1 ((PRDM1) also known as B-lymphocyte-induced maturation protein 1 or BLIMP1) is expressed [10], closely followed by PRDM14, TFAP2C, and IFITM3 (Fragilis) [15]. As PRDM1, positive cells increase and move out to form the primitive streak [10], the cells that express the highest levels of IFITM3 subsequently go on to express DPPA3 (Stella) [16]. The first PGCs are created at E7.25 in this region, marked by their unique expression of TNAP [17–20]. In humans, this corresponds to around the end of the 3rd week of gestation, where they arise in the wall of the yolk sac [16] (Fig. 17.1).

Following specification, PGCs also undergo striking epigenetic changes during their maturation, a process partly initiated by PRDM14 [19]. One important facet of this epigenetic reprogramming of PGCs, is the activity of the X chromosomes in female germ cells before sex determination [21]. In the preimplantation epiblast, both X chromosomes are in an open, active state, while in the surrounding trophectoderm, one of the two X chromosomes is randomly inactivated through repressive epigenetic processes in chromatin structure, which silences gene expression of most genes on one X chromosome [22–24] (Fig. 17.2). This is crucial for obtaining a comparable dosage of X-related gene expression in male and female embryos [23]. Around gastrulation, random X-inactivation also occurs in the postimplantation epiblast, where either the maternal or the paternal X chromosome is inactivated, which is coupled with downregulation of pluripotency genes [23]. It has been shown specifically that this also occurs in mouse PGC precursor cells by E6.5 [25].

At a later stage, from around E9.5 in the mouse, these cells proliferate further and migrate to the emerging gonadal region in the developing embryo. In humans, this occurs at around 5–6 weeks of gestation. Following their induction, PGCs must maintain germ cell potential, without inducing differentiation to somatic lineages. This is made possible by expression of core pluripotency genes POU5F1 and NANOG [26] during migration. Moreover, migrating PGCs express surface receptors that are activated by extracellular molecules that guide the PGCs to the putative gonadal region. For instance, stem cell factor (SCF) stimulates the activation and expression of the CKIT receptor, which is essential to survival and proliferation [27]. Other interactions are created between CXCR4 chemokine receptor and β-integrins [28]. Important markers of early postmigratory PGCs include the RNA-binding proteins DAZL (regulator of meiosis) and VASA [28].

In fact at this stage, PGC precursors undergo extensive nuclear reprogramming that includes genome-wide epigenetic changes. These involve chromatin remodeling and complete erasure of DNA methylation in single copy and imprinted genes, which culminates in the mouse around E13.5. This germline demethylation process is crucial for the resetting of epigenetic memory, which prevents the transfer of epigenetic mutations to the offspring [29,30]. Next to changes in DNA methylation, early germ cells are characterized by specific chromatin signatures [31]. In migrating mouse germ cells, the H3K9me2 chromatin mark is

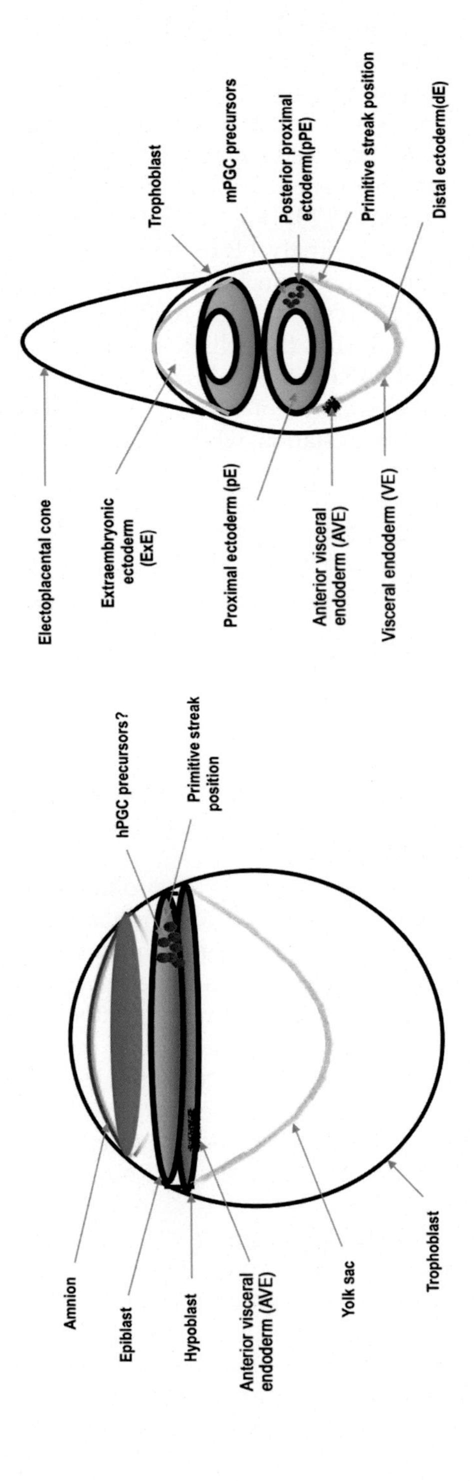

FIG. 17.1 Schematic representation of a human embryo at day 17 and a comparable mouse embryo at day 6.25–6.5 postfertilization. *Adapted from C.G. Extavour, M. Akam, Mechanisms of germ cell specification across the metazoans: epigenesis and preformation, Development 130 (2003) 5869–5884, https://doi.org/10.1242/dev.00804.*

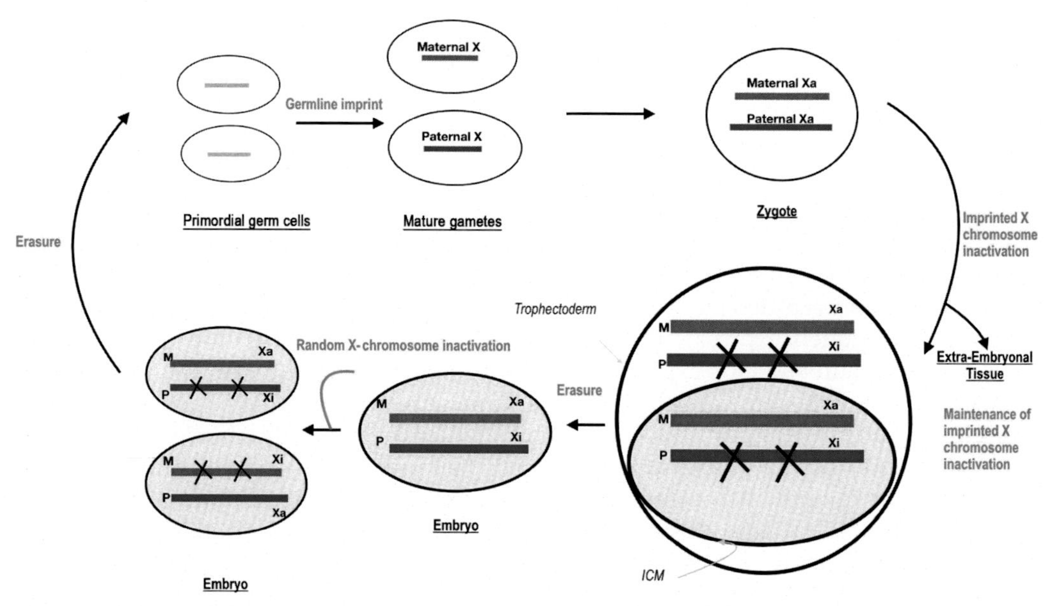

FIG. 17.2 Schematic representation of X chromosome inactivation in mammals. *Red bars* represent the X chromosome of maternal origin (M), *blue bars* represent the X chromosome of paternal origin (P) and *gray bars* represent X chromosomes in primordial germ cells, which have undergone complete parental epigenetic imprint erasure. The activated and inactivated X chromosomes are indicated by Xa and Xi, respectively. In the zygote, both maternal and paternal X chromosomes remain active. However, in the preimplantation blastocyst, inactivation of the paternal X chromosome is established (represented by crosses). While this is maintained in extraembryonic tissues such as the placenta, it is erased in the embryonic tissue. Random X chromosome inactivation takes place in all embryonic cells and is maintained throughout the mammalian life cycle, except in the germline where they are reprogrammed. *Adapted from W. Reik, A. Lewis, Co-evolution of X-chromosome inactivation and imprinting in mammals, Nat. Rev. Genet. 6 (5) (2005) 403–410, https://doi.org/10.1038/nrg1602.*

reduced, followed by an increase in the restrictive H3K27me3 chromatin mark, which results in transcriptional quiescence [32]. At a later stage, upon migration into the gonadal ridges, erasure of genomic imprints is initiated, coupled with further chromatin changes that in turn again generate a transcriptionally more permissive chromatin state [29,33]. Recent studies of porcine PGCs showed similar heterochromatin rearrangements, suggesting that germline epigenetic reprogramming mechanisms are conserved also in nonrodent species [34,35]. It has also been shown that hypomethylation of DNA occurs in early human PGCs [36]. Moreover, they share similar chromatin remodeling and imprinting erasure mechanisms as in mouse and pig PGCs [36]. In female embryos, the silenced X chromosome is also reactivated once again, a process unique to PGC precursor cells [11,24,25]. Even though PGCs are specified from the postimplantation epiblast, which is primed to a somatic fate, epigenetic modifications allow them to regain their developmental totipotency and erase genomic imprinting, protecting the germline cells from induction of somatic cell lineage differentiation [11,37]. Following this epigenetic overhaul toward an ultimately less restricted state, PGCs will enter a phase of global remethylation, ultimately resulting in a highly methylated sperm or a partially methylated oocyte genome [38].

In PGCs in a human fetus, DNA methylation is progressively removed from the genome within 10–11 weeks, when only 5%–7% remains [39–42]. Strikingly, during this epigenomic reprogramming period, the transcriptome remains relatively unchanged [40]. At 11 weeks postfertilization, germ cell precursor cells are generally called oogonia and gonocytes, for female and male, respectively. As a general term for all germ cell stages during human development, fetal germ cells (FGCs) are also used. By week 17, the majority of FGCs have initiated meiotic division. However, strong gene expression heterogeneity exists between individual FGCs [40]. Considerable gene expression differences exist as well between human and mouse FGCs. In early mouse PGCs, the pluripotency marker SOX2 is reexpressed, while this has not been detected in early human PGCs. Instead, early human PGCs have been shown to uniquely express SOX15 and SOX17 in vivo, as well as in vitro [40,43,44]. Both female and male human PGCs express the X chromosome inactivation regulator XIST, even after inactivation of the female X chromosome, which does not occur in mouse PGCs [39].

Recently, Li and colleagues analyzed the transcriptomes of 2000 human FGCs from 4- to 26-week fetuses and their surrounding gonadal niches, using single-cell RNA sequencing to categorize the developmental stages in both males and females [45]. Samples collected from male embryos revealed three distinct populations: migrating FGCs, mitotic FGCs, and mitotic arrest FGCs. Female embryo samples revealed a similar mitotic FGC population, however did not reveal a corresponding migrating or mitotic arrest group. Instead, different populations were distinguished and revealed to be a retinoic acid-responsive group, a meiotic process group, and an oogenesis process group. Furthermore, specific signaling pathway mechanisms were discovered when comparing FGCs and their gonadal niche cells. Male gonadal niche cells showed expression of anti-Mullerian hormone (AMH), while subpopulations of male FGCs showed expression of its receptor BMPR1B. Male FGCs showed expression of Notch ligand DLL3, while their gonadal niche cells showed expression of the Notch receptor NOTCH2 and specific expression of a major Notch target gene HES1 [45,46]. These findings uncovered important signaling pathway differences between male and female germ cell differentiation during embryogenesis. However, a wide range of time points from which the samples were collected in this study assert that larger, more robust studies need to be conducted to derive a higher resolution picture of molecular germ cell differentiation stages and mechanisms.

Spermatogenic Lineage Development

Once specified and matured, primordial germ cells are ready to undergo sex-specific differentiation to form mature germ cells, giving rise to the spermatozoa in the male. In the embryonic stages, PGCs undergo development to form gonocytes, which enclose in testicular cords formed by Sertoli precursor cells and peritubular myoid cells [47,48]. In rodents, gonocytes proliferate for a few days, after which they become quiescent. Together, these cells form the seminiferous epithelium, the tight junctions between adjacent Sertoli cells making it into a basal compartment exposed to many lymph- and blood-borne substances and an adluminal compartment, to which blood-borne substances have limited direct access [49,50]. This blood-testis barrier is an immune privileged site, eventually containing late meiotic germ cells, postmeiotic spermatids, and spermatozoa, and relies on a specific stem cell population to replenish [49]. These are called spermatogonial stem cells (SSCs) and arise from the gonocytes, after they have migrated into the basement membrane during early postnatal development. This coincides with maturation of the surrounding Sertoli cells [49]. After the formation of a functional SSC niche, they divide into their daughter cells, the spermatogonia [49]. In humans, three kinds of spermatogonia are known: type A dark spermatogonia that form the reserve stem cells, type A pale spermatogonia that are considered self-renewing stem cells, and type B spermatogonia that will undergo another round of mitosis and develop further into spermatocytes [51].

The second phase of male germ cell development is initiated in the male testis at the beginning of puberty and is termed spermatogenesis. All its stages are dependent on the intricate interaction between the germ cells and the microenvironment provided by the somatic Sertoli cells [52]. The process initiates when the solid gonadal cord in the juvenile testis develops a lumen [53]. At puberty, spermatogonia at the periphery of the seminal canal advance their proliferation toward the lumen and form primary spermatocytes (spermatocytes I). Through a first meiotic division (meiosis I), primary spermatocytes can form two haploid secondary spermatocytes (spermatocytes II). A second meiotic division (meiosis II) then gives rise to four haploid spermatids. During a subsequent process called spermiogenesis, spermatids undergo morphological and structural alterations, such as acrosome and tail formation, chromosome condensation, and removal of the majority of the cytoplasm, to finally form mature haploid spermatozoa [52–54].

Male Infertility

Disruption at any stage of male germ cell development, either affecting the developing germ cells or their somatic niche cells, may put men at risk for infertility. Impaired germ cell development will result in the clinical presentation of abnormal sperm production, such as aspermia (the absence of semen), oligozoospermia (low sperm count in semen), or azoospermia (the absence of sperm in semen) [52,55]. Developmental errors affecting germline specification, spermatogonial lineage development, or damages to the gonadal niche may result in infertility. Chromosomal factors such as Y chromosome deletions, Klinefelter's syndrome, or specific genetic anomalies are some of the likely causative factors underlying these errors. Furthermore, endocrine disorders (e.g., Kallmann syndrome) or hereditary cases of cancer can also affect fertility. Nongenetic causes of infertility exist and include physiological problems, erectile dysfunction, infections, medical treatments (androgens, chemotherapy, and radiation), and immunologic factors (e.g., antisperm antibodies). However, frequently, the origin of male infertility cases are of unknown origin (idiopathic) [56].

In cases of predictive fertility loss owing to radiation and chemotherapy during cancer treatment, postpubescent men have the option to cryopreserve a semen sample before initiating treatment. This frozen sample can be thawed later to achieve pregnancy by intrauterine insemination, in vitro fertilization (IVF), or IVF with intracytoplasmic sperm injection (ICSI) [56]. For those who do not cryopreserve and have persistent azoospermia post cancer, retrieval of the rare sperm directly from the testis by testicular sperm extraction (TESE) followed by ICSI is an option [56,57]. However, for prepubescent patients and those who fail to show motile sperm after TESE, alternative strategies to create biological offspring need to be devised.

IN VITRO DERIVATION OF GERM CELL PRECURSORS FROM STEM CELLS

Stem Cells as Progenitors for Germ Cell Induction

Weissman's theory proposes that hereditary information moves only from germline to somatic cells and never in reverse [58]. However, recent advancements in the field of somatic cell nuclear transfer and induced pluripotent stem cells challenge the "Weissman barrier" [43,59]. Human germline differentiation has been traditionally modeled based on mouse, due to ethical considerations and limited availability of human samples. Assuming that both mice and human share some mechanisms for gametogenesis [60], human germline cells have been induced based on the principles of the orderly sequence of PGC formation in mice. However, the process has highlighted significant differences between factors involved in murine and human PGC specification, leaving many key regulators of the human germline yet to be identified.

Pluripotent stem cells, in general, can be programmed to induce germline fate. To serve as progenitors, the stem cells can be of embryonic origin (embryonic stem cells, ESCs), induced from mature somatic cells (induced pluripotent stem cells, iPSCs), or derived from adult stem cells designated to form germ cells (e.g., spermatogonial stem cells, SSCs) [4]. Access to surplus embryos from IVF clinics has allowed the study of embryonic development, germline specification, and subsequent differentiation of gametes from embryonic cells. However, reprogramming somatic cells to a pluripotent state would allow the creation of patient-specific gametes and is deemed as a more clinically translational approach to combat infertility. Alternatively, for patients who retain SSCs, induction of spermatogenesis in vitro may be possible (Fig. 17.3).

Embryonic Stem Cells in Different Pluripotent States

Mammalian development entails the regulated proliferation of a population of developmentally plastic, pluripotent stem cells making up the inner cell mass (ICM) of the blastocyst, which ultimately differentiates to give rise to the mature organism [61].

In mice embryos, during in vivo development, the E5.5–E6.0 epiblast progresses from a preimplantation epiblast to a postimplantation epiblast. Both the pre- and postimplantation epiblast can give rise to distinct pluripotent stem cells in vitro, ESCs and epiblast stem cells (EpiSCs), respectively [62,63]. ESCs represent the ground-state pluripotency of the ICM, at least in controlled media conditions including 2 inhibitors (2i, a Mek inhibitor and a GSK3β inhibitor) and human leukemia inhibitory factor (hLIF), and have been shown to contribute to all lineages when injected into blastocysts [64,65]. EpiSCs are also still pluripotent but reside in a more advanced transcriptional state toward differentiation [62,63,65]. To distinguish the two states of pluripotency, the terms "naive" and "primed" are applied [65]. Interestingly, although human ESCs are derived from the same preimplantation stage as mouse ESCs, they adopt the primed state of pluripotency during derivation and share more characteristics with EpiSCs than with mouse ESCs [65].

HESCs have a flat epithelial-like morphology, unlike the domed shape of naive mESCs [64,66,67]. This polar distribution of cells in hESC colony growth is sensitive to apoptosis, which causes them to have low single-cell survival after passaging, which severely hampers

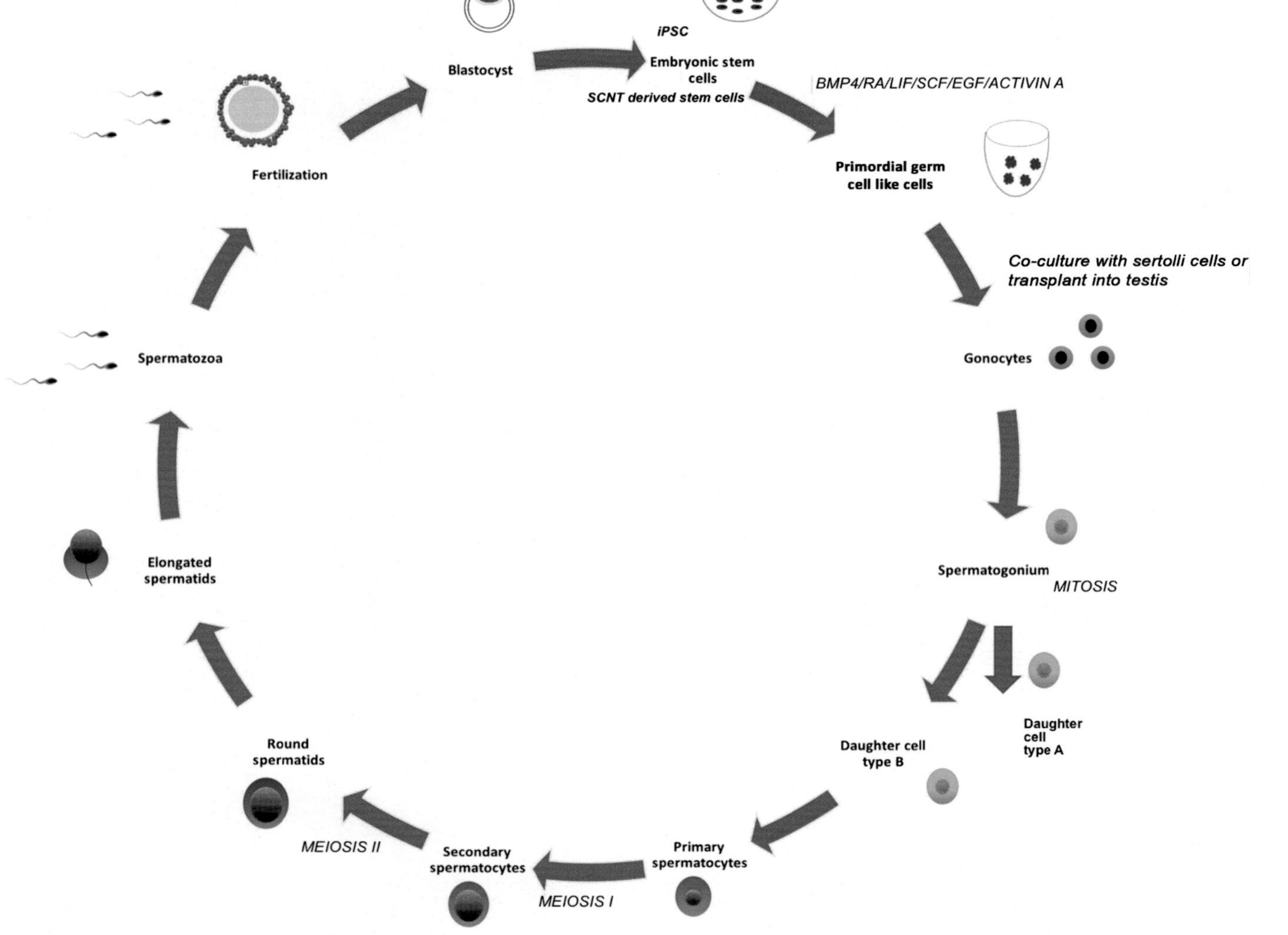

FIG. 17.3 Cyclic flowchart depicting in vitro derivation of spermatozoa from stem cells in mammals. Embryonic, induced pluripotent, or SCNT-derived stem cells can be differentiated to form PGC-LCs in the presence of factors such as BMP4, SCF, LIF, and EGF, with or without activin A or retinoic acid. The derived cells can be cocultured with male gonadal somatic cells to trigger the progress of the germline to derive healthy and viable spermatozoa. These SCD spermatozoa can be functionally validated by allowing them to fertilize oocytes. The blastocyst derived from this fertilization can be transplanted to generate healthy offspring or utilized in a cyclic fashion to derive embryonic stem cells.

efficient transgene transfer [65,68–70]. Different hESC lines also show various differentiation biases toward certain lineages, which is more controllable for naive ESCs [64,71–73]. Epigenetically, DNA from hESCs is hypermethylated, and chromatin accessibility is severely reduced through repressive H3K27me3 chromatin marks, compared with naive ESCs [68,69]. Female hESCs almost always have completed inactivation of one X chromosome, while both X chromosomes in naive ESCs remain active [69]. Primed and naive ESCs also differ in their culture environment requirements. Naive ESCs maintain self-renewal through LIF/Stat3 stimulation, and FGF/Erk signaling is not active, while primed hESCs depend on FGF/Erk signaling, combined with Nodal/activin signaling [62,63,74].

To counter some of the problematic properties inherent to primed pluripotency, with relation to future clinical application, different groups have tried and succeeded in adapting hESCs to a more mouse-like pluripotent state (without the need for transfection/transduction), obtaining variable degrees of naive characteristics [68,75–83].

O'Leary and colleagues identified a transient structure that arises during the derivation of primed hESCs from ICM cells, which they termed the post-ICM intermediate (PICMI). This structure could be isolated consistently, cryopreserved, and reintroduced into culture to give rise to hESCs. Gene expression comparison between hESCs, PICMIs, and ICMs revealed that the PICMI shared similarities with both the ICM and hESCs by expressing both early and late epiblast markers, validating it as a true intermediary. Interestingly, the PICMI also expressed germ cell-related marker genes, such as KIT and IFITM1. Though this suggests that germ cell lineage expression may be active during the derivation of primed hESCs from naive ICM cells, more robust investigation is warranted here [84,85].

Patient Specific PSCs: iPSCs or SCNT

To generate germ cells in vitro from stem cells for infertile patients, pluripotent stem cells need to be patient-specific. Two strategies are known to reprogram somatic cells into pluripotent stem cells; the creation of induced pluripotent stem cells (iPSCs) and somatic cell nuclear transfer (SCNT). The first study elucidating creation of "induced" pluripotent cells or iPSCs from somatic cells was based on the introduction of four reprogramming factors: Oct4, Sox2, Klf4, and cMyc [86]. While Oct4 (POU homeodomain factor1) is identified in ICM cells and ESCs as an essential marker of pluripotency [87], Sox2 is dispensable and can be replaced by other factors such as TGFβ inhibitors in mice models [88]. Kruppel-like zinc finger transcription factor 4 (Klf4) regulates the expression of Nanog and Sox2, while cMyc is an important oncogene that most prominently binds to promoters for cell cycle activating genes and pluripotent state-specific microRNAs [89]. Exogenous induction of these factors using viral or nonviral vectors can induce reprogramming. The first changes occurring are the downregulation of somatic cell markers, followed by upregulation of pluripotency markers, which assure stable pluripotency of the lines independent of the exogenous factors [89]. These reprogrammed pluripotent lines could then be directed to differentiate into germ cells. Alternatively, the patient's somatic cells may be directly programmed to overexpress germ cell markers in addition to pluripotency genes to trigger the cells to induce a germ cell fate to the cells [90].

Another route to derive patient-specific stem cells is through somatic cell nuclear transfer (SCNT). For this, the nucleus of a somatic cell is transferred to the cytoplasm of an enucleated oocyte. With very low efficiency, the somatic nucleus may then be reprogrammed by maternal factors within the oocyte to continue embryonic development toward the formation of a blastocyst. From this SCNT-blastocyst, ESCs could eventually be derived. These ESCs would contain the same genetic background as the somatic cell donor and can be induced to form SCD gametes specific to the donor. The possibility of immune rejection of these SCD gametes when transplanted to the donor is greatly reduced since the nuclear DNA would be identical to the donor [91]. If done stably, this process could provide a safer alternative to iPSCs by eliminating the need for designing reprogramming vectors. This may also help solve the issue of incomplete epigenetic reprogramming as observed more frequently in iPSC lines than in stem cells of embryonic origin [92–94].

Progress in Mouse

Differentiation From SSCs

Many groups have focused on the generation of spermatozoa from adult spermatogonial stem cells (SSCs) present in neonatal mouse testes [95–110] (Table 17.1). Most prominently, Sato and colleagues succeeded in culturing these SSCs in vitro starting from neonatal mouse testicular fragments and differentiating them further to functional spermatids and sperm [95]. They obtained testis tissue fragments from 2.5–3.5-day-old mice that contained only undifferentiated SSCs. Spermatogenesis was induced and maintained for more than 2 months in these testes fragments, through positioning in a gas-liquid interphase on agarose gel, originally devised to promote organ survival in vitro [136]. Haploid round spermatids or sperm cells were generated after 23 or 42 days of culture, respectively. The obtained spermatids and sperm cells were transferred by microinsemination to female mice and generated healthy and fertile offspring for the first time [95]. Neonatal testis tissues were also cryopreserved and successfully thawed, showing complete spermatogenesis in vitro, although the fertilization potential of that sperm was not investigated [95]. Efforts to isolate and differentiate human SSCs toward sperm cells are considerably more scarce and have not yet generated comparable results comparable to reports in mouse (Table 17.1).

Differentiation From Mouse Pluripotent Stem Cells

GENERATION OF GERM CELL PRECURSORS BY SPONTANEOUS DIFFERENTIATION

Preliminary efforts to generate PGCs in vitro from mouse ESCs have relied on germline-specific (e.g., VASA) or meiotic markers (e.g., SCP3) to identify potential PGC cells after spontaneous differentiation. Cell populations positive for these germ cell markers have been obtained, although at a very low efficiency (±1%) [13]. Nevertheless, it was established that in response to BMP4 presence, all E5.5–E6.0 epiblast cells are competent to express PRDM1 and PRDM14 and even hold the capacity to form functional sperm [117].

EPISCS AS A STARTING POINT

There have been attempts to generate PGCs from EpiSCs as well, since they retain properties of the in vivo epiblast. A small subpopulation of EpiSCs even expresses PRDM1 and DPPA3, which already indicates progression toward germ cell fate. However, even when a BMP4 signal is provided, these cells only constitute a tiny fraction of the total cell population (±1.5%) [21]. This may suggest that the primed state of EpiSCs has progressed too far, in order to allow efficient PGC differentiation [117].

TABLE 17.1 Overview of Literature Describing In Vitro Derivation of Male Germ Cells

	First Author	Year	Differentiation Conditions	Early PGC Induction	Late PGC Stage	Meiotic Marker(s)	Haploid Cells (1N)	Spermatozoa	Live Offspring	Fertile Offspring	Reference
In vitro differentiation of SSCs											
Mouse	Nagano	2000	Transplantation	/	/	–	–	X	–	–	[96]
	Nagano	2001	Transplantation	/	/	–	–	X	X	–	[97]
	Kanatsu-Shinohara	2003	Culture + transplantation	/	/	–	–	X	X	X	[99]
	Kubota	2004	Culture + transplantation	/	/	–	–	X	X	X	[106]
	Kanatsu-Shinohara	2005a	Culture + transplantation	/	/	–	–	X	X	X	[100]
	Kanatsu-Shinohara	2005b	Culture + transplantation	/	/	–	–	X	X	X	[101]
	Kanatsu-Shinohara	2006	Culture + transplantation	/	/	–	–	X	X	X	[102]
	Kita	2007	Culture + transplantation	/	/	–	X	–	X	–	[108]
	Kubota	2009	Culture + transplantation	/	/	–	X	X	X	X	[107]
	Ohta	2009	Culture + transplantation	/	/	–	–	X	X	–	[109]
	Sato	2011	Culture	/	/	X	X	X	X	X	[95]
	Kanatsu-Shinohara	2011	Culture + transplantation	/	/	–	–	X	X	X	[104]
	Shiura	2013	Culture + transplantation	/	/	X	–	X	X	X	[110]
Human	Riboldi	2012	Coculture	/	/	X	X	–	/	/	[111]
	Yang	2014	Adherent culture	/	/	X	X	–	/	/	[112]
In vitro differentiation of ESCs											
Mouse	Geijsen	2003	Embryoid bodies + transplantation	–	–	–	X	X	–	–	[113]
	Toyooka	2003	Embryoid bodies + transplantation	–	–	–	–	X	–	–	[114]
	Nayernia	2006	Adherent culture + transplantation	–	X	X	X	X	X	–	[93]
	Kerkis	2007	Embryoid bodies + transplantation	–	X	X	X	X	–	–	[115]

	Yu	2009	Adherent culture	X	X	X	X	X	–	–	[116]
	Ohinata	2009	Embryoid body + coculture + transplantation	X	–	–	X	X	X	X	[117]
	Hayashi	2011	Adherent culture + coculture + transplantation	X	X	–	–	X	X	X	[118]
	Nakaki	2013	Adherent culture + coculture + transplantation	X	–	–	–	X	X	X	[119]
	Murakami	2016	Adherent culture	X	X	–	–	–	–	–	[120]
	Zhou	2016	Adherent culture + coculture + transplantation	X	X	X	X	–	X	X	[121]
Human	Clark	2004	Embryoid body	–	X	X	–	–	/	/	[122]
	Park	2009	Adherent culture	X	X	–	–	–	/	/	[123]
	Aflatoonian	2009	Embryoid body	–	X	X	X	–	/	/	[124]
	Kee	2009	Adherent culture	X	X	X	X	–	/	/	[125]
	Panula	2011	Adherent culture	X	X	X	–	–	/	/	[126]
	West	2011	Adherent culture	–	X	X	X	–	/	/	[127]
	Eguizabal	2011	Adherent culture	–	X	X	X	–	/	/	[92]
	Easley	2012	Adherent culture	–	X	X	X	–	/	/	[128]
	Medrano	2012	Adherent culture	X	X	X	–	–	/	/	[129]
	Duggal	2013	Embryoid body	–	X	–	–	–	/	/	[130]
	Irie	2013	Adherent culture	X	–	–	–	–	/	/	[43]
	Gkountela	2013	Adherent culture + embryoid body + coculture	X	X	X	–	–	/	/	[131]
	Duggal	2014	Embryoid body	–	X	X	–	–	/	/	[132]
	Sugawa	2015	Adherent culture	X	–	–	–	–	/	/	[133]

Continued

TABLE 17.1 Overview of Literature Describing In Vitro Derivation of Male Germ Cells—cont'd

	First Author	Year	Differentiation Conditions	Early PGC Induction	Late PGC Stage	Meiotic Marker(s)	Haploid Cells (1N)	Spermatozoa	Live Offspring	Fertile Offspring	Reference
In vitro differentiation of iPSCs											
Human	Park	2009	Adherent culture + coculture	X	X	–	–	–	/	/	[123]
	Panula	2011	Adherent culture	X	X	X	–	–	/	/	[126]
	Eguizabal	2011	Adherent culture	–	X	X	X	–	/	/	[92]
	Easley	2012	Adherent culture	–	X	X	X	–	/	/	[128]
	Medrano	2012	Adherent culture	X	X	X	–	–	/	/	[129]
	Irie	2013	Adherent culture	X	–	–	–	–	/	/	[43]
	Sugawa	2015	Adherent culture	X	–	–	–	–	/	/	[133]
	Sasaki	2015	Adherent culture	X	–	–	–	–	/	/	[134]
	Ge	2015	Embryoid body	–	X	X	X	–	/	/	[135]
In vitro differentiation of somatic cells											
Human	Medrano	2016	Adherent culture	X	X	X	X	–	/	/	[90]

X, achieved; –, not achieved; and /, not applicable. Early PGC induction = study PGC specification toward PGC-LCs, showing expression of early germ cell markers, such as PRDM1 (Blimp1), PRDM14, NANOS3, and TFAP2C (AP2y). Late PGC stage = VASA/DAZL expression shown or later stages shown. Meiotic markers = SYCP1 and SYCP3.

ESCS AS A STARTING POINT

Other groups have focused on directed differentiation from mouse ESCs [93,113–115,117–119,137] (Table 17.1). Nayernia and colleagues were the first ones to obtain live offspring after germ cell differentiation from mESCs in vitro. Their strategy depended upon the use of reporter mESC lines bearing fluorescent reporter genes GFP and DsRed under the control of spermatogonia (Stra8) and postmeiotic spermatid-specific (Prm1) promotors, respectively. MESCs that were initially maintained in LIF and serum conditions were differentiating upon removal of LIF and addition of retinoic acid. Eventually, the team of Nayernia succeeded in the differentiation of mESCs toward functional germ cell precursors, including spermatogenesis and meiosis, ultimately generating cells with a spermatid-like morphology. The resulting sperm-like cells were used to fertilize donor oocytes and resulted in several pregnancies. However, most offspring died before birth and the remainder a few months after birth. They showed that certain imprinting regions were incompletely unmethylated in the male spermatid-like cells, which is a likely explanation for the observed birth complications [93].

INDUCTION OF PGC-LIKE CELLS IN VITRO

Hayashi and colleagues were the first group to establish a culture system in which they could control the advancement from naive state mouse ESCs to an early pregastrulating epiblast-like state, before PGC induction [118]. For this, they cultured ESCs in low-serum culture conditions, with addition of activin A and basic fibroblast growth factor (bFGF). This effectively converted their ESCs into pregastrulating epiblast-like cells, bearing competence for the PGC fate. Analysis of gene expression showed an expression profile similar to pregastrulating epiblast, while distinct from EpiSCs. Putative epiblast-like cells (Epi-LCs) were consequently differentiated to PGC-like cells (PGC-LCs), based on evaluation of previously introduced PRDM1- and DPPA3-reporter constructs. They found that the highest reporter expression could be obtained primarily through BMP4 induction from day 2 Epi-LCs onward, enhanced by combined presence of growth factors LIF, SCF, and EGF, as well as BMP8b. Gene expression analysis validated the profile of PGC-LCs as being similar to what is expected for PGC specification, with upregulation of PGC induction markers, elevation of pluripotency genes, and downregulation of mesodermal marker genes. At a slightly later stage of development, they confirmed that the PGC-LCs could be fully considered established PGC-LCs.

On a more global transcriptomic scale, they analyzed gene expression comparatively between ESCs, Epi-LCs, EpiSCs, epiblasts, and PGC-LCs, through microarray analysis. Principle component analysis (PCA) indicated the directional progression from ESC, through Epi-LC induction and toward PGC specification. Unsupervised hierarchical clustering of amplified samples showed clear similarities between Epi-LCs and epiblast samples and separately between PGC-LCs and in vivo E9.5 PGCs. These findings soundly verified established PGC-LC formation from ESCs, through Epi-LCs. Epigenetically, the PGC-LCs were shown to have increased HRK27me3, reduced H3K9me2 and cytosine methylation (5mC) levels.

Finally, they succeeded in recapitulating PGC-LC induction from iPSCs and successfully induced spermatogenesis after introduction of PRDM1-positive PGC-LCs into germ cell-depleted neonatal mouse testes in vivo. Spermatozoa were generated, the resultant sperm was used to fertilize mouse oocytes through ICSI and generated live and fertile offspring. Some ESC lines were unable to reach up to this stage, others formed teratomas, and some of the offspring died prematurely from tumor formation in the neck region. Nonetheless, they showed that iPSCs can be induced to form PGC-LCs with fertilization potential for fertile offspring.

UNCOVERING THE MOLECULAR MECHANISMS GOVERNING PGC INDUCTION

Despite successful induction of germ cell fate from mice stem cells, the precise mechanisms of the cytokines involved in PGC induction and transcription factors that were used to validate establishment of PGC-LC cells remained unclear. Nakaki and colleagues demonstrated that without the presence of cytokines, like BMP4, simultaneous overexpression of three transcription factors (either in ESCs or in Epi-LCs), PRDM1, PRDM14, and TFAP2C (also known as AP2γ), directly induces Epi-LCs (but not ESCs) efficiently into a PGC state in a synergistic way. Moreover, PRDM14 overexpression alone, without PRDM1 or TFAP2C, was shown to be sufficient for the induction of the PGC-LCs. PRDM1-positive PGC-LCs obtained by overexpression of PRDM1, PRDM14, and TFAP2C were also transplanted into germ cell-depleted neonatal mouse testes and gave rise to functional spermatozoa. Viable fertile offspring was generated from these spermatozoa as well [119]. The authors outlined a basic transcriptional model in which PRDM14 and PRDM1 play a central cooperative role in inducing core PGC genes and maintaining pluripotency genes. Furthermore, PRDM1 plays a pivotal role in the inhibition of the mesoderm differentiation program, while PRDM14 downregulates genes associated with neural differentiation. PRDM1 and PRDM14 both enhance expression of each other and expression of TFAP2C, which role seems to be supplementary, compared to the other two, for PGC fate induction.

THE ROLE OF PLURIPOTENCY MARKERS IN PGC INDUCTION

Murakami and colleagues expanded our understanding of germ cell formation further by uncovering molecular transitions from pluripotency maintenance toward PGC specification [120]. They showed that the core pluripotency factor Nanog alone can induce PGC-LCs from Epi-LCs, independently of BMP4 presence. Specifically, Nanog binds to enhancer elements of PRDM1 and PRDM14, thereby indicating epigenetic regulatory changes. In the changing molecular environment from the inner cell mass (ICM) in the preimplantation blastocyst to the late postimplantation epiblast, the role of Nanog seems to transform from, respectively, being indispensable for pluripotency maintenance to inducing PGC-specification genes epigenetically. These epigenetic changes were not only found to influence selected PGC marker genes but also uncovered on a genome-wide scale. Interestingly, SOX2, which is another core pluripotency maintenance gene, seems to contrast the role of Nanog by repressing PGC-LC specification. In vivo, Nanog is undetectable in postimplantation mouse embryos from E6.25 to E7.75, just after the time where PGC specification has occurred and until the time where PGC precursor cells are migrating to the pregonadal region [26], suggesting Nanog is downregulated after its role in PGC specification is fulfilled.

NECESSARY REQUIREMENTS FOR MEIOSIS TO OCCUR

One of the major undefined parts of the germ cell specification protocol, as outlined by Hayashi and colleagues, was the necessity for maturation of transplanted PGC-LCs in neonatal testes lacking endogenous germ cells. Potential translation of this method to human ESCs will require a defined in vitro differentiation system from ESCs to functional spermatozoa. Not least challenging is the recapitulation of meiosis in vitro, which is a unique process to germ cells. A panel of gold standards has been defined to reflect key features of meiosis that need to be attained to warrant successful in vitro derived gametes [138]. For successful meiosis to occur, several processes must be established: correct nuclear DNA content at different meiotic stages (for male cells, premeiotic, meiotic, S-phase, first reductional, and second meiotic division stages), normal chromosome number and organization, appropriate nuclear and chromosomal localization of proteins involved in homologous synapsis and recombination, and capacity of the in vitro-produced germ cells to produce viable euploid offspring.

IN VITRO GENERATION OF MEIOTIC SPERMATID-LIKE CELLS

In 2016, Zhou and colleagues succeeded in in vitro recapitulation of these key elements of meiosis during male germ cell differentiation [121]. Furthermore, their resulting spermatid-like cells (SLCs) were shown to be capable of producing viable fertile offspring. Their strategy involved the generation of transgenic ESC lines containing fluorescent reporters under the control of key germ cell specification genes, from different time points of the development process. PRDM1, DPPA3, SSEA1, and B3-integrin promotors were used to identify PGC-LCs; STRA8 promotors were used to distinguish early-stage spermatogonia; and PRM1 promotors were used to identify postmeiotic spermatids. The authors based their initial differentiation approach on a PGC-LC formation protocol starting from naive ESCs, equivalent to what was described by Hayashi and colleagues, validated by expression of the PRDM1/DPPA3 reporter system, and through gene expression analysis and histone modification analysis. Expanding upon Hayashi and colleagues, the authors then FACS-sorted PGC-LCs double positive for SSEA1 and B-integrin and brought them into coculture with neonatal somatic testicular cells from germ cell-deficient mice. These were subjected to a combination of activin A, BMPs, and retinoic acid (RA), which resulted in cell populations positive for STRA8-promotor activity, suggesting the initiation of meiosis. Follow-up replacement of morphogens by the sex hormones FSH, BPE, and testosterone consequently generated the formation of postmeiotic Prm1-expressing SLCs. Key elements of successful meiosis were shown to have occurred in these cells, such as erasure of genetic imprinting, chromosomal synapsis and recombination, and single chromosomal DNA content, which solidifies their identity of male haploid SLCs. Although fully mature sperm was not obtained, these SLCs were also able to generate fertile normal offspring, after ICSI and embryo transfer was performed with donor mouse oocytes [121].

Progress in Human

Differentiation From SSCs

Due to ethical considerations and limited access to research material, direct study of in vivo PGC development in early human embryos is not possible. Some groups did study the differentiation of human SSCs, obtained from male infertile patients, either by coculture with Sertoli cells [111] or through retinoic acid-induced differentiation [112]. In both cases, haploid sperm precursor cells that expressed meiotic markers were obtained [111,112].

Differentiation From Human Pluripotent Stem Cells

SPONTANEOUS DIFFERENTIATION TOWARD HUMAN GERM CELL PRECURSORS

To date, most research into human germ cell differentiation has focused on directed differentiation of human pluripotent stem cell lines, either embryonic (hESCs) or induced (hiPSCs), toward germ cell-like cells [43,90,92,122–135,139,140] (Table 17.1). As with mouse, spontaneous differentiation from human ESCs toward PGC precursor cells was initially described but with low efficiency and limited understanding of underlying molecular mechanisms [122,131,141,142]. These studies did provide a proof of concept that demonstrates differentiation toward human germ cell precursors in vitro is possible [4,122,131].

ECTOPIC EXPRESSION OF GERM CELL MARKERS

Other groups succeeded in reaching later stages of PGC development through ectopic expression of late PGC marker genes, such as members of the DAZ gene family. Kee and colleagues established the direct differentiation of hESCs toward haploid germ-cell-like cells through ectopic expression of DAZL, DAZ, and BOULE genes [125]. The resulting cells were shown to be VASA positive and enriched for DAZL, PRDM1, and DPPA3. However, specific meiotic markers were absent, except for low levels of SYCP3, suggesting the resulting cells to reside in a premeiotic stage. Moreover, global epigenetic methylation levels were shown to be reduced, which is concordant with germ cell development from an earlier hypermethylated state.

MEIOTIC PROGRESSION BY MINIMAL OVEREXPRESSION OF GERM CELL MARKER GENES

A number of other studies have similarly demonstrated that meiotic germ cell-like cells can be obtained from hESCs or iPSCs from varying sources, through ectopic overexpression of germline marker genes [126,129]. Medrano et al. further demonstrated that haploid germ cell-like cells can also be obtained from adult somatic cells, provided the right germ cell factors are ectopically expressed. They identified six essential germ cell development factors by introducing them into lentiviral vectors for transduction. Ectopic expression of PRDM1, PRDM14, LIN28A, DAZL, VASA, and SYCP3 was minimally sufficient to induce germ cell fate from somatic human foreskin fibroblasts [135]. Induced germ cell-like cells demonstrated a transcriptomic switch toward germ cell expression profile, with expression of multiple postmeiotic germline marker genes. Moreover, they showed meiotic progression and occurrence of epigenetic reprogramming. Roughly 1% of germ cell-like cells achieved haploid status, validating complete meiosis. Moreover, they were able to xenotransplant a subset of induced cells, showing successful colonization of the spermatogonial niche in nude mice [90].

HAPLOID SPERMATID-LIKE CELLS FROM IPSCS BY RETINOIC ACID INDUCED DIFFERENTIATION

Eguizabal and colleagues used induced pluripotent stem cell lines and spontaneous differentiation of monolayers by removal of bFGF, followed by addition of retinoic acid. They were able to isolate $VASA^{+}/SSEA\text{-}1^{-}$ cells by FACS analysis, which indicated progression from the PGC state toward premeiotic cells. Surrounding cells also indicated expression of Sertoli and Leydig cell marker genes, suggesting possible reconstruction of a testicular niche. A subset of cells showed expression of spermatogonial markers, such as CD9 and CD49f, of which a fraction was VASA positive, which indicated that their protocol could generate spermatogonia-like cells. Further culture of their spermatogonial-like cell population in the presence of hLIF, FRSK, bFGF, and R115866, a CYP26 inhibitor yielded a subset of cells positive for the meiotic markers SCP3 and γH2AX. Further confirmation of meiosis was provided by detecting meiotic prophase I and expression of the premeiotic gene STRA8, confirming haploidy and validation of imprinting establishment. Although functional spermatozoa cells were not yet obtained, Eguizabal and colleagues thus described a protocol for the generation of haploid spermatid-like cells, from iPSCs, over a spontaneous, but stepwise guided differentiation period of 10 weeks [92].

GENERATION OF POSTMEIOTIC CELLS USING SSC CONDITIONS

Easley and colleagues described a method to directly differentiate pluripotent stem cells into haploid germ-like cells in vitro, in only 10 days of time. Pluripotent cells developed into spermatogonia-like cells, spermatocyte-like cells, and spermatid-like cells. To achieve this, they cultured hESCs and iPSCs in SSC conditions, inspired from mouse SSC proliferation conditions [99]. The resulting cells were shown to be mostly positive for VASA and DAZL. High levels of nuclear PLZF also indicated that these cells generated progenitor spermatogonia. A small percentage of cells was shown to be haploid and harbored polar acrosin localization, suggesting their postmeiotic identity. Furthermore, few cells also expressed meiotic marker SYCP3 and postmeiotic marker HIWI, indicative of a bottleneck for meiotic progression. Acrosin-positive cells were isolated and showed nuclear/perinuclear presence of protamine 1 and transition protein 1, which are markers for round spermatid identity. The haploid cells also showed paternal imprints for H19 and IGF2 [128]. Although this study generated putative round spermatid-like cells, complete spermatogenesis was not attained. Furthermore, 10 days of differentiation is extremely short, compared with the classical length of spermatogenesis in human (74 days). Longer culture time did also not result in more haploid cells, and the percentage even decreased after 20 days of culture. Nevertheless, Easley successfully generated spermatid-like cells, coming relatively close to deriving mature spermatozoa.

CHEMICALLY DEFINED STEP-WISE GERM CELL FORMATION

Other groups have focused on a more defined, stepwise approach to differentiate hESCs toward PGC-LCs. Initial steps toward this end were taking by Ge and colleagues, who successfully generated early germ-like cells from human fetus skin-derived stem cells, which, after culture in porcine follicle fluid (PFF) conditioned media, showed expression of VASA and DAZL. They also showed expression of the meiotic marker SCP3 [142a].

Irie and colleagues started from a pluripotent condition resembling the one described by Gafni et al. that was shown to maintain hESCs and iPSCs in a naive state [43,68]. To follow the differentiation process, they used reporter hESC lines for the PGC-specific marker gene NANOS3. These cells were subjected to a PGC-LC differentiation protocol, similar to the approach described by Hayashi et al. for mESCs. Cells positive for the NANOS3 reporter and for the pluripotency and germ cell marker nonspecific alkaline phosphatase (TNAP) were obtained. Other key PGC genes were also expressed in these cells, among which were PRDM1, TFAP2C, DPPA3, and CKIT. They were also able to identify strong upregulation, compared with hESCs, of the endodermal lineage marker SOX17.

Time-course analysis showed that the expression of SOX17 occurs before the expression of PRDM1 during PGC specification from hESCs. Moreover, they identified SOX17 as a key inducer of PRDM1. SOX17 was shown to be both necessary and sufficient to induce PGC-LC differentiation from hESCs on its own, suggesting it to be a major regulator of overall hPGC-LC differentiation. This is not the case during mouse PGC-LC formation where SOX17 does not appear to play a role during PGC differentiation [139,143]. This underlines the occurrence of interspecies molecular differences and similarities in developmental events and their recapitulation in vitro.

When hESCs lacking SOX17 expression were subjected to the same PGC-LC differentiation protocol, they were unable to identify the molecular signature indicative of hPGC-LCs. Conversely, introduction of ectopic SOX17 expression into these SOX17-negative cells could rescue the successful generation of PGC-LCs, as validated by expression of PGC-LC marker genes. Transcriptomic comparison among hPGC-LCs, embryonic gonadal hPGCs (week 7 male), and a seminoma line validated a likely PGC-LC identity for their cells. As another hallmark of PGC specification, hPGC-LCs also showed evidence for the initiation of epigenetic changes and DNA demethylation, comparable with E8 mouse PGCs. Overall, Irie et al. greatly advanced our understanding of human PGC development and provided a framework for the study of in vitro germ cell differentiation.

SUPPLEMENTATION WITH ACTIVIN A PROMOTES GERM CELL INDUCTION

Research published by Duggal and colleagues showed an increased propensity of embryonic stem cells derived in activin A toward germ cell lineage. Additional supplementation of activin A in the in vitro germ cell differentiation medium upregulated early germ-cell markers such as STELLA/DPPA3. Supplementation of BMP4 along with activin A provided an additional boost to the culture allowing the expression of late germ cell markers such as STELLA and DAZL. The obtained PGC-LCs were further matured in an in vitro maturation (IVM) medium resulting in the formation of germ cell-like clusters and induction of meiotic gene expression. VASA expression was also shown to occur, with both cytoplasmic and nuclear localization [130,132].

MESODERMAL LINEAGE INDUCTION PRECEDING GERM CELL INDUCTION

Another study, by Sugawa and colleagues, used a slightly different but ultimately quite similar strategy compared with Irie et al., to differentiate hESCs to PGC-LCs in vitro [133]. Their approach was based on pre-differentiation of hESCs to primitive streak/mesodermal-like precursor cells, which has been shown to occur in mice prior to PGC specification. The authors argued that identification of early marker genes that are expressed before PGC specification is necessary to truly understand PGC differentiation. To this end, they pre-differentiated hESCs in serum-free culture conditions, with added activin A, bFGF, and BMP4. Contrary to earlier studies, they first evaluated the expression of mesodermal lineage marker gene Brachyury (T). T was seen to be rapidly upregulated after just one day, followed by upregulation of PRDM1 on the second day. Subsequently, the mesodermal-like precursor cells were cocultured as aggregates in PGC-LC differentiation media, with additional BMP4, LIF, and ROCK inhibitor. To isolate PGC-LCs, the authors FACS sorted the cells that were double positive for CKIT and TRA-1-81, by which they acquired a high fraction of PGC-LCs at day 4 and 6 of differentiation. Transcriptomic comparison revealed transcriptional similarities with 16-week prenatal human germline cells. Similar to Irie et al., they also saw gradual increase of NANOS3 expression during PGC-LC differentiation. Unlike mouse ESCs, the human PGC-LCs had very low levels of PRDM14, suggesting that PRDM14 might be dispensable for human germline development [133]. To validate this notion, knockdown experiments were performed on day 2 precursor cells for the genes PRDM1 and PRDM14. For PRDM1, this caused a significant reduction in the number of PGC-LCs that could be generated comparatively, while knockdown of PRDM14 did not change the induction efficiency of

PGC-LCs at all. This serves as another example of species-specific differences between human and mouse PGC-LC induction, demonstrating that extrapolation is not always possible.

INCIPIENT MESODERM-LIKE CELLS VERSUS EPIBLAST-LIKE CELLS AS PRECURSORS FOR PGC-LIKE CELLS

Another approach was described by Sasaki and colleagues [134]. They used a transgenic hiPSC line bearing a double reporter system, using promotors of PRDM1 and TFAP2C. The authors expanded on the epiblast-like cell differentiation section of the PGC-LC differentiation protocol described by Hayashi and colleagues. They added the Wnt agonist CHIR99021 to the EpiLC induction medium and found that hiPSCs could be induced into flat epithelial-like cells with distinct cell-to-cell boundaries. The flat epiblast-like cells were shown to harbor modest mesodermal lineage marker gene expression, such as T, EOMES, SP5, and MIXL1, which prompted the authors to designate these cells as incipient mesoderm-like cells (iMeLCs). After progression through the coculture aggregate formation and appropriate cytokine stimulation, these epiblast-like cells efficiently generated cells positive for both PRDM1 and TFAP2C as early as day 2 and around 30%–40% of all cells, which increased up to 60% in the following days. Global transcriptional and epigenetic analysis validated that the PRDM1/TFAP2C-positive cells were very similar to early hPGCs and showed competence for the germ cell fate.

The authors also compared their hPGC-LC expression profiles to the publications of Sugawa et al. and Irie et al. They observed major fold expression differences when comparing their PGC-LC transcriptome with the study of Sugawa et al. This was much less the case when comparing to Irie et al. This prompted the authors to argue that the hPGC-LCs obtained by Sugawa et al. might not be properly sorted or induced. More similar expression profiles were demonstrated with the hPGC-LCs obtained by Irie et al. However, the authors also pointed out that the hESCs of Irie et al. do not show proper ground-state pluripotency gene expression profiles. A possible uniting explanation could be that the hESC condition used in Irie et al. has been adapted from the original condition described by Gafni et al. in such a way that it is no longer fully supporting ground-state pluripotency and is in fact more similar to a pregastrulating epiblast-like state, such as the iMeLCs described by Sasaki et al. However, ultimately, their hPGC-LC cells did not show expression of late PGC markers DAZL and VASA and thus also correspond to early hPGCs [134].

DIFFERENCES IN PGC-LC GENERATION BETWEEN HUMAN AND MOUSE

A detailed comparative analysis of DNA methylation landscapes in PGC-LCs of both mouse and human origin was laid out in a study by von Meyenn et al. [144]. Human PGC-LCs were induced from naive hESCs using the strategy as defined by Hayashi et al. [118], while a GFP reporter for the OCT4-dPE promoter allowed tracking of the progression of hPGC-LCs development in culture. The same protocol was adopted for mice, using a Blimp1-Venus reporter to track the mPGC-LC progression. RNA sequencing analysis comparing CKiT or SSEA positive PGC-LCs with naive ESC, primed ESC and EpiLCs, confirmed that the PGC-LCs clustered separately from the stem cells. While PGC-LCs from mice showed a preference to cluster separately by time points, hPGC-LCs had a more mixed population indicating smaller changes. This was confirmed again by bisulfite sequencing, which showed that demethylation in

PGC-LCs is a much slower and gradual process in human compared with mice, depicting different developmental timings. A similar profile was observed in the in vivo PGCs as well, although contrary to mouse, some subtle differences were observed with hPGC-LCs, particularly for the repression of UHRF1. Similarly, an increased expression of KDM3A and KDM3B was observed in hPGC's but not in hPGC-LC's. Overall, the study compared early germline development in mice and human in both in vivo and in vitro models and concluded that global epigenetic reprogramming is about five times faster in mice than in humans.

Safety and Ethical Concerns of Stem Cell-Derived Gametes

While the prospect of deriving patient-specific gametes from pluripotent stem cells offers a new alternative route to genetic parenthood, current state of the art in vitro germ cell derivation techniques are still insufficient. Without efficient and complete protocols to derive human gametes and stringent guidelines to oversee and assess their safety and quality, the risks and ethical implications are high. Moreover, robust functional assays need to be designed to ascertain the viability and fertility of these gametes before any clinical applications can be tested.

In principle, healthy and fertile SCD gametes should exhibit the following properties: (a) a normal chromosomal and genetic profile; (b) a normal epigenetic profile, assuring complete imprint erasure; (c) the ability to proliferate in culture and colonize gonads on transplantation; (d) no risk of accumulation of chromosomal abnormalities or a tumorigenic profile while in culture; (e) no risk of inducing an immunogenic reaction if used in allogenic transplants; and, most importantly, (f) the ability to fertilize and result in the birth of normal, healthy, and fertile offspring. A secure method for ascertaining these properties in SCD gametes is to compare their profiles to normal and potent natural gametes. By estimate, the closer they are to their in vivo counterparts, the safer they should be.

Studies in mice, although more advanced than the human model, have reported a very low efficiency in deriving healthy and fertile progeny. Nayernia and colleagues, who published the first paper demonstrating that mESC-derived gametes could lead to viable mice, tested the functional aspects of their SCD spermatozoa by ICSI, presumably because their sperm cells exhibited low motility unable to fertilize oocytes on their own [93,145]. Progeny obtained from this fertilization was smaller or larger than the controls, unhealthy, and died between 5 days and 5 months after birth. At this point, it was unknown if epigenetic reprogramming occurred adequately when deriving the gametes from stem cells in culture. To test whether erasure of genomic imprinting was manifested, they tested the methylation status of three imprinted genes with well-defined methylated regions (imprinting control region (ICR) of H19, the differentially methylated region 2 (DMR) of Igf2r, and DMR1 of the small nuclear ribonucleoprotein N (Snrpn) genes). The results showed that imprints were efficiently erased at many, but not all alleles of two of the three analyzed genes. It was postulated that this incomplete imprint erasure was possibly the reason for the improper growth and poor viability of the mice [145]. Hayashi and colleagues reported birth of fertile progeny from transplanted mPGC-LCs derived from induced Epi-LCs, but many of the offspring would die later mostly due to the emergence of tumors around their neck region, a phenomenon also observed in a study by Okita et al. [94,118]. Teratomas were also observed in most of the embryos obtained. Both studies reported abnormal methylation patterns in the induced cell lines.

In humans, in an attempt to derive haploid postmeiotic cells from hESCs and hiPSCs, Eguizabal and colleagues tested six imprinted genes: the maternally inherited expressed H19, the pleckstrin homology-like domain family A member 2 (PHLDA2) and cyclin-dependent kinase inhibitor 1C (CDKN1C) along with the paternally inherited mesoderm-specific transcript homologue (MEST), insulin-like growth factor 2 (IGF2), neuronatin (NNAT), and small nuclear ribonucleoprotein polypeptide N (SNRPN) [92]. A decrease in H19 methylation in one of the lines depicted faulty imprinting establishment, while in others, although a propensity toward monoparental imprinting was observed, the process was deemed incomplete. Another study by Park et al. reported that SCD gametes derived in hiPSC lines were incapable of initiating imprinting erasures, while hESC lines had inconsistent methylation patterns in a few loci, although they initiated the process [123].

Ethical constraints on creating embryos for research in most countries imply that functional aspects of SCD-derived human gametes may not be as clearly tested as in mice. However, in certain countries, such as Belgium, functional testing is permitted for diagnostic purposes. The mouse oocyte activation test (MOAT) is a test offered to patients with low fertility to test the capability of sperm to induce oocyte activation [146]. Human "test" sperm is injected by ICSI into mouse oocytes, and the percentage of oocyte activation is measured by analyzing the increase in the oocyte activation marker, that is, the intracellular calcium concentration. Thus, SCD sperm derived in the laboratory could be functionally tested using MOAT. Additionally, where donor oocytes are available for research, the potential of SCD sperm to activate human oocytes could be analyzed, thus providing a more robust functional assay [147].

Continued progress in deriving human SCD gametes will inevitably endure safety concerns and ethical debates. However, creation of SCD gametes for the purpose of reproduction, like most treatments for infertility, is grounded on the high value attributed to genetic parenthood [148]. While this does not ensure "good" parenthood, it is generally perceived as a basic human right, granting a valuable opportunity to both heterosexual and same-sex couples [148].

CONCLUSION

In vitro germ cell differentiation studies lay the groundwork for future human cell-based germ cell models that go beyond PGC-LC differentiation, which can potentially generate functional gametes for further research and possibly for therapeutic use. New advancements in molecular research progress may grant the possibility to fine-tune protocols for differentiation toward the germ-cell lineage, starting from different developmental starting points. The efficiency of PGC-LC from human pluripotent stem cell lines has become increasingly robust, but to fully understand molecular signaling pathway mechanisms during germ cell formation, more defined approaches will be required. The nature of correct epigenetic reprogramming remains mainly uncharted, and future research will need to focus on clarification of the dynamic epigenome states during germ cell specification, migration, and maturation. Yet, it is now possible to successfully generate fertile normal offspring after in vitro differentiation of stem cell lines in the mouse model, with or without the use of supporting cells. Coupled with our ever-widening understanding of human sperm cell development, where meiotic haploid sperm cells can already be generated in vitro, this seems to suggest that the generation of fertile human gametes in vitro is realistically attainable. Therefore, it is paramount that discussion of policies for the generation of embryos for research purposes (not for reproductive purposes) takes place within the near future in order to advance our understanding. The ability to functionally test in vitro differentiated gametes for research alone is fundamental to truly know the quality and capacity of any generated in vitro gamete. By achieving this, the development of safe and functional gametes suitable for reproduction should be possible, providing an essential cure to infertility.

References

[1] J. Kashir, B. Heindryckx, C. Jones, P. De Sutter, J. Parrington, K. Coward, Oocyte activation, phospholipase C zeta and human infertility, Hum. Reprod. Update 16 (2010) 690–703.

[2] J. Kashir, C. Jones, T. Child, S.A. Williams, K. Coward, Viability assessment for artificial gametes: the need for biomarkers of functional competency, Biol. Reprod. 87 (2012) 114.

[3] B. Dale, M. Wilding, G. Botta, M. Rasile, M. Marino, L. Di Matteo, G. De Placido, A. Izzo, Pregnancy after cytoplasmic transfer in a couple suffering from idiopathic infertility: case report, Hum. Reprod. 16 (2001) 1469–1472.

[4] I. Moreno, J.M. M?guez-Forjan, C. Sim?n, Artificial gametes from stem cells, Clin. Exp. Reprod. Med. 42 (2015) 33, https://doi.org/10.5653/cerm.2015.42.2.33.

[5] E. Borges, A.S. Setti, D.P. Braga, R.D.C.S. Figueira, A. Iaconelli, Decline in semen quality among infertile men in Brazil during the past 10 years, Int. Braz. J. Urol. 41 (2015) 757–763, http://www.ncbi.nlm.nih.gov/pubmed/26401870, Accessed July 31, 2017.

[6] H. Levine, N. Jørgensen, A. Martino-Andrade, J. Mendiola, D. Weksler-Derri, I. Mindlis, R. Pinotti, S.H. Swan, J.W. Levy Library, Temporal trends in sperm count: a systematic review and meta-regression analysis, Hum. Reprod. Update (2017) 1–14, https://doi.org/10.1093/humupd/dmx022/0000-0002-5597-4916.

[7] M. Rolland, J. Le Moal, V. Wagner, D. Royère, J. De Mouzon, Decline in semen concentration and morphology in a sample of 26 609 men close to general population between 1989 and 2005 in France, Hum. Reprod. 28 (2013) 462–470, https://doi.org/10.1093/humrep/des415.

[8] A.J. Newson, A.C. Smajdor, Artificial gametes: new paths to parenthood? J. Med. Ethics 31 (2005) 184–186.

[9] S.M. Chuva de Sousa Lopes, B.A.J. Roelen, On the formation of germ cells: the good, the bad and the ugly, Differentiation 79 (2010) 131–140, https://doi.org/10.1016/j.diff.2009.11.003.

[10] M. De Felici, Origin, migration, and proliferation of human primordial germ cells. in: Oogenesis, Springer, London, 2013, pp. 19–37, https://doi.org/10.1007/978-0-85729-826-3_2.

[11] C.G. Extavour, M. Akam, Mechanisms of germ cell specification across the metazoans: epigenesis and preformation, Development 130 (2003) 5869–5884, https://doi.org/10.1242/dev.00804.

[12] M. Saitou, M. Yamaji, Germ cell specification in mice: signaling, transcription regulation, and epigenetic consequences, Reproduction 139 (2010) 931–942, https://doi.org/10.1530/REP-10-0043.

[13] K.A. Lawson, N.R. Dunn, B.A.J. Roelen, L.M. Zeinstra, A.M. Davis, C.V.E. Wright, J.P.W.F.M. Korving, B.L.M. Hogan, Bmp4 is required for the generation of primordial germ cells in the mouse embryo, Genes Dev. 13 (1999) 424–436, https://www.ncbi.nlm.nih.gov/pmc/articles/PMC316469/pdf/x11.pdf, .

[14] Y. Ying, X. Qi, G.-Q. Zhao, Induction of primordial germ cells from murine epiblasts by synergistic action of BMP4 and BMP8B signaling pathways, Proc. Natl. Acad. Sci. 98 (2001) 7858–7862, https://doi.org/10.1073/pnas.151242798.

[15] S. Weber, D. Eckert, D. Nettersheim, A.J.M. Gillis, S. Schäfer, P. Kuckenberg, J. Ehlermann, U. Werling, K. Biermann, L.H.J. Looijenga, H. Schorle, Critical function of AP-2gamma/TCFAP2C in mouse embryonic germ cell maintenance, Biol. Reprod. 82 (2010) 214–223, https://doi.org/10.1095/biolreprod.109.078717.

[16] M. Saitou, S.C. Barton, M.A. Surani, A molecular programme for the specification of germ cell fate in mice, Nature 418 (2002) 293–300, https://doi.org/10.1038/nature00927.

[17] Y. Ohinata, B. Payer, D. O'Carroll, K. Ancelin, Y. Ono, M. Sano, S.C. Barton, T. Obukhanych, M. Nussenzweig, A. Tarakhovsky, M. Saitou, M.A. Surani, Blimp1 is a critical determinant of the germ cell lineage in mice, Nature 436 (2005) 207–213, https://doi.org/10.1038/nature03813.

[18] S.D. Vincent, N.R. Dunn, R. Sciammas, M. Shapiro-Shalef, M.M. Davis, K. Calame, E.K. Bikoff, E.J. Robertson, The zinc finger transcriptional repressor Blimp1/Prdm1 is dispensable for early axis formation but is required for specification of primordial germ cells in the mouse, Development 132 (2005) 1315–1325, https://doi.org/10.1242/dev.01711.

[19] M. Yamaji, Y. Seki, K. Kurimoto, Y. Yabuta, M. Yuasa, M. Shigeta, K. Yamanaka, Y. Ohinata, M. Saitou, Critical function of Prdm14 for the establishment of the germ cell lineage in mice, Nat. Genet. 40 (2008) 1016–1022, https://doi.org/10.1038/ng.186.

[20] K. Hayashi, M.A. Surani, Self-renewing epiblast stem cells exhibit continual delineation of germ cells with epigenetic reprogramming in vitro, Development 136 (2009) 3549–3556, https://doi.org/10.1242/dev.037747.

[21] W. Reik, A. Lewis, Co-evolution of X-chromosome inactivation and imprinting in mammals, Nat. Rev. Genet. 6 (5) (2005) 403–410, https://doi.org/10.1038/nrg1602.

[22] E. Heard, Recent advances in X-chromosome inactivation, Curr. Opin. Cell Biol. 16 (2004) 247–255, https://doi.org/10.1016/j.ceb.2004.03.005.

[23] I. Okamoto, C. Patrat, D. Thépot, N. Peynot, P. Fauque, N. Daniel, P. Diabangouaya, J.-P. Wolf, J.-P. Renard, V. Duranthon, Eutherian mammals use diverse strategies to initiate X-chromosome inactivation during development, Nature 472 (2011) 370–374.

[24] V. Pasque, J. Tchieu, R. Karnik, M. Uyeda, A. Sadhu Dimashkie, D. Case, B. Papp, G. Bonora, S. Patel, R. Ho, R. Schmidt, R. McKee, T. Sado, T. Tada, A. Meissner, K. Plath, X chromosome reactivation dynamics reveal stages of reprogramming to pluripotency, Cell 159 (2014) 1681–1697, https://doi.org/10.1016/j.cell.2014.11.040.

[25] S.M. Chuva de Sousa Lopes, K. Hayashi, T.C. Shovlin, W. Mifsud, M.A. Surani, A. McLaren??, X chromosome activity in mouse XX primordial germ cells, PLoS Genet. 4 (2008) e30, https://doi.org/10.1371/journal.pgen.0040030.

[26] S. Yamaguchi, H. Kimura, M. Tada, N. Nakatsuji, T. Tada, Nanog expression in mouse germ cell development, Gene Expr. Patterns 5 (2005) 639–646, https://doi.org/10.1016/j.modgep.2005.03.001.

[27] C. Runyan, K. Schaible, K. Molyneaux, Z. Wang, L. Levin, C. Wylie, Steel factor controls midline cell death of primordial germ cells and is essential for their normal proliferation and migration, Development 133 (2006) 4861–4869, https://doi.org/10.1242/dev.02688.

[28] A.K. Lim, C. Lorthongpanich, T.G. Chew, C.W.G. Tan, Y.T. Shue, S. Balu, N. Gounko, S. Kuramochi-Miyagawa, M.M. Matzuk, S. Chuma, D.M. Messerschmidt, D. Solter, B.B. Knowles, The nuage mediates retrotransposon silencing in mouse primordial ovarian follicles, Development 140 (2013) 3819–3825, https://doi.org/10.1242/dev.099184.

[29] P. Hajkova, K. Ancelin, T. Waldmann, N. Lacoste, U.C. Lange, F. Cesari, C. Lee, G. Almouzni, R. Schneider, M.A. Surani, Chromatin dynamics during epigenetic reprogramming in the mouse germ line, Nature 452 (2008) 877–881, https://doi.org/10.1038/nature06714.

[30] Y. Seki, K. Hayashi, K. Itoh, M. Mizugaki, M. Saitou, Y. Matsui, Extensive and orderly reprogramming of genome-wide chromatin modifications associated with specification and early development of germ cells in mice, Dev. Biol. 278 (2005) 440–458, https://doi.org/10.1016/j.ydbio.2004.11.025.

[31] M.A. Surani, K. Hayashi, P. Hajkova, Genetic and epigenetic regulators of pluripotency, Cell 128 (2007) 747–762, https://doi.org/10.1016/j.cell.2007.02.010.

[32] J.A. Hackett, J.J. Zylicz, M.A. Surani, Parallel mechanisms of epigenetic reprogramming in the germline, Trends Genet. 28 (2012) 164–174, https://doi.org/10.1016/j.tig.2012.01.005.

[33] S. Kagiwada, K. Kurimoto, T. Hirota, M. Yamaji, M. Saitou, Replication-coupled passive DNA demethylation for the erasure of genome imprints in mice, EMBO J. 32 (2012) 340–353, https://doi.org/10.1038/emboj.2012.331.

[34] S.M. Hyldig, N. Croxall, D.A. Contreras, P.D. Thomsen, R. Alberio, Epigenetic reprogramming in the porcine germ line, BMC Dev. Biol. 11 (2011) 11, https://doi.org/10.1186/1471-213X-11-11.

[35] S.M.W. Hyldig, O. Ostrup, M. Vejlsted, P.D. Thomsen, Changes of DNA methylation level and spatial arrangement of primordial germ cells in embryonic day 15 to embryonic day 28 pig embryos, Biol. Reprod. 84 (2011) 1087–1093, https://doi.org/10.1095/biolreprod.110.086082.

[36] C. Eguizabal, L. Herrera, L. De Oñate, N. Montserrat, P. Hajkova, J.C. Izpisua Belmonte, Characterization of the epigenetic changes during human gonadal primordial germ cells reprogramming, Stem Cells 34 (2016) 2418–2428, https://doi.org/10.1002/stem.2422.

[37] I. Cantone, A.G. Fisher, Epigenetic programming and reprogramming during development, Nat. Publ. Gr. 20 (2013), https://doi.org/10.1038/nsmb.2489.

[38] Z.D. Smith, M.M. Chan, T.S. Mikkelsen, H. Gu, A. Gnirke, A. Regev, A. Meissner, A unique regulatory phase of DNA methylation in the early mammalian embryo, Nature 484 (2012) 339–344, https://doi.org/10.1038/nature10960.
[39] S. Gkountela, K.X. Zhang, T.A. Shafiq, W.-W. Liao, J. Hargan-Calvopiña, P.-Y. Chen, A.T. Clark, DNA demethylation dynamics in the human prenatal germline, Cell 161 (2015) 1425–1436, https://doi.org/10.1016/j.cell.2015.05.012.
[40] F. Guo, L. Yan, H. Guo, L. Li, B. Hu, Y. Zhao, J. Yong, Y. Hu, X. Wang, Y. Wei, W. Wang, R. Li, J. Yan, X. Zhi, Y. Zhang, H. Jin, W. Zhang, Y. Hou, P. Zhu, J. Li, L. Zhang, S. Liu, Y. Ren, X. Zhu, L. Wen, Y.Q. Gao, F. Tang, J. Qiao, The transcriptome and DNA methylome landscapes of human primordial germ cells, Cell 161 (2015) 1437–1452, https://doi.org/10.1016/j.cell.2015.05.015.
[41] W.W.C. Tang, S. Dietmann, N. Irie, H.G. Leitch, V.I. Floros, C.R. Bradshaw, J.A. Hackett, P.F. Chinnery, M.A. Surani, A unique gene regulatory network resets the human germline epigenome for development, Cell 161 (2015) 1453–1467, https://doi.org/10.1016/j.cell.2015.04.053.
[42] F. von Meyenn, W. Reik, N. Irie, H.G. Leitch, V.I. Floros, C.R. Bradshaw, J.A. Hackett, P.F. Chinnery, M.A. Surani, H.R. Schöler, et al., Forget the parents: epigenetic reprogramming in human germ cells, Cell 161 (2015) 1248–1251, https://doi.org/10.1016/j.cell.2015.05.039.
[43] N. Irie, L. Weinberger, W.W.C. Tang, T. Kobayashi, S. Viukov, Y.S. Manor, S. Dietmann, J.H. Hanna, M.A. Surani, SOX17 is a critical specifier of human primordial germ cell fate, Cell 160 (2015) 253–268, https://doi.org/10.1016/j.cell.2014.12.013.
[44] R.M. Perrett, L. Turnpenny, J.J. Eckert, M. O'Shea, S.B. Sonne, I.T. Cameron, D.I. Wilson, E.R.-D. Meyts, N.A. Hanley, The early human germ cell lineage does not express SOX2 during in vivo development or upon in vitro culture, Biol. Reprod. 78 (2008) 852–858, https://doi.org/10.1095/biolreprod.107.066175.
[45] L. Li, J. Dong, L. Yan, J. Yong, X. Liu, Y. Hu, X. Fan, X. Wu, H. Guo, X. Wang, Single-cell RNA-seq analysis maps development of human germline cells and gonadal niche interactions, Cell Stem Cell (2017).
[46] X. Song, G.B. Call, D. Kirilly, T. Xie, Notch signaling controls germline stem cell niche formation in the Drosophila ovary, Development 134 (2007) 1071–1080, https://doi.org/10.1242/dev.003392.
[47] B.T. Phillips, K. Gassei, K.E. Orwig, Spermatogonial stem cell regulation and spermatogenesis, Philos. Trans. R. Soc. B Biol. Sci. 365 (2010) 1663–1678.
[48] M. Culty, Gonocytes, the forgotten cells of the germ cell lineage, Birth Defects Res. Part C Embryo Today Rev. 87 (2009) 1–26.
[49] J.M. Oatley, R.L. Brinster, Spermatogonial stem cells, Methods Enzymol. 419 (2006) 259–282, https://doi.org/10.1016/s0076-6879(06)19011-4.
[50] N. Ahmed, H. Yufei, P. Yang, W. Muhammad Yasir, Q. Zhang, T. Liu, C. Hong, H. Lisi, C. Xiaoya, Q. Chen, Cytological study on Sertoli cells and their interactions with germ cells during annual reproductive cycle in turtle, Ecol. Evol. 6 (2016) 4050–4064.
[51] M. Dym, M. Kokkinaki, Z. He, Spermatogonial stem cells: mouse and human comparisons, Birth Defects Res. Part C Embryo Today Rev. 87 (2009) 27–34, https://doi.org/10.1002/bdrc.20141.
[52] T. Ogawa, I. Dobrinski, M.R. Avarbock, R.L. Brinster, Transplantation of male germ line stem cells restores fertility in infertile mice, Nat. Med. 6 (2000) 29–34.
[53] R.C. Gupta, Reproductive and Developmental Toxicology, Academic Press as an Imprint of Elsevier, San Diego, 2017.
[54] E.M. Eddy, Male germ cell gene expression, Recent Prog. Horm. Res. 57 (2002) 103–128.
[55] S.H. Greenberg, L.I. Lipshultz, A.J. Wein, Experience with 425 subfertile male patients, J. Urol. 119 (1978) 507–510.
[56] H. Valli, K. Gassei, K.E. Orwig, Stem cell therapies for male infertility: where are we now and where are we going? in: Biennial Review of Infertility, Springer International Publishing, Switzerland, 2015, pp. 17–39.
[57] W. Hsiao, P.J. Stahl, E.C. Osterberg, E. Nejat, G.D. Palermo, Z. Rosenwaks, P.N. Schlegel, Successful treatment of postchemotherapy azoospermia with microsurgical testicular sperm extraction: the Weill Cornell experience, J. Clin. Oncol. 29 (2011) 1607–1611.
[58] M.A. Surani, Breaking the germ line-soma barrier, Nat. Rev. Mol. Cell Biol. 17 (3) (2016) 136.
[59] J. Solana, Closing the circle of germline and stem cells: the primordial stem cell hypothesis, EvoDevo 4 (2013) 2.
[60] D. Chen, A.T. Clark, Human germline differentiation charts a new course, EMBO J. (2015). e201591447.
[61] S.J. Rodda, S.J. Kavanagh, J. Rathjen, P.D. Rathjen, Embryonic stem cell differentiation and the analysis of mammalian development, Int. J. Dev. Biol. 46 (2002) 449–458, http://www.ncbi.nlm.nih.gov/pubmed/12141431, Accessed July 31, 2017.

[62] I.G.M. Brons, L.E. Smithers, M.W.B. Trotter, P. Rugg-Gunn, B. Sun, S.M. Chuva de Sousa Lopes, S.K. Howlett, A. Clarkson, L. Ahrlund-Richter, R.A. Pedersen, L. Vallier, Derivation of pluripotent epiblast stem cells from mammalian embryos, Nature 448 (2007) 191–195, https://doi.org/10.1038/nature05950.

[63] P.J. Tesar, J.G. Chenoweth, F.A. Brook, T.J. Davies, E.P. Evans, D.L. Mack, R.L. Gardner, R.D.G. McKay, New cell lines from mouse epiblast share defining features with human embryonic stem cells, Nature 448 (2007) 196–199, https://doi.org/10.1038/nature05972.

[64] Q.-L. Ying, J. Wray, J. Nichols, L. Batlle-Morera, B. Doble, J. Woodgett, P. Cohen, A. Smith, The ground state of embryonic stem cell self-renewal, Nature 453 (2008) 519–523, https://doi.org/10.1038/nature06968.

[65] J. Nichols, A. Smith, J. Nichols, L. Batlle-Morera, B. Doble, J. Woodgett, P. Cohen, A. Smith, S. Yamanaka, C.L. Hsieh, et al., Naive and primed pluripotent states, Cell Stem Cell 4 (2009) 487–492, https://doi.org/10.1016/j.stem.2009.05.015.

[66] M. Ohgushi, M. Matsumura, M. Eiraku, K. Murakami, T. Aramaki, A. Nishiyama, K. Muguruma, T. Nakano, H. Suga, M. Ueno, T. Ishizaki, H. Suemori, S. Narumiya, H. Niwa, Y. Sasai, Molecular pathway and cell state responsible for dissociation-induced apoptosis in human pluripotent stem cells, Cell Stem Cell 7 (2010) 225–239, https://doi.org/10.1016/j.stem.2010.06.018.

[67] K. Watanabe, M. Ueno, D. Kamiya, A. Nishiyama, M. Matsumura, T. Wataya, J.B. Takahashi, S. Nishikawa, S. Nishikawa, K. Muguruma, Y. Sasai, A ROCK inhibitor permits survival of dissociated human embryonic stem cells, Nat. Biotechnol. 25 (2007) 681–686, https://doi.org/10.1038/nbt1310.

[68] O. Gafni, L. Weinberger, A.A. Mansour, Y.S. Manor, E. Chomsky, D. Ben-Yosef, Y. Kalma, S. Viukov, I. Maza, A. Zviran, Y. Rais, Z. Shipony, Z. Mukamel, V. Krupalnik, M. Zerbib, S. Geula, I. Caspi, D. Schneir, T. Shwartz, S. Gilad, D. Amann-Zalcenstein, S. Benjamin, I. Amit, A. Tanay, R. Massarwa, N. Novershtern, J.H. Hanna, Derivation of novel human ground state naive pluripotent stem cells, Nature 504 (2013) 282–286, https://doi.org/10.1038/nature12745.

[69] J. Hanna, A.W. Cheng, K. Saha, J. Kim, C.J. Lengner, F. Soldner, J.P. Cassady, J. Muffat, B.W. Carey, R. Jaenisch, Human embryonic stem cells with biological and epigenetic characteristics similar to those of mouse ESCs, Proc. Natl. Acad. Sci. U. S. A. 107 (2010) 9222–9227, https://doi.org/10.1073/pnas.1004584107.

[70] T.W. Theunissen, M. Friedli, Y. He, E. Planet, R.C. O'Neil, S. Markoulaki, J. Pontis, H. Wang, A. Iouranova, M. Imbeault, J. Duc, M.A. Cohen, K.J. Wert, R. Castanon, Z. Zhang, Y. Huang, J.R. Nery, J. Drotar, T. Lungjangwa, D. Trono, J.R. Ecker, R. Jaenisch, Molecular criteria for defining the naive human pluripotent state, Cell Stem Cell 19 (2016) 502–515, https://doi.org/10.1016/j.stem.2016.06.011.

[71] C. Bock, E. Kiskinis, G. Verstappen, H. Gu, G. Boulting, Z.D. Smith, M. Ziller, G.F. Croft, M.W. Amoroso, D.H. Oakley, A. Gnirke, K. Eggan, A. Meissner, Reference maps of human ES and iPS cell variation enable high-throughput characterization of pluripotent cell lines, Cell 144 (2011) 439–452, https://doi.org/10.1016/j.cell.2010.12.032.

[72] K. Osafune, L. Caron, M. Borowiak, R.J. Martinez, C.S. Fitz-Gerald, Y. Sato, C.A. Cowan, K.R. Chien, D.A. Melton, Marked differences in differentiation propensity among human embryonic stem cell lines, Nat. Biotechnol. 26 (2008) 313–315, https://doi.org/10.1038/nbt1383.

[73] S.R. Hough, A.L. Laslett, S.B. Grimmond, G. Kolle, M.F. Pera, A continuum of cell states spans pluripotency and lineage commitment in human embryonic stem cells, PLoS One 4 (2009) e7708, https://doi.org/10.1371/journal.pone.0007708.

[74] L. Vallier, M. Alexander, R.A. Pedersen, Activin/nodal and FGF pathways cooperate to maintain pluripotency of human embryonic stem cells, J. Cell Sci. 118 (2005) 4495–4509, https://doi.org/10.1242/jcs.02553.

[75] H. Chen, I. Aksoy, F. Gonnot, P. Osteil, M. Aubry, C. Hamela, C. Rognard, A. Hochard, S. Voisin, E. Fontaine, M. Mure, M. Afanassieff, E. Cleroux, S. Guibert, J. Chen, C. Vallot, H. Acloque, C. Genthon, C. Donnadieu, J. De Vos, D. Sanlaville, J.-F. Guérin, M. Weber, L.W. Stanton, C. Rougeulle, B. Pain, P.-Y. Bourillot, P. Savatier, Reinforcement of STAT3 activity reprogrammes human embryonic stem cells to naive-like pluripotency, Nat. Commun. 6 (2015) 7095, https://doi.org/10.1038/ncomms8095.

[76] Y.-S. Chan, J. Göke, J.-H. Ng, X. Lu, K.A.U. Gonzales, C.-P. Tan, W.-Q. Tng, Z.-Z. Hong, Y.-S. Lim, H.-H. Ng, Induction of a human pluripotent state with distinct regulatory circuitry that resembles preimplantation epiblast, Cell Stem Cell 13 (2013) 663–675, https://doi.org/10.1016/j.stem.2013.11.015.

[77] G. Duggal, S. Warrier, S. Ghimire, D. Broekaert, M. Van der Jeught, S. Lierman, T. Deroo, L. Peelman, A. Van Soom, R. Cornelissen, B. Menten, P. Mestdagh, J. Vandesompele, M. Roost, R.C. Slieker, B.T. Heijmans, D. Deforce, P. De Sutter, S.C. De Sousa Lopes, B. Heindryckx, Alternative routes to induce naïve pluripotency in human embryonic stem cells, Stem Cells 33 (2015) 2686–2698, https://doi.org/10.1002/stem.2071.

[78] Y. Takashima, G. Guo, R. Loos, J. Nichols, G. Ficz, F. Krueger, D. Oxley, F. Santos, J. Clarke, W. Mansfield, W. Reik, P. Bertone, A. Smith, Resetting transcription factor control circuitry toward ground-state pluripotency in human, Cell 158 (2014) 1254–1269, https://doi.org/10.1016/j.cell.2014.08.029.

[79] T.W. Theunissen, B.E. Powell, H. Wang, M. Mitalipova, D.A. Faddah, J. Reddy, Z.P. Fan, D. Maetzel, K. Ganz, L. Shi, T. Lungjangwa, S. Imsoonthornruksa, Y. Stelzer, S. Rangarajan, A. D'Alessio, J. Zhang, Q. Gao, M.M. Dawlaty, R.A. Young, N.S. Gray, R. Jaenisch, Systematic identification of culture conditions for induction and maintenance of naive human pluripotency, Cell Stem Cell 15 (2014) 471–487, https://doi.org/10.1016/j.stem.2014.07.002.

[80] B. Valamehr, M. Robinson, R. Abujarour, B. Rezner, F. Vranceanu, T. Le, A. Medcalf, T.T. Lee, M. Fitch, D. Robbins, P. Flynn, Platform for induction and maintenance of transgene-free hiPSCs resembling ground state pluripotent stem cells, Stem Cell Rep. 2 (2014) 366–381, https://doi.org/10.1016/j.stemcr.2014.01.014.

[81] C.B. Ware, A.M. Nelson, B. Mecham, J. Hesson, W. Zhou, E.C. Jonlin, A.J. Jimenez-Caliani, X. Deng, C. Cavanaugh, S. Cook, P.J. Tesar, J. Okada, L. Margaretha, H. Sperber, M. Choi, C.A. Blau, P.M. Treuting, R.D. Hawkins, V. Cirulli, H. Ruohola-Baker, Derivation of naive human embryonic stem cells, Proc. Natl. Acad. Sci. U. S. A. 111 (2014) 4484–4489, https://doi.org/10.1073/pnas.1319738111.

[82] M.G. Carter, B.J. Smagghe, A.K. Stewart, J.A. Rapley, E. Lynch, K.J. Bernier, K.W. Keating, V.M. Hatziioannou, E.J. Hartman, C.C. Bamdad, A primitive growth factor, NME7AB, is sufficient to induce stable naïve state human pluripotency; reprogramming in this novel growth factor confers superior differentiation, Stem Cells 34 (2016) 847–859, https://doi.org/10.1002/stem.2261.

[83] Y. Yang, B. Liu, J. Xu, J. Wang, J. Wu, C. Shi, Y. Xu, J. Dong, C. Wang, W. Lai, J. Zhu, L. Xiong, D. Zhu, X. Li, W. Yang, T. Yamauchi, A. Sugawara, Z. Li, F. Sun, X. Li, C. Li, A. He, Y. Du, T. Wang, C. Zhao, H. Li, X. Chi, H. Zhang, Y. Liu, C. Li, S. Duo, M. Yin, H. Shen, J.C.I. Belmonte, H. Deng, Derivation of pluripotent stem cells with in vivo embryonic and extraembryonic potency, Cell 169 (2017) 243–257.e25, https://doi.org/10.1016/j.cell.2017.02.005.

[84] T. O'leary, B. Heindryckx, S. Lierman, D. Van Bruggen, J.J. Goeman, M. Vandewoestyne, D. Deforce, S.M.C. de Sousa Lopes, P. De Sutter, Tracking the progression of the human inner cell mass during embryonic stem cell derivation, Nat. Biotechnol. 30 (2012) 278–282.

[85] T. O'Leary, B. Heindryckx, S. Lierman, M. Van der Jeught, G. Duggal, P. De Sutter, S.M. Chuva de Sousa Lopes, Derivation of human embryonic stem cells using a post-inner cell mass intermediate, Nat. Protoc. 8 (2013) 254–264, https://doi.org/10.1038/nprot.2012.157.

[86] K. Takahashi, K. Tanabe, M. Ohnuki, M. Narita, T. Ichisaka, K. Tomoda, S. Yamanaka, Induction of pluripotent stem cells from adult human fibroblasts by defined factors, Cell 131 (2007) 861–872, https://doi.org/10.1016/j.cell.2007.11.019.

[87] J. Nichols, B. Zevnik, K. Anastassiadis, H. Niwa, D. Klewe-Nebenius, I. Chambers, H. Schöler, A. Smith, Formation of pluripotent stem cells in the mammalian embryo depends on the POU transcription factor Oct4, Cell 95 (1998) 379–391, http://www.ncbi.nlm.nih.gov/pubmed/9814708, Accessed July 31, 2017.

[88] S. Masui, Y. Nakatake, Y. Toyooka, D. Shimosato, R. Yagi, K. Takahashi, H. Okochi, A. Okuda, R. Matoba, A.A. Sharov, M.S.H. Ko, H. Niwa, Pluripotency governed by Sox2 via regulation of Oct3/4 expression in mouse embryonic stem cells, Nat. Cell Biol. 9 (2007) 625–635, https://doi.org/10.1038/ncb1589.

[89] T. Teramura, J. Frampton, Induced pluripotent stem cells in reproductive medicine, Reprod. Med. Biol. 12 (2013) 39–46, https://doi.org/10.1007/s12522-012-0141-x.

[90] J.V. Medrano, A.M. Martínez-Arroyo, J.M. Míguez, I. Moreno, S. Martínez, A. Quiñonero, P. Díaz-Gimeno, A.I. Marqués-Marí, A. Pellicer, J. Remohí, C. Simón, Human somatic cells subjected to genetic induction with six germ line-related factors display meiotic germ cell-like features, Sci. Rep. 6 (2016) 24956, https://doi.org/10.1038/srep24956.

[91] W.S. Hwang, S. Il Roh, B.C. Lee, S.K. Kang, D.K. Kwon, S. Kim, S.J. Kim, S.W. Park, H.S. Kwon, C.K. Lee, J.B. Lee, J.M. Kim, C. Ahn, S.H. Paek, S.S. Chang, J.J. Koo, H.S. Yoon, J.H. Hwang, Y.Y. Hwang, Y.S. Park, S.K. Oh, H.S. Kim, J.H. Park, S.Y. Moon, G. Schatten, Patient-specific embryonic stem cells derived from human SCNT blastocysts, Science 308 (2005) 1777–1783, http://science.sciencemag.org/content/308/5729/1777, Accessed July 31, 2017.

[92] C. Eguizabal, N. Montserrat, R. Vassena, M. Barragan, E. Garreta, L. Garcia-Quevedo, F. Vidal, A. Giorgetti, A. Veiga, J.C.I. Belmonte, Complete meiosis from human induced pluripotent stem cells, Stem Cells 29 (2011) 1186–1195, https://doi.org/10.1002/stem.672.

[93] K. Nayernia, J. Nolte, H.W. Michelmann, J.H. Lee, K. Rathsack, N. Drusenheimer, A. Dev, G. Wulf, I.E. Ehrmann, D.J. Elliott, In vitro-differentiated embryonic stem cells give rise to male gametes that can generate offspring mice, Dev. Cell 11 (2006) 125–132.

[94] K. Okita, T. Ichisaka, S. Yamanaka, Generation of germline-competent induced pluripotent stem cells, Nature 448 (2007) 313–317, https://doi.org/10.1038/nature05934.
[95] T. Sato, K. Katagiri, A. Gohbara, K. Inoue, N. Ogonuki, A. Ogura, Y. Kubota, T. Ogawa, In vitro production of functional sperm in cultured neonatal mouse testes, Nature 471 (2011) 504–507.
[96] M. Nagano, T. Shinohara, M.R. Avarbock, R.L. Brinster, Retrovirus-mediated gene delivery into male germ line stem cells, FEBS Lett. 475 (2000) 7–10, http://www.ncbi.nlm.nih.gov/pubmed/10854847, .
[97] M. Nagano, C.J. Brinster, K.E. Orwig, B.-Y. Ryu, M.R. Avarbock, R.L. Brinster, Transgenic mice produced by retroviral transduction of male germ-line stem cells, Proc. Natl. Acad. Sci. 98 (2001) 13090–13095, https://doi.org/10.1073/pnas.231473498.
[98] J. Lee, M. Kanatsu-Shinohara, H. Morimoto, Y. Kazuki, S. Takashima, M. Oshimura, S. Toyokuni, T. Shinohara, Genetic reconstruction of mouse spermatogonial stem cell self-renewal in vitro by Ras-cyclin D2 activation, Cell Stem Cell 5 (2009) 76–86, https://doi.org/10.1016/j.stem.2009.04.020.
[99] M. Kanatsu-Shinohara, N. Ogonuki, K. Inoue, H. Miki, A. Ogura, S. Toyokuni, T. Shinohara, Long-term proliferation in culture and germline transmission of mouse male germline stem cells, Biol. Reprod. 69 (2003) 612–616, https://doi.org/10.1095/biolreprod.103.017012.
[100] M. Kanatsu-Shinohara, H. Miki, K. Inoue, N. Ogonuki, S. Toyokuni, A. Ogura, T. Shinohara, Long-term culture of mouse male germline stem cells under serum-or feeder-free conditions, Biol. Reprod. 72 (2005) 985–991, https://doi.org/10.1095/biolreprod.104.036400.
[101] M. Kanatsu-Shinohara, N. Ogonuki, T. Iwano, J. Lee, Y. Kazuki, K. Inoue, H. Miki, M. Takehashi, S. Toyokuni, Y. Shinkai, M. Oshimura, F. Ishino, A. Ogura, T. Shinohara, Genetic and epigenetic properties of mouse male germline stem cells during long-term culture, Development 132 (2005), http://dev.biologists.org/content/132/18/4155, .
[102] M. Kanatsu-Shinohara, K. Inoue, J. Lee, H. Miki, N. Ogonuki, S. Toyokuni, A. Ogura, T. Shinohara, Anchorage-independent growth of mouse male germline stem cells in vitro, Biol. Reprod. 74 (2006) 522–529, https://doi.org/10.1095/biolreprod.105.046441.
[103] M. Kanatsu-Shinohara, S. Takashima, T. Shinohara, Transmission distortion by loss of p21 or p27 cyclin-dependent kinase inhibitors following competitive spermatogonial transplantation, Proc. Natl. Acad. Sci. 107 (2010) 6210–6215, https://doi.org/10.1073/pnas.0914448107.
[104] M. Kanatsu-Shinohara, K. Inoue, N. Ogonuki, H. Morimoto, A. Ogura, T. Shinohara, Serum- and feeder-free culture of mouse germline stem cells, Biol. Reprod. 84 (2011) 97–105, https://doi.org/10.1095/biolreprod.110.086462.
[105] K. Takahashi, S. Yamanaka, Induction of pluripotent stem cells from mouse embryonic and adult fibroblast cultures by defined factors, Cell 126 (2006) 663–676, https://doi.org/10.1016/j.cell.2006.07.024.
[106] H. Kubota, M.R. Avarbock, R.L. Brinster, Growth factors essential for self-renewal and expansion of mouse spermatogonial stem cells, Proc. Natl. Acad. Sci. 101 (2004) 16489–16494, https://doi.org/10.1073/pnas.0407063101.
[107] H. Kubota, M.R. Avarbock, J.A. Schmidt, R.L. Brinster, Spermatogonial stem cells derived from infertile Wv/Wv mice self-renew in vitro and generate progeny following transplantation, Biol. Reprod. 81 (2009) 293–301, https://doi.org/10.1095/biolreprod.109.075960.
[108] K. Kita, T. Watanabe, K. Ohsaka, H. Hayashi, Y. Kubota, Y. Nagashima, I. Aoki, H. Taniguchi, T. Noce, K. Inoue, H. Miki, N. Ogonuki, H. Tanaka, A. Ogura, T. Ogawa, Production of functional spermatids from mouse germline stem cells in ectopically reconstituted seminiferous tubules, Biol. Reprod. 76 (2007) 211–217, https://doi.org/10.1095/biolreprod.106.056895.
[109] H. Ohta, Y. Ohinata, M. Ikawa, Y. Morioka, Y. Sakaide, M. Saitou, O. Kanagawa, T. Wakayama, Male germline and embryonic stem cell lines from NOD mice: efficient derivation of GS cells from a nonpermissive strain for ES cell derivation, Biol. Reprod. 81 (2009) 1147–1153, https://doi.org/10.1095/biolreprod.109.079368.
[110] H. Shiura, R. Ikeda, J. Lee, T. Sato, N. Ogonuki, M. Hirose, A. Ogura, T. Ogawa, K. Abe, Generation of a novel germline stem cell line expressing a germline-specific reporter in the mouse, Genesis 51 (2013) 498–505, https://doi.org/10.1002/dvg.22391.
[111] M. Riboldi, C. Rubio, A. Pellicer, M. Gil-Salom, C. Simón, In vitro production of haploid cells after coculture of CD49f+ with Sertoli cells from testicular sperm extraction in nonobstructive azoospermic patients, Fertil. Steril. 98 (2012). 580–590.e4, https://doi.org/10.1016/j.fertnstert.2012.05.039.
[112] S. Yang, P. Ping, M. Ma, P. Li, R. Tian, H. Yang, Y. Liu, Y. Gong, Z. Zhang, Z. Li, Generation of haploid spermatids with fertilization and development capacity from human spermatogonial stem cells of cryptorchid patients, Stem Cell Rep. 3 (2014) 663–675.

[113] N. Geijsen, M. Horoschak, K. Kim, J. Gribnau, K. Eggan, G.Q. Daley, Derivation of embryonic germ cells and male gametes from embryonic stem cells, Nature 427 (2004) 148–154, https://doi.org/10.1038/nature02247.

[114] Y. Toyooka, N. Tsunekawa, R. Akasu, T. Noce, Embryonic stem cells can form germ cells in vitro, Proc. Natl. Acad. Sci. 100 (2003) 11457–11462, https://doi.org/10.1073/pnas.1932826100.

[115] A. Kerkis, S.A.S. Fonseca, R.C. Serafim, T.M.C. Lavagnolli, S. Abdelmassih, R. Abdelmassih, I. Kerkis, *In vitro* differentiation of male mouse embryonic stem cells into both presumptive sperm cells and oocytes, Cloning Stem Cells 9 (2007) 535–548, https://doi.org/10.1089/clo.2007.0031.

[116] Z. Yu, P. Ji, J. Cao, S. Zhu, Y. Li, L. Zheng, X. Chen, L. Feng, Dazl promotes germ cell differentiation from embryonic stem cells, J. Mol. Cell Biol. 1 (2009) 93–103, https://doi.org/10.1093/jmcb/mjp026.

[117] Y. Ohinata, H. Ohta, M. Shigeta, K. Yamanaka, T. Wakayama, M. Saitou, A signaling principle for the specification of the germ cell lineage in mice, Cell 137 (2009) 571–584, https://doi.org/10.1016/j.cell.2009.03.014.

[118] K. Hayashi, H. Ohta, K. Kurimoto, S. Aramaki, M. Saitou, Reconstitution of the mouse germ cell specification pathway in culture by pluripotent stem cells, Cell 146 (2011) 519–532, https://doi.org/10.1016/j.cell.2011.06.052.

[119] F. Nakaki, K. Hayashi, H. Ohta, K. Kurimoto, Y. Yabuta, M. Saitou, Induction of mouse germ-cell fate by transcription factors in vitro, Nature 501 (2013), https://doi.org/10.1038/nature12417.

[120] K. Murakami, U. Günesdogan, J.J. Zylicz, W.W.C. Tang, R. Sengupta, T. Kobayashi, S. Kim, R. Butler, S. Dietmann, M.A. Surani, NANOG alone induces germ cells in primed epiblast in vitro by activation of enhancers, Nature (2016).

[121] Q. Zhou, M. Wang, Y. Yuan, X. Wang, R. Fu, H. Wan, M. Xie, M. Liu, X. Guo, Y. Zheng, G. Feng, Q. Shi, X.-Y. Zhao, J. Sha, Q. Zhou, Complete meiosis from embryonic stem cell-derived germ cells in vitro, Cell Stem Cell 18 (2016) 330–340, https://doi.org/10.1016/j.stem.2016.01.017.

[122] A.T. Clark, M.S. Bodnar, M. Fox, R.T. Rodriquez, M.J. Abeyta, M.T. Firpo, R.A.R. Pera, Spontaneous differentiation of germ cells from human embryonic stem cells in vitro, Hum. Mol. Genet. 13 (2004) 727–739, https://doi.org/10.1093/hmg/ddh088.

[123] T.S. Park, Z. Galic, A.E. Conway, A. Lindgren, B.J. van Handel, M. Magnusson, L. Richter, M.A. Teitell, H.K.A. Mikkola, W.E. Lowry, K. Plath, A.T. Clark, Derivation of primordial germ cells from human embryonic and induced pluripotent stem cells is significantly improved by coculture with human fetal gonadal cells, Stem Cells 27 (2009) 783–795, https://doi.org/10.1002/stem.13.

[124] B. Aflatoonian, L. Ruban, M. Jones, R. Aflatoonian, A. Fazeli, H.D. Moore, In vitro post-meiotic germ cell development from human embryonic stem cells, Hum. Reprod. 24 (2009) 3150–3159, https://doi.org/10.1093/humrep/dep334.

[125] K. Kee, V.T. Angeles, M. Flores, H.N. Nguyen, R.A. Reijo Pera, Human DAZL, DAZ and BOULE genes modulate primordial germ-cell and haploid gamete formation, Nature 462 (2009) 222–225, https://doi.org/10.1038/nature08562.

[126] S. Panula, J.V. Medrano, K. Kee, R. Bergstr?m, H.N. Nguyen, B. Byers, K.D. Wilson, J.C. Wu, C. Simon, O. Hovatta, R.A. Reijo Pera, Human germ cell differentiation from fetal- and adult-derived induced pluripotent stem cells, Hum. Mol. Genet. 20 (2011) 752–762, https://doi.org/10.1093/hmg/ddq520.

[127] F.D. West, J.L. Mumaw, A. Gallegos-Cardenas, A. Young, S.L. Stice, Human haploid cells differentiated from meiotic competent clonal germ cell lines that originated from embryonic stem cells, Stem Cells Dev. 20 (2011) 1079–1088, https://doi.org/10.1089/scd.2010.0255.

[128] C.A. Easley, B.T. Phillips, M.M. McGuire, J.M. Barringer, H. Valli, B.P. Hermann, C.R. Simerly, A. Rajkovic, T. Miki, K.E. Orwig, G.P. Schatten, G.P. Schatten, Direct differentiation of human pluripotent stem cells into haploid spermatogenic cells, Cell Rep. 2 (2012) 440–446, https://doi.org/10.1016/j.celrep.2012.07.015.

[129] J.V. Medrano, C. Ramathal, H.N. Nguyen, C. Simon, R.A. Reijo Pera, Divergent RNA-binding proteins, DAZL and VASA, induce meiotic progression in human germ cells derived in vitro, Stem Cells 30 (2012) 441–451, https://doi.org/10.1002/stem.1012.

[130] G. Duggal, B. Heindryckx, S. Warrier, T. O'Leary, M. Van der Jeught, S. Lierman, L. Vossaert, T. Deroo, D. Deforce, S.M. Chuva de Sousa Lopes, P. De Sutter, Influence of activin A supplementation during human embryonic stem cell derivation on germ cell differentiation potential, Stem Cells Dev. 22 (2013) 3141–3155, https://doi.org/10.1089/scd.2013.0024.

[131] S. Gkountela, Z. Li, J.J. Vincent, K.X. Zhang, A. Chen, M. Pellegrini, A.T. Clark, The ontogeny of cKIT+ human primordial germ cells proves to be a resource for human germ line reprogramming, imprint erasure and in vitro differentiation, Nat. Cell Biol. 15 (2013) 113–122, https://doi.org/10.1038/ncb2638.

[132] G. Duggal, B. Heindryckx, S. Warrier, J. Taelman, M. Van Der Jeught, D. Deforce, S.C. De Sousa Lopes, P. De Sutter, Exogenous supplementation of activin A enhances germ cell differentiation of human embryonic stem cells, Mol. Hum. Reprod. 21 (2014), https://doi.org/10.1093/molehr/gav004.

[133] F. Sugawa, M.J. Arauzo-Bravo, J. Yoon, K.-P. Kim, S. Aramaki, G. Wu, M. Stehling, O.E. Psathaki, K. Hubner, H.R. Scholer, Human primordial germ cell commitment in vitro associates with a unique PRDM14 expression profile, EMBO J. 34 (2015) 1009–1024, https://doi.org/10.15252/embj.201488049.

[134] K. Sasaki, S. Yokobayashi, T. Nakamura, I. Okamoto, Y. Yabuta, K. Kurimoto, H. Ohta, Y. Moritoki, C. Iwatani, H. Tsuchiya, S. Nakamura, K. Sekiguchi, T. Sakuma, T. Yamamoto, T. Mori, K. Woltjen, M. Nakagawa, T. Yamamoto, K. Takahashi, S. Yamanaka, M. Saitou, Robust in vitro induction of human germ cell fate from pluripotent stem cells, Cell Stem Cell 17 (2015) 178–194, https://doi.org/10.1016/j.stem.2015.06.014.

[135] W. Ge, H.-G. Ma, S.-F. Cheng, Y.-C. Sun, L.-L. Sun, X.-F. Sun, L. Li, P. Dyce, J. Li, Q.-H. Shi, W. Shen, Differentiation of early germ cells from human skin-derived stem cells without exogenous gene integration, Sci. Rep. 5 (2015) 13822, https://doi.org/10.1038/srep13822.

[136] O.A. TROWELL, The culture of mature organs in a synthetic medium, Exp. Cell Res. 16 (1959) 118–147, http://www.ncbi.nlm.nih.gov/pubmed/13639945, Accessed July 31, 2017.

[137] J. Yu, K. Hu, K. Smuga-Otto, S. Tian, R. Stewart, I.I. Slukvin, J.A. Thomson, Human induced pluripotent stem cells free of vector and transgene sequences, Science 324 (2009) 797–801, https://doi.org/10.1126/science.1172482.

[138] M.A. Handel, J.J. Eppig, J.C. Schimenti, Applying "gold standards" to in-vitro-derived germ cells, Cell 157 (2014) 1257–1261, https://doi.org/10.1016/j.cell.2014.05.019.

[139] K. Hara, M. Kanai-Azuma, M. Uemura, H. Shitara, C. Taya, H. Yonekawa, H. Kawakami, N. Tsunekawa, M. Kurohmaru, Y. Kanai, Evidence for crucial role of hindgut expansion in directing proper migration of primordial germ cells in mouse early embryogenesis, Dev. Biol. 330 (2009) 427–439, https://doi.org/10.1016/j.ydbio.2009.04.012.

[140] S. Yang, J. Bo, H. Hu, X. Guo, R. Tian, C. Sun, Y. Zhu, P. Li, P. Liu, S. Zou, Y. Huang, Z. Li, Derivation of male germ cells from induced pluripotent stem cells in vitro and in reconstituted seminiferous tubules, Cell Prolif. 45 (2012) 91–100, https://doi.org/10.1111/j.1365-2184.2012.00811.x.

[141] N. Fukunaga, T. Teramura, Y. Onodera, T. Takehara, K. Fukuda, Y. Hosoi, Leukemia inhibitory factor (LIF) enhances germ cell differentiation from primate embryonic stem cells, Cell. Reprogram. 12 (2010) 369–376. (Formerly "Cloning Stem Cells"), https://doi.org/10.1089/cell.2009.0097.

[142] K. Tilgner, S.P. Atkinson, A. Golebiewska, M. Stojković, M. Lako, L. Armstrong, Isolation of primordial germ cells from differentiating human embryonic stem cells, Stem Cells 26 (2008) 3075–3085, https://doi.org/10.1634/stemcells.2008-0289.

[142a] W. Ge, H.-G. Ma, S.-F. Cheng, Y.-C. Sun, L.-L. Sun, X.-F. Sun, L. Li, P. Dyce, J. Li, Q.-H. Shi, W. Shen, Differentiation of early germ cells from human skin-derived stem cells without exogenous gene integration, Sci. Rep. 5 (1) (2015) 13822, https://doi.org/10.1038/srep13822.

[143] M. Kanai-Azuma, Y. Kanai, J.M. Gad, Y. Tajima, C. Taya, M. Kurohmaru, Y. Sanai, H. Yonekawa, K. Yazaki, P.P.L. Tam, Y. Hayashi, Depletion of definitive gut endoderm in Sox17-null mutant mice, Development 129 (2002) 2367–2379, http://www.ncbi.nlm.nih.gov/pubmed/11973269, Accessed June 11, 2017.

[144] F. von Meyenn, R.V. Berrens, S. Andrews, F. Santos, A.J. Collier, F. Krueger, R. Osorno, W. Dean, P.J. Rugg-Gunn, W. Reik, et al., Comparative principles of DNA methylation reprogramming during human and mouse in vitro primordial germ cell specification, Dev. Cell 39 (2016) 104–115, https://doi.org/10.1016/j.devcel.2016.09.015.

[145] D. Lucifero, W. Reik, Artificial sperm and epigenetic reprogramming, Nat. Biotechnol. 24 (2006) 1097–1098, https://doi.org/10.1038/nbt0906-1097.

[146] A. Rybouchkin, D. Dozortsev, M.J. Pelinck, P. De Sutter, M. Dhont, Analysis of the oocyte activating capacity and chromosomal complement of round-headed human spermatozoa by their injection into mouse oocytes, Hum. Reprod. 11 (1996) 2170–2175, http://www.ncbi.nlm.nih.gov/pubmed/8943524, Accessed July 31, 2017.

[147] D. Nikiforaki, F. Vanden Meerschaut, C. Qian, I. De Croo, Y. Lu, T. Deroo, E. Van den Abbeel, B. Heindryckx, P. De Sutter, Oocyte cryopreservation and in vitro culture affect calcium signalling during human fertilization, Hum. Reprod. 29 (2014) 29–40, https://doi.org/10.1093/humrep/det404.

[148] S. Segers, H. Mertes, G. Pennings, G. de Wert, W. Dondorp, Using stem cell-derived gametes for same-sex reproduction: an alternative scenario, J. Med. Ethics 43 (2017) 688–691, https://doi.org/10.1136/medethics-2016-103863.

CHAPTER

18

Computational Approaches in Reproductomics

Eva Vargas, Francisco J. Esteban*, Signe Altmäe*[†,‡]

*Department of Experimental Biology, Faculty of Experimental Sciences, University of Jaén, Jaén, Spain [†]Department of Biochemistry and Molecular Biology, Faculty of Sciences, University of Granada, Granada, Spain [‡]Competence Centre on Health Technologies, University of Tartu, Tartu, Estonia

ABBREVIATIONS

APP	amyloid precursor protein
ART	assisted reproductive techniques
DAVID	Database for Annotation, Visualization, and Integrated Discovery
dbPTB	database for preterm birth
EMBL-EBI	European Molecular Biology Laboratory-European Bioinformatics Institute
FDR	false discovery rate
GEO	Gene Expression Omnibus database
GWAS	genome-wide association studies
HGEx-ERdb	Human Gene Expression Endometrial Receptivity database
IPA	Ingenuity Pathway Analysis
IVF	in vitro fertilization
lncRNA	long noncoding RNA
MS	mass spectrometry
miRNA	microRNAs
NCBI	National Center for Biotechnology Information
NGS	next-generation sequencing
PCA	principal component analysis
PCOS	polycystic ovary syndrome
RC	raw count
RMA	robust multiarray average
RSEM	RNA-seq by expectation maximization

Reproductomics
https://doi.org/10.1016/B978-0-12-812571-7.00019-8

RVM	relevance vector machine
RNA-seq	RNA sequencing
SNP	single nucleotide polymorphism
SVM	support vector machine
UQ	upper quantile

INTRODUCTION

Over the last few decades, the development of the omics technologies, which include genomics, transcriptomics, epigenomics, proteomics, metabolomics, and microbiomics, has led to a better understanding of the molecular basis of several physiological and pathological processes.

Omics technologies imply the study of specific events and interactions from structures and biological processes to the biological functions in a complex and global manner. Therefore, it is possible to analyze in one experiment several biological compounds, epigenetic biomarkers, genes, mRNA, proteins, or metabolites simultaneously [1].

A decade ago, it was complicated to study multifactorial human reproductive diseases. Due to the emergence of omics technologies, it has been possible to improve knowledge in this field and obtain a broader view of complex biological systems [2]. One of the main advantages of these technologies is the large amount of information they provide, which can also be obtained with relatively little cost and effort [1]. However, this advantage is sometimes an obstacle, since powerful tools are needed to extract conclusions with biological significance from the large amount of information generated. Understanding and analyzing the omics data is further complicated by the complexity of the human body. The field of reproductomics, for instance, is challenged by the cyclically regulated hormones and other various factors and together with the individual's genetic background result in different biological responses (Fig. 18.1).

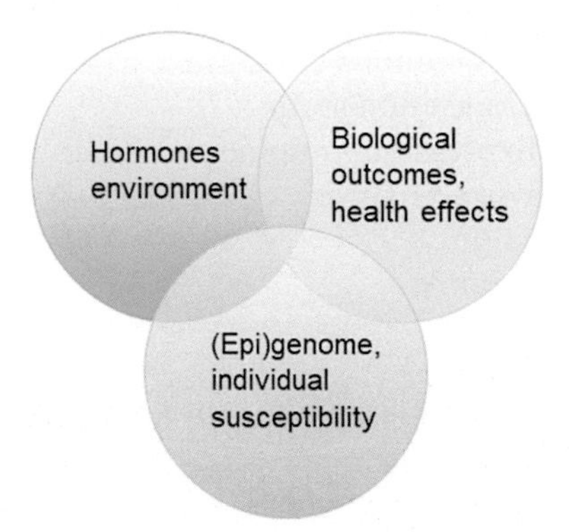

FIG. 18.1 Reproductomics is the study of interactions between an individual's hormonal regulation and their environment, individual genetic susceptibility (DNA makeup and epigenome) and health effects, and biological outcomes. *Adapted from J. Vlaanderen, L.E. Moore, M.T. Smith, Q. Lan, L. Zhang, C.F. Skibola, N. Rothman, R. Vermeulen, Application of OMICS technologies in occupational and environmental health research; current status and projections, Occup. Environ. Med. 67 (2) (2010) 136–143.*

The mass of data generated by these high-throughput technologies is clearly underexplored, representing a large challenge for biomedical research, since all of this information needs to be processed and analyzed carefully using powerful bioinformatic applications. The fact is that we have reached a bottleneck in data management—we have far more data than we are able to analyze and fully interpret. Indeed, millions of gene expression datasets are publicly available in repositories (e.g., *Gene Expression Omnibus, GEO*, and *ArrayExpress*), but few researchers fully utilize these data to explore new information; rather, they focus only on a small set of the data to compare with their own finding [3].

The current chapter summarizes and discusses the different computational approaches/options for analyzing the generated omics data, presents developed databases and analysis tools in the field, and provides an example of an analysis application in reproductomics.

IN SILICO DATA MINING

One of the most straightforward computational approaches in the omics data analysis is the in silico data mining approach, where different studies using the same omics approach and research questions are analyzed together (e.g., transcriptomics of endometrial receptivity and genes involved in endometriosis).

In the reproductomics field, despite the mass of omics data generated for the endometrial transcriptome, there are only three studies published to date, where the in silico data mining approach using previously published gene expression data has been utilized [4–6]. Bhagwat et al. created a *human gene expression receptivity database* (HGEx-ERdb) that consists of 19,285 genes expressed in the human endometrium, and they have identified 179 receptivity-associated genes [6]. Zhang et al.'s study analyzed raw microarray data from three previous microarray studies [7–9] and proposed 148 potential biomarkers for the receptive endometrium [5], while the study by Tapia et al. integrated gene lists from seven previous microarray studies and presented a list of 61 endometrial receptivity biomarkers [4]. However, these three in silico analysis studies only share nine genes in common. This clearly highlights the shortcomings of omics analyses (discussed in Ref. [10]), where the large variation in the study designs, analysis methods, and data processing lead to varying results. It is obvious that the large mass of data generated within endometrial transcriptomics studies, which is the most studied approach in the reproductomics field, is underexplored. Therefore, future studies are needed to analyze huge sets of data simultaneously in order to raise the power, credibility, and reliability of the findings.

In regard to endometriosis, these data mining approaches have also been applied in order to determine the role of certain genes in the pathogenesis of this widespread condition. Wang et al. constructed a decision tree, which is a widely applied methodology of data mining technology for performing classification and prediction, to analyze clinical data in 178 endometriosis patients. With this computational approach, they were able to identify a threshold based on the number and size of endometriomas that determined the effectiveness of the ultrasound-guided aspiration treatment [11]. More recently, Mathew et al.'s study mined publically available human endometrial microarray data from the *GEO* database and indicated the *FOXD3* as a key transcription factor involved in the hormonal regulation of the endometrium and the regulation of some aberrantly expressed genes associated with endometriosis.

Subsequently, a further wet-lab approach confirmed the results of the bioinformatic analysis, suggesting a differential expression of *FOXD3* in women with endometriosis [12]. Using a different data mining-related strategy, Liu and Zhao performed a text mining analysis using 19,904 papers published in the PubMed database in order to extract a list of endometriosis-related genes, with the aim of analyzing them and providing insights into the molecular mechanisms underlying this disorder. Their analysis retrieved a list of 1531 genes, 121 of which were significantly associated with endometriosis and were therefore submitted to an enrichment analysis, showing their implication in signal transduction, cell communication, or multicellular organism development, among others [13].

CORRELATION ANALYSIS

Correlation analysis is an interesting computational approach in reproductomics, although it is not always easy to perform and interpret. Epigenomics is becoming increasingly applied in reproductomics, since it has been established that these modifications can have a broad impact in gene expression and thereby in the underlying biological processes. Among all of the epigenetic modifications described so far, DNA methylation is probably the most studied, and many attempts have been made to unravel its effect in reproductive tissues. It is widely considered that when hypermethylation takes place in promoter regions, a transcriptional repression is observed, whereas hypomethylation is associated with enhanced gene activity [14].

As DNA methylation can be considered a dynamic process, it is interesting to point out its effect on such a dynamic and hormonal-dependent tissue such as the endometrium. It is known that DNA methylation plays a key role in gene expression regulation and influences functional changes in normal endometrial tissue [15]. Thus, recent studies have tried to investigate the endometrial methylome changes throughout the menstrual cycle.

Houshdaran et al. analyzed 18 eutopic endometrial tissue samples, 6 in each of the proliferative, early secretory, and midsecretory phases of the menstrual cycle in healthy women, and used the Spearman correlation test to investigate the relationship between DNA methylation and gene expression. They found that the methylome of the human endometrium differed throughout the cycle, underscoring the importance of epigenetic regulation of gene expression in different hormonal milieu in the human endometrium [16]. Saare et al. went a step further, and they analyzed the endometrial DNA methylome signature in both healthy women and endometriosis patients in correlation with the menstrual cycle. Their results demonstrated that the endometrial methylation profile of women with and without endometriosis was very similar, as the differences observed were too small to discern among both experimental groups. Therefore, they concluded that endometrial epigenetic changes are probably not behind the altered expression of the genes involved in the pathogenesis of this disorder [17].

Recently, Kukushkina et al. hypothesized that transcriptomic changes observed in the endometrium during the window of implantation may be caused by changes in the global DNA methylation pattern. Their correlation analysis of methylation and gene expression allowed them to confirm that methylation can activate and/or repress gene expression, as they found both positive and negative correlations depending on the genomic region [18].

However, other studies have proposed a nonlineal relationship between the epigenome and transcriptome [19,20], which is even more difficult to ascertain in reproductive processes due to the complexity [21]. Thus, it is not yet clear how DNA methylation correlates with gene expression, and therefore, there is a need for further studies to discern the real relationship.

META-ANALYSIS

Meta-analysis is another next-level computational approach in omics analyses. A meta-analysis comprises statistical methods for contrasting and combining results from different studies in the hope of identifying patterns among studies in order to achieve a higher statistical power and increased reliability/credibility of the findings [22].

In the reproductomics field, one of the most studied omics approaches is the transcriptome analysis of endometrial receptivity. As mentioned above, the perceived limitations of the omics approaches lie in the differences in experimental design, timing and conditions of endometrial sampling, sample selection criteria, array/sequencing platforms and annotation versions used, pipelines for data processing, and a lack of consistent standards for data presentation [10]. Together, these restrictions have made it nearly impossible to perform a meta-analysis of similar studies on endometrial receptivity.

The preferred meta-analysis method requires analysis of the raw expression datasets; nevertheless, such a rigorous approach is often not possible due to the unavailability of the raw data. Variation in the number of gene transcripts known at a given moment together with the technological platform employed makes the proper integration of raw datasets rather complex. Furthermore, the limited sample size and noisiness of microarray data have resulted in inconsistency of biological conclusions [23]. In order to overcome these limitations, Altmäe et al. applied a meta-analysis method, utilizing a robust rank aggregation method, which has been specifically designed for comparison of different gene lists and identification of commonly overlapping genes among studies [24]. With this approach, Altmäe et al. has directly analyzed differentially expressed gene lists from nine published studies, comprising a total of 96 endometrial biopsy samples from healthy normally cycling women. This work has provided an up-to-date metasignature of endometrial receptivity biomarkers (Fig. 18.2) [25]. With their elegant meta-analysis, they identified 57 genes that could serve as potential biomarkers of endometrial receptivity. Furthermore, genes such as *SPP1, PAEP, GPX3, GADD45A, MAOA, CLDN4, IL15, CD55, DP44, ANXA4*, and *S100P* were indicated to be of special interest as they have also been identified as putative biomarkers of endometrial receptivity in previous data mining and review studies [4–6,26,27]. This is the first transcriptome-based meta-analysis study in reproductomics and could serve to inspire new analytic approaches to unravel other research questions in reproductomics.

Another important meta-analysis in the field has been performed on genome-wide association studies (GWAS) for endometriosis [28]. Rahmioglu et al. included eight previously published studies in order to test both the consistency and heterogeneity of the obtained results. They were able to demonstrate that there was remarkable consistency in the endometriosis GWAS, with little evidence of population-based heterogeneity [28].

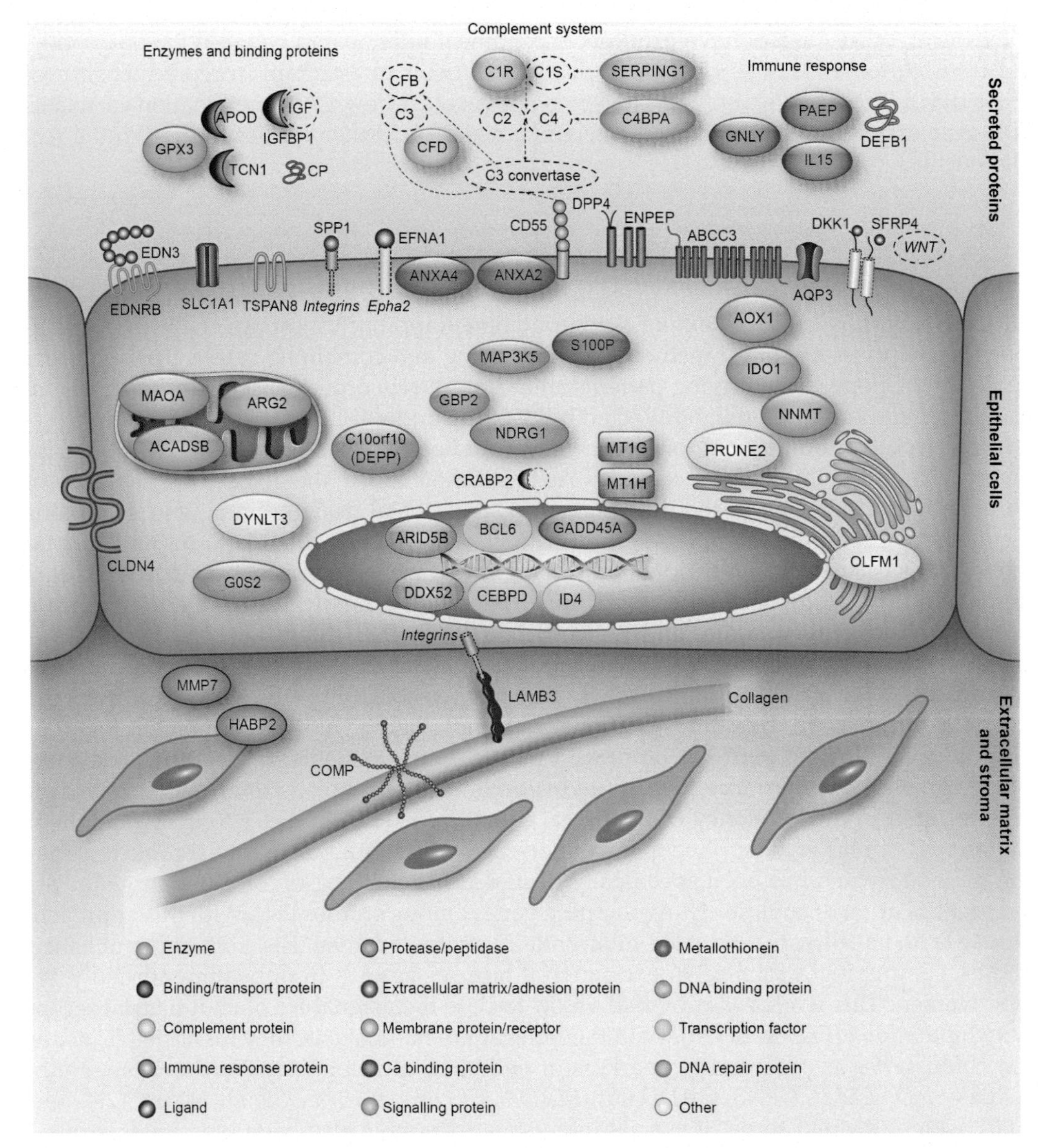

FIG. 18.2 Schematic overview of the 57 metasignature genes and their localization and involvement in the receptive phase in the endometrium. Different membrane-associated proteins (ABCC3, ANXA2, ANXA4, AQP3, CD55, DKK1, DPP4, EDN3, EFNA1, ENPEP, SFRP4, SLC1A1, SPP1, and TSPAN8), epithelial cell tight junction protein (CLDN4), secreted enzymes and binding proteins (APOD, CP, GPX3, IGFBP1, and TCN1), secreted immune response proteins (DEFB1, GLNY, IL15, and PAEP), extracellular matrix-associated proteins (COMP, HABP2, LAMB3, and MMP7), different enzymes (ACADSB, AOX1, ARG2, IDO1, MAOA, and NNMT), signaling proteins (C10orf10, GBP2, G0S2, MAP3K5, and NDRG1), metallothioneins (MT1G and MT1H), DNA binding and repair proteins (ARID5B, DDX52, and GADD45A), transcription factors (BCL6, CEBPD, and ID4), and other intracellular proteins (CRABP2, DYNLT3, OLFM1, PRUNE2, and S100P) are indicated. Additionally, the enriched KEGG pathway of complement cascade with the identified genes *C1R*, *SERPING1*, *CD55*, *C4BPA*, and *CFD* is presented [25]. *Reproduced with permission from Nature Publishing Group. This work is licensed under the Creative Commons Atrribution 4.0 International License. To view a copy of this license, visit http://creativecommons.org/license/by/4.0/ or send a letter to Creative Commons, PO Box 1866, Mountain View, CA 94942, USA.*

SYSTEMS BIOLOGY

Systems biology is an advanced way to analyze the omics data. Systems biology refers to the integration of the generated omics data, compiling information from genomics, epigenomics, transcriptomics, proteomics, and metabolomics (among others), and analyzes it using computational models that infer the behavior of a cell, tissue, or organism (Fig. 18.3). The global analysis of the omics under an interdisciplinary approach provides us with a wealth of information with a high potential for unraveling the interaction of the molecular processes underlying the function of an organism in both health and disease [10].

It is not easy to extract the true correlations with biological meaning, as true system interactions are nonlinear and a number of external factors have to be taken into consideration [30]. The main reason why systems biology has become a successful approach is the fact that it has been realized that the traditional reductionist approaches focusing only on a few molecules at a time cannot describe the complexity of the molecular environments across a whole system [31]. Therefore, a holistic approach is needed. However, the challenge of systems biology resides in the compilation of data derived from very different areas such as biology, chemistry, statistics, physics, mathematics, and computational engineering. As systems biology tries to provide a complete interpretation of all this knowledge, the high-throughput omics platforms have to be integrated for the analysis, display, and recording of information in order to guarantee compatibility and accessibility to these datasets [32]. In short, systems biology represents an integrative approach to understand biology, by allowing the functional analysis of the structure and dynamics of cells, as it focuses on complex interactions (i.e., a *holistic approach*) rather than on the isolated characteristics of biological systems (i.e., a *traditional approach*) [10].

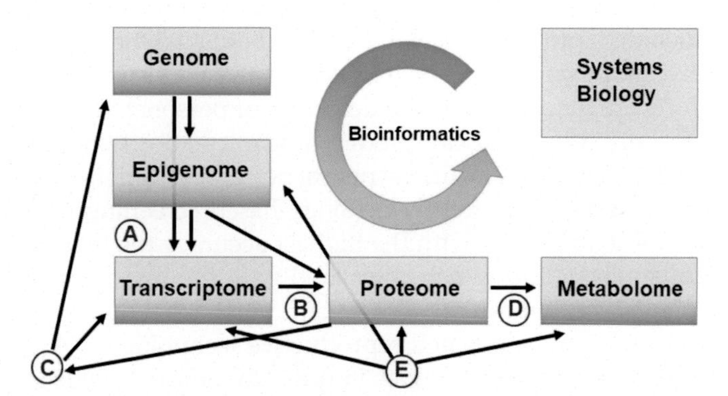

FIG. 18.3 Bioinformatic analysis tools unite/integrate the different omics levels into a systems biology approach. The complexity of omics fields in biological processes that contribute to the study and understanding of biological systems is illustrated. The human genome consists of more than ~25,000 genes that encode for ~100,000–200,000 gene transcripts and over 1 million different proteins, whereas there are as few as ~3000 different metabolites that make up the human metabolome [29]. The genome is essentially invariant among cells and tissues, where the epigenome has low/moderate temporal variance and can influence both the transcriptome and proteome. The transcriptome has a high variance and is translated into the proteome tissue specifically and physiological state specifically, thus affecting the metabolome in a tissue-specific manner. This "simple" model is further modulated by multiple factors: (A) differential splicing that can be affected by proteome, (B) posttranslational modification of proteins, (C) transcription factor binding, (D) receptor ligand binding, and (E) environmentally induced factors. *From S. Altmäe, F.J. Esteban, A. Stravreus-Evers, C. Simón, L. Giudice, B.A. Lessey, J.A. Horcajadas, N.S. Macklon, T. D'Hooghe, C. Campoy, B.C. Fauser, L.A. Salamonsen, A. Salumets, Guidelines for the design, analysis and interpretation of 'omics' data: focus on human endometrium, Hum. Reprod. Update 20 (1) (2014) 12–28; published with permission from Human Reproduction Update, Oxford publishing group.*

The systems biology approach in reproductomics has been applied in different areas, such as endometrial receptivity analysis, male sperm analysis, placenta analysis, and preterm birth analysis [33–36]. One of the first systems biology approach was performed by Ghosh et al. on endometrial omics analysis in order to identify molecules important for blastocyst implantation [37]. They performed a comparative analysis of the transcriptomic profile in the implantation stage endometrium using cDNA-based arrays and employed endometrial samples from rhesus monkeys as a model. Subsequently, they constructed expression networks to represent the transcriptomic profile of the endometrium in both normal hormonal milieu and one without progesterone starvation, which has an insufficiency effect in endometrial receptivity. They concluded that the presence of an embryo modifies the transcriptomic network of the endometrium, thus resulting in a physiological process to support blastocyst implantation [37].

Another study applying a systems biology approach aimed to integrate proteomics, transcriptomics, and interactomics datasets in order to construct a dynamic interactome map of the human sperm flagella microtubulome [35]. They defined a set of 116 gene products associated with the human sperm microtubulome by applying an integrated genomic workflow. This analysis, together with the construction of a disease-interaction network allowed them to find some novel factors potentially associated with altered sperm motility and male fertility. Among these molecules, *CUL3* and *DCDC2C* seemed to have important roles in the sperm flagellum, despite the fact that their association with male germ-line differentiation has not been described thus far [35].

Systems Medicine

When systems biology approaches are applied to biomedical research, this is called "systems medicine." In other words, systems medicine is the application of systems biology with medical purposes with the ultimate goal of transition to a predictive, preventive, and personalized medicine [38]. This interesting approach is becoming the trend in biomedical research, although it is still far from being included into the clinical practice, as both analytic and computational technologies have to evolve in the same direction in order to solve unanswered biological questions.

To date, some efforts are made toward the detection of personalized biomarkers in reproductive biology, especially those that can predict outcomes after treatments. Kyrgiou et al. recently presented a system based on artificial neural networks for personalized management of women with cervical issues detected by cytology-based screening [39]. They concluded that their method was able to predict with the highest accuracy the histological diagnosis in women with cervical abnormalities when compared with the use of traditional tests alone, which may contribute toward a more preventive and personalized care [39]. Another active field in systems medicine is in the assisted reproductive technologies (ART), where focus is being put on improving the actual techniques in order to increase yields in transferable embryos and to develop novel tools for selecting the best embryos [40].

However, this is still a new research area in reproductomics, with a high potential in the future to develop personalized treatment protocols for improving fertility/infertility/gynecologic complications.

Interactomics

In general, the systems biology approach is focused on the global analysis of multiple interactions at different levels, and the main strategy usually employs networks as a representation of interacting molecules. Modules are built around each separate regulated function, with the interrelations among modules finally arising as complex networks. This interaction network approach has become known as "interactomics."

In the reproductomics field, there have been a number of well-conducted studies of the molecular interactome between the embryo and endometrium in an attempt to reveal the initial steps of the implantation process [33,41,42].

The study by Haouzi et al. presented the global gene expression comparison in exogenous gonadotrophin-stimulated endometrium and blastocyst trophectoderm cells during the implantation period in women undergoing infertility treatment by in vitro fertilization (IVF). They identified a number of interacting molecules, including *LAMA1* in embryo and *ITGB8*, *LAMA2*, *DCN*, *CEACAM1*, *CD44*, and collagens 1 and 4 that have also been detected in other studies as having role in implantation [41]. Another study analyzed the transcriptome of mural trophectoderm cells from human blastocysts and compared the pattern with human embryonic stem cell-derived trophoblasts, offering a new view into the players in the very early stages of the human implantation process [42]. Interestingly, several of the detected molecules known to be involved in the implantation process were identified, such as CXCL12, HBEGF, inhibin A, DKK3, WNT5A, follistatin, IL-6, IL-6ST, LEPR, OCLN, SERPINE1, TGFB1, and VCAN [42]. Altmäe et al. published the first true interactome study of molecular networks between the human embryo and endometrium [33]. They integrated the embryonic and endometrial transcriptomic profiles with protein-protein interactions and applied a novel network profiling algorithm (HyperModules), which combines topological module identification and enrichment analysis [43]. Their curated embryo-endometrium interactome highlighted the importance of cell adhesion molecules and cytokine-cytokine receptor interactions in the implantation process (Fig. 18.4). The methodology presented could inspire new analytic approaches to untangle complex networks in human physiology and pathophysiology.

Interactome analysis has also been applied in male reproduction, where the amyloid precursor protein (APP) interaction network in human testis was studied. A unique interactome reflecting the connections between APP and key molecules for male reproduction, particularly in the sperm-oocyte interaction, was identified [44].

COMPUTATIONAL TOOLS

Databases

The large amount of data derived from omics studies has to be stored in order to provide researchers with useful information. For this purpose, there has been an increasing development of databases (public and private) in which the results of the omics experiments can be stored.

The genome variation databases *GWASdb* (http://jjwanglab.org/gwasdb), *GWAS central* (http://www.gwascentral.org/), and *GWAS catalog* developed by the *European Bioinformatics Institute* (EMBL-EBI) (https://www.ebi.ac.uk/gwas/) represent excellent examples of repositories that are being updated frequently and in which human genetic variant data identified by genome-wide association studies are stored [45,46]. In general, they combine collections of traits/diseases associated with *single nucleotide polymorphisms* (SNPs), which are the most common genetic variation.

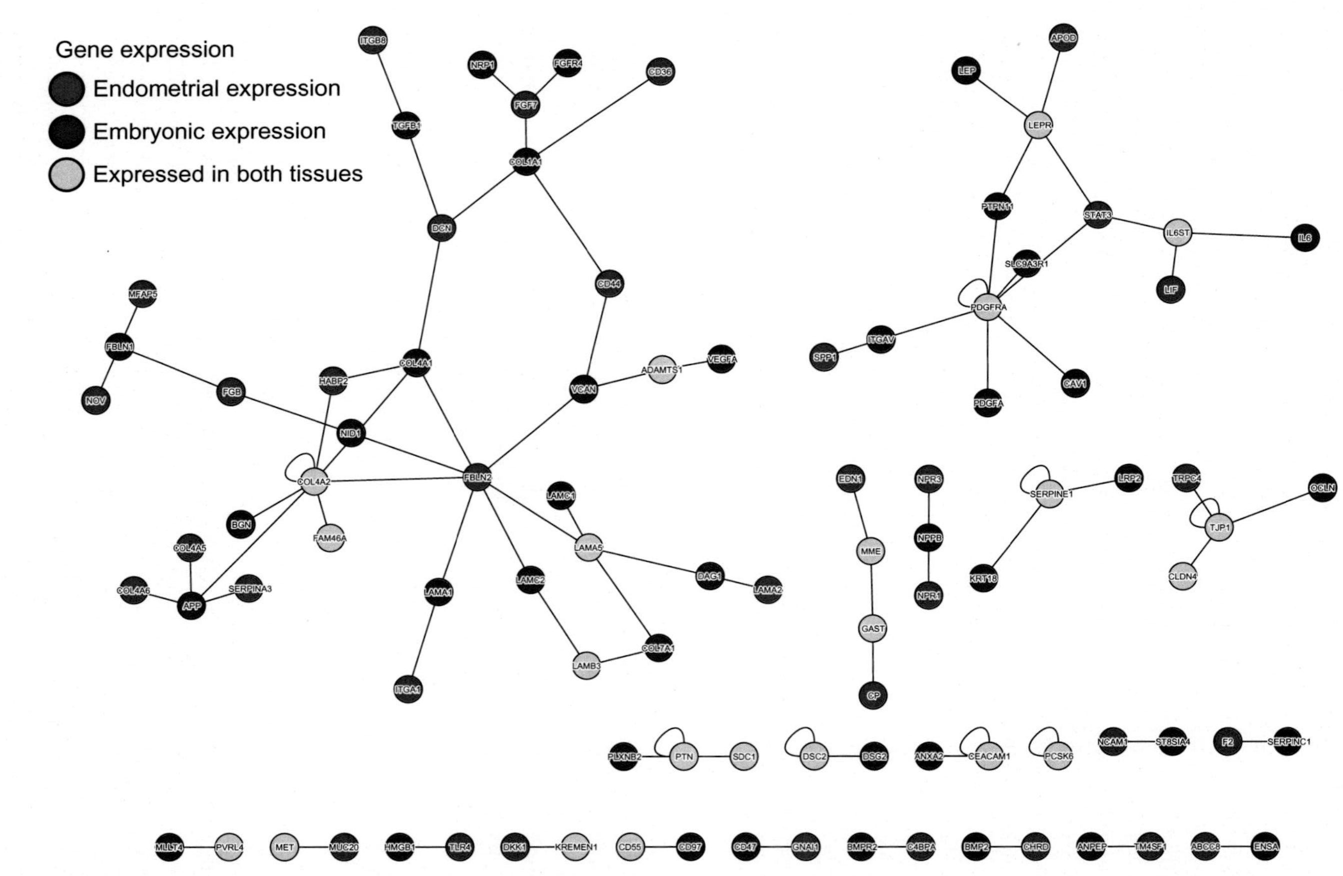

FIG. 18.4 Human embryo-endometrium interactome. The interaction network analysis from protein-protein interaction data and literature curation. Node color represents tissue-specific differential gene expression: *blue*, expressed in the embryo; *red*, expressed in the endometrium; and *gray*, expressed in both tissues *From S. Altmäe, J. Reimand, O. Hovatta, P. Zhang, J. Kere, T. Laisk, M. Saare, M. Peters, J. Vilo, A. Stavreus-Evers, A. Salumets, Research resource: interactome of human embryo implantation: identification of gene expression pathways, regulation, and integrated regulatory networks, Mol. Endocrinol. 26 (1) (2012) 203–217; published with permission from Molecular Endocrinology.*

In regard to methylome data, *MethBase* (http://smithlabresearch.org/software/methbase/) was created from public bisulfite-sequencing datasets to provide information about the methylome of well-studied species at varying methylation levels at individual sites [47]. The human disease methylation database *DiseaseMeth version 2.0* (http://www.bio-bigdata.com/diseasemeth/) contains a collection of aberrant DNA methylation in human diseases and provides lists of disease-gene associations based on high-throughput methylome experiments [48].

The study of microRNA (miRNA) expression profile generates a large amount of information that is likewise stored in public databases. *miRBase* (http://www.mirbase.org/) provides researchers with published miRNA sequences, annotation, and target prediction and represents a useful way to analyze the miRNome [49]. The *microRNA.org* resource (http://www.microrna.org) is a comprehensive platform including miRNA target predictions based on the miRanda algorithm and expression profiles [50].

One of the most important repositories of gene expression data is the one developed by the *National Center for Biotechnological Information (NCBI)*, the so-called *Gene Expression Omnibus* (https://www.ncbi.nlm.nih.gov/geo/), a public functional genomics database that supports minimal-information-about-an-experiment-compliant data submissions. This information is composed in minimum of (1) the raw data from an experiment, (2) the normalized data, (3) the samples annotation, and (4) the experimental design. This project started in 2002, and now (June 2017), it contains >4000 expression datasets, and this number is increasing regularly [51]. Although *ArrayExpress* (http://www.ebi.ac.uk/arrayexpress/) was developed later, it is one of the most recommended repositories to archive functional genomics data from omics platforms such as microarrays and next-generation sequencing (NGS) techniques such as RNA sequencing (RNA-seq) to support reproducible research [52].

Lately, the study of the proteome has been attracting increasing attention, and there has been a development of some platforms to store the generated data. Some examples of proteome repositories are *PeptideAtlas* (http://www.peptideatlas.org/), which archives a collection of peptides identified by tandem mass spectrometry (MS) proteomics experiments [53]; the *PRIDE database* (https://www.ebi.ac.uk/pride/archive/), a centralized database for proteomics data such as protein and peptide identifications or posttranslational modifications [54]; and the *Plasma Proteome Database* (http://www.plasmaproteomedatabase.org/), one of the largest resources on proteins reported in plasma and serum [55]. The *Human Proteome Map* (http://www.humanproteomemap.org/) is an interactive resource that integrates results from different proteomics experiments with genetic data for several tissues [56]. Finally, the information available in *Proteomics DB* (https://www.proteomicsdb.org/) covers around 80% of the human proteome and allows researchers to explore it protein by protein [57].

Apart from this variety of software, several web-based tools have been created in order to host information about genotypes related to a given disease state, not only in reproduction but also in other biomedical areas. *Phenopedia* (https://phgkb.cdc.gov/HuGENavigator/startPagePhenoPedia.do) offers a disease-centered view of genetic association studies, and both genes and diseases can be used as a starting point for the study of some reproductive conditions such as polycystic ovary syndrome, menopause, azoospermia, or cryptorchidism, just to name a few [58].

DiseaseConnect (http://disease-connect.org/) is a web server for the analysis and visualization of diseases with shared molecular mechanisms [59]. This resource was applied in 2015 to detect how infertility etiologies such as premature ovarian failure or testicular cancer share particular genes and pathways with other pathologies [60].

Finally, and in a more integrative way, a manually curated metadatabase called *OMICtools* (https://omictools.com/) was developed. This open-access directory provides an overview of some of the accessible tools related to omics data [61].

In the area of reproductomics, different web resources and bioinformatic tools including specific information for regular or disease situations or clinical reproductive outcomes have been generated. We will focus on some of the most representative reproductome databases:

- *Follicle Online* (http://mcg.ustc.edu.cn/bsc/follicle/index.php) is a centralized database in which follicle assembly, development, and ovulation information are integrated in order to make the understanding of folliculogenesis functions and related genes and proteins easier [62].
- *Spermatogenesis Online database* (http://mcg.ustc.edu.cn/bsc/spermgenes/index.php) was created to offer a comprehensive resource of genes reported to participate in spermatogenesis [63].
- *GermlncRNA* (http://germlncrna.cbiit.cuhk.edu.hk/) is a catalog of long noncoding RNAs (lncRNAs) and associated regulations in male germ cell development. It represents a useful platform for researchers to visualize, analyze, and download annotated or novel lncRNAs expressed in key stages of male germ cell development such as type A spermatogonia, pachytene spermatocytes, and round spermatids [64].
- *The ReproGenomics Viewer* (http://rgv.genouest.org) is a website in which data generated from genomics studies in mammals are stored. Although it currently hosts datasets mainly related to male reproductive traits (such as testis biology or spermatogenesis), in the future, these will extend to other aspects of reproduction [65].
- *Gametogenesis Epigenetic Database, GED* (http://gametsepi.nwsuaflmz.com/), is a manually curated resource that includes information about the effects of epigenetic processes such as DNA methylation in different phases of gametogenesis process [66].
- *Ovarian Kaleidoscope Database* (http://okdb.appliedbioinfo.net/) provides a large amount of information about biological functions, expression patterns, and regulation of genes expressed in the ovary. Although it is one of the oldest reproduction-related databases (it was created in 2000), it is being updated weekly, and it will continue to facilitate research by ovarian researchers [67–69].
- *Endometrial Database* (http://www.endometrialdatabase.com/edb/) includes endometrium-related publications and gathers information about the molecular characteristics of this fascinating tissue.
- *Human Gene Expression Endometrial Receptivity database (HGEx-ERdb)* (http://resource.ibab.ac.in/HGEx-ERdb/) is a comprehensive database that stores manually curated data from studies related to the receptive phase endometrium from *GEO* and peer-reviewed journals. The aim of this catalog is to offer a broad vision of the transcriptomic profile of the human endometrium, including information such as the expression status or the relative expression level of any gene of interest previously reported as expressed in the endometrium [6].
- *GEneSTATION* (http://www.genestation.org/) was developed with the aim of integrating diverse types of omics data, mainly functional genomics, across placental mammals (including humans), in order to advance the understanding of the molecular basis of gestation. Moreover, for each gene, this database includes evolutionary, organismal, and molecular information and some links to other public databases, which can be pregnancy disease-specific or not [70]. Fig. 18.5 shows the different categories into which available gene expression studies in *GEneSTATION* are classified.

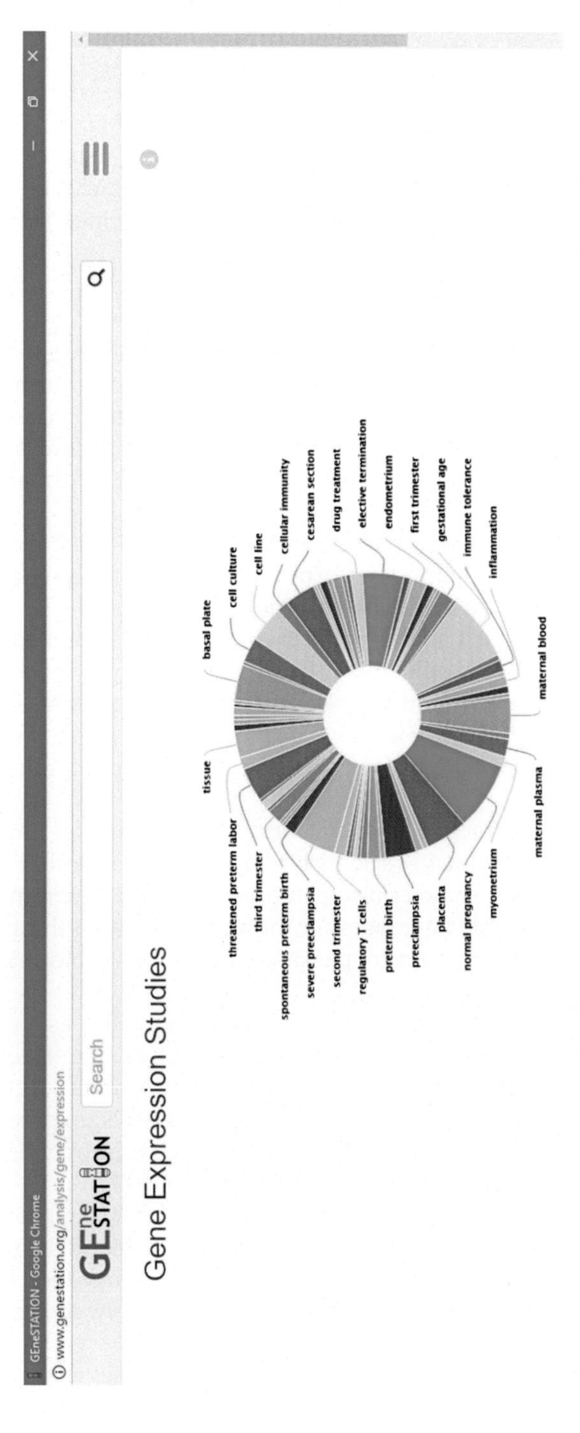

FIG. 18.5 Screenshot of *GEneSTATION* [70] showing the different categories into which gene expression studies included in this platform are divided.

- *Database for Preeclampsia, dbPEC* (http://ptbdb.cs.brown.edu/dbpec/), in which published literature and associated genes for preeclampsia are stored [71]. Uzun et al. has recently developed a database for the preeclampsia phenomena, one of the most common causes of maternal and fetal mortality and morbidity.
- *The database for preterm birth* (*dbPTB*) (http://ptbdb.cs.brown.edu/dbPTBv1.php) collects data of genomic variants related to preterm birth from public databases, archives of expression arrays, and curated literature, leading to a better understanding of the genetic architecture of this process whose causes have yet to be well defined [72].

Computational Tools for Analyzing Omics Data

Regardless of their origin, once the data are either generated or accessed via public databases, it has to be analyzed in order to extract as much biomedical information as possible. Although most of the omics high-throughput technology corporations usually provide researchers with specific software for the analysis of their products, different open-access software resources have been created to facilitate the integration of the information derived from these powerful tools.

The *microarray software suite TM4* (http://mev.tm4.org) is a great example of how a unique and free accessible tool can help to perform visualization, analysis, and integration of microarray-derived data in a user-friendly environment [73].

With the emergence of NGS such as RNA-seq, there was a need for a specific software to analyze the data derived from its application. Thus, software as *Chipster* (http://chipster.csc.fi/) was developed. It is a user-friendly platform for analyzing RNA-seq and microarray data [74]. Similarly, *ReadXplorer* (http://www.readxplorer.org) is a free Java-based software for the visualization and extensive analysis of genomic and transcriptomics next-generation sequencing data [75].

Despite that it is not usually attractive for researchers to deal with large amounts of data, the emergence of omics technologies is forcing scientists to move their research from laboratories to computational systems. Although some critical computational and programming knowledge is required, two of the most versatile languages for the analysis of high-throughput data are *R* (https://www.r-project.org/), its *Bioconductor* packages (http://www.bioconductor.org/), and *Python* (https://www.python.org/).

Bioconductor (http://www.bioconductor.org/) was created in 2001 as an open source and open development project for the analysis and comprehension of high-throughput genomic data [76], based on *R* programming language (https://www.r-project.org/). The main advantage of this program is its low computational cost [77]. *R* allows preprocessing, statistical testing, and downstream analysis, and it can be applied to each type of omics data. Thus, it is a particularly useful tool in bioinformatics and systems biology [31].

Python (https://www.python.org/) is an open, usable, and distributable resource widely applied in data analysis. It has become a popular language in biomedicine whose success mainly lies in the fact that it provides some libraries and third-party toolkits that extend its functionality to biological domains such as sequence analysis or the interpretation of omics data [78].

The processing steps carried out using computational approaches depend on both the nature of the data and the way to get knowledge from them. However, the typical pipeline

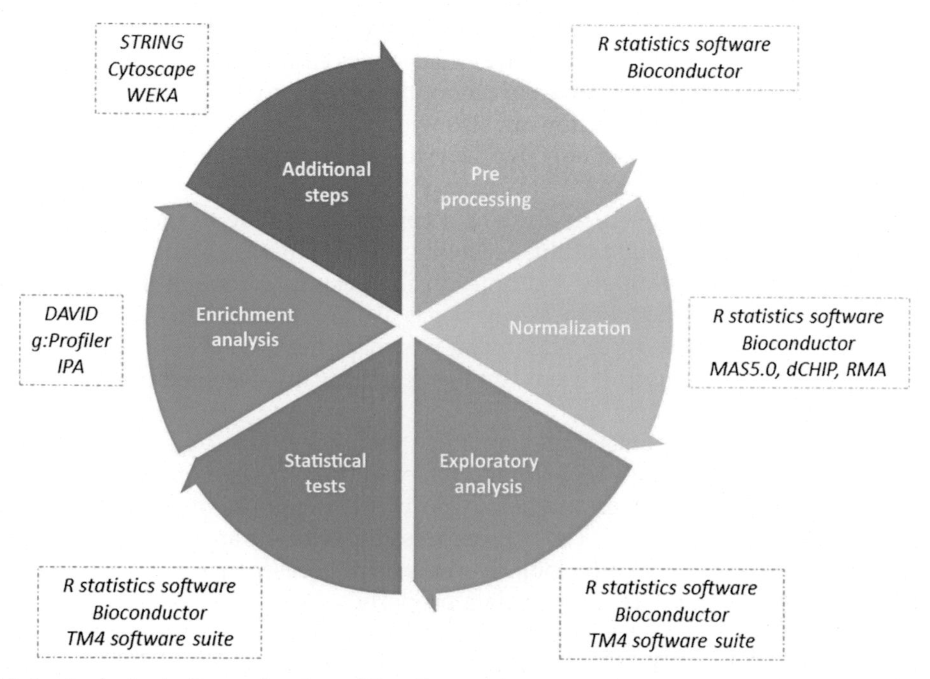

FIG. 18.6 Analysis pipeline and tools traditionally used by systems biologists. After data collection, data are preprocessed and normalized using different approaches. These include an exploratory analysis that is carried out in order to find possible patterns among the data; statistical tests that allow the detection of differentially expressed molecules in different situations; and finally, an enrichment analysis and some other steps that are applied to the data to further elucidate biological related issues.

analysis of omics technology data is composed of a preprocessing step, followed by a normalization, an exploratory analysis, some statistical tests, and finally an enrichment analysis (although some additional steps can be performed) (Fig. 18.6).

Preprocessing

It is evident that biological data show high variability, which is further increased when we have to deal with data related to complex situations. Thus, there is a need for a preprocessing step to discard technical outliers. This step is highly variable and predominantly depends on the criteria chosen by the researchers, with the chosen criteria being influenced by both the employed technology and the actual data [79]. Thus, its description in scientific papers has to be as detailed as possible, in order to guarantee the reproducibility of the experiments. Below, we discuss some of the approaches most widely followed for preprocessing omics data.

Although a number of approaches are used for this purpose, it depends on the type of data that is being analyzed. However, the bases of all approaches are similar, as they all remove the potential confounding factors that can affect the reliability of the results. One of the most popular and easiest methods to correct these disturbances is the logarithmic transformation, which makes data more homogeneous by unifying the scale in the expression rates [80]. For instance, raw downregulations are normally found between zero and one and upregulations between one and infinity. This is a widely used approach in managing transcriptomics data from microarrays, because this procedure scales upregulation and downregulation equally around zero, making the interpretation of the fold changes easier [31]. Other procedures such as the background correction, the summarization, the quality control, the missing value estimation, or the background subtraction are also widely employed [81]. Discretization of the numerical features is also another effective way to preprocess numerical features such as hormone levels [82].

Sequencing data (i.e., RNA-seq) analysis is a complex task due to the numerous biases that can arise from the application of this technology. These biases should be identified and removed in order to obtain accurate and reproducible results. Although there is not a gold standard for this step so far, the preprocessing step is usually divided into the following steps: quality control, filtering procedures, alignment, postmapping quality control and counting, and several pipelines, with most of them run using specifically designed programming language [83–85].

Methylome data usually show a high variability, therefore making the preprocessing step necessary. The most widely applied methods for preprocessing genome-wide methylation data try to remove technical biases and batch effects due to the use of arrays by usually removing failed and/or SNP probes. These preprocessing methods are applied to ensure reliable and high-quality data generation [86]. However, further research is needed to provide an optimal approach to analyze this complex data properly.

The development of algorithms for preprocessing proteomics data has been an active area of research. No standard method nor order for the preprocessing steps for proteomics data such as those derived from MS has been established so far. However, it is widely accepted that the preprocessing procedure is divided into several subtasks, which include alignment, filtering, peak detection, and quantification [87].

Normalization

Any variation in the data with no biological interest should be removed in order to make them comparable—which is the major aim of the normalization step. Some sources of variation in omics high-throughput technology data are the preparation of the biological sample, the manufacturing processes, or the biases associated to some molecules needed for the performance of the analysis, such as some fluorescent dyes [88]. Moreover, the data generated by the high-throughput technologies are usually relative (i.e., the abundance of a peptide measured by mass spectrometry), thus making the application of normalization methods necessary in order to help in the removal of these nonbiological confounding factors that affect the expression value of a given molecule (genes, proteins, miRNA, metabolites, etc.) [89].

It is important to carefully choose the appropriate method. This selection is mainly based on the type of platform and the aim of the analysis, since it could predetermine the subsequent results. Some methods such as loess, quantiles, MAS5.0, or dCHIP have been developed to help in the removal of the systematic inherent errors and are becoming increasingly popular in the microarray data analysis [90], with others such as quantiles becoming widely used in RNA-seq data analysis [91]. However, the most popular procedure for microarray data analysis is the robust multiarray average (RMA), a comprehensive tool that includes background adjustment, quantile normalization, log transformation, and summarizing from the individual probe values down to probe set values [31,92].

Normalization is a critical step for RNA-seq data analysis. Therefore, several methods have been proposed to correct biases between and within samples, among which raw count (RC), upper quantile (UQ), and median (Med) are the most popular. Furthermore, a new trend is the development of novel methods based on machine learning algorithms such as RNA-seq by expectation maximization (RSEM) and Sailfish [91]. Recently, Yang et al. proposed an integrated approach for RNA-seq data normalization based on an appropriate alignment of the signal intensity. They obtained a higher detection of differentially expressed genes than other methods using either the raw data or the data normalized using traditional methods [93]. However, the optimal method to normalize these complex data has yet to be determined.

In regard to proteome data, normalization procedures are widely applied in order to correct for variation due to the differential amount of protein among samples or its degradation. Therefore, different algorithms that include both preprocessing and normalization steps have been developed [87].

It is widely considered that a suitable method for normalizing the data is one that does not greatly modify the data. Some authors propose to repeat the analysis with several normalization methods and consider the results obtained [89]. Given the difficulty in choosing the appropriate method, some tools such as the *R* package for proteomics data, *Normalyzer* (http://quantitativeproteomics.org/normalyzer) have been designed to cope with this issue [94].

Exploratory Analysis

It is interesting to identify the unknown patterns in the data in order to reveal unexpected relationships among them. The objective of the multivariate analysis is to explore the large amount of information enclosed in the datasets and to summarize it into a few variables that capture the real structure of the data [95]. This can be achieved through the so-called exploratory techniques, which can be defined as a set of procedures with the potential to retrieve some groups of data with a similar behavior from the whole dataset. More precisely, the exploratory analysis of omics data relies on unsupervised methods such as hierarchical clustering and principal component analysis (PCA). Other popular approaches include k-means clustering, consensus clustering, nonnegative matrix factorization, and mixture models [96].

Clustering techniques classifies data objects into a set of disjoint groups depending on their similarities. Therefore, objects within a class have high similarity to each other, while objects in separate classes are supposed to be less similar [97]. The degree of similarity can be measured through the so-called distances, of which the *Euclidean distance*, the *Pearson's correlation coefficient*, the *Jaccard index*, and the *Manhattan distance* are examples. The exact choice depends on both the quantitative (linear/nonlinear) and qualitative nature of the data [80]. A complete review about clustering techniques can be found in [97].

Due to the high multidimensionality of the data derived from the omics high-throughput experiments, there is also a need for a dimension reduction through methods that rely on the projection of the data onto a new basis of lower dimensionality [95]. As mentioned earlier, one of the most versatile unsupervised methods for data reduction is PCA [96]. It allows analysts to obtain an overview of the structures in the data by projecting high dimensional data into a lower dimensional span, where objects with similar levels for a feature tend to cluster together [98].

Statistical Tests

Statistical tests are applied in order to detect those molecules (genes, methylation sites, miRNAs, etc.) that show a specific behavior when different situations are compared, thus allowing for the finding of differences among experimental groups.

Firstly, studies in the field have focused on comparing the expression levels of the given molecules for each experimental group, looking for statistically significant increases or reductions between them, namely, *fold changes*. However, this simple approach appeared not to have as much statistical signification as expected, and it was therefore supplemented with some additional approaches [99], which are discussed below.

The first approaches were based on simple statistical methods that tested the association using models based on the inherent nature of the data and the dimensionality of the outcomes of interest [95]. Besides that, these methods were mainly based on the selection of molecules over a threshold according to their *P*-values. Although it depends on the criteria of the researchers, a *P*-value below 0.01 or 0.05 was usually considered as statistically significant.

Data generated using omics high-throughput technologies are characterized by high dimensionality, which means a number of measurements per subject, much more than the actual number of subjects in the study [100]. Moreover, due to the fact that thousands of characteristics are measured, this provides results of poor biological interpretability and reliability, with a high rate of false positives, which are only due to chance. Furthermore, as the different omics profiles can be measured through a wide variety of analytic methods, the type and correlation structure of the generated data are likely to be very heterogeneous [95]. Altogether, these special features make the data biased, which therefore justifies the application of multiple testing correction strategies to control for error rates [30]. A broad description of several methods designed for multiple testing corrections can be found in [95]. Among the variety of methods to address this issue, the most commonly used methods are discussed below.

The *false discovery rate* (FDR) control is defined as the expected number of false positives among all associations detected as significant [95]. Its interpretation is simple, which makes it a more suitable tool for the analysis of results than *raw* (noncorrected) *P*-values. There are different methods to perform this FDR correction, with the one developed by Benjamini-Hochberg being the most popular.

Another method widely used (especially when dealing with microarray data) is the so-called Bonferroni correction, which proposes that the corrected *P*-value is obtained by the quotient between the raw *P*-value and the number of genes in the array [80]. Despite its simplicity, it is stringent when a large number of tests have been made [31].

Nonparametric tests such as the *rank product* can be applied to high-throughput data analysis, especially when the number of biological replicates is low [101]. The statistical significance is calculated based on ranks of fold changes and on the assumptions about the data that is generated [31].

At this point, it has to be pointed out that the omics technologies are evolving and specializing even faster than the specific analytic tools and sophisticated platforms. Therefore, it is still necessary to improve data analysis through the development of integrative, specific, and friendly software in order to adequately deal with this complex amount of information.

Enrichment Analysis

Once the molecules with differential behavior are detected, a subsequent step is to perform an enrichment analysis to detect the biological processes carried out by them or those in which they are overrepresented.

Traditionally, this process has been addressed following a reductionist approach, by trying to elucidate the molecular function or the biological processes carried out by a molecule of interest (e.g., a gene, a protein, or a metabolite). Some specific tools such as *GeneCards* (http://www.genecards.org/), *Entrez Gene* (https://www.ncbi.nlm.nih.gov/gene), *UniProt* (http://uniprot.org/), or *the Human Metabolome Database* (http://www.hmdb.ca/) have been designed for this purpose.

However, it is difficult to follow this classical approach when the number of molecules is high, which usually happens when dealing with omics high-throughput data. This type of analysis manages high volumes of data, which causes a large number of molecules of interest to arise. Therefore, it makes the enrichment analysis both a tedious and a biased task [80]. To address this gap, a series of tools have been developed:

- One popular tool is the *Database for Annotation, Visualization, and Integrated Discovery* (DAVID) (https://david.ncifcrf.gov/), which is an integrative platform to extract information about functional annotation from gene lists [102].
- *Gene ontology* (http://www.geneontology.org/) is another framework for the integration of biology with a high potential for performing enrichment analysis [103].
- *g:Profiler* (http://biit.cs.ut.ee/gprofiler/) was developed in order to help systems biologists to unravel the biological processes underlying some molecules of interest [104].
- *Enrichr* (http://amp.pharm.mssm.edu/Enrichr/) is a web-based tool for the analysis of the collective functions of gene or protein lists in an intuitive way [105].
- *REVIGO* (http://revigo.irb.hr/) is a tool for reducing the long list of enriched biological processes into clusters of similar gene ontology terms by allowing a small amount of similarity [106].

Apart from these publicly available tools for enrichment analysis, a number of commercial platforms have been developed. *Ingenuity Pathway Analysis* (IPA, from QIAGEN Inc., https://www.qiagenbioinformatics.com/products/ingenuitypathway-analysis) is the most commonly used private software, although other platforms such as *Pathway Studio* (https://www.pathwaystudio.com/) have an important role in the enrichment analysis [107].

Network Analysis

An elegant way to represent interactions widely applied in systems biology is the construction of networks. Groups of elements (nodes) that are related by connections (edges) compose the structure of a network. This straightforward approach displays a smart manner of representing protein-protein and protein-DNA interactions when transcriptomics and proteomics data have to be integrated [30]. Furthermore, it has been stated that this is a very useful way to show biological data related to gene regulation and cellular and signaling pathways, since quantitative information can be extracted from the nodes in order to infer their importance in the system they represent [108]. This quantitative information is gathered in the form of network parameters such as the centrality indicators, which allow for the identification of the most important nodes in a system. Some of the aforementioned parameters are the *degree*, the *betweenness*, the *closeness*, or the *Eigenvector centralities*. These analyses allow researchers to highlight some highly connected nodes in the structure of a network, which are called "hubs" in network theory-based language [109].

As the visualization and interpretation of complex data in the context of a pathway or network usually provide more insightful information than other types of representation [110], many computational tools for network analysis have been developed to date. The most commonly used tools are discussed:

- *EGAN* (http://akt.ucsf.edu/EGAN/) is a free software tool that allows the visualization and interpretation of the results of high-throughput exploratory assays in an interactive graph of genes, interactions, and metadata. Moreover, this Java desktop application provides a comprehensive and automated calculation of metadata coincidence for gene lists and provides links to literature and web resources such as *PubMed, Gene Ontology,* or *NCBI Entrez Gene* sites [111]. When this powerful tool was applied to the study of the inflammasome of different populations of Sertoli cells, it was concluded that a group of differentially expressed genes was related to the activation of autophagy and innate immunity pathways. Thus, suggesting that Sertoli cells could participate in male infertility pathogenesis via inflammatory cytokine induction [112].
- *STRING* (http://string-db.org) is a database of known and predicted protein-protein interactions, including physical and functional associations from high-throughput lab experiments, text mining, and previous knowledge, among others [113]. The information provided by this platform is further required when the aim of the research is to elucidate the function of a gene in a complex context such as the structure of a network. This tool was applied to study the interaction among some genes that showed a differential expression in the endometrium during the window of implantation when comparing natural with controlled ovarian stimulation cycles [114].
- *Cytoscape* (http://www.cytoscape.org/) is an open-source software platform designed for network visualization and analysis [115], and it is run under a user-friendly interface. Furthermore, it extends its functionality by utilizing a wide variety of plugins designed by the scientific community [31]. Due to its versatility, its application has revealed the interaction among several molecules related to reproduction. This is the case for the network pharmacology approach carried out to explore the molecular mechanisms of a herbal medicine on anovulatory infertility [116]. Another example is represented by Sabetian and Shamsir's work, who were the first to study the protein-protein interaction network related to azoospermia, in order to identify the complex mechanisms and candidate genes involved in this major cause of male infertility [117]. A novel embryo-endometrium interaction study was published by Altmäe et al., where they demonstrated the first systems biology approach into the complex molecular network of the implantation process in humans [33].

Classifiers and Predictors

Another step in systems biology is the detection of classifiers or predictors of a given situation, that is, an algorithm that selects a (group of) molecule(s) that can discriminate among a normal and a pathological sample or predict future data trends, respectively. Several algorithms belonging to data mining and machine learning fields have been developed to aid in this phase, whose explanation deserves an entire book to itself.

Two main steps compose the data classification process. In the first step, the classifier or the model is built (i.e., this is the learning step), whereas in the second step, it is used for the classification itself.

In the first learning phase, the classifier is built using a set of classification algorithms from a training set, made from part of the data (a set of tuples) and their associated class labels. Subsequently, the test data is applied using the estimation of the accuracy of classification rules previously established. A high accuracy means that the classification rules can be used with new data tuples. The most popular classifiers include *support vector machine* (SVM), *random forest*, *relevance vector machine* (RVM), and *logistic curves*, just to name a few. These algorithms can be run using specific software such as *WEKA* (http://www.cs.waikato.ac.nz/ml/weka/), which is a suite of data mining and machine language software platforms written in the Java programming language, and it is one of the most common tools for classification and prediction tasks [118].

Although these approaches are typically used only in gene expression and/or genomic profiling, some other factors such as metabolic pathways are starting to be considered [119]. In practice, Liu et al. was recently able to develop a method based on the random forest classifier that was capable of discriminating between pathogenic and normal SNPs [120]. It is becoming clear that these validated classifiers and predictors have the power to be included as part of the healthcare plan, leading to the development of targeted therapies and personalized medicine [121], which can also be applied to reproductive research purposes.

EXAMPLE OF DATA ANALYSIS APPLICATIONS

This section gives two examples of how to carry out a typical data analysis using the data derived from the high-throughput technologies. To achieve this objective, datasets were accessed from some of the previously described public repositories for two study cases.

Example 1: Microarray Data Analysis of Female Infertility Due to PCOS

Polycystic ovary syndrome (PCOS) is the most common endocrine disorder among women during reproductive age and is characterized by anovulatory dysfunction and impaired androgenic hormone secretions. Moreover, it has been stated that women with PCOS present a number of systemic symptoms apart from those related to the reproductive system such as β-cell dysfunction, metabolic syndrome, cardiovascular diseases, or derangements in adipose tissue [122,123]. The latter has attracted scientists' attention, and several groups have tried to elucidate the relationship between the adipose tissue and PCOS. For that reason, we chose a public dataset focused on this topic and analyzed it following the steps described below.

Raw intensity signal data were obtained from the *NCBI GEO* website, under accession number GSE5090 (https://www.ncbi.nlm.nih.gov/geo/query/acc.cgi?acc=GSE5090) [124]. The dataset includes gene expression values for nine PCOS samples and eight control samples obtained from morbidly obese women that were biopsied with the aim of studying the effect of the omental adipose tissue in the pathogenesis of PCOS [124].

After a preprocessing step that included a logarithmic transformation and the adjustment of those expression values below 2 using the *impute* package in R [125], the data were

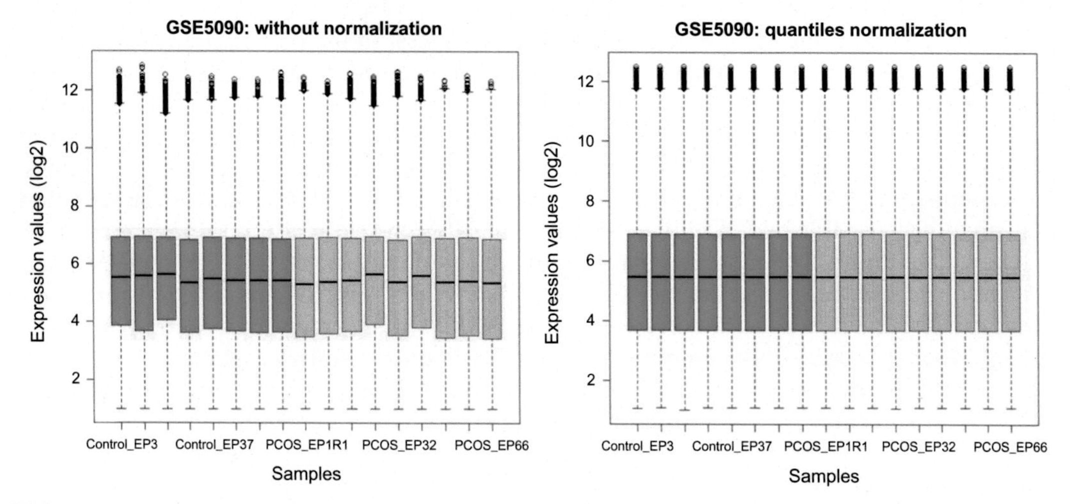

FIG. 18.7 Effect of the normalization procedure in the expression dataset GSE5090, *blue* is for control group and *yellow* for the PCOS group. Data were normalized between arrays using the quantile method in R.

quantile normalized using the quantile method implemented in the *limma* package in R [126]. The results are shown in Fig. 18.7.

After this preliminary analysis and with the aim of detecting common expression patterns among samples, the *TM4* software was employed to perform some exploratory analysis techniques. The hierarchical clustering (Fig. 18.8) did not reveal a specific expression pattern among groups, and neither did the PCA analysis (Fig. 18.9).

Statistically significant differences in gene expression between groups were evaluated by using the rank product method, implemented in the R package *RankProd* (http://www.bioconductor.org/packages/release/bioc/html/RankProd.html) [127]. This approach applies a multiple test hypothesis for raw *P*-value corrections to ascertain false positive rates, and its use is recommended when the number of samples in each group is small [101]. Taking all of this into account, a *P*-value <0.05 and a 1.5-fold change in gene expression were selected as the criteria for considering differentially expressed genes in the dataset. Of the 22,284 genes evaluated in the array, a group of 179 genes exhibited a differential expression when both experimental groups were compared. More specifically, 54 of them increased their expression, while 135 were downregulated. In an attempt to elucidate the biological processes affected by this differential expression, we performed a gene enrichment analysis using the platform *DAVID* (https://david.ncifcrf.gov/). The obtained results indicated that inflammatory and immune responses were the major functions in which downregulated genes were enriched, while enrichment analysis of upregulated genes did not yield any significant biological process (FDR < 0.05) (Table 18.1).

In conclusion, following an alternate approach as that of the one used by the authors allowed us to identify some new biological processes that have not been related to PCOS thus far. This is important as these biological processes may be playing an important role in the development of this commonly occurring syndrome in women.

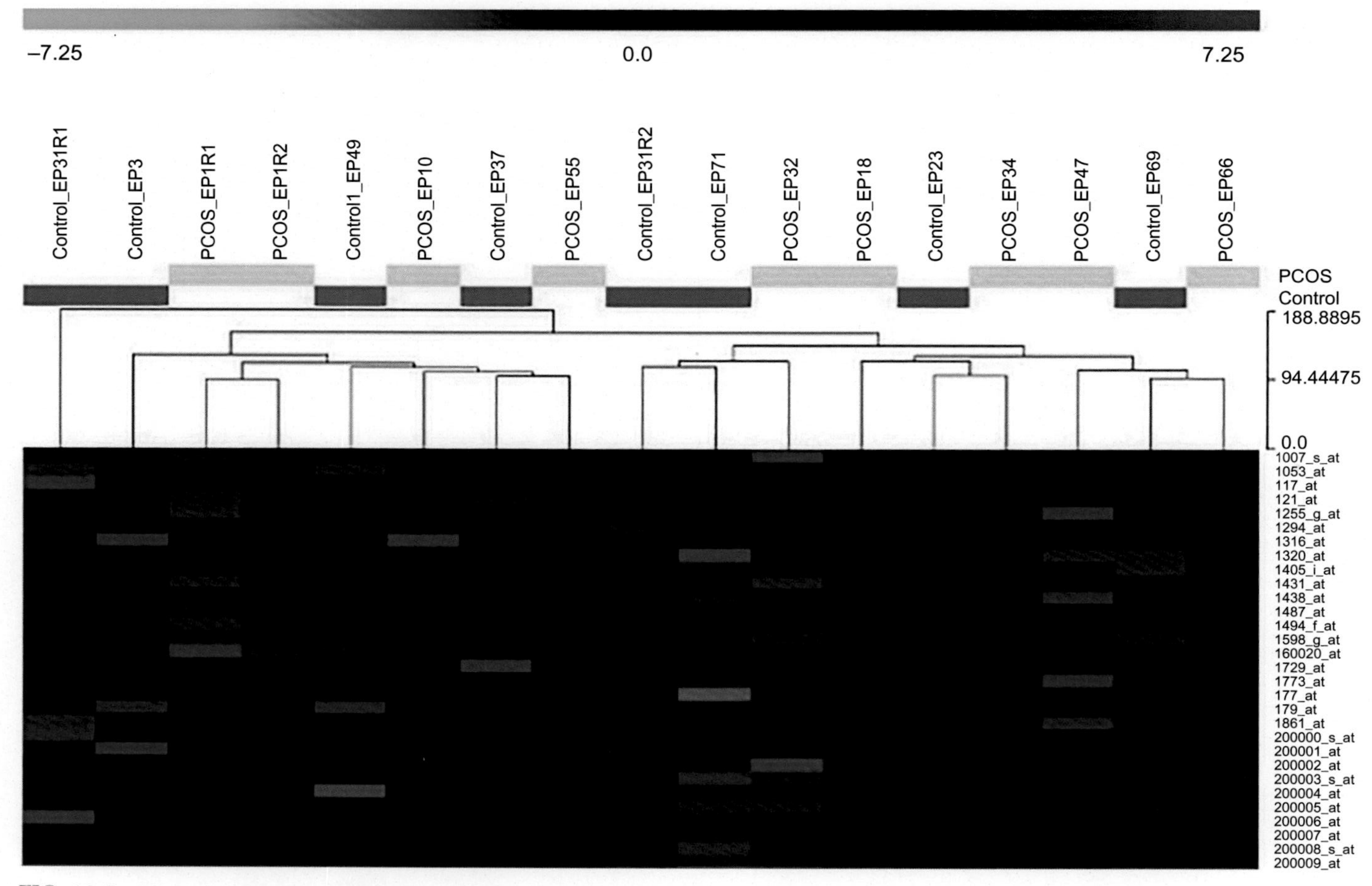

FIG. 18.8 Hierarchical clustering that shows the absence of a specific gene expression pattern among the control group *(blue)* and the PCOS group *(yellow)*. It was built using Euclidean distance as a similarity measure and the complete linkage as a method for clustering. Note: the image has been shortened for a better visualization.

FIG. 18.9 PCA analysis demonstrates the absence of a specific gene expression pattern between the PCOS (*yellow*) and control (*blue*) groups.

TABLE 18.1 Enrichment Analysis for the Genes Differentially Expressed in PCOS Samples

Upregulated Genes			
GO Term	**Genes**	***P*-Value**	**FDR**
Neuron recognition	CNTNAP2, NTM	0.01	0.13
Antigen processing and presentation of peptide or polysaccharide antigen via MHC class II	HLA-DQA1, HLA-DQB1	0.04	0.45

Downregulated Genes			
GO Term	**Example of Genes**	***P*-Value**	**FDR**
Inflammatory response	CCL2, CCL20, CXCL1, TNFAIP6, CHI3L1	2.07×10^{-11}	3.31×10^{-8}
Immune response	CCL2, CCL220, CXCL1, IGKC, IL6, IL17A, LIF	3.60×10^{-7}	5.76×10^{-4}
Chemotaxis	CCL2, CCL20, CXCL1, FOSL1, FPR2, PLAUR	1.14×10^{-5}	0.02
Response to molecule of bacterial origin	CXCL2, CXCL8, CD24, MALT1	1.94×10^{-5}	0.03

GO: gene ontology terms. FDR: false discovery rate. Note that the biological processes in which upregulated genes are involved do not show statistical significance (FDR < 0.05).

Example 2: Network Construction Applied to Study Male Infertility Due to Azoospermia

Azoospermia is the term referred to ejaculates that lack spermatozoa without implying a specific underlying cause. Although its definition is ambiguous, the diagnosis of the absence of spermatozoa in a semen sample is an important criterion for the diagnosis of male infertility [128]. A list of several genes that have been related with this widespread condition is publicly available through the *Phenopedia* database (https://phgkb.cdc.gov/HuGENavigator/startPagePhenoPedia.do). This list of genes can be transformed into a network using the platform *STRING* (Fig. 18.10), together with some information regarding the connections between the genes.

As stated before, an enrichment analysis gives information about the biological processes carried out by a group of genes. Using *DAVID*, we were able to detect a list of statistically significant biological processes, which are shown in Table 18.2. Most of them are related to male germ cell development.

This information can be further analyzed using a more sophisticated software such as *Cytoscape*, which will offer a variety of options for the visualization, some network centrality parameters and specific plugins to perform additional analysis. In this application example, we decided to select those genes that had been related to azoospermia in at least three publications. This reduced our original gene list composed of 229 genes to one with only 42 genes.

After this step, we decided to follow two different approaches for the visualization of this network: (1) a traditional analysis, in which some topology parameters such as the degree or the betweenness were assessed (Fig. 18.11), and (2) an analysis using the *Cytoscape* plugin *GeneMANIA* [129,130] (Fig. 18.12).

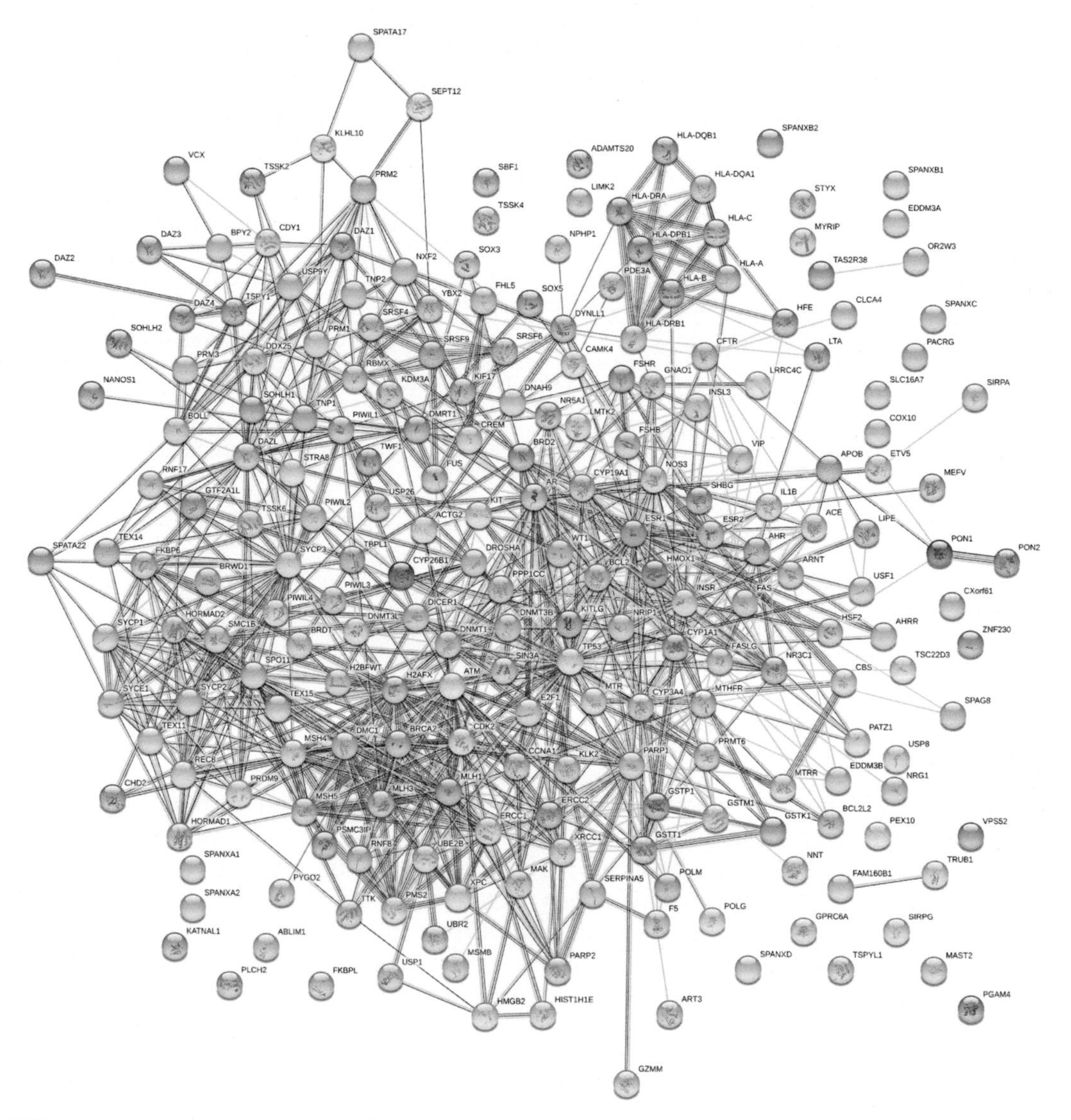

FIG. 18.10 *STRING* network that shows the connections among the group of 229 genes involved in azoospermia (this list was obtained from *Phenopedia* database).

The first approach resulted in the network that is displayed in Fig. 18.11. Two topology parameters were taken into account for the visualization of the network: (i) the degree, defined as the number of connections that a node has to other nodes, and (ii) the betweenness, a measure of the way in which a node acts as a bridge connecting two a priori separated nodes, that is, the way in which a node acts as a bottleneck between two given nodes (see the bottom of Fig. 18.11). It is interesting to point out that some of the most connected nodes such as the *androgen receptor*

TABLE 18.2 Enrichment Analysis of the Azoospermia-Related Genes

Biological Term	Example of Genes Involved	P-Value	FDR
Spermatogenesis	AR, DAZL, E2F1, BRD2, INSL3, STYX, MLH1, SOHLH1, SOHLH2, YBX2	4.15×10^{-54}	6.90×10^{-51}
Spermatid development	KIT, TNP1, SPO11, KIT, RNF8, REC8, SPANXB1, DDX25, PACRG, PRM2	7.08×10^{-13}	1.18×10^{-9}
Synaptonemal complex assembly	REC8, TEX15, HORMAD1, SYCE1, MLH3, SPO11, SYCP2, SYCP1, TEX11	1.00×10^{-10}	1.66×10^{-7}
Male gonad development	HMGB2, ESR1, KITLG, KIT, FSHR, WT1, BCL2, PATZ1, KLHL10, SRD5A2	1.76×10^{-10}	2.92×10^{-7}
Male meiosis I	REC8, SYCP3, BRCA2, BRDT, UBR2, SPO11, CCNA1, DMC1	1.31×10^{-9}	2.18×10^{-6}
Reciprocal meiotic recombination	REC8, STRA8, MSH4, MLH3, SPO11, DMC1, SYCP1, ATM, TEX11	2.61×10^{-9}	4.34×10^{-6}
Cell differentiation	DAZ3, DAZ4, DAZ1, TSPY1, DAZ2, FKBP6, MAK, CREM, TP53, SOHLH1	4.66×10^{-9}	7.76×10^{-6}
Meiotic nuclear division	REC8, FKBP6, HORMAD1, HORMAD2, DMC1, BOLL, CDK2, SMC1B	1.34×10^{-8}	2.22×10^{-5}
Multicellular organism development	DAZ3, DAZ4, DAZ1, RNF17, DAZ2, MAK, CREM, TP53, NXF2, TNP1	4.70×10^{-8}	7.85×10^{-5}

FDR: false discovery rate.

gene (AR) or the *deleted-in-azoospermia-like* (DAZL) gene are involved in the most significant biological processes obtained by the enrichment analysis as previously performed (Table 18.2).

The second approach, a finer approach, was conducted using the *Cytoscape* plugin *GeneMANIA*, which uses previously published studies to predict gene-gene interactions among a query gene set [129,130]. The input list of the 42 genes related to azoospermia that were found in at least three publications was analyzed, and the results are shown in Fig. 18.12. The output network from GeneMANIA has 62 nodes and 249 edges due to the fact that this plugin grows the network by adding some nodes that have been predicted as genes that interact with those from the input list. Among the 42 nodes and their interacting genes, it was found that 48.05% shared some coexpression characteristics and 31.72% displayed similar physical interactions. Other results including colocalization, pathways, and shared protein domains are also shown in Fig. 18.12.

GeneMANIA has proved to be a powerful tool not only to display different correlations among nodes but also to perform an enrichment analysis. When the genes related to azoospermia were submitted to an enrichment analysis, they were mainly involved in a wide variety of processes related to male reproduction, which may indicate that changes in their expression would result in disturbances in the reproductive success. Top biological processes are shown in Table 18.3.

Although further research is required, the two approaches carried out in this section are just an example of how network analysis can help to unravel the interaction among biological molecules and decipher the underlying biological processes.

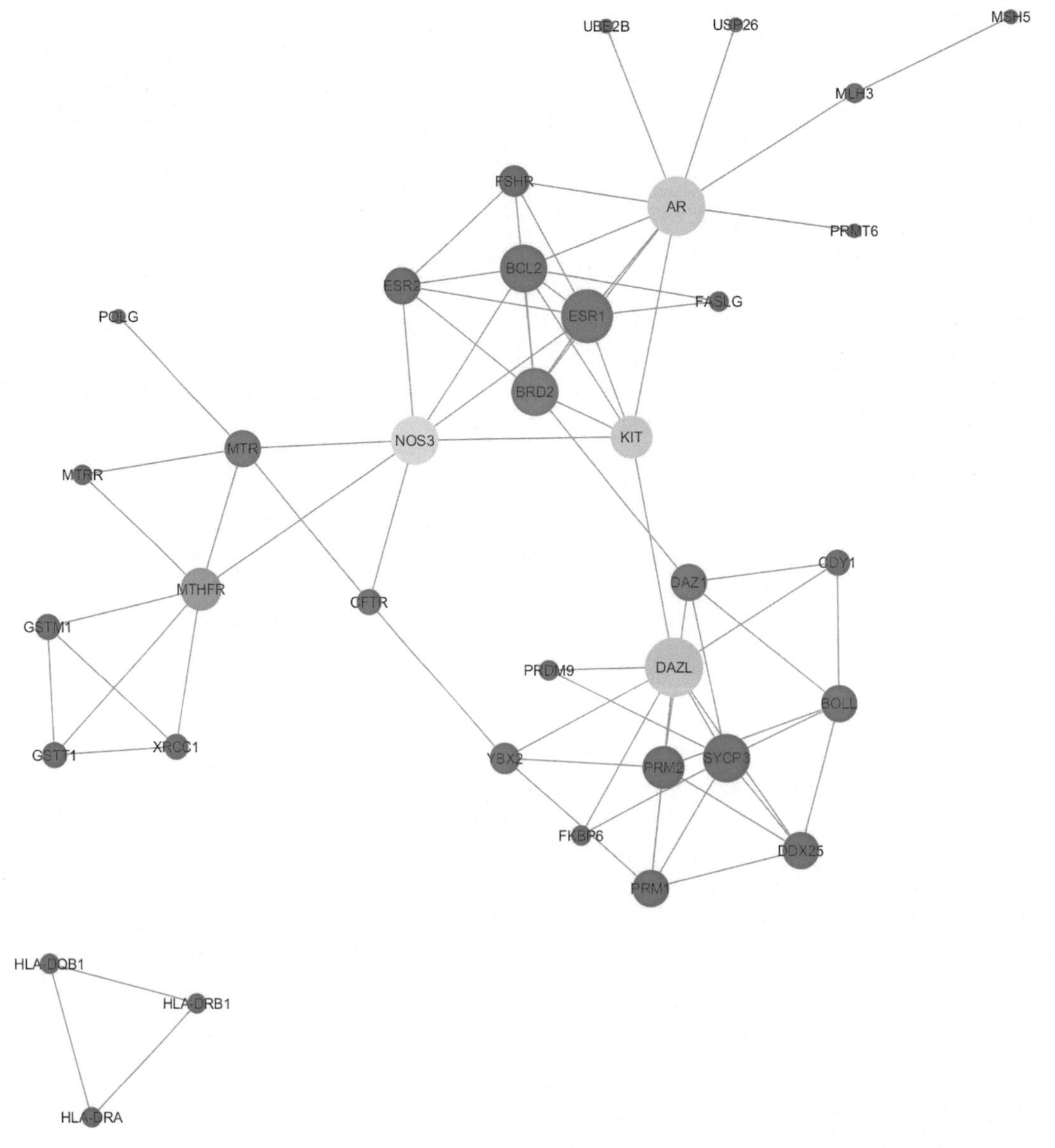

FIG. 18.11 Azoospermia network built under a traditional gene network analysis using Cytoscape. Two topology parameters were considered, namely, the degree and the betweenness; more specifically, the size of the nodes increases as the degree does, and the genes with the highest betweenness have lighter colors. Note: some of the original genes were lost in the analysis due to the absence of interactions with other nodes, thus reducing the initial list from 42 genes to 36.

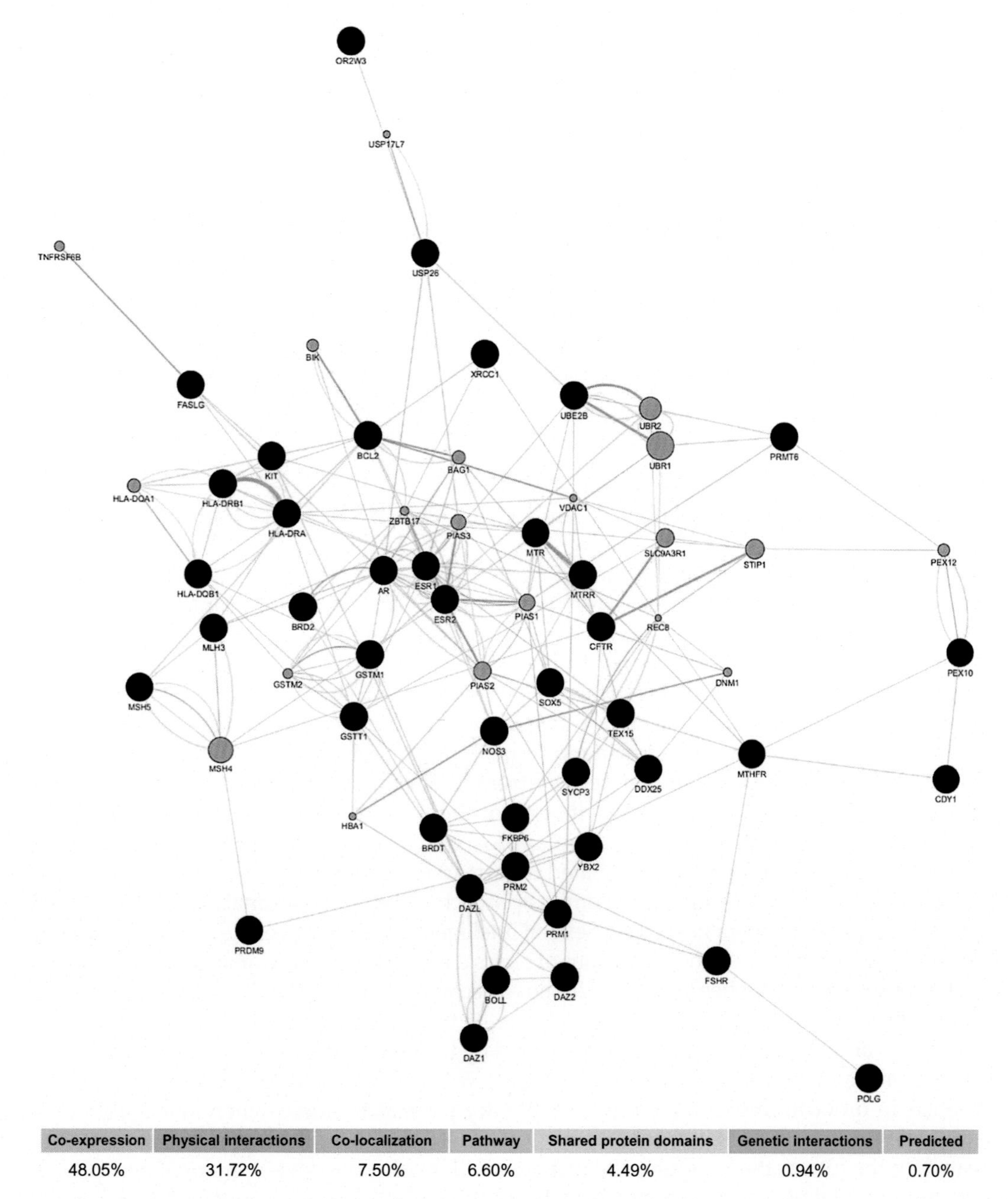

FIG. 18.12 GeneMANIA network of azoospermia-related genes. *Black nodes* match with the 42 genes submitted from the original list, and different edge colors represent different correlations. *Gray nodes* indicate genes associated with query genes.

TABLE 18.3 GeneMANIA Enrichment Analysis of the Genes Involved in Azoospermia

GO id	Description	*q*-Value	Number of Genes
GO:0048610	Cellular process involved in reproduction	7.54×10^{-9}	12
GO:0019953	Sexual reproduction	7.54×10^{-9}	13
GO:0007276	Gamete generation	7.54×10^{-9}	12
GO:0007283	Spermatogenesis	7.54×10^{-9}	11
GO:0048232	Male gamete generation	7.54×10^{-9}	11
GO:0048609	Multicellular organismal reproductive process	2.07×10^{-8}	12
GO:0032504	Multicellular organism reproduction	5.19×10^{-8}	12
GO:0007126	Meiotic nuclear division	4.51×10^{-7}	8

GO id: gene ontology identifier.

CONCLUSIONS AND FUTURE PERSPECTIVES

With the emergence of omics technologies, there has been improved knowledge regarding human reproductive health and disease by providing a large amount of information about the biological processes involved in reproductive processes. It is clear that the analysis of data obtained from the application of omics technologies has provided a broader view of complex biological systems in both physiological and pathological situations, so that data integration through systems biology approaches offers users a unified view of such data.

Today, computational approaches in reproductomics are advancing reproductive medicine and technology and the information derived from it will change the way ART techniques are performed [1]. Even though it is clear that all of these computer-based analyses should be complimented with experimental wet-lab approaches, systems biology represents a useful way to deal with the complexity associated to human reproduction.

One of the most critical steps in the omics approach is the adequate analysis of the complex high-throughput omics data generated. Several intelligent tools have been developed in order to overcome different obstacles and limitations of omics data, such as in silico analyses. Nevertheless, several consensus approaches together with novel techniques need to be developed.

Reproductive data heterogeneity is still one of the largest challenges in biological data integration, and this challenge could be handled by the standardization of the methods used. This requires a greater involvement of biologists with bioinformaticians and computational scientists in order to improve the interdisciplinary projects required when managing omics data [131]. As a result of this collaborative way of research, new databases and tools will be created. In the future, the technological progress will involve the development of new powerful computational platforms of analysis that can deal with the large amount of data generated and, in addition, with a user-friendly interfaces that enable noncomputational scientists to analyze their data.

The ultimate goal of the omics analyses is to understand human reproduction in health and disease and to identify new biomarkers for improving fertility/infertility and gynecologic disorders. Preferably, the identified biomarkers should be noninvasive, which in fact is one of the benefits of the high-throughput omics techniques. Despite these encouraging perspectives, well-designed, sufficiently powered, and adequately analyzed as well as interpreted studies are needed in order to achieve successful outcomes.

References

[1] R.R. Egea, N.G. Puchalt, M.M. Escrivá, A.C. Varghese, OMICS: current and future perspectives in reproductive medicine and technology, J. Hum. Reprod. Sci. 7 (2) (2014) 73–92.

[2] E. Silvestri, A. Lombardi, P. de Lange, D. Glinni, R. Senese, F. Cioffi, A. Lanni, F. Goglia, M. Moreno, Studies of complex biological systems with applications to molecular medicine: the need to integrate transcriptomic and proteomic approaches, J Biomed Biotechnol 2011 (2011). 810242.

[3] M. Baker, Gene data to hit milestone, Nature 487 (7407) (2012) 282–283.

[4] A. Tapia, C. Vilos, J.C. Marín, H.B. Croxatto, L. Devoto, Bioinformatic detection of E47, E2F1 and SREBP1 transcription factors as potential regulators of genes associated to acquisition of endometrial receptivity, Reprod. Biol. Endocrinol. 9 (2011) 14.

[5] D. Zhang, C. Sun, C. Ma, H. Dai, W. Zhang, Data mining of spatial-temporal expression of genes in the human endometrium during the window of implantation, Reprod. Sci. 19 (10) (2012) 1085–1098.

[6] S.R. Bhagwat, D.S. Chandrashekar, R. Kakar, S. Davuluri, A.K. Bajpai, S. Nayak, S. Bhutada, K. Acharya, G. Sachdeva, Endometrial receptivity: a revisit to functional genomics studies on human endometrium and creation of HGEx-ERdb, PLoS One 8 (3) (2013) e58419.

[7] S. Talbi, A.E. Hamilton, K.C. Vo, S. Tulac, M.T. Overgaard, C. Dosiou, N. Le Shay, C.N. Nezhat, R. Kempson, B.A. Lessey, N.R. Nayak, L.C. Giudice, Molecular phenotyping of human endometrium distinguishes menstrual cycle phases and underlying biological processes in normo-ovulatory women, Endocrinology 147 (3) (2006) 1097–1121.

[8] R.O. Burney, S. Talbi, A.E. Hamilton, K.C. Vo, M. Nyegaard, C.R. Nezhat, B.A. Lessey, L.C. Giudice, Gene expression analysis of endometrium reveals progesterone resistance and candidate susceptibility genes in women with endometriosis, Endocrinology 148 (8) (2007) 3814–3826.

[9] A. Hever, R.B. Roth, P. Hevezi, Human endometriosis is associated with plasma cells and overexpression of B lymphocyte stimulator, Proc. Natl. Acad. Sci. U. S. A. 104 (30) (2007) 12451–12456.

[10] S. Altmäe, F.J. Esteban, A. Stravreus-Evers, C. Simón, L. Giudice, B.A. Lessey, J.A. Horcajadas, N.S. Macklon, T. D'Hooghe, C. Campoy, B.C. Fauser, L.A. Salamonsen, A. Salumets, Guidelines for the design, analysis and interpretation of 'omics' data: focus on human endometrium, Hum. Reprod. Update 20 (1) (2014) 12–28.

[11] Y.F. Wang, M.Y. Chang, R.D. Chiang, L.J. Hwang, C.M. Lee, Y.H. Wang, Mining medical data: a case study of endometriosis, J. Med. Syst. 37 (2) (2013) 9899.

[12] D. Mathew, J.A. Drury, A.J. Valentijn, O. Vasieva, D.K. Hapangama, In silico, in vitro and in vivo analysis identifies a potential role for steroid hormone regulation of FOXD3 in endometriosis-associated genes, Hum. Reprod. 31 (2) (2016) 345–354.

[13] J.L. Liu, M. Zhao, A PubMed-wide study of endometriosis, Genomics 108 (3–4) (2016) 151–157.

[14] Z.L. Vincent, C.M. Farquhar, M.D. Mitchell, A.P. Ponnampalam, Expression and regulation of DNA methyltransferases in human endometrium, Fertil. Steril. 95 (4) (2011) 1522–1525.

[15] V. Caplakova, E. Babusikova, E. Blahovcova, T. Balharek, M. Zelieskova, J. Hatok, DNA methylation machinery in the endometrium and endometrial cancer, Anticancer Res. 36 (9) (2016) 4407–4420.

[16] S. Houshdaran, Z. Zelenko, J.C. Irwin, L.C. Giudice, Human endometrial DNA methylome is cycle-dependent and is associated with gene expression regulation, Mol. Endocrinol. 28 (7) (2014) 1118–1135.

[17] M. Saare, V. Modhukur, M. Suhorutshenko, B. Rajashekar, K. Rekker, D. Sõritsa, H. Karro, P. Soplepmann, A. Sõritsa, C.M. Lindgren, N. Rahmioglu, A. Drong, C.M. Becker, K.T. Zondervan, A. Salumets, M. Peters, The influence of menstrual cycle and endometriosis on endometrial methylome, Clin. Epigenetics 8 (2016) 2.

[18] V. Kukushkina, V. Modhukur, M. Suhorutšenko, M. Peters, R. Mägi, N. Rahmioglu, A. Velthut-Meikas, S. Altmäe, F.J. Esteban, J. Vilo, K. Zondervan, A. Salumets, T. Laisk-Podar, DNA methylation changes in endometrium and correlation with gene expression during the transition from pre-receptive to receptive phase, Sci. Rep. 7 (1) (2017) 3916.

[19] J.R. Wagner, S. Busche, B. Ge, T. Kwan, T. Pastinen, M. Blanchette, The relationship between DNA methylation, genetic and expression inter-individual variation in untransformed human fibroblasts, Genome Biol. 15 (2) (2014) R37.

[20] J. Wan, V.F. Oliver, G. Wang, H. Zhu, D.J. Zack, S.L. Merbs, J. Qian, Characterization of tissue-specific differential DNA methylation suggests distinct modes of positive and negative gene expression regulation, BMC Genomics 16 (2015) 49.

[21] N. Bunkar, N. Pathak, N.K. Lohiya, P.K. Mishra, Epigenetics: a key paradigm in reproductive health, Clin. Exp. Reprod. Med. 43 (2) (2016) 59–81.

[22] E. Walker, A.V. Hernandez, M.W. Kattan, Meta-analysis: its strengths and limitations, Cleve. Clin. J. Med. 75 (6) (2008) 431–439.

[23] U. Võsa, T. Vooder, R. Kolde, J. Vilo, A. Metspalu, T. Annilo, Meta-analysis of microRNA expression in lung cancer, Int. J. Cancer 132 (12) (2013) 2884–2893.

[24] R. Kolde, S. Laur, P. Adler, J. Vilo, Robust rank aggregation for gene list integration and meta-analysis, Bioinformatics 28 (4) (2012) 573–580.

[25] S. Altmäe, M. Koel, U. Võsa, P. Adler, M. Suhorutšenko, T. Laisk-Podar, V. Kukushkina, M. Saare, A. Velthut-Meikas, K. Krjutškov, L. Aghajanova, P.G. Lalitkumar, K. Gemzell-Danielsson, L. Giudice, C. Simón, A. Salumets, Meta-signature of human endometrial receptivity: a meta-analysis and validation study of transcriptomic biomarkers, Sci. Rep. 7 (1) (2017) 10077.

[26] J.A. Horcajadas, A. Pellicer, C. Simón, Wide genomic analysis of human endometrial receptivity: new times, new opportunities, Hum. Reprod. Update 13 (1) (2007) 77–86.

[27] L.H. Tseng, I. Chen, M.Y. Chen, H. Yan, C.N. Wang, C.L. Lee, Genome-based expression profiling as a single standardized microarray platform for the diagnosis of endometrial disorder: an array of 126-gene model, Fertil. Steril. 94 (1) (2010) 114–119.

[28] N. Rahmioglu, D.R. Nyholt, A.P. Morris, S.A. Missmer, G.W. Montgomery, K.T. Zondervan, Genetic variants underlying risk of endometriosis: insights from meta-analysis of eight genome-wide association and replication datasets, Hum. Reprod. Update 20 (5) (2014) 702–716.

[29] L. Botros, D. Sakkas, E. Seli, Metabolomics and its application for non-invasive embryo assessment in IVF, Mol. Hum. Reprod. 14 (12) (2008) 679–690.

[30] A. Alyass, M. Turcotte, D. Meyre, From big data analysis to personalized medicine for all: challenges and opportunities, BMC Med. Genet. 8 (2015) 33.

[31] S.W. Robinson, M. Fernandes, H. Husi, Current advances in systems and integrative biology, Comput. Struct. Biotechnol. J. 11 (18) (2014) 35–46.

[32] S.A. Chervitz, E.W. Deutsch, D. Field, H. Parkinson, J. Quackenbush, P. Rocca-Serra, S.A. Sansone, C.J. Stoeckert Jr., C.F. Taylor, R. Taylor, C.A. Ball, Data standards for Omics data: the basis of data sharing and reuse, Methods Mol. Biol. 719 (2011) 31–69.

[33] S. Altmäe, J. Reimand, O. Hovatta, P. Zhang, J. Kere, T. Laisk, M. Saare, M. Peters, J. Vilo, A. Stavreus-Evers, A. Salumets, Research resource: interactome of human embryo implantation: identification of gene expression pathways, regulation, and integrated regulatory networks, Mol. Endocrinol. 26 (1) (2012) 203–217.

[34] S. Gracie, C. Pennell, G. Ekman-Ordeberg, S. Lye, J. McManaman, S. Williams, L. Palmer, M. Kelley, R. Menon, M. Gravett, PREBIC "-Omics" Research group, An integrated systems biology approach to the study of preterm birth using "-omic" technology—a guideline for research, BMC Pregnancy Childbirth 11 (2011) 71.

[35] F. Jumeau, F. Chalmel, F.J. Fernandez-Gomez, C. Carpentier, H. Obriot, M. Tardivel, M.L. Caillet-Boudin, J.M. Rigot, N. Rives, L. Buée, N. Sergeant, V. Mitchell, Defining the human sperm microtubulome: an integrated genomics approach, Biol. Reprod. 96 (1) (2017) 93–106.

[36] T.M. Mayhew, Morphomics: an integral part of systems biology of the human placenta, Placenta 36 (4) (2015) 329–340.

[37] D. Ghosh, J. Sengupta, A systems biology approach to elucidate the process of blastocyst implantation, Indian J. Physiol. Pharmacol. 54 (5) (2010) 41–50.

[38] L. Díaz-Beltrán, C. Cano, D.P. Wall, F.J. Esteban, Systems biology as a comparative approach to understand complex gene expression in neurological diseases, Behav. Sci. (Basel) 3 (2) (2013) 253–272.

[39] M. Kyrgiou, P.J.G. Pouliakis a, N. Margari, P. Bountris, G. Valasoulis, M. Paraskevaidi, E. Bilirakis, M. Nasioutziki, A. Loufopoulos, M. Haritou, D.D. Koutsouris, P. Karakitsos, E. Paraskevaidis, Personalised management of women with cervical abnormalities using a clinical decision support scoring system, Gynecol. Oncol. 141 (1) (2016) 29–35.

[40] E. Seli, C. Robert, M.A. Sirard, OMICS in assisted reproduction: possibilities and pitfalls, Mol. Hum. Reprod. 16 (8) (2010) 513–530.

[41] D. Haouzi, H. Dechaud, S. Assou, C. Monzo, J. de Vos, S. Hamamah, Transcriptome analysis reveals dialogues between human trophectoderm and endometrial cells during the implantation period, Hum. Reprod. 26 (6) (2011) 1440–1449.

[42] L. Aghajanova, S. Shen, A.M. Rojas, S.J. Fisher, J.C. Irwin, L.C. Giudice, Comparative transcriptome analysis of human trophectoderm and embryonic stem cell-derived trophoblasts reveal key participants in early implantation, Biol. Reprod. 86 (1) (2012) 1–21.

[43] A. Leung, G.D. Bader, J. Reimand, HyperModules: identifying clinically and phenotypically significant network modules with disease mutations for biomarker discovery, Bioinformatics 30 (15) (2014) 2230–2232.

[44] J.V. Silva, S. Yoon, S. Domingues, S. Guimarães, A.V. Goltsev, E. da Cruz, E.F. Silva, J.F. Mendes, E. da Cruz, O.A. Silvia, M. Fardilha, Amyloid precursor protein interaction network in human testis: sentinel proteins for male reproduction, BMC Bioinf. 16 (2015) 12.

[45] M.J. Li, P. Wang, X. Liu, E.L. Lim, Z. Wang, M. Yeager, M.P. Wong, P.C. Sham, S.J. Chanock, J. Wang, GWASdb: a database for human genetic variants identified by genome-wide association studies, Nucleic Acids Res. 40 (Database issue) (2012) D1047–1054.

[46] T. Beck, R.K. Hastings, S. Gollapudi, R.C. Free, A.J. Brookes, GWAS central: a comprehensive resource for the comparison and interrogation of genome-wide association studies, Eur. J. Hum. Genet. 22 (7) (2014) 949–952.

[47] Q. Song, B. Decato, E.E. Hong, M. Zhou, F. Fang, J. Qu, T. Garvin, M. Kessler, J. Zhou, A.D. Smith, A reference methylome database and analysis pipeline to facilitate integrative and comparative epigenomics, PLoS One 8 (12) (2013) e81148.

[48] Y. Xiong, Y. Wei, Y. Gu, S. Zhang, J. Lyu, B. Zhang, C. Chen, J. Zhu, Y. Wang, H. Liu, Y. Zhang, DiseaseMeth version 2.0: a major expansion and update of the human disease methylation database, Nucleic Acids Res. 45 (D1) (2017) D888–D895.

[49] S. Griffiths-Jones, H.K. Saini, S. van Dongen, A.J. Enright, miRBase: tools for microRNA genomics, Nucleic Acids Res. 36 (Database issue) (2008) D154–158.

[50] D. Betel, M. Wilson, A. Gabow, D.S. Marks, C. Sander, The microRNA.org resource: targets and expression, Nucleic Acids Res. 36 (Database issue) (2008) D149–153.

[51] R. Edgar, M. Domrachev, A.E. Lash, Gene Expression Omnibus: NCBI gene expression and hybridization array data repository, Nucleic Acids Res. 30 (1) (2002) 207–210.

[52] H. Parkinson, M. Kapushesky, M. Shojatalab, N. Abeygunawardena, R. Coulson, A. Farne, E. Holloway, N. Kolesnykov, P. Lilja, M. Lukk, R. Mani, T. Rayner, A. Sharma, E. William, U. Sarkans, A. Brazma, ArrayExpress—A public database of microarray experiments and gene expression profiles, Nucleic Acids Res. 35 (Database issue) (2007) D747–750.

[53] F. Desiere, E.W. Deutsch, N.L. King, A.I. Nesvizhskii, P. Mallick, J. Eng, S. Chen, J. Eddes, S.N. Loevenich, R. Aebersold, The PeptideAtlas project, Nucleic Acids Res. 34 (Database issue) (2006) D655–658.

[54] J.A. Vizcaíno, A. Csordas, N. del-Toro, J.A. Dianes, J. Griss, I. Lavidas, G. Mayer, Y. Perez-Riverol, F. Reisinger, T. Ternent, Q.W. Xu, R. Wang, H. Hermjakob, 2016 update of the PRIDE database and its related tools, Nucleic Acids Res. 44 (D1) (2016) D447–456.

[55] V. Nanjappa, J.K. Thomas, A. Marimuthu, B. Muthusamy, A. Radhakrishnan, R. Sharma, A. Ahmad Kjam, L. Balakrishnan, N.A. Sahasrabuddhe, S. Kumar, B.N. Jhaveri, K.V. Sheth, R. Kumar Khatana, P.G. Shaw, S.M. Srikanth, P.P. Mathur, S. Shankar, D. Nagaraja, R. Christopher, S. Mathivanan, R. Raju, R. Sirdeshmukh, A. Chatterjee, R.J. Simpson, H.C. Harsha, A. Pandey, T.S. Prasad, Plasma proteome database as a resource for proteomics research: 2014 update, Nucleic Acids Res. 42 (Database issue) (2014) D959–965.

[56] M.S. Kim, S.M. Pinto, D. Getnet, R.S. Nirujogi, S.S. Manda, R. Chaerkady, A.K. Madugundu, D.S. Kelkar, R. Isserlin, S. Jain, J.K. Thomas, B. Muthusamy, P. Leal-Rojas, P. Kumar, N.A. Sahasrabuddhe, L. Balakrishnan, J. Advani, B. George, S. Renuse, L.D. Selvan, A.H. Patil, V. Nanjappa, A. Radhakrishnan, S. Prasad, T. Subbannayya, R. Raju, M. Kumar, S.K. Sreenivasamurthy, A. Marimuthu, G.J. Sathe, S. Chavan, K.K. Datta, Y. Subbannayya, A. Sahi, S.D. Yelamanchi, S. Jayaram, P. Rajagopalan, J. Sharma, K.R. Murthy, N. Syed, R. Goel, A.A. Khan, S. Ahmad, G. Dey, K. Mudgal, A. Chatterjee, T.C. Huang, J. Zhong, X. Wu, P.G. Shaw, D. Freed, M.S. Zahari, K.K. Mukherjee, S. Shankar, A. Mahadevan, H. Lam, C.J. Mitchell, S.K. Shankar, P. Satishchandra, J.T. Schroeder, R. Sirdeshmukh, A. Maitra, S.D. Leach, C.G. Drake, M.K. Halushka, T.S. Prasad, R.H. Hruban, C.L. Kerr, G.D. Bader, C.A. Iacobuzio-Donahue, H. Gowda, A. Pandey, A draft map of the human proteome, Nature 509 (7502) (2014) 575–581.

[57] M. Wilhem, J. Schlegl, H. Hahne, A.M. Gholami, M. Lieberenz, M.M. Savitski, E. Ziegler, L. Butzmann, S. Gessulat, H. Marx, T. Mathieson, S. Lemeer, K. Schnatbaum, U. Reimer, H. Wenschuh, M. Mollenhauer, J. Slotta-Huspenina, J.H. Boese, M. Bantscheff, A. Gerstmair, F. Faerber, B. Kuster, Mass-spectrometry-based draft of the human proteome, Nature 509 (7502) (2014) 582–587.

[58] W. Yu, M. Clyne, M.J. Khoury, M. Gwinn, Phenopedia and Genopedia: disease-centered and gene-centered views of the evolving knowledge of human genetic associations, Bioinformatics 26 (1) (2010) 145–146.

[59] C.C. Liu, Y.T. Tseng, W. Li, C.Y. Wu, I. Mayzus, A. Rzhetsky, F. Sun, M. Waterman, J.J. Chen, P.M. Chaudhary, J. Loscalzo, E. Crandall, X.J. Zhou, DiseaseConnect: a comprehensive web server for mechanism-based disease-disease connections, Nucleic Acids Res. 42 (Web Server issue) (2014) W137–146.

[60] J.J. Tarín, M.A. García-Pérez, T. Hamatani, A. Cano, Infertility etiologies are genetically and clinically linked with other diseases in single meta-diseases, Reprod. Biol. Endocrinol. 13 (2015) 31.

[61] V.J. Henry, A.E. Bandrowski, A.S. Pepin, B.J. Gonzalez, A. Desfeux, OMICtools: an informative directory for multi-omic data analysis, Database (Oxford) 2014 (2014). bau069.

[62] J. Hua, B. Xu, Y. Yang, R. Ban, F. Igbal, H.J. Cooke, Y. Zhang, Q. Shi, Follicle Online: an integrated database of follicle assembly, development and ovulation, Database (Oxford) 2015 (2015). bav036.

[63] Y. Zhang, L. Zhong, B. Xu, Y. Yang, R. Ban, J. Zhu, H.J. Cooke, Q. Hao, Q. Shi, SpermatogenesisOnline 1.0: a resource for spermatogenesis based on manual literature curation and genome-wide data mining, Nucleic Acids Res. 41 (Database issue) (2013) D1055–1062.

[64] A.C. Luk, H. Gao, S. Xiao, J. Liao, D. Wang, J. Tu, O.M. Rennert, W.Y. Chan, T.L. Lee, GermlncRNA: a unique catalogue of long non-coding RNAs and associated regulations in male germ cell development, Database (Oxford) 2015 (2015). bav044.

[65] T.A. Darde, O. Sallou, E. Becker, B. Evrard, C. Monjeaud, Y. Le Bras, B. Jégou, O. Collin, A.D. Rolland, F. Chalmel, The ReproGenomics viewer: an integrative cross-species toolbox for the reproductive science community, Nucleic Acids Res. 43 (W1) (2015) W109–116.

[66] W. Bai, W. Yang, W. Wang, Y. Wang, C. Liu, Q. Jiang, J. Hua, M. Liao, GED: a manually curated comprehensive resource for epigenetic modification of gametogenesis, Brief. Bioinform. 18 (1) (2017) 98–104.

[67] C.P. Leo, U.A. Vitt, A.J. Hsueh, The ovarian kaleidoscope database: an online resource for the ovarian research community, Endocrinology 141 (9) (2000) 3052–3054.

[68] I. Ben-Shlomo, U.A. Vitt, A.J. Hsueh, Perspective: the ovarian kaleidoscope database-II. Functional genomic analysis of an organ-specific database, Endocrinology 143 (6) (2002) 2041–2044.

[69] A.J. Hsueh, R. Rauch, Ovarian kaleidoscope database: ten years and beyond, Biol. Reprod. 86 (6) (2012) 192.

[70] M. Kim, B.A. Cooper, R. Venkat, J.B. Phillips, H.R. Eidem, J. Hirbo, S. Nutakki, S.M. Williams, L.J. Muglia, J.A. Capra, K. Petren, P. Abbot, A. Rokas, K.L. McGary, GEneSTATION 1.0: a synthetic resource of diverse evolutionary and functional genomic data for studying the evolution of pregnancy-associated tissues and phenotypes, Nucleic Acids Res. 44 (Database issue) (2016) D908–D916.

[71] A. Uzun, E.W. Triche, J. Schuster, A.T. Dewan, J.F. Padbury, dbPEC: a comprehensive literature-based database for preeclampsia related genes and phenotypes, Database (Oxford) 2016 (2016). baw006.

[72] A. Uzun, A. Laliberte, J. Parker, C. Andrew, E. Winterrowd, S. Surendra, S. Istrail, J.F. Padbury, dbPTB: a database for preterm birth, Database (Oxford) 2012 (2012). bar069.

[73] A.I. Saeed, N.K. Bhagabati, J.C. Braisted, W. Liang, V. Sharov, E.A. Howe, J. Li, M. Thiagarajan, J.A. White, J. Quackenbush, TM4 microarray software suite, Methods Enzymol. 411 (2006) 134–193.

[74] M.A. Kallio, J.T. Tuimala, T. Hupponen, P. Klemelä, M. Gentile, I. Scheinin, M. Koski, J. Käki, E.I. Korpelainen, Chipster: user-friendly analysis software for microarray and other high-throughput data, BMC Genomics 12 (2011) 507.

[75] R. Hilker, K.B. Stadermann, D. Doppmeier, J. Kalinowski, J. Stoye, J. Straube, J. Winnebald, A. Goesmann, ReadXplorer-visualization and analysis of mapped sequences, Bioinformatics 30 (16) (2014) 2247–2254.

[76] M. Reimers, V.J. Carey, Bioconductor: an open source framework for bioinformatics and computational biology, Methods Enzymol. 411 (2006) 119–134.

[77] Y. Zhang, J. Szustakowski, M. Schinke, Bioinformatics analysis of microarray data, Methods Mol. Biol. 573 (2009) 259–284.

[78] B. Ekmekci, C.E. McAnany, C. Mura, An introduction to programming for bioscientists: a python-based primer, PLoS Comput. Biol. 12 (6) (2016) e1004867.

[79] T.K. Rausch, A. Schillert, A. Ziegler, A. Lüking, H.D. Zucht, P. Schulz-Knappe, Comparison of pre-processing methods for multiplex bead-based immunoassays, BMC Genomics 17 (1) (2016) 601.

[80] F.J. Esteban, C. Cano, I. de la Haza, A. Cano-Ortiz, N.V. Mendizábal, J. Goñí, J.A. Horcajadas, Análisis bioinformático de datos: aplicación en microarrays, Cuadernos de medicina reproductiva 51 (14) (2008) 87–96.

[81] Y. Sui, X. Zhao, T.P. Speed, Z. Wu, Background adjustment for DNA microarrays using a database of microarray experiments, J. Comput. Biol. 16 (11) (2009) 1501–1515.

[82] S.A. Mirroshandel, F. Ghasemian, S. Monji-Azad, Applying data mining techniques for increasing implantation rate by selection best sperms for intracytoplasmic sperm injection treatment, Comput. Methods Prog. Biomed. 137 (2016) 215–229.

[83] A. Goncalves, A. Tikhonov, A. Brazma, M. Kapushesky, A pipeline for RNA-seq data processing and quality assessment, Bioinformatics 27 (6) (2011) 867–869.

[84] C.W. Law, M. Alhamdoosh, S. Su, G.K. Smyth, M.E. Ritchie, RNA-seq analysis is easy as 1-2-3 with limma, Glimma and edgeR. Version 2, F1000 Res. 5 (2016) 1408.

[85] D.J. McCarthy, K.R. Campbell, A.T. Lun, Q.F. Wills, Scater: pre-processing, quality control, normalization and visualization of single-cell RNA-seqdata in R, Bioinformatics 33 (8) (2017) 1179–1186.

[86] E. Cazaly, R. Thomson, J.R. Marthick, A.F. Holloway, J. Charlesworth, J.L. Dickinson, Comparison of pre-processing methodologies for Illumina 450k methylation array data in familial analyses, Clin. Epigenetics 8 (2016) 75.

[87] A. Cruz-Marcelo, R. Guerra, M. Vannucci, Y. Li, C.C. Lau, T.K. Man, Comparison of algorithms for pre-processing of SELDI-TOF mass spectrometry data, Bioinformatics 24 (19) (2008) 2129–2136.

[88] J.H. Do, D.K. Choi, Normalization of microarray data: single-labeled and dual-labeled arrays, Mol. Cells 22 (3) (2006) 254–261.

[89] S.P. Borgaonkar, H. Hocker, H. Shin, M.K. Markey, Comparison of normalization methods for the identification of biomarkers using MALDI-TOF and SELDI-TOF mass spectra, OMICS 14 (1) (2010) 115–126.

[90] M. Reimers, Statistical analysis of microarray data, Addict. Biol. 10 (1) (2005) 23–35.

[91] P. Li, Y. Piao, H.S. Shon, K.H. Ryu, Comparing the normalization methods for the differential analysis of Illumina high-throughput RNA-Seq data, BMC Bioinf. 16 (2015) 347.

[92] R.A. Irizarry, B. Hobbs, F. Collin, Y.D. Beazer-Barclay, K.J. Antonellis, U. Scherf, T.P. Speed, Exploration, normalization, and summaries of high density oligonucleotide array probe level data, Biostatistics 4 (2) (2003) 249–264.

[93] S. Yang, D.E. Mercante, K. Zhang, Z. Fang, An integrated approach for RNA-seq data normalization, Cancer Informat. 15 (2016) 129–141.

[94] A. Chawade, E. Alexandersson, F. Levander, Normalyzer: a tool for rapid evaluation of normalization methods for omics data sets, J. Proteome Res. 13 (6) (2014) 3114–3120.

[95] M. Chadeau-Hyam, G. Campanella, T. Jombart, L. Bottolo, L. Portengen, P. Vineis, B. Liquet, R.C. Vermeulen, Deciphering the complex: methodological overview of statistical models to derive OMICS-based biomarkers, Environ. Mol. Mutagen. 54 (7) (2013) 542–557.

[96] F. Wagner, GO-PCA: an unsupervised method to explore gene expression data using prior knowledge, PLoS One 10 (11) (2015) e0143196.

[97] D. Jiang, C. Tang, A. Zhang, Cluster analysis for gene expression data: a survey, IEEE Trans. Knowl. Data Eng. 16 (11) (2004) 1370–1386.

[98] S. Altmäe, K. Tamm-Rosenstein, F.J. Esteban, J. Simm, L. Kolberg, H. Peterson, M. Metsis, K. Haldre, J.A. Horcajadas, A. Salumets, A. Stavreus-Evers, Endometrial transcriptome analysis indicates superiority of natural over artificial cycles in recurrent implantation failure patients undergoing frozen embryo transfer, Reprod. Biomed. Online 32 (6) (2016) 597–613.

[99] G.W. Hatfield, S.P. Hung, P. Baldi, Differential analysis of DNA microarray gene expression data, Mol. Microbiol. 47 (4) (2003) 871–877.

[100] Y. Guo, A. Graber, R.N. McBurney, R. Balasubramanian, Sample size and statistical power considerations in high-dimensionality data settings: a comparative study of classification algorithms, BMC Bioinf. 11 (2010) 447.

[101] D. Franco, F. Bonet, F. Hernández-Torres, E. Lozano-Velasco, F.J. Esteban, A.E. Aranega, Analysis of microRNA microarrays in cardiogenesis, Methods Mol. Biol. 1375 (2016) 207–221.

[102] W. Huang da, B.T. Sherman, R.A. Lempicki, Systematic and integrative analysis of large gene lists using DAVID bioinformatics resources, Nat. Protoc. 4 (1) (2009) 44–57.

[103] M. Ashburner, C.A. Ball, J.A. Blake, D. Botstein, H. Butler, J.M. Cherry, A.P. Davis, K. Dolinski, S.S. Dwight, J.T. Eppig, M.A. Harris, D.P. Hill, L. Issel-Tarver, A. Kasarskis, S. Lewis, J.C. Matese, J.E. Richardson, M. Ringwald, G.M. Rubin, G. Sherlock, Gene ontology: tool for the unification of biology, Nat. Genet. 25 (1) (2000) 25–29.

[104] J. Reimand, T. Arak, P. Adler, L. Kolberg, S. Reisberg, H. Peterson, J. Vilo, G:Profiler—a web server for functional interpretation of gene lists (2016 update), Nucleic Acids Res. 44 (W1) (2016) W83–89.

[105] E.Y. Chen, C.M. Tan, Y. Kou, Q. Duan, Z. Wang, G.V. Meirelles, N.R. Clark, A. Ma'ayan, Enrichr: integrative and collaborative HTML5 gene list enrichment analysis tool, BMC Bioinf. 14 (2013) 128.

[106] F. Supek, M. Bošnjak, N. Škunca, T. Šmuc, REVIGO summarizes and visualizes long lists of gene ontology terms, PLoS One 6 (7) (2011) e21800.

[107] A. Nikitin, S. Egorov, N. Daraselia, I. Mazo, Pathway studio—The analysis and navigation of molecular networks, Bioinformatics 19 (16) (2003) 2155–2157.

[108] C.H. Chin, S.H. Chen, H.H. Wu, C.W. Ho, M.T. Ko, C.Y. Lin, cytoHubba: identifying hub objects and subnetworks from complex interactome, BMC Syst. Biol. 8 (Suppl. 4) (2014) S11.

[109] J.A. Santiago, V. Bottero, J.A. Potashkin, Dissecting the molecular mechanisms of neurodegenerative diseases through network biology, Front. Aging Neurosci. 9 (2017) 166.

[110] N. Gehlenborg, S.I. O'Donoghue, N.S. Baliga, A. Goesmann, M.A. Hibbs, H. Kitano, O. Kohlbacher, H. Neuweger, R. Schneider, D. Tenenbaum, A.C. Gavin, Visualization of omics data for systems biology, Nat. Methods 7 (3 Suppl) (2010) S56–68.

[111] J. Paquette, T. Tokuyasu, EGAN: exploratory gene association networks, Bioinformatics 26 (2) (2010) 285–286.

[112] S. Hayrabedyan, K. Todorova, A. Jabeen, G. Metodieva, S. Toshkov, M.V. Metodiev, M. Mincheff, N. Fernández, Sertoli cells have a functional NALP3 inflammasome that can modulate autophagy and cytokine production, Sci. Rep. 6 (2016) 18896.

[113] D. Szklarczyk, A. Franceschini, S. Wyder, K. Forslund, D. Heller, J. Huerta-Cepas, M. Simonovic, A. Roth, A. Santos, K.P. Tsafou, M. Kuhn, P. Bork, L.J. Jensen, C. von Mering, STRING v10: protein-protein interaction networks, integrated over the tree of life, Nucleic Acids Res. 43 (Database issue) (2015) D447–452.

[114] J.A. Horcajadas, P. Mínguez, J. Dopazo, F.J. Esteban, F. Domínguez, L.C. Giudice, A. Pellicer, C. Simón, Controlled ovarian stimulation induces a functional genomic delay of the endometrium with potential clinical implications, J. Clin. Endocrinol. Metab. 93 (11) (2008) 4500–4510.

[115] P. Shannon, A. Markiel, O. Ozier, N.S. Baliga, J.T. Wang, D. Ramage, N. Amin, B. Schwikowski, T. Ideker, Cytoscape: a software environment for integrated models of biomolecular interaction networks, Genome Res. 13 (11) (2003) 2498–2504.

[116] H. Liu, L. Zeng, K. Yang, G. Zhang, A network pharmacology approach to explore the pharmacological mechanism of Xiaoyao powder on anovulatory infertility, Evid. Based Complement. Alternat. Med. 2016 (2016) 2960372.

[117] S. Sabetian, M.S. Shamsir, Systematic analysis of protein interaction network associated with azoospermia, Int. J. Mol. Sci. 17 (2016) 1857.

[118] E. Frank, M.A. Hall, I.H. Witten, The WEKA Workbench. Online Appendix for "Data Mining: Practical Machine Learning Tools and Techniques.", fourth ed., Morgan Kaufmann, 2016.

[119] F.H. Hung, H.W. Chiu, Cancer subtype prediction from a pathway-level perspective by using a support vector machine based on integrated gene expression and protein network, Comput. Methods Prog. Biomed. 141 (2017) 27–34.

[120] Q. Liu, M. Gan, R. Jiang, A sequence-based method to predict the impact of regulatory variants using random forest, BMC Syst. Biol. 11 (Suppl 2) (2017) 7.

[121] G.P. Way, R.J. Allaway, S.J. Bouley, C.E. Fadul, Y. Sanchez, C.S. Greene, A machine learning classifier trained on cancer transcriptomes detects NF1 inactivation signal in glioblastoma, BMC Genomics 18 (1) (2017) 127.

[122] A.P. Delitala, G. Capobianco, G. Delitala, P.L. Cherchi, S. Dessole, Polycystic ovary syndrome, adipose tissue and metabolic syndrome, Arch. Gynecol. Obstet. 296 (3) (2017) 405–419.

[123] D. Macut, J. Bjekić-Macut, D. Rahelić, M. Doknić, Insulin and the polycystic ovary syndrome, Diabetes Res. Clin. Pract. 130 (2017) 163–170.

[124] M. Cortón, J.I. Botella-Carretero, A. Benguría, G. Villuendas, A. Zaballos, J.L. San Millán, H.F. Escobar-Morreale, B. Peral, Differential gene expression profile in omental adipose tissue in women with polycystic ovary syndrome, J. Clin. Endocrinol. Metab. 92 (1) (2007) 328–337.

[125] T. Hastie, R. Tibshirani, B. Narasimhan, G. Chu, Impute: imputation for microarray data, 2017. R package version 1.50.1.

[126] M.E. Ritchie, B. Phipson, D. Wu, Y. Hu, C.V. Law, W. Shi, G.K. Smyth, Limma powers differential expression analyses for RNA-sequencing and microarray studies, Nucleic Acids Res. 43 (7) (2015) e47.

[127] F.D. Carratore, A. Janckevics, F. Hong, B. Wittner, R. Breitling, F. Battke, RankProd: Rank Product method for identifying differentially expressed genes with application in meta-analysis, 2016. R package version 3.2.0.
[128] N. Aziz, The importance of semen analysis in the context of azoospermia, Clinics (Sao Paulo) 68 (Suppl 1) (2013) 35–38.
[129] J. Montojo, K. Zuberi, H. Rodriguez, F. Kazi, G. Wright, S.L. Donaldson, Q. Morris, G.D. Bader, GeneMANIA Cytoscape plugin: fast gene function predictions on the desktop, Bioinformatics 26 (22) (2010) 2927–2928.
[130] D. Warde-Farley, S.L. Donaldson, O. Comes, K. Zuberi, R. Badrawi, P. Chao, M. Franz, C. Grouios, F. Kazi, C.T. Lopes, A. Maitland, S. Mostafavi, J. Montojo, Q. Shao, G. Wright, G.D. Bader, Q. Morris, The GeneMANIA prediction server: biological network integration for gene prioritization and predicting gene function, Nucleic Acids Res. 38 (Web Server issue) (2010) W214–220.
[131] V. Lapatas, M. Stefanidakis, R.C. Jimenez, A. Via, M.V. Schneider, Data integration in biological research: an overview, J. Biol. Res. (Thessalon) 22 (1) (2015) 9.

Further Reading

[132] J. Vlaanderen, L.E. Moore, M.T. Smith, Q. Lan, L. Zhang, C.F. Skibola, N. Rothman, R. Vermeulen, Application of OMICS technologies in occupational and environmental health research; current status and projections, Occup. Environ. Med. 67 (2) (2010) 136–143.

CHAPTER

19

Conclusions—Year 2020: The "-Omics" Technologies and Personalized Assisted Reproduction

José A. Horcajadas, Jaime Gosálvez*[†]

[*]Department of Molecular Biology and Biochemical Engineering, University Pablo de Olavide, Sevilla, Spain [†]Department of Biology, University Autónoma of Madrid, Madrid, Spain

The completion of the Human Genome Project and the development of new technologies have brought the "omics" era to many research fields with possible clinical impact. The application of omics in the field of reproduction, under the common name of "reproductomics," allows the analysis of every single step in the reproductive process. Bioinformatics, which is the unique nonomics matter in the omics scenario, remains a key element to collect, order, and analyze and provides understanding to the large amount of data that omics generates.

The term reproductomics covers the use of genomics, functional genomics or transcriptomics, metabolomics, and proteomics to investigate, evaluate, or diagnose men and women at reproductive age, gametes, zygotes, or fetuses in order to help improve clinical outcomes.

The following is a summary of how the editors of this book envision the future from an omics era perspective. This information is summarized in Table 19.1.

GENOMICS

Undoubtedly, genomics is the omics with more applications in reproduction. Genomics can be applied to determine the carrier status of the couple prior to conception. This is crucial in the case of monogenic diseases, especially with recessive diseases, in which the absence of symptoms makes it impossible to detect who are carriers of mutations that can affect the future baby.

Nowadays, the high cost of the tests prevents large-scale use; however, it is expected that the reduction of the cost of the technology will allow for generalized use for all couples, both fertile and infertile. Furthermore, we believe that the use of genomics for the detection of mutations related with infertility such as chromosome Y microdeletions in male infertility or

https://doi.org/10.1016/B978-0-12-812571-7.00029-0

TABLE 19.1 Use of Omics Technologies in Human Reproduction and the Expectations of Their Use in Clinic in 2025

Year 2025					
Omics	**Test**		**Applied to Patients**	**Availability in Clinics**[a]	**Utility**
Genomics	Carrier screening		100%	>80%	Carrier status
	Genetic variations of infertility		100%	>80%	Medical guide
	Pharmacogenetics		100%	~100%	Personalized treatment
	Nutrigenetics		On demand	<50%	Diet recommendations
	Polar body screening		~0%	Negligible	Oocyte quality
	PGS	Embryo biopsy	>90%	100%	Embryo quality
		Blastocele	Unpredictable	100%	Embryo quality
		Spent media	Unpredictable	100%	Embryo quality
	Fetal cells in uterus		By clinical indication	100%	Fetus status
	NIPT		On demand	100%	Fetus status
	Uterine metagenomics		By clinical indication?	Unpredictable	Unexplained infertility
	Sperm DNA damage		Unpredictable	Unpredictable	Sperm quality
Transcriptomics	Endometrial transcriptomics		By clinical indication	>80%	Endometrial evaluation
	Sperm		Unpredictable	Unpredictable	Sperm quality
	Cumulus cells		On demand	<10%	Oocyte quality
Proteomics and Metabolomics	Uterine lipidomics		By clinical indication?	Unpredictable	Embryo quality
	Sperm		Unpredictable	Unpredictable	Sperm quality
	Spent media		On demand	>80%	Embryo quality

The type of tests, the percentage of patients that can have access to this tests, the availability of tests in clinics, and their clinical utility are indicated.
[a] *US and EU clinics.*

genetic variations in hormone receptors such as FSH or LHR for female disorders related with infertility such as recurrent miscarriage or premature ovarian failure will also be accessible for general populations before treatment.

The reproductive pharmacogenomics and pharmacogenetics, which are poorly applied for personalized treatment today, will advance current knowledge in the next few years and will very likely be the most important key for the personalization of treatments for ovarian stimulation, that is, the right drug, at the right dose, for the right woman, at the right time, and will provide the right outcome. Additionally, even the periconceptional nutrigenomics and nutrigenetics will play a role in the reproductive scenario in the future.

In relation with gametes, the only use of genomics has been with the polar body. It does not look like the analysis of polar body DNA will serve for oocyte or embryo selection in the future. The absence of genetic material from the other partner of the embryo, the sperm, greatly hampers the utility of the information obtained from the polar body. We have to add also the errors that occur during the second division to distinguish a good embryo, genetically talking. Furthermore, it looks like there is no more controversies regarding the use of genomics to diagnose embryos from embryo biopsy at day 5 using NGS or even CGH. NGS has also the ability to identify and screen for embryos with reduced viability such as mosaic embryos and those with partial aneuploidies or triploidy. Finally, the approaches based on noninvasive or semi-invasive methods for genetic embryo screening such as blastocentesis or free DNA from spent media have currently not shown the expected results. Improvements in DNA collection, amplification, and testing may allow for PGS without biopsy in the future. In any case, the original developments that involve noninvasive methods and the use of metabolomics or proteomics in spent media represent a threat to the current PGS of embryos.

Prenatal testing also has different genomics approaches, which are used to determine chromosomal or genetic abnormalities early on in pregnancy. Classically, these tests were carried out by taking a tissue sample of the chorionic villi or amniotic fluid from the placenta and performing a classic cytogenetic analysis to show the chromosomes at metaphase. Nowadays, the same tissue samples can be used for genomics analysis by CGH or NGS. These techniques allow for knowing not only the chromosomal dotation of the cells but also the presence of punctual mutations or small deletions related with monogenic diseases. Another advantage is that these methods do not require the cells to be cultured, as is needed in classic cytogenetic testing; therefore, the results can be obtained in a few days. The prenatal testing revolution has allowed for the development of NIPT, noninvasive prenatal testing. NIPT is a safe, simple, and highly accurate prenatal screening blood test that was first introduced into clinical practice around 6 years ago. It was created to detect trisomies or monosomies of a few chromosomes: 13, 18, 21, X, and Y, but nowadays, it can detect any aneuploidy and a limited number of microdeletions related with some syndromes. In 2011, the first NIPT was launched in the United States at a cost of over $2000. That figure has dropped to a few hundred dollars today, and millions of tests are performed every year in the United States, Europe, and China.

Another emerging genomics application has appeared under the study of the uterine microbiome. It is known that an abnormal endometrial microbiota has been associated with implantation failure, pregnancy loss, and other gynecologic and obstetric conditions. The knowledge of the uterine microbiome and how some infections can affect endometrial receptivity or induce endometrial disorders will allow for the application of this specific omics in clinic's daily routine.

With regard to sperm, one of the most common analyses currently performed is the checking of the DNA integrity. This started to be investigated thanks to the previously described methods such as single-cell gel electrophoresis (COMET) assay, sperm chromatin structure assay (SCSA), acridine orange test (AOT), terminal deoxynucleotidyl transferase-mediated deoxyuridine (TdT) triphosphate (dUTP) nick end labeling (TUNEL) assay, and sperm chromatin dispersion (SCD) test. However, studying sperm's DNA might be very complex because the sperm DNA differs from the somatic cell DNA with its unique structure. The clinical value of these tests remains to be elucidated. In spite of half a century of research within the area, this analysis is not routinely implemented into the fertility clinics. Therefore, the future application of genomics in DNA integrity analysis is hard to imagine.

TRANSCRIPTOMICS

Since the development of gene expression microarray technology, many groups of researchers working in different areas of human reproduction started to apply it for research purposes. The first and most important results came from the analysis of the endometrium at different phases of the menstrual cycle, and in many aspects, it was related with pathologies and infertility. The large number of publications regarding this topic and the abundant amount of data available in gene expression allowed the development of different methods of endometrial evaluation. Although there are four established molecular methods of endometrial evaluation, only two can be considered transcriptomics tools. These are the endometrial receptivity array (ERA) and the endometrial receptivity map (ER Map), also called ER Grade. Both of these methods use a limited number of genes to classify the endometrium as prereceptive, receptive, postreceptive, and nonreceptive. The current utility of these methods is soon to be demonstrated; the first data from a randomized clinical trial suggests that the evaluation of the endometrium is very important in substitute cycles in order to personalize the embryo transfer, especially in those women with implantation failure. It is expected that with time, the use of these methods will be applied, under clinical demand, at any IVF clinic.

The use of transcriptomics' technology to find molecular differences between the spermatozoa transcriptomes in infertile and fertile men with similar sperm counts started more than 10 years ago. The finding that the over- or underexpressed biological functions that were different between groups involving spermatozoa differentiation, spermatid development, gametogenesis, spermatid differentiation, and male gamete generation processes suggests that sperm infertility markers may not be related to sperm production in terms of sperm count but are probably related to sperm function. The use in clinic of this information in the future remains unpredictable.

Transcriptomics microarray analysis can also identify the profiles of oocytes at various stages of growth and maturation. This has provided a better understanding of the genes expressed during oocyte development. Somatic cells associated with the oocytes' function, such as CCs, are also related to embryo competence and pregnancy outcomes. In this sense, it has been defined that the transcriptomic signature of CC included hundreds of genes associated with pregnancy outcome, being those that were differentially expressed mainly upregulated. However, the use in clinic of this information has not had the appropriated impact, and the benefits of its use for oocyte/embryo selection will have to be elucidated in the future.

PROTEOMICS AND METABOLOMICS

Proteomics and metabolomics have been used to know the bases of many processes of human reproduction, since oocyte and sperm maturation to molecular factors related with women infertility. Proteomics and especially metabolomics are the closest omics to phenotype. The identification of changes in proteins and metabolites related with a specific process, although related with the underlying genomics and transcriptomics, is the result of the interaction of the genome with a specific environment giving a more complete and nearer image of a particular biological situation. The humble opinion of these two editors is that both omics are going to contribute in the development of noninvasive methods for embryo selection, endometrial evaluation, and detection of gynecologic disorders related with infertility.

Using lipidomics, some researchers have identified the presence of several important lipids whose variations are very well regulated during the menstrual cycle, specifically at the time of embryo implantation. The profile of lipids, particularly prostaglandins, from uterine fluid can serve as a noninvasive method for endometrial evaluation or as a guide to diagnose infertility of endometrial origin.

Unexplained male infertility has been a definition traditionally employed for those patients whose analysis results were within normal values and their physical and endocrine profiles and an explanation for the infertility that can't be justified. As technology has been developed and incorporated to the study of infertility, new tools as proteomics and metabolomics have provided new data and information for professionals. Proteomics, after the drafting of the genome, tries to understand target proteins that could explain infertility and metabolomics that has provided new insights into the understanding of dysregulation in reproductive organ metabolism using seminal plasma from a noninvasive perspective. Maybe both technologies will highlight in the near future some unexplained infertility and offer to clinicians additional clues for patient treatment prior to IVF and also useful to both proteomic and metabolomic information for sperm selection within the assisted reproduction techniques.

Finally, one of the most promising approaches in metabolomics is the analysis of the spent media; it means the media in which the embryo has been cultured in the IVF lab, to establish metabolomics profile related with pregnancy outcomes or aneuploidy. It has been used to detect also variations in obese women undergoing in vitro fertilization. We strongly believe that the future of embryo selection passes through the discovery of metabolites related with embryo quality to avoid the invasive methods such as embryo biopsy, currently used today.

In Fig. 19.1, a graphic summary of reproductomics' clinical applications, currently used in 2018, is indicated. This figure tries to represent the state of the art in the year of publication of the first edition of *REPRODUCTOMICS*.

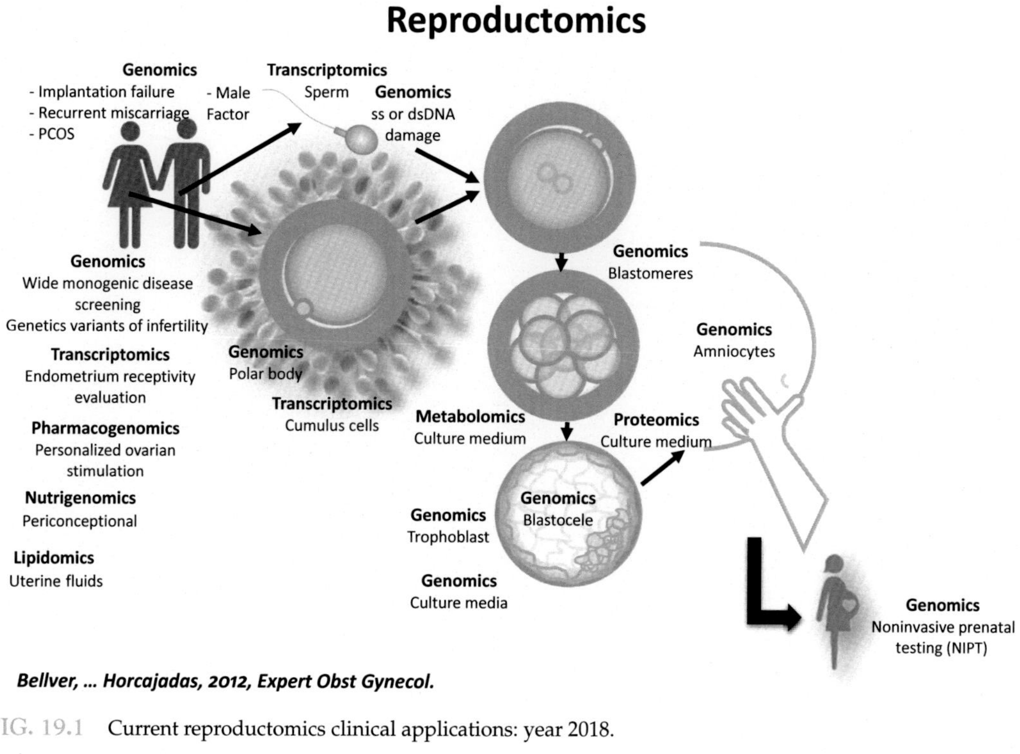

FIG. 19.1 Current reproductomics clinical applications: year 2018.

Index

Note: Page numbers followed by *f* indicate figures and *t* indicate tables.

Printed in the United States
By Bookmasters